ASTROCHEMISTRY OF COSMIC PHENOMENA

INTERNATIONAL ASTRONOMICAL UNION
UNION ASTRONOMIQUE INTERNATIONALE

ASTROCHEMISTRY OF COSMIC PHENOMENA

PROCEEDINGS OF THE 150TH SYMPOSIUM OF THE
INTERNATIONAL ASTRONOMICAL UNION,
HELD AT CAMPOS DO JORDÃO, SÃO PAULO, BRAZIL, AUGUST 5–9, 1991

EDITED BY

P. D. SINGH

Instituto Astronômico e Geofísico,
Universidade de São Paulo, Brazil

KLUWER ACADEMIC PUBLISHERS
DORDRECHT / BOSTON / LONDON

Library of Congress Cataloging-in-Publication Data

```
International Astronomical Union. Symposium (150th : 1991 : Campos do
  Jordão, Brazil)
    Astrochemistry of cosmic phenomena : proceedings of the 150th
  Symposium of the International Astronomical Union, held at Campos do
  Jordão, São Paulo, Brazil, August 5-9, 1991 / edited by P.D. Singh.
       p.   cm.
    Includes bibliographical references and indexes.
    ISBN 0-7923-1824-2 (HB : acid-free paper)
    1. Cosmochemistry--Congresses.  2. Cosmochemistry--Congresses.
  I. Singh, P. D.   II. Title.
  QB450.I58   1991
  523'.02--dc20                                              92-17224
```

ISBN 0-7923-1824-2 (HB)

Published on behalf of
the International Astronomical Union
by
Kluwer Academic Publishers, P.O. Box 17, 3300 AA Dordrecht, The Netherlands.

Kluwer Academic Publishers incorporates
the publishing programmes of
D. Reidel, Martinus Nijhoff, Dr W. Junk and MTP Press.

Sold and distributed in the U.S.A. and Canada
by Kluwer Academic Publishers,
101 Philip Drive, Norwell, MA 02061, U.S.A.

In all other countries, sold and distributed
by Kluwer Academic Publishers Group,
P.O. Box 322, 3300 AH Dordrecht, The Netherlands.

Printed on acid-free paper

All Rights Reserved
© 1992 International Astronomical Union

No part of the material protected by this copyright notice may be reproduced or utilized in any form or by any means, electronic or mechanical including photocopying, recording or by any information storage and retrieval system, without written permission from the publisher.

Printed in the Netherlands

TABLE OF CONTENTS

Preface		xvii
The Organising Committee		xix
List of Participants		xxi
Programme of the Astrochemistry Symposium (RP: Review Paper; CP: Contributed Paper)		
Astrochemistry of Cosmic Phenomena An Introduction(RP)	A.Dalgarno	1

BASIC STUDIES

Chemical reactions in Astrochemistry(RP)	B.R.Rowe	7
Neutral-Neutral Reactions: A new CRESU Study(CP)	C.Rebrion A.Defrance J.L.Queffelec B.R.Rowe D.Travers	13
ISO and Laboratory Astrophysics(CP)	L.J.Allamondola	15
Electrostatic Fragmentation of Dust Particles in Laboratory(CP)	J.Svestka E.Grun	17
Astrochemistry Library with Artificial Intelligence for Quality Control(CP)	Sheo S.Prasad P.Gangopadhaya	19
Contribution of PAHs to the Interstellar Extinction Curve(CP)	C.Joblin A.Leger P.Martin D.Defourneau	21
Cumulene Carbenes in Space and in the Laboratory(CP)	J.M.Vrtilek C.A.Gottlieb T.C.Killian P.Thaddeus J.Cernicharo M.Guelin G.Paubert	23

UV-Visible and Near IR Absorption Characteristics of Interstellar PAHS. I.$C_{10}H_8^+$(CP)	F.Salama L.J.Allamandola	25
Infrared Spectroscopy of Interstellar and Solar System Ice Analogs: Measurement of Optical Constants(CP)	D.M.Hudgins S.A.Sanford A.G.G.M.Tielens L.J.Allamandola	27
Formation of Organic Molecules by Formaldehyde Reactions in Astrophysical Ices at Very Low Temperatures(CP)	W.A.Schutte L.J.Allamondola S.A.Sandford	29
Computational Chemistry Aproach to Space Chemistry(RP)	Y.Ellinger	31
Metamorphism of Cosmic Dust: Diagnostic Infrared Signatures(CP)	Joseph A.Nuth III	39
Positronium in Astrophysical Condition(CP)	V.V.Burdyzha V.L.Kauts N.P.Yudin	41
Astrophysical Problems Involving Carbon Re-appraised(RP)	J.P.Hare H.W.Kroto	47
Advancement of Photoionization and Photodissociation Rates Relevant to Astrochemistry(RP)	S.P.Tarafdar	55
Grain Spectroscopy(RP)	L.J.Allamandola	65

EARLY UNIVERSE

Chemistry in the Early Universe(RP)	Paul R.Shapiro	73
Primordial Molecular Hydrogen Production and the Formation of Population III Objects(CP)	J.C.N.de Araujo R.Opher	83
Deuterium in the Diffuse	R.Ferlet	85

Interstellar Medium(RP)

The D/H Ratio in Molecular Clouds(RP)	A.G.G.M.Tielens	91
Deuterium in the Solar System(RP)	T.Owen	97

EXTERNAL GALAXIES

H_2 Emission from External Galaxies(RP)	Kimiaki Kawara	103
Interstellar Polarization in the Small Magellanic Cloud(CP)	C.Rodrigues A.M.Magalhaes G.Coyne V.Piirola	109
Molecular Abundances in Extragalactic Sources(RP)	C.Henkel R.Mauersberger	111
A 158 μm [CII] Map of NGC 6946: Detection in Extragalactic Atomic and Ionized Gas(CP)	S.C.Madden N.Geis R.Genzel F.Herrmann P.Poglitsch J.Jackson G.J.Stacy C.H.Townes	117
Molecule Formation in External Galaxies(RP)	T.J.Millar E.Herbst	121

DIFFUSE, TRANSLUCENT AND HIGH-LATITUDE CLOUDS

Observations of Diffuse and Translucent Clouds(RP)	Philippe Crane	127
The Interstellar C-H Stretching Band Near 3.4 μm: Constraints on the Composition of Organic Material in the Diffuse ISM(CP)	S.A.Sandford L.J.Allamandola A.G.G.M.Tielens K.Sellgren M.Tapia Y.Pendleton	133

Search for the 4430 ÅDIB in the Spectrum of Coronene(CP)	P.Ehrenfreund L.d'Hendecourt L.Verstraete A.Leger W.Schmidt	135
Theoretical Modelling of the Infrared Fluorescence by Interstellar Polycyclic Aromatic Hydrocarbons(CP)	W.A.Schutte A.G.G.M.Tielens L.J.Allamandola	137
Are Carbon Clusters the Cause of Interstellar Diffuse Bands?(CP)	S.P.Tarafdar K.S.Krishna Swamy C.Badrinathan D.Mathur	139
Diffuse,Translucent & High-Latitude Clouds:Theoretical Considerations(RP)	E.F.van Dishoeck	143
Abundance of CH^+ in Translucent Molecular Clouds:Problems for Shock Models?(CP)	R.Gredel E.F.van Dishoeck J.H.Black	153
Chemical Effects of Shear Alfven Waves in Molecular Clouds(CP)	S.B.Charnley W.G.Roberge	155
Interpretation of the Level Population Distribution of Highly Rotationally Excited H_2 Molecules in Diffuse Clouds(CP)	R.Wagenblast	157

QUIESCENT CLOUDS AND REGIONS OF STAR FORMATION

The Structure of Quiescent Clouds(RP)	E.Falgarone	159
Abundance of DCO^+ in Nearby Molecular Clouds(CP)	Harold M.Butner	169
Molecular Abundance Variations Among and Within Cold, Dark Molecular Clouds(RP)	Masatoshi Ohishi William M.Irvine Norio Kaifu	171

Results From a Three Position Spectral Scan in the Sgr B2 Molecular Cloud Core(CP)	P.Bergman Å.Hjalmarson P.Friberg W.M.Irvine T.J.Millar M.Ohishi S.Saito	179
What Species Remain to be Seen?(RP)	B.E.Turner	181
Observations of HNO(CP)	L.E.Snyder J.M.Hollis L.M.Ziurys Y.-J.Kuan	187
On the Possible Detection of Solid O_2 in Interstellar Grains(CP)	P.Ehrenfreund R.Breukers L.d'Hendecourt J.M.Greenberg	189
Interstellar H_3O^+ (CP)	T.G.Phillips E.F.van Dishoeck Jocelyn B.Keene	191
Carbon Isotopic Chemistry(RP)	W.D.Langer	193
Chemical Models of Active Regions(RP)	S.B.Charnley	199
Evolutionary Models of Interstellar Chemistry(RP)	S.S.Prasad	205
The Formation of Deuterated Molecules in Dense Clouds(RP)	T.J.Millar	211
Molecular Line Studies of Dense Core Motions(RP)	Yuefang Wu	217
Molecular Abundances in the Sgr A Molecular Clouds(CP)	Y.C.Minh W.M.Irvine P.Friberg	223

Molecular Line Observations Towards W58 and GL490(CP)	D.G.Roh H.R.Kim Y.C.Minh B.R.Auh B.C.Koo	225
Nitrogen Sulfide (NS) in Star Forming Regions(CP)	Douglas McGonagle William Irvine Young Minh	227
Radiation Transfer and Photochemical Effects in Inhomogeneous Dense Clouds(CP)	S.Aiello C.Cecchi Pestellini S.Tine	231
Observations of Shocked Regions(RP)	L.M.Ziurys	237
H_2O Maser Pumping by Shock Waves(CP)	J.R.D.Lepine A.Heske	245
The C:O Ratio in Dark Clouds with Cyclic Star Formation(CP)	L.A.M.Nejad D.A.Williams	249
Hot Cores and Cold Grains(RP)	C.M.Walmsley P.Schilke	251
The Environment of High-Mass Young Stellar Objects(RP)	Jean-P.Maillard George F.Mitchell	259
Infrared Molecular Spectroscopy of Orion(RP)	Neal J.Evans II	265
Chemical Gradients in the Orion Molecular Cloud(CP)	H.Ungerechts E.A.Bergin J.Carpenter P.F.Goldsmith W.M.Irvine A.Lovell D.Mcgonagle F.P.Schloerb R.L.Snell	271

The Polarized Water Maser in Orion(CP)	Z.Abraham J.W.S.Vilas Boas	275
OH Masers in Outflow Regions(CP)	M.D.Gray D.Field R.C.Doel	277
Abundances of Refractory Elements in the Orion Nebula(CP)	R.H.Rubin E.F.Erickson M.R.Haas S.W.J.Colgan J.P.Simpson R.J.Dufour	281
Panel Discussion: The CO/H_2 Abundance Ratio	E.F.van Dishoeck A.E.Glassgold M.Guelin D.T.Jaffe D.A.Neufeld A.G.G.M.Tielens C.M.Walmsley	285

CHEMISTRY OF INTERFACE REGIONS

Observations of Atomic Gas in Photodissociation Regions(RP)	Glenn J.White Rachael Padman	297
Observations of CII,CI and CO in Interface Regions(RP)	Jurgen Stutzki	303
CH^+ in Shocks,Cloud-Intercloud Interfaces, and Dense Photodissociation Regions(CP)	W.W.Duley T.W.Hartquist A.Sternberg R.Wagenblast D.A.Williams	309
Warm Molecular Clouds(RP)	D.T.Jaffe	311
Hot Core Chemistry: Gas Phase Molecule Formation in situ(CP)	S.B.Charnley A.G.G.M.Tielens	317

The Physical and Chemical Structure of Warm, Dense Regions:IC 63 and IC 443(CP)	D.J.Jansen E.F.van Dishoeck J.H.Black T.G.Phillips	319
Preliminary Observations of the Galaxy with a 7° Beam by the Cosmic Background Explorer(COBE) (RP)	Edward L.Wright	321
Chemical Composition Variations in the Galaxy(CP)	Ju L.Frantsman	327
The Chemistry of Photon-Dominated Regions(RP)	Amiel Sternberg	329
Near Infrared Emission of Neutral Carbon from Photon-Dominated Regions(CP)	V.Escalante A.Sternberg A.Dalgarno	333
Observations of Water and Molecular Oxygen in the Interstellar Gas(RP)	David A.Neufeld	335
Methanol Masers in W3(OH)(RP)	Q.Zeng	341
VLBI Observations of Amonia(9,6) Masers(CP)	Preethi Pratap Karl M.Menten Mark J.Reid J.M.Moran C.M.Walmsley	345
A Model for the Maser Source NGC 7538 IRS 1 (CP)	Preethi Pratap Lewis E.Snyder Wolfgang Batrla	347
Results of the Monitoring of a Very Strong Water Maser Event in W49N(CP)	E.Scalise,Jr. G.M.Pacheco A.M.Gomes Balboa Z.Abraham	349

NEAR STELLAR ENVIRONMENTS

H Atom Observations in Near-Stellar Environments(RP)	Luis F.Rodriguez	351
Infall in Collapsing Protostars(CP)	J.M.C.Rawlings T.W.Hartquist K.M.Menten D.A.Williams	357
Carbon to Helium Ratio in the Wind of Central Stars of Planetary Nebulae with WC Spectrum(CP)	J.A.de F. Pacheco R.D.D.Costa F.X.de Araujo D.Petrini	359
Chemical Composition of a Southern Planetary Nebulae Sample(CP)	J.A.de F.Pacheco R.D.D.Costa	361
Chemistry in Protoplanetary Nebulae(CP)	D.A.Howe T.J.Millar D.A.Williams	363
Molecules in Novae and Supernovae(RP)	J.M.C.Rawlings	365
Silicon Monoxide in Supernova SN1987A(CP)	Craig H.Smith D.K.Aitken P.F.Roche	371
Molecular Studies of Herbig-Haro Objects(RP)	Salvador Curiel	373
Molecules in Stellar Winds(RP)	A.E.Glassgold	379
Excited Hydrogen Chemistry in Protostellar Outflows(CP)	J.M.C.Rawlings J.E.Drew M.J.Barlow	387
Molecules in the Envelopes of Late-Type-Stars(RP)	R.Lucas	389
Lithium Abundance and Spacial Distribution of T Tauri Stars(CP)	J.Gregorio-Hetem J.R.D.Lepine	395

Sub-Millimetre Observations of SiCC in IRC 10216(CP)	L.W.Avery T.Amano M.B.Bell P.A.Feldman J.W.C.Johns J.M.MacLeod H.E.Mathews D.C.Morton J.K.G.Watson B.E.Turner S.S.Hyashi G.D.Watt A.S.Webster	399
A Molecular Line Survey of Carbon Star IRAS 15194-5115(CP)	L.-Å.Nyman R.S.Booth U.Carlstrom L.E.B.Johansson H.Olofsson R.Wolstencroft	401
Carbon Stars with Oxygen-Rich Circumstellar Envelopes(CP)	S.L.Martins S.J.C.Landaberry	403
G305.8 - 0.2 : A Young Object with a Dust and Gas Envelope(CP)	J.W.S.Vilas Boas E.Scalise,Jr. G.C.Sanzovo G.Mendes Pacheco A.M.Gomes Balboa	405
Bipolar Outflow and Turbulence in Molecular Clouds Due to Protostellar Alfven Waves(CP)	V.Jatenco-Pereira R.Opher	407
Spectral Synthesis of Cool Components of Symbiotic Stars(CP)	C.B.Pereira S.C.Landaberry	409
Radioactive Isotope ^{26}Al in the Interstellar Matter(Resulting From a Mass Loss by AGB Stars) (CP)	Ju L.Frantsman	411

Are Molecules Responsible for Origin of Cold Giants Mass Loss?(CP)	I.K.Shmeld V.S.Strelnitskij A.V.Fedorova O.V.Fedorova	413

SOLAR SYSTEM

Cometary Chemistry(RP)	Michael F.A'Hearn	415
Cometary Origin of Carbon and Water on the Terrestrial Planets(CP)	A.H.Delsemme	421
Interstellar and Meteoritic Organic Matter at 3.4 μm(CP)	P.Ehrenfreund F.Robert L.d'Hendecourt F.Behar	423
Diamagnetic Abundance Differentiation in the Solar System(CP)	R.Steinitz E.Kunoff	425
Cometary Molecules(RP)	Lewis E.Snyder	427
Laboratory Studies of Planetary Molecules and Ices:The Case of IO(CP)	F.Salama S.A.Sandford L.J.Allamandola	435
Type II Clathrate Hydrate Formation in Cometary Ice Analogs in Vacuo(CP)	D.F.Blake L.Allamandola S.Sandford D.Hudgins F.Freund	437
Observations of Parent Molecules in Comets at Radio Wavelengths: HCN,H_2S,H_2CO and CH_3OH(CP)	P.Colom D.Bockelee-Morvan J.Crovisier D.Despois G.Paubert	439
Radio Interferometric Observations of Cometary Molecules(CP)	P.Palmer L.E.Snyder I.de Pater	441

Gas Production Rates in Comets(CP)	A.A.de Almeida	443
Mass Loss Rates of Three Comets(CP)	P.D.Singh W.F.Huebner D.C.Boice I.Konno E.Scalise,Jr.	447
A Model of P/Temple 2 With Dust and Detailed Chemistry(CP)	W.F.Huebner D.C.Boice I.Konno P.D.Singh	449
Solar System - Interstellar Medium A Chemical Memory of the Origins(RP)	D.Despois	451
Formation of Comets: Constraints From the Abundance of Hydrogen Sulfide and Other Sulfur Species(CP)	D.Despois J.Crovisier D.Bockelee-Morvan P.Colom	459
A Diagnostic Spectral Indicator of the Exposure Age of an Asteroidal Surface(CP)	Joseph A.Nuth III	461
Summary	D.A.Williams	463
The Cosmic-Ray Ionization Rate*	S.Lepp	471
Dense Knots in High Latitude Molecular Clouds*	L.Blitz	477
Source Index		483
Index of Chemical Species		489
Subject Index		497
Author Index		505

(*) Received too late for proper incorporation and indexing.

PREFACE

The IAU symposium No. 150 "Astrochemistry of Cosmic Phenomena" was held at the beautiful and scenic town of Campos do Jordao, Sao Paulo, Brazil from August 5 to 9, 1991, and was attended by 111 registered participants with 17 accompanied guests from 19 countries. The symposium had a wide ranging discussion of the chemistry of astronomical environments with an emphasis on the description of molecular processes that critically influence the nature and evolution of astronomical objects and the identification of specific observations that directly address significant astronomical questions. The subject areas of the symposium included atomic and molecular processes at low and high temperatures and photon interactions, the chemical structure of molecular clouds in the Milky Way and in external galaxies, the chemistry of outflows and their interactions with the interstellar medium, the chemical connections between the interstellar medium and the solar system and pregalactic chemistry. The scientific programme comprised of review talks and contributed papers, with a general introduction by Professor A. Dalgarno and a final overview of the whole symposium by Professor D.A.Williams.

Financial supports from the Sao Paulo State Foundation Support (FAPESP), Brazilian National Research Council(CNPq), Finance Company of Studies and Projects (FINEP), Institute of Astronomy and Geophysics of University of Sao Paulo (IAG-USP) and International Astronomical Union (IAU) are greatfully acknowledged.

I am greatly indebted to Professor A. Dalgarno, Chairman of the Scientific Organising Committee, for the success of the conference and for organising the scientific contents of the program. I am thankful to Professors Walter F.Huebner and J.Mayo Greenberg who accepted the invitation to participate in the symposium. I am greatful to the members of Local Organising Committee, especially Dr.A.A. de Almeida who proved himself as the backbone of the committee. Last but not the least, I am thankful to Miss Maria Salete Vaceli who took a great pain in making the conference a great success.

March, 1992   P.D.Singh

Scientific Organising Committee

Chairman: A.Dalgarno

Members:
L.W.Avery (Canada)
M.Guélin (France)
N.Kaifu (Japan)
L.F.Rodriguez (Mexico)
P.L.Smith (USA)
G.Winnewisser (Germany)

L.Blitz (USA)
S.P.Tarafdar (India)
J.R.D.Lépine (Brazil)
P.D.Singh (Brazil)
D.A.Williams (England)

Local Organising Committee

Chairman: P.D.Singh

Members:
Z.Abraham
A.A.de Almeida
E.Scalise, Jr.

S.J.C.Landaberry
J.R.D.Lépine

IAU Symposium No. 150 was sponsored by
IAU Commission No. 34 and co-sponsored
by Commission Nos. 14,22,26,33 and 44.

LIST OF PARTICIPANTS

Abraham, Zulema, Instituto Astronomico e Geofisico, Univ.de Sao Paulo,
   Av. Miguel Stefano 4200, Agua Funda,Cep.04301, Sao Paulo, Brazil.

Aiello, S., Departimento di Astronomia e Scienza/Spazio,
   Via Leone Pancaldo, 3/45-50127, Firenze, Italy.

A'Hearn, Michael F., Astronomy Program, University of Maryland,
   College Park, MD.20742, U.S.A.

Allamandola, Lou, NASA-Ames Research Center, M/S 245-6,
   Moffett Field, CA. 94035, U.S.A.

Araujo, Francisco X.,Observatorio Nacional, Rua General Jose Cristino, 77,
   Sao Cristovao, Rio de Janeiro, Brazil.

Avery, Lorne, Herzberg Institute of Astrophysics,
   100 Sussex Drive, Ottawa, Ontario, Canada K1A 0R6.

Blitz, Leo, Astronomy Program, University of Maryland,
   College Park, MD. 20742, U.S.A.

Boulanger, Francois, Ecole Normale Superieure, Laboratoire
   de Radioastronomie, 24 Rue Lhomond, 75005, Paris, France.

Burdyzha, Vladimir, Astro Space Center, Profsoyuznaya St.
   84/32 Moscow, 117810, USSR.

Butner, Harold M., NASA Ames Research Center, MS 245-6,
   Moffett Field, CA. 94035,U.S.A.

Charnley, Steven, Space Science Division, MS 245-I, NASA
   Ames Research Center, Moffett Field, CA. 94035, U.S.A.

Colom, P., Observatoire de Meudon, DERAD, 5, place J.Jansen,
   F. 92195 Meudon Cedex, France.

Crane, Phil., European Southern Observatory 2/WD8046
   Garching, Germany.

Curiel, Salvador, Center for Astrophysics, 60 Garden St.,
   Cambridge, Mass. 02138, U.S.A.

Dalgarno, Alex., Center for Astrophysics, 60 Garden St.,
   Cambridge, Mass. 02138, U.S.A.

de Almeida, A.A., Instituto Astronomico e Geofisico, Universidade de
   Sao Paulo, Av. Miguel Stefano 4200, Sao Paulo, Cep. 04301, Brazil.

de Freitas Pacheco, J.A., Instituto Astronomico e Geofisico, Universidade de

Sao Paulo, Av. Miguel Stefano 4200, Cep. 04301, Sao Paulo, Brazil.

Delsemme, Armando H., 2509 Meadowwood, Toledo, Ohio 43606, U.S.A.

Despois, Didier, Observatoire de Bordeaux, B.P. 89, 33270, Floirac, France.

d'Hendecourt, Louis, Groupe de Physique des Solides, Universite Paris 6, T23, 4 Place Jussieu, 75251 Paris, Cedex 05, France.

Deutsch,L.K., MS245-6, NASA Ames Research Center, Moffett Field, CA. 94035, U.S.A.

Ellinger,Yves, Ecole Normale Superieure, Laboratoire de Radioastronomie, 24 Rue Lhomond, 75005, Paris, France.

Ehrenfreund,Pascale, Huygens Laboratory of Astrophysics, University of Leiden, Niels Bohrweg 2, 2333 RA Leiden, The Netherlands.

Escalante,V., Instituto de Astronomia, Apdo. Postal 70-264, Mexico D.F. 04510, Mexico.

Evans, Neal, Astronomy Department, University of Texas, Austin, Texas 78712, U.S.A.

Falgarone, Edit, Radioastronomie-Laboratoire de Physique E.N.S. 24 rue Lhomond 75005, Paris, France.

Ferlet, R., Institut d'Astrophysique, 98 bis Bd. Arago 75014, Paris, France.

Frantsman, Ju, Latvian Academy of Sciences, Radioastrophysical Observatory, 226524 Riga, Turgeneva 19, USSR.

Glassgold, A.E., Astronomy Department, 601 Campbell Hall, University of California, Berkeley, CA. 94720, U.S.A.

Gomez Balboa, A., INPE, DAS, P.O.Box 515, 12201 S.J.Campos, SP, Brazil.

Gray, Malcolm, School of Chemistry, University of Bristol, Cantock's Close, Bristol, BS8 1TS, England.

Greenberg, J.Mayo, Huygens Laboratory of Astrophysics, University of Leiden, Niels Bohrweg 2, 2333 RA Leiden, The Netherlands.

Gredel,R., European Southern Observatory, Casilla 19001, Santiago 19, Chile.

Guelin, Michel, Institut de Radioastronomie Millimetrique, 300 Rue de La Piscine, Domaine Universitaire, F. 38406 St. Martin d'Heres, France.

Henkel,C., MPIFR, Auf dem Hugel 69, 5300 Bonn 1, Germany.

Hjalmarson, Ake, Onsala Space Observatory, S-43900 Onsala, Sweden.

Howe, David, Maths Department, UMIST, Manchester M60 1QD, England.

Huebner, Walter F., Southwest Research Institute, 6220 Culebra Road,

Post Office Drawer 28510 San Antonio, Texas, U.S.A.

Hudgins, Douglas, 1228 W.McKinley Ave.,
Apt.#3, Sunnyvale, CA. 94086, U.S.A.

Irvine, W.M., FCRAO, 619 Lederle GRC,
University of Mass., Amherst, MA.01003, U.S.A.

Jaffe, D.T., Department of Astronomy, RL Moore Hall,
Univ. of Texas, Austin, Texas 78712, U.S.A.

Jansen, David, Sterrewacht Leiden, P.O.Box 9513,
2300 RA Leiden, The Netherlands.

Joblin, C., G-P-S tour 23 - Universite Paris 7 -2,
Place Jussieu, 75251 Paris, Cedex 05, France.

Kawara, Kimiaki, National Astronomical Observatory,
Mitaka, Tokyo, 181, Japan.

Khare, Bishun N., 306 Space Sciences Building,
Cornell University, Ithaca, New York 14853, U.S.A.

Kim, Hyo-Ryoung, Daeduk Radio Astronomy Observatory, ISSA,
San 36-1, Whaam-dong, Yousung-gu, Daejeon 305-348, Korea.

Kroto, Harold, School of Chemistry and Molecular Sciences,
University of Sussex, Brighton, BN1 9QJ, U.K.

Kunoff, Estelle, Physics Department, Ben-Gurian University,
84105, Beer Sheva, Israel.

Landaberry, S.J.C., Observatorio Nacional, Rua General Bruce
526, Sao Cristovao, Rio de Janeiro, RJ, Brazil.

Leger, Alain, GPS-Tour 23, Universite Paris 7, 4 Pl. Jussieu
75251 Paris Cedex 05, France.

Langer, William, MS 169-327, Jet Propulsion Laboratory,
Pasadena, CA. 91109, U.S.A.

Lepine, J.R.D., Instituto Astronomico e Geofisico, Univ. de Sao Paulo,
Av. Miguel Stefano 4200, Agua Funda, Sao Paulo, Cep 04301, Sao Paulo, Brazil.

Lepp, Stephen, Center for Astrophyscs, 60 Garden Street,
Cambridge, Mass.02138, U.S.A.

Liszt, Harvey S., NRAO, Edgemont Road, Charllotesville, VA 22901, U.S.A.

Lucas, Robert, IRAM, 300, Rue de La Piscine, Domaine
Universitaire, F. 38406, St.Martin d'Heres, France.

Madden S., Max Planck Institut fur Extraterrestrische Physik,
Giessenbachstrasse, D 8046, Garching Bei Muchen, Germany.

Maillard, J-P, Institut d'Astrophysique de Paris,
   98 bis Blvd Arago, Paris 75014, France.

Martins, S.Lorenz, Rua Jose Cristino 77, Sao Cristovao,
   Cep. 20921, Rio de Janeiro, Brazil.

Melnick, G.J., Center for Astrophysics, 60 Garden St.,
   Cambridge, Mass. 02138, U.S.A.

Mendes Pacheco,G., INPE/DAS, P.O.Box 515, 12201 S.J.Campos, SP., Brazil.

Millar,T.J., Mathematics Department, UMIST,P.O. Box 88,
   Manchester M60 1QD, England.

Minh, Young,Institute of Space Science and Astronomy,
   Whaam-dong Yusong Daejon 305-348, Korea.

Natta, Antonella, Observatorio di Arcetri, Largo Fermi 5,
   50125 Firenze, Italy.

Neufeld, David, Department of Physics and Astronomy, Johns Hopkins
   University, Homewood Campus, Baltimore, MD. 21218, U.S.A.

Nuth, Joseph, Code 691, NASA-Goddard Space Flight Center,
   Greenbelt, MD. 20771, U.S.A.

Nyman, Lars, ESO/La Silla, Casilla 19001, Santiago 19,Chile.

Ohishi, Masatoshi, Nobeyama Radio Observatory, Nobeyama,
   Minamimaku, Minamisaku, 384-13, Nagano, Japan.

Opher, R., Instituto Astronomico e Geofisico,Univ. de Sao Paulo,
   Av. Miguel Stefano 4200, Agua Funda, Cep. 04301, Sao Paulo, Brazil.

Owen, Tobias, Institute for Astronomy,
   2680 Woodlawn Drive, Honolulu, HI 96822, U.S.A.

Pecker, J.-C., College of France, Annexe, 3 Rue d'Ulm,
   75231 Paris Cedex 05, France.

Pereira, Claudio Bastos, Rua Jose Cristino 77, Cep. 20921
   Sao Cristovao, Rio de Janeiro, Brazil.

Pereira, Vera J., Instituto Astronomico e Geofisico, Av.Miguel Stefano
   4200, Agua Funda, Cep. 04301, Sao Paulo, Brazil.

Prasad, S.S., Lockheed Palo Alto Research Laboratory, (0/91-20, B255)
   3251 Hanover Street, Palo Alto, CA 94304, U.S.A.

Rawlings, J.M.C., Department of Physics, Nuclear Physics Laboratory,
   Keble Road, Oxford, OX1 3RH United Kingdom.

Robertry, Heloisa M.B., Observatorio de Valongo, UFRJ,
   Ladeira Pedro Antonio 43, Cep.20080, Rio de Janeiro, Brazil.

Rodriguez, Claudia V., Instituto Astronomico e Geofisico, Univ. de Sao Paulo,
   Av. Miguel Stefano 4200, Agua Funda, Cep. 04301, Sao Paulo, Brazil.

Rodriguez, Luis F., Instituto de Astronomia, UNAM,
   Apdo. Postal 70-264, Mexico, D.F. 04510, Mexico.

Roh, Duk-Gyoo, Daeduk Radio Astronomy Observatory, ISSA,
   San 36-1, Whaam-dong, Yousung-Gu, Daejeon 305-348, Korea.

Rowe, B., Departement de Physique Atomique et Moleculaire, Universite
   de Rennes I, Campus de Beaulieu, Rennes 35042, France.

Rubin, Robert, NASA Ames Research Center, Mail Stop 245-6,
   Moffett Field, CA 94035-1000, U.S.A.

Sabalisck, Nancy S.P., Instituto Astronomico e Geofisico, Univ. de Sao Paulo,
   Av. Miguel Stefano 4200, Agua Funda, Cep. 04301, Sao Paulo, Brazil.

Salama, Farid, NASA-Ames Research Center, Space Science
   Division, MS:245-6, Moffett Field, CA 940-1000, U.S.A.

Sandford, Scott, NASA/Ames Research Center, Mail Stop 245-6
   Moffett Field, CA. 94035, U.S.A.

Sanzovo, Gilberto Carlos, Inst. Astronomico e Geofisico, Univ. de Sao Paulo,
   Av. Miguel 4200, Agua Funda, Sao Paulo, Cep 04301, Brazil.

Scalise, Jr., E., Departamento de Astropfisica, INPE,
   Av. dos Astronauticas, P.O.Box 515, S.J.Campos,SP,Brazil.

Shapiro, Paul, Department of Astronomy,
   University of Texas, Austin, TX. 78712, U.S.A.

Shmeld, I., Latvian Academy of Sciences,
   Radioastrophysical Observatory, 226524 Riga, Turgeneva 19 USSR.

Singh, P.D., Instituto Astronomico e Geofisico, Univ. de Sao Paulo,
   Av. Miguel Stefano 4200, Agua Funda, Sao Paulo, Cep 04301, Brazil.

Singh, M., Department of Physics, University of Gorakhpur,
   Gorakhpur, U.P., India.

Snyder, Lewis E., 103 Astronomy Building,
   1002 W.Green Street, Urbana, Illinois 61801, U.S.A.

Steinitz, Raphael, Physics Department, Ben Gurien University
   84105 Beer Sheva, Israel.

Sternberg, Amiel, School of Physics & Astronomy,
   Tel Aviv University, Ramat Aviv, Israel.

Stief, Louis J., NASA/Goddard Space Flight Center,
   Code 690, Greenbelt, MD. 20771, U.S.A.

Stutzki, J., I.Phisikalisches Institut der Universitat zu Koln, Zulpicher
Strasse 77, Universitatsstrasse 14 D-5000 Koln 41, Germany.

Smith, Craig H., Department of Physics, University College, Australian
Defense Force Academy, Campbell ACT 2601, Australia.

Svestka, Jiri, Prague Observatory, Petrin 205,
118 46 Prague 1, Czechoslovakia.

Tarafdar, S.P., Theoretical Astrophysics Group,
TIFR, Colaba, Bombay-400005, India.

Tielens, Xander, MS 245-3, NASA Ames Research Center,
Moffett Field, CA. 94035, U.S.A.

Turner, Barry E., NRAO, Edgemont Road,
Charlottesville, VA 22901, U.S.A.

Vaceli, M. Salete, Inst. Astronomico e Geofisico, Univ. de Sao Paulo,
Av. Miguel Stefano 4200 Agua Funda, Cep. 04301, Sao Paulo, Brazil.

van Dishoeck, E.F., Leiden Observatory, P.O.Box 9513,
2300 RA Leiden, The Netherlands.

Vilas Boas, Jose Williams S., CRAAE-Departamento de Engenharia Civil,
Escola Politecnica da USP, Caixa Postal 8174, Cep. 05508 Sao Paulo, Brazil.

Vrtilek, J.M., Divison of Applied Sciences, Harvard Univ.,
29 Oxford Street, Cambridge, Mass. 02138, U.S.A.

Williams, D.A., Department of Mathematics,
University of Manchester, M60 1QD, England.

Wagenblast, Ralf, UMIST, Maths Department, P.O.Box 88,
Manchester, M60 1QD, England.

Walmsley, C.M., Max-Planck Institut Fur Radioastronomie,
Auf em Hugel 69, 53 Bonn, Germany.

White, G.J., Physics Department, Queen Marry and Westfield College,
University of London, Mile End Road, London, England.

Wright, Ned, 10541 Seabury Lane, 474-3633, Los Angeles,
CA. 90077,U.S.A.

Wu, Yuefang, 410 Building 16, Wei Xiu Yuan, Peking University,
Beijing 100871, People Republic of China.

Zeng, Quin, Purple Mountain Observatory,
Nanjing 210008, P.R.China.

Ziurys, L.M., Department of Chemistry,
Arizona State University, Tempe, AZ. 85287-1604, U.S.A.

ASTROCHEMISTRY OF COSMIC PHENOMENA: AN INTRODUCTION

A. DALGARNO
Harvard-Smithsonian Center for Astrophysics
60 Garden Street
Cambridge, MA 02138
USA

ABSTRACT. A general introduction is given to the subject matter of the symposium.

The first IAU Symposium on Astrochemistry took place in Goa, India in December 1985 (Tarafdar and Vardya 1987). It provided an overview of the subject matter of Astrochemistry which I take to be the formation, destruction and excitation processes of molecules in astronomical environments and the role of molecules in astronomical phenomena as diagnostic probes of the ambient physical conditions and, more fundamentally, as influences, often decisive, on the structure, dynamics and evolution of astronomical objects.

The subject has matured rapidly in the intervening years, driven like most of astronomy primarily by observational advances but helped by a developing understanding of the chemistry to the point where we can begin to use the chemical composition to draw conclusions about astronomical events, particularly those associated with star formation. We understand qualitatively that the chemical composition of the gas in regions of high and low mass star formation is modified in varying degrees by the direct effects of radiation from the stars and by the impact of material inflows and outflows accompanying the birth, evolution and death of stars. The radiation warms the grains and releases volatiles from the grain surfaces into the gas phase. The radiation also heats the gas and may change the chemistry into one in which endothermic reactions with molecular hydrogen became effective on short time scales. The outflows drive shocks which compress, heat and accelerate the gas. If rapid enough, the shocks erode and destroy the grains and dissociate and ionize the gas. The ionized gas recombines to produce a precursor radiation field that can photodissociate the molecules as they form in the cooling gas. The chemical composition of a cooling gas is quite different from that of a cold gas and quite different from that of a gas warmed by a slower non-dissociative shock.

The Orion K-L outflow region around the infrared source IRC 2 is a familiar example of a chemistry profoundly affected by massive star formation. Much higher abundances of HCO^+, SO and SO_2 occur in the plateau region than in the hot core. The relatively small abundance of HCO^+ in the hot core may reflect a low level of ionization in dense gas or it may be due to a high abundance of H_2O which converts HCO^+ to H_3O^+ by the reaction

$HCO^+ + H_2O \rightarrow H_3O^+ + CO$. The high abundances of H_2O, and of H_2S, may arise from their release from grain surfaces but in a warm non-dissociated gas they are produced by reactions initiated by $O+H_2 \rightarrow OH + H$ and $S + H_2 \rightarrow SH + H$. The reactions producing sulfur compounds are very slow in cold clouds.

The molecules H_2O and H_2O and H_2S are powerful coolants so that the temperature and the chemistry are intimately related.

Persuasive evidence that these regions have been modified by shock or radiation processing of grains are the observations of large abundances of deuterated species HDO, NH_2D, CH_3OD, DCN and D_2CO in the hot core and ridge regions which suggest that the molecules were produced in a cold gas, underwent fractionation at the low temperatures, were condensed on to grains and then released less than 10^4 years ago by the action of the newly formed star.

The effects of intense radiation fields on the chemical composition and cooling of interstellar gas in the absence of shocks have been explored quantitatively beginning with a study by Tielens and Hollenbach in 1985. We know from earlier work on diffuse clouds and a model of the dense cloud L134N that in the envelopes of molecular clouds there is a transition from a gas of atomic hydrogen to a gas of molecular hydrogen in which molecular hydrogen forms and shields itself from the destructive effects of the radiation field. The regions in clouds subjected to intense radiation fields have been called photodissociation regions (PDR) by Tielens and Hollenbach and photochemical regions by van Dishoeck. Because they are regions in which molecular formation is driven by photoionization and molecular destruction is determined by photodissociation, I suggest the term "photon-dominated regions". The acronym PDR can be retained.

Photon-dominated regions occur on a larger scale in starburst galaxies and infrared emission from excited rotation-vibration levels of molecular hydrogen, produced by ultraviolet pumping by the radiation field, have been detected in external galaxies. As in objects in our galaxy, the energy source for the excitation may come also from shocks associated with star formation and supernovae. A further source of excitation is pumping by photoelectrons ejected in the absorption of X-rays. There occur differences in the emission spectra resulting from the three mechanisms and they can be distinguished by observations of transitions from high-lying levels.

Photon-dominated regions are regions in which C^+ is converted to CO and in them the C^+/O and C/CO ratios are anomalously high. If the gas density is high enough to quench vibrationally excited H_2 molecules, the regions are warm and they are characterized by strong emissions of the oxygen fine-structure lines and of high rotational transitions of CO. The enhanced C^+ leads to infrared emission from metastable levels of neutral carbon. The emission line intensities provide a direct measure of the incident ultraviolet fluxes and the gas densities. Similar photon-dominated regions exist around hot stars and 21 cm radiation from the transition zones has been detected. The chemistry of PDR's has unique features that have yet to be fully explored.

Analysis of PDR's often indicates the existence of high density clumps with densities between 10^6 cm^{-3} and 10^7 cm^{-3}. Clumpy structure may be a common phenomenon in

PDR's and more generally in interstellar clouds.

Evidence has accumulated that even in the absence of active star formation, interstellar clouds have structure on small and large scales that may be reflected in a varying chemical composition and different spatial distributions of individual atomic and molecular species. Some part of the inferred chemical variations may be attributable to changing excitation conditions. Particularly sensitive are molecules with large dipole moments which are usually regarded as probes of high density regions. They are also the molecules most readily excited by electron impacts. Multifrequency multilevel studies are particularly valuable in separating out the possible excitation mechanisms.

On a large scale, the chemical compositions of interstellar clouds with similar densities and temperatures like TMC-1 and L134N differ, suggesting that one cloud is at an earlier stage in its evolution than the other. It may not be clear which is younger and which is older but the comparison raises again the possibility that properly interpreted the molecular composition can serve as a chemical clock. However there occur changes in composition in regions of the same clouds over scales as small as 0.2 pc. Swade has suggested the differences in L134N are due to an oxygen abundance gradient across the core of the cloud but conceivably they indicate the presence of a low-luminosity protostar that has caused some local heating. A heated gas cools more rapidly than its chemistry relaxes and a region though cold may maintain a warm chemical composition over a considerable length of time.

Such complexities make it difficult to address the important question of the distribution of a given element into its atomic, molecular and ionic forms. Gas phase oxygen that is not taken up into CO and OH is mostly in the forms O, H_2O and O_2 and gas phase nitrogen that is not taken up into NH_3 is mostly in the form N and N_2. None of these is detectable in cold clouds. A coordinated study of different molecular species may permit some progress. For example, the abundance ratio CH_2CN/CH_2CO must be related to the atomic abundance ratio N/O.

Of cosmological significance is the D/H ratio which is a measure of the baryon density in the early Universe. Of galactic significance is the cosmic ray flux and its variations. The D/H ratio may be inferred from observations of deuterated molecules and the cosmic ray flux from observations of molecular ions, provided that realistic models of the chemistry can be constructed.

Chemical models of a quiescent slowly varying interstellar cloud have been valuable in defining the processes that enter into interstellar chemistries but they were recognized to be no more than idealizations if only because they did not take account of the loss of gas atoms and molecules by depletion on to grains. Severely depleted regions are rare at best so that there must be mechanisms for returning the material to the gas phase or depletion must be inhibited at high densities or depleted gas has a short lifetime perhaps because it rapidly collapses.

Interstellar chemistry takes place in a complex, continuously changing, dynamically active medium in which gas forms into clouds containing clumps moving in an interclump gas and on a smaller scale dense cores moving in an intercore gas subjected to and disrupted by radiation, winds and shocks arising from star formation and during stellar evolution. The chemical response and the dynamical events are closely coupled and together control the continuing evolution of star-forming regions. It is one task of astrochemistry to turn this qualitative description into quantitative models and the first steps have been taken.

Chemistry may influence the clump size distributions in clouds through its control of the ionization level. Turbulence provides support against gravitational collapse. If turbulence is a superposition of Alfven waves, it will be dissipated at a rate which is sensitive to the fractional ionization. The fractional ionization varies with depth into clouds from a value of about 10^{-4} at the surface due to C^+ to about 10^{-5} at a visual extinction of unity due to S^+ followed by narrow zones of Si^+ and Fe^+ until the molecular region is reached with a fractional ionization of 10^{-7}. Still deeper the ionization may become atomic again or the positive charge may reside on grains. The zones of diminishing ionization have different dissipation rates and different ambipolar diffusion rates. The turbulent motions may also affect the chemistry directly and perhaps be dissipated by chemical reactions.

Star formation is accompanied by a protostellar wind that collides with the remnant core material. The winds may be mostly neutral. If the density is high, the wind cools and molecular synthesis can occur. The chemistry is hydrogen rich and molecular hydrogen can be formed by reactions catalysed by electrons and protons as in the early Universe. The heart of the chemistry is the hydroxyl radical which leads to other species, particularly CO, as in the chemistry of a dissociative shock and in photon-dominated regions.

A similar chemistry operates in the envelopes of supernovae, where H_3^+ may have been detected. CO and SiO have been identified in the core of SN 1987A but there it is hydrogen poor and the major source of CO is the radiative association of C and O atoms.

In circumstellar shells, molecular observations are important in the determination of mass loss rates, but they also pose interesting questions of chemistry in a region where grains and photodissociation play major roles. The study of circumstellar shells may help in understanding the influence of grains in interstellar chemistry. Circumstellar chemistry has some similarities to the chemistry of cometary materials where processing by the Sun may be minimal and the connection between the solar system and the interstellar medium can be investigated. Comparing the chemical compositions, the connection is not immediately apparent and more observational data are needed to make progress.

Diffuse and translucent cloud studies and high latitude cloud studies are probably the major areas where the chemical schemes can be tested quantitatively. Problems still remain concerning the abundances of CH^+, CO and CN that may not be resolvable by postulating shocks and a reduced ultraviolet radiation field. Perhaps dense clumps are present even in diffuse clouds. Perhaps polycyclic aromatic hydrocarbons, suggested as major repositories of carbon in the interstellar medium, participate in the chemistry.

Our galaxy provides a test bed for the effectiveness of molecules as indicators of density, temperature, radiation field intensities and cosmic ray fluxes that can then be used in the study of molecular regions of external galaxies. Many species have been detected in addition to vibrationally excited H_2 and the ubiquitous CO. Isotopic variants have been found though as yet no deuterated species.

The carbon monoxide is usually taken to be a quantitative measure of the total mass. It is important to identify other species that might test this assumption. In any event, the observations of other molecules promise to provide much insight into the nature of external galaxies and greatly extend the domain of astrochemistry.

The early Universe is an interesting area of astrochemistry which despite its chemical simplicity may not be fully explored. The formation of H_2 was a determinant of the relict electron density after recombination and H_2 played a crucial role in the initial mass spectrum of the first distinct astronomical entities. Other species may have been important in galaxy formation.

Basic data underly all the theoretical and observational advances. Here much progress has been made in quantum-chemical calculations of transition frequencies, dipole moments and excitation cross sections, in measurements at low temperatures of ion-molecule reactions, in measurements of the end-products of dissociative recombination, in measurements of the ultraviolet absorption spectrum of CO and the resulting calculations of the photodissociation rate of CO and in measurements of the rate coefficient for dissociative recombination of H_3^+ which show it to be again (probably) a fast reaction. Exciting developments have taken place in interstellar and laboratory studies of polycyclic aromatic hydrocarbons and carbon clusters and in spectroscopy of molecules on surfaces. A quantitative assessment has been made of the photodissociation rates induced by cosmic rays by the Prasad-Tarafdar mechanism. Much remains to be done in all the areas of basic studies.

I hope that IAU Symposium #150 will show that astrochemistry addresses significant astronomical issues. I expect to hear specific discussions that demonstrate the utility of astrochemistry and that identify the critical questions that need to be answered if astrochemistry is to develop further as a major branch of astronomy.

The ideas presented in this introduction were advanced by many different individuals. I have given no references. They can be found in the Invited and Contributed Papers that follow.

My research in astronomy is supported by the National Science Foundation under grant AST 89-21939 and the National Aeronautics and Space Administration under grant NAGW-1516.

QUESTIONS AND ANSWERS

V.Burdyuzha: Why molecules with deuterium are more abundant in Ori region than in other regions of our galaxy?

A.Dalgarno: The interesting question is why they are abundant in a warm gas. Deuterated molecules are observed to be abundant in cold interstellar clouds.

D.A.Williams: Are we sometimes led astray by our terminology: e.g. what is a quiescent cloud? Such objects as TMC-1 may in fact be transient?

A.Dalgarno: I tried to use the term "apparently quiescent" because of consideration such as you present.

J.C.Pecker: Obvious "clumps" in ISM give place to some differences in the observed chemical molecular composition. Could it be due (in part?) to their fine structure, and the local unresolved physical inhomogeneities? And could the observed differences be then a somewhat spurious effect of the diagnosis difficulties?

A.Dalgarno: I think it is due to physical inhomogeneities. The question is what then is the origin of the inhomogeneities.

CHEMICAL REACTIONS IN ASTROCHEMISTRY

B.R. ROWE
Département de Physique Atomique et Moléculaire, URA 1203 du CNRS,
Université de Rennes I, 35042 Rennes Cedex, France.

ABSTRACT. This paper is devoted to chemistry in the gas phase dealing firstly with ion-molecule reactions at extremely low temperature. The experimental techniques that have been used in this field are shortly presented and the reactions that have been studied using the CRESU(S) method reviewed. In the second part, the most recent measurements concerning dissociative recombination are discussed, including studies of branching ratio and new determination of the rate coefficient for H_3^+ ions.

1. Introduction

As shown throughout these proceedings, it is now apparent that the physics and chemistry of interstellar clouds is far more complicated than the simple picture described by the first gas phase models of a quiescent, stationnary cloud. Chemistry on grain surfaces and in shocked regions cannot be neglected in the processes that lead to molecule formation and the evolution of the cloud itself has to be considered. However gas phase chemistry is certainly involved in the formation of a major part of the molecules observed in interstellar space. In the thousands of chemical and physical processes that have been considered in various models there is always a lack of experimental data, leading to major uncertainties concerning the predicted abundances of various species. The last three years have brought remarquable breakthroughs in two fields : ion-molecule reaction rate coefficients have been measured down to extremely low temperatures and, in a few cases, the product channels and branching ratio in the dissociative recombination (hereafter DR) of polyatomic ions have been determined. On an other hand a large controversy broke out concerning the value of the DR rate coefficient of H_3^+, which plays a key role in cloud chemistry. The most recent measurements yield a large coefficient in contradiction with the extremely small value reported by Adams and Smith [1].

2. Ion-molecule reactions at extremely low temperature

2.1. EXPERIMENTAL TECHNIQUES

The study of rate coefficients at extremely low temperature is very challenging from an experimental point of view and only four research groups have succeeded in such measurements. Cryogenic cooling has a great limitation since neutral reactants other than helium and hydrogen condense on the walls of the reaction cells. However, it has been used in the static drift tube of H. Böhringer et al [2] and in the ion trap of G. Dunn and co-workers [3]. These experiments were difficult to operate and are no longer used. Another cooling strategy is to generate the very low temperature with a supersonic expansion. Then it is possible to use neutral reactants in supersaturation conditions. The CRESU (and CRESUS which allows mass selection of the ions) apparatus has been described in detail elsewhere [4], [5], [6]. The most important feature of this experiment is that true thermal conditions exist locally (excepted for H_2 and D_2 rotational levels). The rotationnal levels are in thermal equilibrium with a kinetic temperature which corresponds to a local Maxwell-Boltzmann distribution of velocity. This is not the case in the free jet experiment recently developed in Tucson (USA) by M. Smith et al [7]. In this last case, the jet is obtained by expansion of premixed gases through a small orifice and there are strong density gradients in the flow. Consequently there is not a real local thermodynamical equilibrium in this experiment and the mean effective estimated rotational "temperature" is often far greater than the effective kinetic "temperature".

2.2. MAINS RESULTS CONCERNING ION MOLECULE REACTIONS

The rate coefficient of many ion-molecule reactions can be calculated as a capture rate coefficient, i.e. using only the long range part of the potential energy surface on which the reaction takes place.

For molecules having no permanent dipole moments, the ion-induced dipole potential leads to the Langevin formula which predicts a rate coefficient independent of temperature. For polar molecules the calculation of the capture rate coefficient is far more complicated and can been performed using various methods which have been reviewed elsewhere [8]. Table 1 summarizes the CRESU results obtained for reactions between ions and polar molecules and shows that a strong increase of the rate coefficient is obtained at low temperature, which is well reproduced by theoretical calculations. An interesting feature pointed out by theory is the strong influence of rotational state on the reactivity. This effect can be highlighted when considering the measurements of the rate coefficient of $NH_4^+ + NH_3$ at 2.5 K by Hawley et al. [7] using the Tucson apparatus, yielding a much lower value than expected on theoretical basis. This can be explained if one considers that rotation in this experiment is not fully relaxed, and corresponds to a temperature that could be ten times larger than the kinetic one.

CRESU(S) measurements have shown that, when the rate coefficient of an ion-molecule reaction is much smaller than the capture rate it often increases at lower temperatures, sometimes approaching k_c at temperatures close to 0 K [9], [10], [11]. However this is not a general rule as shown by the fact that the $He^+ + H_2$ reaction remains extremely slow at low temperature [12].

A very interesting reaction from an astrochemical point of view is the reaction of N^+ with H_2. It has been studied down to 8 K using normal and para hydrogen as well as HD and D_2 and exhibits a strong decrease of the rate coefficient with temperature. Analysis of the results shows that the endothermicity of the reaction $N^+ + H_2 \longrightarrow NH^+ + H$ for ground state reactants and products is 18 ± 2 meV [13]. Implication of this finding for interstellar ammonnia formation has been discussed by Herbst et al. [14].

An important class of ion-molecule reaction is radiative association. Unfortunately the few results available to date are often contradictory as in the case of $CH_3^+ + H_2$. The results obtained by Barlow et al. [15] and Gerlich and Kaefer [16] at respectively 13 and 80 K would imply a strong temperature dependence which is extremely unlikely on the basis of theoretical considerations [17] and of the CRESUS study of the analogous ternary association between 20 and 80 K [18].

3. Dissociative recombination of molecular ions

3.1. BRANCHING RATIOS IN DISSOCIATIVE RECOMBINATION

A good knowledge of the branching ratios in dissociative recombination is extremely important for interstellar chemistry. Only recently a few measurements have been performed concerning the yield of OH radicals or H atoms in DR reactions ([19], [20], [21]) using either laser induced fluorescence or VUV resonance absorption techniques. One of the major findings of these studies was that the channel leading to H_2O in H_3O^+ DR has a branching ratio lower than 0.34. Also they showed that H_3O^+ and $OCSH^+$ cannot be an important source of H_2S and OCS respectively in interstellar clouds, since H atoms production is small in recombination of these ions.

3.2. RATE COEFFICIENTS IN DISSOCIATIVE RECOMBINATION

Although a wide variety of reliable techniques now allows α values to be determined, some important contradictions remain as in the various studies concerning the DR rate coefficient of H_3^+ (see table 2).

A new Flowing Afterglow apparatus has been used by the author and his coworkers to study the DR of several ions including H_3^+. The new feature of this apparatus is that a movable mass spectrometer allows the ion densities to be measured in the recombination zone. These measurements confirmed Amano's results, i.e. that dissociative recombination of H_3^+ ions with low vibrationnal levels (including v=o) is in the range $1.1 - 1.5 \ 10^{-7}$ cm^3 s^{-1} ([22], [23]). Other ions studied with this apparatus include HCO^+ and HCS^+ [24], [25] with $\alpha = 2.4 \ 10^{-7}$cm^{-3} s^{-1} and $7 \ 10^{-7}$cm^{-3} s^{-1} respectively.

A detailed discussion of the problems associated with the vibrational states of the ions is given in the references above.

Table1: reaction rates for ion-polar molecules ($10^{-9} cm^3 s^{-1}$), uncertainty is $\pm$ 30%

reactants	T=27K	T=30K	T=68K	T=300K
$He^+ + HCl$	11.0		4.6	3.3
$He^+ + SO_2$	8.2		6.5	4.3
$He^+ + H_2S$	5.5		4.6	2.8
$He^+ + NH_3$	4.5		3.0	1.65
$He^+ + H_2O$	4.3		1.8	0.48
$C^+ + HCl$	3.8		1.9	1.0
$C^+ + SO_2$	5.7		4.1	2.3
$C^+ + H_2S$	4.8		3.1	1.7
$C^+ + NH_3$	4.6		3.2	2.3
$C^+ + H_2O$	12.0		5.2	2.5
$H_3^+ + SO_2$		11.0		
$H_3^+ + H_2S$		6.5		3.4
$H_3^+ + NH_3$		9.1		4.2
$N^+ + NH_3$	5.2		3.2	2.4
$N^+ + H_2O$	9.9		6.0	2.8

Table2: Summary of the main results for $\alpha(H_3^+)$ at 300K

$\alpha(H_3^+)$ ($cm^3 s^{-1}$)	method	reference
$2.3 \cdot 10^{-7}$	μwa-ms[a]	[26]
$2.5 \cdot 10^{-7}$	inclined beam	[27]
$\approx 2.1 \cdot 10^{-7}$	merged beam	[28]
$1.5 \cdot 10^{-7}$	ion trap	[29]
$2.1 \cdot 10^{-7}$	merged beam	[30]
$\leq 2 \cdot 10^{-8}$	FALP	[31]
$1.5 \cdot 10^{-7}$	μwa-ms[a]	[32]
$\leq 10^{-11}$	FALP	[1]
$\approx 2 \cdot 10^{-8}$	merged beam	[33]
$1.8 \cdot 10^{-7}$	I.R. spectroscopy	[34]
$1.5 \cdot 10^{-7}$	FALP	[23]

[a]: μwa-ms = microwave afterglow mass spectrometer

[1] N.G. Adams, D. Smith, In: Vardya, M.S. & Tarafdar, S.P. (eds), IAU Symposium 120, Astrochemistry, Reidel, Dordrecht, 1987.
[2] H. Bohringer, F. Arnold, Int. J. Mass Spectrom. Ion Proc. **49**,61, 1983.
[3] S.E. Barlow, G.H. Dunn, M. Schauer, Phys. Rev. Let. **52**, 902, 1984
[4] B.R.Rowe, G.Dupeyrat, J.B.Marquette, P. Gaucherel, J. Chem. Phys. **80**,4915, 1984
[5] G. Dupeyrat, J.B. Marquette, B.R. Rowe, Phys. Fluids, **28**, 1273,1985
[6] B.R.Rowe, J.B.Marquette, C.Rebrion, J.Chem.Soc. Faraday Trans. 2, **85**, 1631, 1989
[7] M. Hawley, T.L.Mazely, L.K. Randenyia, R.S. Smith, X.K. Zeng, M.A. Smith, Int. J. Mass Spectrom. Ion Proc. **97**, 55, 1990
[8] C. Rebrion, J.B. Marquette, B.R. Rowe and D.C. Clary, Chem. Phys. Lett. **143**, 130, 1988
[9] B.R. Rowe, G. Dupeyrat, J.B. Marquette, D. Smith, N.G. Adams, E.E. Ferguson, J. Chem. Phys. **80**, 241, 1984
[10] P. Gaucherel, J.B. Marquette, C. Rebrion, G. Poissant, G. Dupeyrat, B.R. Rowe, Chem. Phys. Lett. **132**, 63, 1986
[11] C. Rebrion, B.R. Rowe, J.B. Marquette, J. Chem. Phys. **91**, 6142, 1989
[12] M.M.Schauer, S.R.Jefferts, S.E. Barlow, G.H. Dunn, J. Chem. Phys. **91**, 4593, 1989
[13] J.B. Marquette, C. Rebrion, B.R. Rowe, J. Chem. Phys. **89**, 2041, 1988
[14] E. Herbst, D.J. Defrees, A.D. Meslean, Astrophys. J. **321**, 898, 1987.
[15] S.E. Barlow, G.H. Dunn, M. Schauer, Phys. Rev. Letters **52**, 902, 1984
[16] D. Gerlich, G. Kaefer, Astrophys. J. **347**, 849, 1989
[17] J.W.M. Smith, Astrophys J. **347**, 289, 1989
[18] B.R.Rowe, J.B.Marquette, C.Rebrion, J.Chem.Soc. Faraday Trans. 2, **85**, 1631, 1989
[19] N.G. Adams, C.R. Herd, D. Smith, J. Chem. Phys. **91**, 963, 1989
[20] N.G. Adams, C.R. Herd, M. Geoghegan, D. Smith, A. Canosa, J.C. Gomet, B.R. Rowe, J.L. Queffelec, M. Morlais, J. Chem. Phys. **94**, 4852, 1991
[21] C.R. Herd, N.G. Adams, D. Smith, Astrophys. J, **349**, 388, 1990
[22] A. Canosa, B.R. Rowe, J.B.A. Mitchell, J.C. Gomet and C. Rebrion, Astron. Astrophys. **248**, L21, 1991
[23] A. Canosa, B.R.Rowe, J.B.A. Mitchell, J.L.Queffelec, J.C. Gomet, J. Chem. Phys. to be submitted
[24] B.R. Rowe, J.C. Gomet, A. Canosa, C. Rebrion, J.B.A. Mitchell, J. Chem. Phys, in press
[25] H. Abouelaziz, J.C. Gomet, B.R. Rowe, C. Rebrion, D. Travers, J.L. Queffelec, Chem. Phys. Lett. to be submitted
[26] M.T. Leu, M.A. Biondi, R. Johnsen, Phys. Rev. A, **8**, 413, 1973
[27] B. Peart, K.T. Dolder, J. Phys. B **7**, 1948, 1974
[28] D. Auerbach, R. Cacak, R. Caudano, T.D. Gaily, C.J. Keyser, J.Wm McGowan, J.B.A. Mitchell, S.F.J. Wilk, J. Phys. B **10**, 3797, 1977
[29] D. Mathur, S.U. Khan, J.B. Hasted, J. Phys. B **11**, 3615, 1978
[30] J. Wm McGowan, P.M. Mul, V.S. D'Angelo, J.B.A. Mitchell, P. Defrance, H.R. Froelich, Phys. Rev. Lett. **42**, 373, 1978
[31] N.G. Adams, D. Smith, E. Alge, J. Chem. Phys. **81**, 1778, 1984
[32] J.A. McDonald, M.A. Biondi, R. Johnsen, Planet. Space Sci. **32**, 651, 1984
[33] H. Hus, F.B. Yousif, A. Sen, J.B.A. Mitchell, Phys. Rev. A **38**, 658, 1988
[34] T. Amano, J. Chem. Phys. **92**, 6492, 1990

QUESTIONS AND ANSWERS

D.A.Williams: How many H-atoms per reaction arise from CH_5^+ recombination?

B.Rowe: This is one of the cases where all the measurements have a very good internal consistency (between Rennes and Birmingham experiments and with various precursor ions). The yield of H atoms is close to 1.2.

A.Dalgarno: Will measurements be possible of reactions of ions like He^+ with reactive species such as atomic oxygen and atomic nitrogen?

B.Rowe: There is in principle no difficulties in doing reactions with oxygen and nitrogen atoms (carbon is much more difficult). However, if the reaction of the ion with the corresponding diatomic molecules is fast and the reaction with the atom slow, the measurement could be extremely difficult.

M.Guelin: The new recombination rate value you derive for H_3^+ restores the very low limits as the electron density derived in the dark cloud cores from DCO^+ and $H^{13}CO^+$ observations (e.g. Guelin et al. 1982, Astron.Astrophys. 107,107). These limits ($10^{-9} < x_e < few \times 10^{-7}$) have important implications as metal depletion, ambipolar diffusion time scales, cosmic ray ionization rate, etc.

B.Rowe: I thank you for this comment and I am glad to see that laboratory works are really useful in astrochemistry.

NEUTRAL-NEUTRAL REACTIONS: A NEW CRESU STUDY

C. Rebrion, A. Defrance, J.L. Queffelec, B.R. Rowe and D. Travers,
Département de Physique Atomique et Moléculaire
Université de Rennes 1
35042 RENNES Cedex
FRANCE

ABSTRACT: A new CRESU experiment devoted to neutral-neutral chemistry at low temperature is being built. The first reactions to be studied are $N+NO$ and $CN+O_2$.

1. Introduction

Models of interstellar clouds usually emphasize the importance of ion chemistry in spite of the fact that recent theoretical studies [1] [2] have stressed out that many neutral reactions may have no barrier on the potential surface. The dynamics of such reactions would be dominated by long-range forces and in some cases the reaction could be very fast.

In a comparison of two current models, Millar et al. [3] have shown that the calculated abundances of several molecules depend critically on the choice of the rate coefficient for some neutral reactions. However, models have to rely on functional forms chosen either for lack of better information or from theoretical guidance. Even at liquid nitrogen temperature there is a scarcity of experimental data concerning gas phase neutral chemistry [4] and no data at temperatures relevant for interstellar chemistry.

2. The experiment

The CRESU method, based on the supersonic expansion of a buffer gas through a Laval nozzle, is now well known in the astrochemical community. The first version, located in Meudon (France), is devoted to ion-molecule reactions at very low temperature. As ions make very easily clusters and also because of termolecular reactions, a very low pressure is needed. Such problems are much less important in neutral chemistry and higher pressure in the flow can be used. The DPAM is building a new CRESU experiment working in a higher pressure range which will be used to study neutral chemistry down to 10K.

Two glass nozzles are available: one helium nozzle working at $T=140$ K, and one nitrogen nozzle working at $T=170$ K (lower temperatures can be reached when the reservoir of the nozzles is cooled with liquid nitrogen: $T=37$ K and $T=45$ K respectively). Their characteristics have been tested using two independent methods (impact pressure measurements and time of flight measurements) and both sets of results are in good agreement.

3. N + NO → N_2 + O

The data available in the 160-400 K range agree to find its rate coefficient temperature independent. However, theoretical calculations by Jaffe et al. [5] show that an energy barrier of several meV could occur on the reaction path, which should induce a reactivity variation at 45 K.

The cooled nitrogen nozzle will be used for this study. A mixture of N_2 and NO will be injected through the nozzle and a micro-wave discharge will dissociate N_2 into $N(^4S)$, $N(^2D)$ and $N(^2P)$. The population of the metastable states of N will be checked, in order to take into account the deexcitation processes and the different reaction rates with NO.

4. CN (v=0) + O_2 → NCO + O

The sketch of the experiment needed for this study in drawn below. The reaction will be studied with the helium nozzle. The CN radical is obtained by laser photolysis of NOCN (synthetized itself from NO, Cl_2 and AgCN) at 532 nm. Its decay will be followed by laser induced fluorescence tuned to a rotational line in the B-X(0,0) band.

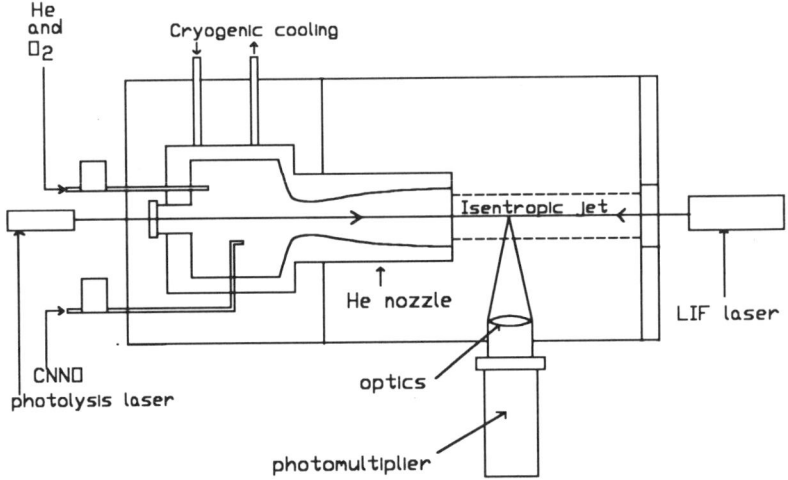

[1] D.C. Clary, Mol. Phys. **53**, 3, 1984
[2] M.M. Graff, Astrophys. J. **339**, 239, 1989
[3] T.J. Millar, C.M. Leung, E. Herbst, Astron. Astrophys. **183**, 109, 1987
[4] M.J. Frost, P. Sharkey, I.W.M. Smith, Faraday Discuss. Chem. Soc. 1991, in press
[5] S.P. Walch, R.L. Jaffe, J. Chem. Phys. **86**, 6946, 1987

ISO AND LABORATORY ASTROPHYSICS

L.J. ALLAMANDOLA
NASA-Ames Research Center
Mountain View, CA 94035
USA

ISO, the Infrared Space Observatory is capable of measuring spectra across most of the mid-infrared, the region from 4000 to 500 cm^{-1} (2.5-20 μm). Of these 3500 cm^{-1}, about 670 have been inaccessible to astronomers due to strong atmospheric absorptions and about 750 cm^{-1} require an airborne platform, making access limited. ISO will provide high quality spectra in the 19% of the celestial mid-infrared which has never been studied and dramatically increase the number of spectra in the 21% with limited access. Thus ISO has the potential to nearly double our knowledge of the mid-IR spectroscopic properties of the cosmos. This knowledge, in turn, will revolutionize our understanding of its chemical make-up because of the mid-infrared's powerful compositional diagnostic capabilities.

In order to fully exploit this potential requires that laboratory data are available with which one can analyze the spectra ISO measures. To maximize the scientific return from this limited lifetime mission it is particularly important to have as much data available as possible before launch in order to plan an intelligent follow-up program before the cryogen runs out. This dictates that laboratory studies focus on the spectral properties of astrophysically relevant materials with important bands in the new regions and regions of limited access.

As I am most familiar with the IR absorption spectra of dense and diffuse clouds and with IR emission spectra associated with aromatics, I will summarize the work needed in order to be prepared for the discoveries in these areas. The table summarizes what potentially will be probed. As with all pioneering spectroscopic studies unexpected features will be found. These can be recognized only if archives of the appropriate laboratory data are available.

To interpret the absorption spectra infrared spectroscopic studies should be carried out on realistic astrophysical ice analogs containing the species and subgroups listed in the third column of the table. For the emission bands, studies are needed on isolated, neutral and ionized polycyclic aromatic hydrocarbons which are hydrogenated (PAHs) and deuterated (PADs).

With the appropriate laboratory data, the absorption spectra ISO obtains will shed light on D/H ratios, composition, abundances, and chemical processing of ices and dust in dense molecular clouds, the diffuse interstellar medium and in cold disks. In many cases gas phase column densities of the same molecules can be measured along the same line of sight. Emission spectra will shed light on PAH D/H ratios and compositional variations as a function of object type along the H-R diagram and the relationship of PAHs (if any) to the IR cirrus.

TABLE OF WAVELENGTH REGIONS AND TARGET BANDS RELATED TO INTERSTELLAR DUST WHICH WILL BE PIONEERED BY ISO.

OPAQUE REGION		TARGET BAND CARRIERS	
Wavelength (μm)	Obscurer	Absorption	Emission
2.5 - 2.75	[H_2O]	Free H_2O	?
4.1 - 4.4	[CO_2]	CD, OD, CO_2, C≡N stretches	PAD:CD stretch
*5.1 - 9	H_2O, CH_4, N_xO_y,...	Organics >C=O	PAH:C-C stretch (most intense)
9.5 - 10	[O_3]	O_3	?
14 - 16	[H_2O, CO_2]	CD, OD, O_3, CO_2, =C-H bends	PAD:CD bend
>50	[H_2O]	Ice phonons	PAH plane bends

* Limited access - requires airborne platform.

ELECTROSTATIC FRAGMENTATION OF DUST PARTICLES IN LABORATORY

J. SVESTKA[1] and E. GRÜN[2]
[1]*Prague Observatory,*
11846 Prague 1, Czechoslovakia
[2]*Max-Planck-Institut für Kernphysik,*
6900 Heidelberg 1, Germany

ABSTRACT. Experimental laboratory work on simulation of the electrostatic fragmentation was started with loosely bound Al_2O_3 particles of 1 to 10 micrometers size. These particles were suspended in an electrodynamic quadrupole inside a vacuum chamber and electrically charged by ion beams of energies up to 5 keV. The electrostatic fragmentation was observed and derived tensile strengths of the particles range from 10^3 to 10^5 Pa what is compatible with theoretical estimates. A dependence of the tensile strength on the size of particles has been found. This dependence can be well fitted by inverse square power law.

The electrostatic fragmentation of dust particles, i.e., disruption of particles by repulsive electrostatic forces, is one of important destruction mechanisms for cosmic dust particles which leads to conversion of the solid phase of interstellar and interplanetary matter to the gas phase. Many phenomena connected to dust particles within the solar system can be explained by their electric charging with subsequent electrostatic fragmentation - see, e.g., Fechtig et al. (1979), Grün et al. (1984) or Boehnhardt and Fechtig (1987).

Repulsive electrostatic forces within a charged particle produce the electrostatic stress which is proportional to the second power of the particle surface electric field strength. The electrostatic fragmentation should occur when this stress exceeds the tensile strength of the particle material. Measured tensile strengths of materials relevant for cosmic dust particles range from 10^3 Pa for fluffy aggregates to 2×10^9 Pa in the case of metals. Corresponding critical electric field strengths which should result in electrostatic fragmentation range from 3×10^7 to 2×10^{10} $V.m^{-1}$ (for references see, e.g., Grün et al., 1984).

Cosmic dust particles are probably highly non-spherical, and their surfaces are often not smooth. Any roughness, that is usually much smaller than the grain itself, will be removed, however, due to a local enhancement of the electric field strength, before fragmentation of a particle can occur. Hill and Mendis (1981) calculated electrostatic stress for prolate and oblate spheroids and showed that the stress was increasing with distance towards the centre of prolate spheroids with possible chipping off its ends. As a result, the grain can become more and more spherical during its electric charging.

Tensile strengths of cosmic dust particles are influenced by the size of particles, by the irradiation history, etc. It is known that tensile strengths of carbon whiskers, carbonized organic fibers, or ion microscope tips of many materials including graphite are higher than 10^{10} Pa. It was also found that tensile strength of graphite increased during irradiation from 5×10^7 to 10^8 Pa (for references see Draine and Salpeter, 1979).

Trying to improve our knowledge about electrostatic fragmentation we started an experimental laboratory work on simulation of this process.

For test measurements we used loosely bound Al_2O_3 particles of 1 to 10 micrometers size. These particles were suspended in an electrodynamic quadrupole inside a vacuum chamber and electrically charged by ion beams of energies up to 5 keV. The stress at which fragmentation takes place was determined from the charge-to-mass ratio of a particle at the moment of fragmentation and its size. The size was derived from the time during which the amplitude of the particle motion in vertical direction decreased by a certain factor due to particle collisions with atoms of the rest gas at known pressure and temperature. For more details about the suspension system see Svestka et al. (1987) and Pinter et al. (1990).

The derived values of tensile strengths of the Al_2O_3 particles range from 10^3 to 10^5 Pa and there is an obvious dependence of the derived tensile strengths on the size of the agglomerate which can be well fitted by an inverse square power law. In case of such a dependence, the fragmentation condition involves, instead of the electric field strength, the particle electrostatic potential. The electrostatic potential of cosmic particles of sizes within the range we studied is often expected to be practically independent of the size of particles. The same would then apply to the fragmentation condition.

These results are, however, still somewhat uncertain mainly because of probable non-sphericity of the particles. In near future we plan to build a smaller vacuum chamber with a smaller quadrupole inside which much better vacuum will be reached. As a result, instrumental effects on charging processes should then be then minimized. In the present apparatus these effects lead to a substantial reduction of the maximum attainable surface field strength of charged particles to about 10^8 V.m^{-1} (Svestka and Grün, 1991). With the new apparatus we should be able to fragment spherical compact particles, and the comparison of experimental results with theoretical estimates will be extended over a larger range of particle materials and sizes.

References

Boehnhardt, H. and Fechtig, H. (1987) 'Electrostatic charging and fragmentation of dust near P/Giacobini-Zinner and P/Halley', Astron. Astrophys. 187, 824-828.

Draine, B.T. and Salpeter, E.E. (1979) 'On the physics of dust grains in hot gas', Astrophys. J. 231, 77-94.

Fechtig, H., Grün, E. and Morfill, G.E. (1979) 'Micrometeoroids within ten Earth radii', Planet. Space Sci. 27, 511-531.

Grün, E., Morfill, G.E. and Mendis, D.A. (1984) 'Dust-magnetosphere interactions', in R. Greenberg and A. Brahic (eds.), Planetary Rings, Univ. of Arizona Press, Tuscon, pp. 275-332.

Hill, J.R. and Mendis, D.A. (1981) 'Electrostatic disruption of a charged conducting spheroid', Can. J. Physics 59, 897-901.

Pinter, S., Svestka, J. and Grün, E. (1990) 'Interaction of dust particles with electrons and ions', in E. Bussoletti and A.A. Vittone (eds.), Dusty Objects in the Universe, Kluwer Academic Publishers, Dordrecht, pp. 139-146.

Svestka, J., Grün, E., Pinter, S. and Schumacher, S. (1987) 'Laboratory charging of dust by electrons and ions', Publ. of Astron. Inst. of Czechoslovak Academy of Sci. 67, 277-280.

Svestka, J. and Grün, E. (1991) 'Methods, difficulties and first results in laboratory simulation of cosmic dust electric charging' in A.C. Levasseur-Regourd (ed.), Origin and Evolution of Interplanetary Dust, in press.

ASTROCHEMISTRY LIBRARY WITH ARTIFICIAL INTELLIGENCE FOR QUALITY CONTROL

SHEO S. PRASAD [1,2] & PRADIP GANGOPADHAYA [2]
[1]Lockheed Palo Alto Research Laboratory (O/91-20, B255)
Hanover Street, Palo Alto, CA 94304, USA

[2]Space Science Center (SHS-274) & Department of Physics
University of Southern California, Los Angeles, CA 90089-1341, USA

ABSTRACT. Libraries of reactions used in astrochemistry modeling have seen an explosive increase in size in recent years. Their quality control by manual effort is almost impossible. Expert systems with artificial intelligence are now needed to ensure the quality of large scale astrochemistry libraries.

1. PROBLEM DEFINITION AND SOLUTION CONCEPT:

Two decades ago, in early 70s, interstellar chemistry modeling used only modestly large reaction sets (e.g., 100 reactions involving 30 species). The beginning of the next decade saw a quantum leap in the size of the network with the publication in 1980 of Prasad and Huntress (1980) model. This model used 1423 reactions and 137 species. The new direction became popular, and the trend towards large scale models continued throughout the 1980s. The most recent publication in this class is the UMIST Ratefile (Millar et al 1991) with 2880 reactions and 313 species.

The problems now are: (i) How to guard against unavoidable human errors in the traditional way of assembling a library? or, (ii) How to deal with the difficulties of manually searching the literature adequately? "Expert systems" with artificial intelligence for quality control constitute the solution to these problems.

The quality control capabilities of "expert system" would originate from the current knowledge (both scientific and empirical) of human experts that are stored in computer subprograms queried by the master quality control program. Such "expert systems" are quite common in numerous fields (e.g., medical diagnosis). It is now time to consider its application in astrochemistry.

2. OUTLINE OF MINIMAL CAPABILITY "EXPERT SYSTEMS":

Quality controllers can be as sophisticated as we may wish them to be. But the toll on time and effort increases steeply with the intended sophistication. To begin with we may try to have at least the following capability for our systems: (i) ability to make charge and mass balance check, (ii) detect repetitions, (iii) consistency between forward and backward reactions, (iv) ensuring that endothermic reaction do have activation energies and that the reaction rates do not have alarming values due to human errors during data entry, (v) ensuring maximum scan of the published work, (vi) use of evaluated data when more than one study exist for a given reaction.

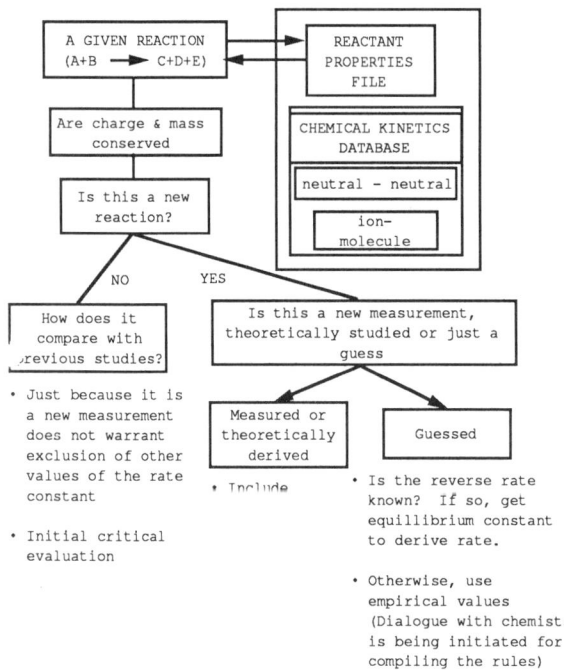

These capabilities can be accomplished with an approach outlined in the Figure 1. We need to create a property file containing the following data for each reactant: charge, mass, atomic composition, heat of formation, and entropy. For the neutral partners of ion-molecule reactions, we must also include their rotational constants, dipole moments and polarizabilities. These would be needed to get the reaction rates at the very low temperatures (~10 - 20K). Preparation of the property file is now rendered easier due to the availability of computerized data bases for the various chemical and physical properties of atoms and molecules from national standards institutions. The required information may not be in the data base for many of the astrochemistry species. For them empirical rule may be used.

Chemical kinetics data bases on computer readable media are also available for both ion-molecule and neutral-neutral reactions. These and the property file become input to chemical reaction scrutinizer outlined in the figure 1.

Figure 1: Reaction scrutinizer expert system: an outline of its possible functions.

3. CONCLUDING REMARKS:

Building of "intelligent" expert systems is not easy. It certainly cannot be done by a single astrochemistry modeler. The purpose of the presentation at this symposium has been to expose the new opportunity due to the recent advances in data base technology, and to invite collaboration from interested modelers.

ACKNOWLEDGEMENTS:

This research has been supported by NASA's ISR Astronomy program through a grant to the University of Southern California and a contract to the Lockheed Missile and Space Company.

REFERENCES:

Prasad, S.S. & Huntress, W.T., Jr. 1980, ApJS, **43**, 1.
Millar, T.J., Rawlings, J.M.C., Bennett, A., Brown, P.D., & Charnley, S.B., 1991, *A &AS*, **87**, 585.

CONTRIBUTION OF PAHS TO THE INTERSTELLAR EXTINCTION CURVE

C. JOBLIN[1], A. LEGER[1], P. MARTIN[2], D. DEFOURNEAU[1]
[1] *Groupe de Physique des Solides, Univ. Paris 7, Tour 23, 2, place Jussieu, 75251 Paris Cédex 05, France*
[2] *LURE, bât. 209d, Univ. Paris 11, 91405 Orsay Cédex, France*

ABSTRACT. Absorption spectra of some gaseous PAHs, either pure species or natural mixtures, have been obtained in the VUV-visible region and compared to the interstellar extinction curve.
The assumption that free PAHs are ubiquitous in the ISM cannot be rejected by incompatibility between the interstellar extinction curve and the absorption spectra of such molecules. PAHs absorb in the FUV rise and may give an important contribution to the bump at 2200 Å. We have derived that about 15% of the cosmic carbon is involved in these molecules.

1. Introduction

The astrophysical presence of PAHs has been suggested by the similarity between the unidentified IR bands, seen in emission at 3.3 - 6.2 - 7.7 and 11.3 μm in several astrophysical environments, and vibrational features of such molecules. If the whole energy emitted in the 12 μm IRAS band is due to PAHs then these molecules contribute to almost 20% of the absorption by dust. The PAH model has therefore to be tested by comparing absorption spectra of these molecules to the extinction curve. However, published spectra have been obtained in solutions and, restricted to wavelengths larger than 2500 Å, do not permit to conclude.

2. Results

We have measured the absolute absorption cross-sections of various gaseous PAHs on a large spectral range 400 Å - 7000 Å with a resolution of 20 Å, using the Orsay synchrotron radiation facility (LURE). Studied molecules are pure species and natural mixtures. These mixtures are certainly different from the interstellar ones, but they can help to understand the behaviour of the absorption spectrum when a great deal of species is involved.
  Figure 1 shows the absorption spectra of coronene and of two mixtures obtained by evaporation of some natural mixtures: coal pitch and catalytic deposit extracts. The composition of the gaseous mixtures, computed with an evaporation model, is given in fig.1d and 1e.
  Each laboratory spectrum is compared to the interstellar absorption curve attributed to small grains. This curve has been derived from the total extinction one, by substracting the contribution of big grains calculated in Désert et al. (1990) and by assuming a zero albedo for small grains. The main result is that laboratory spectra and the interstellar curve are rather similar in first approximation. All spectra show a rise in the FUV. In the near UV, individual molecules exhibit prominent features but these features merge into some wider structures when a sufficient number of species are present. However, some accumulation points may appear like the point at 2100 Å in some of our mixtures. An absorption tail stretches further in the visible as molecules become larger.

3. Discussion

Studied molecules are different from interstellar ones that are expected to be larger, partly ionised and possibly dehydrogenated. However, from the analysis of our spectra, we conclude that there is no incompatibility between the extinction curve and the presence of PAHs in the ISM. Moreover, we propose that no individual PAH is dominant in the ISM and that interstellar mixtures have a wide mass distribution necessary to obtain a smooth structure. An accumulation point may exist at the 2200 Å bump position. In summary, interstellar PAHs absorb the energy they reemit in the IR, in the FUV rise, in a UV-visible continuum and possibly in the bump. From the energetic budget, we can derive that 15% of the cosmic carbon is implied in PAHs, if these molecules have to account for the whole 12 μm IRAS band emission.

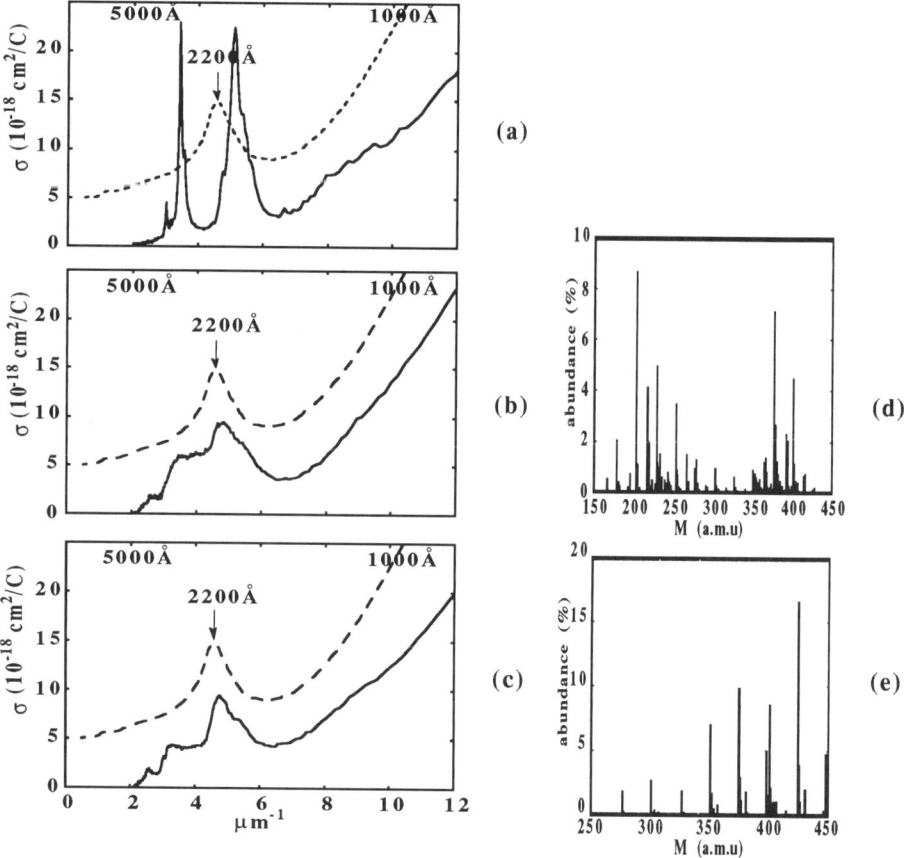

Figure 1. Comparison of the interstellar absorption curve of small grains (dashed line; offset, arbitrary units) with absorption spectra of PAHs (full line): (a) coronene, (b) gaseous mixture from coal pitch extract and (c) from catalytic deposit extract. (d) and (e) are mass spectra associated respectively with mixtures (b) and (c).

REFERENCES
Désert F.-X., Boulanger F., Puget J.L. (1990), A.& A. **237**, 215-236

CUMULENE CARBENES IN SPACE AND IN THE LABORATORY

J. M. VRTILEK, C. A. GOTTLIEB,
T. C. KILLIAN, and P. THADDEUS
Division of Applied Sciences
Harvard University
Cambridge, MA 02138, U.S.A.

J. CERNICHARO, M. GUÉLIN,
and G. PAUBERT
IRAM
Domaine Universitaire de Grenoble
F-38406 St. Martin D'Hères, France

ABSTRACT. Astronomical searches for H_2CCC and H_2CCCC, based on frequencies from our laboratory identifications, have resulted in detections toward TMC–1 and IRC+10216. These new interstellar species are possibly the first of a new family of highly polar carbon chains; they are only the second and third carbenes (carbon molecules with two nonbonded electrons) known in space.

1. Introduction

As part of a long-term investigation of rotational spectra of transient and reactive molecules of astrophysical importance, we recently undertook a laboratory search for several members of a family of highly polar hydrocarbons with linear double-bonded carbon backbones and terminal nonbonded carbene electrons. Two of these, H_2CCC and H_2CCCC, have now been successfully observed in the same cooled acetylene discharge that has been a source of numerous other reactive hydrocarbons (Vrtilek *et al.* 1990; Killian *et al.* 1990). They are remarkable for several reasons: both are isomers of interstellar species, H_2CCC of cyclic-C_3H_2 and H_2CCCC of diacetylene (almost certainly abundant though not directly detected); both are highly polar, with the two nonbonded electrons characteristic of carbenes; and, like a number of other transient interstellar species, both are unsaturated carbon chains.

With highly accurate (~1 km s^{-1} or better in equivalent velocity) frequency predictions in hand, we initiated successful astronomical searches for H_2CCC and H_2CCCC, using principally the IRAM 30 m telescope but also data from the Effelsberg 100 m and the NRAO 43 m (Cernicharo *et al.* 1991a,b). Observations were also obtained at Nobeyama (Kawaguchi *et al.* 1991).

Although the smallest H_2C_n molecule, vinylidene, H_2CC, has proven elusive, larger members of this sequence, predicted to be stable and even more polar, are excellent candidates for laboratory and radioastronomical detection, and searches for them are now underway.

2. Rotational Spectra of H₂CCC and H₂CCCC

H_2CCC and H_2CCCC are calculated (DeFrees and McLean 1986; Dykstra, Parsons, and Oates 1979) to be the isomers next in energy above their ground configurations. Very nearly prolate symmetric top molecules, with rotational spectra characterized by strong, harmonically related *R*-

branch lines, both species have many lines well situated for observation with mm-wave and cm-wave radio telescopes.

Our prediction of a very specific nearly harmonic sequence of doublets in the millimeter-wave rotational spectra of H_2CCC and H_2CCCC was the basis for a systematic laboratory search for these species. In ladders with $K_a > 0$ the slight deviation from symmetry splits energy levels degenerate in the symmetric top case (K-doubling). For intermediate values of K_a, here $K_a = 3$, the doublet splitting takes on a distinctive and convenient signature of a few MHz.

Support for our identifications comes from a number of chemical and physical assays, in addition to agreement of the observed spectrum with that expected for molecules with C_{2v} symmetry and the geometry of H_2CCC and H_2CCCC. For H_2CCC we now have conclusive evidence: a full substitution structure (Killian et al. 1991, in prep.).

3. Measurements in Space

Our initial identification of H_2CCC in TMC–1 rested on three lines found at the expected frequencies (Cernicharo et al. 1991a), from which we derived a rotational temperature of 4–6 K, consistent with expectations from other molecules, and a column density of $(2.5 \pm 0.5) \times 10^{12}$ cm^{-2}, smaller by a factor of ~70 than that of the lowest energy isomer, cyclic-C_3H_2. Three lines detected toward IRC+10216 imply a rotational temperature of 25 K and a column density of $\sim 2.6 \times 10^{12}$ cm^{-2}.

Eight mm-wave transitions of H_2CCCC have been assigned toward IRC+10216; all are fit by a single rotational temperature, 20 ± 3 K; the implied column density is $(1.6 \pm 0.4) \times 10^{13}$ cm^{-2}, 6 times that of H_2CCC toward IRC+10216 (Cernicharo et al. 1991b). Several lines of H_2CCCC have now been found in TMC–1, as for H_2CCC at precisely the expected frequencies (Kawaguchi et al. 1991). The column density of H_2CCCC in TMC–1, $\sim 8 \times 10^{12}$ cm^{-2}, is greater, as in IRC+10216, than that of the smaller molecule H_2CCC.

The presence of three closely spaced, low-lying lines, two *ortho* ($2_{12} - 1_{11}$ at 17.79 GHz and $2_{11} - 1_{10}$ at 17.94 GHz) and one *para* ($2_{02} - 1_{01}$ at 17.86 GHz), in the spectrum of H_2CCCC at frequencies conveniently located for radio astronomy permits accurate measurement of the *ortho-para* ratio, hence a possible constraint on proton density, and a search in the two *ortho* lines for the Townes-Cheung effect found in H_2CO.

4. References

Cernicharo, J., Gottlieb, C. A., Guélin, M., Killian, T. C., Paubert, G., Thaddeus, P., and Vrtilek, J. M. 1991a, *Ap. J. Letters*, **368**, L39.

Cernicharo, J., Gottlieb, C. A., Guélin, M., Killian, T. C., Thaddeus, P., and Vrtilek, J. M. 1991b, *Ap. J. Letters*, **368**, L43.

DeFrees, D. J., and McLean, A. D. 1986, *Ap. J. Letters*, **308**, L31.

Dykstra, C. E., Parsons, C. A., and Oates, C. L. 1979, *J. Am. Chem. Soc.*, **101**, 1962.

Kawaguchi, K., Kaifu, N., Ohishi, M., Ishikawa, S., Hirahara, Y., Yamamoto, S., Saito, S., Takano, S., Murakami, A., Vrtilek, J. M., Gottlieb, C. A., Thaddeus, P., and Irvine, W. M., 1991, *Pub. A. S. J.*, in press.

Killian, T. C., Vrtilek, J. M., Gottlieb, C. A., Gottlieb, E. W., and Thaddeus, P. 1990, *Ap. J. Letters*, **365**, L89.

Vrtilek, J. M., Gottlieb, C. A., Gottlieb, E. W., Killian, T. C., and Thaddeus, P. 1990, *Ap. J. Letters*, **364**, L53.

UV-VISIBLE AND NEAR IR ABSORPTION CHARACTERISTICS OF INTERSTELLAR PAHS. I. $C_{10}H_8^+$

F. SALAMA and L.J. ALLAMANDOLA
NASA-Ames Research Center. Space Science Division, MS: 245-6
Moffett Field, CA 94035, U.S.A

ABSTRACT. We have initiated a systematic and detailed study of the spectroscopy of *neutral* and *ionized* Polycyclic Aromatic Hydrocarbons (PAHs). We report here, the results obtained for the smallest PAH ($C_{10}H_8$) [1] and discuss their astrophysical applications.

1. Introduction

Polycyclic Aromatic Hydrocarbons (PAHs) in their neutral and ionized states have been postulated as an important, ubiquitous component of the interstellar material. They are considered as the best candidates for the carriers of the UIR features, and they may also play a role in other critical astrophysical issues such as: (i) The FUV part of the interstellar extinction curve and the 2200 Å hump. (ii) The carriers of the diffuse interstellar bands DIBs. (iii) The radiation and energy balance in space (UV to IR conversion). (iv) The extended red emission (ERE) seen in reflection nebulae. *There is therefore a crucial need for quantitative UV-Visible-NIR spectroscopic data on isolated, neutral and ionized PAHs to test the existing theories and their implications.*

2. Experimental Approach and Results:

A computer-controlled UV-Visible-NIR spectrometer system, coupled to a cryogenic cell, was designed to measure the absorption and emission spectra of neutral, ionic, or radical species in the 1800 - 9000 Å range under conditions relevant to astrophysical environments. *Matrix Isolation Spectroscopy is particularly well adapted here because it involves the trapping of the species of interest in a chemically inert, rigid cage at low temperature.*
2.1: Neutral naphthalene ($C_{10}H_8$): The UV-Visible absorption spectra of $C_{10}H_8$ isolated in Ar and Ne matrices [1] indicate that neutral naphthalene does not absorb in the visual and therefore cannot contribute to the DIBs. The UV spectrum of naphthalene consists of three systems peaking at 3115, 2792, and 2116 Å respectively in neon. The entire spectrum is dominated by a very strong band at 2116 Å in Ne and 2162 Å in Ar. The Ne matrix results show that the strongest absorption of $C_{10}H_8$ *does not peak at the interstellar 2175 Å absorption* but, if present, can contribute as substructure to the hump of the extinction curve. The cross section of the 2116 Å band (4 10^{-16} cm^2/molecule), requires a $C_{10}H_8$ abundance of 1.3 10^{-7} with respect to hydrogen in order to be detectable as substructure on the 2200 Å hump (at the 10 % level). The effect of the matrix material on the band positions has been studied [1] and it has been shown that *it is mandatory to utilize neon as the matrix material for astrophysical applications* (Ne is the closest to the gas phase conditions).

2.2: Ionized naphthalene: The UV-Visible absorption spectrum of $C_{10}H_8^+$, formed by direct photoionization of the neutral molecule isolated in a Ne matrix, can be separated into two components (Fig. 1): the discrete features and the continuum. The spectrum indicates that contrary to its neutral precursor, $C_{10}H_8^+$ does absorb in the visual, raising a possible connection with the DIBs. The small number of DIBs (~ 15%) which could be accounted for by $C_{10}H_8^+$ shows, however, that *$C_{10}H_8^+$ alone can not explain the DIBs in this wavelength range.* This comparison hints, however, to the fact that *a family of different PAH ions* may contribute to the DIBs. *The answer to this puzzling problem can only be obtained through a systematic comparison of the DIBs with appropriate laboratory data.* The complete UV-Visible spectrum (not shown) of the naphthalene cation consists of 7 systems with different intensities. Another striking aspect of these experiments is the production of a broad and strong CONTINUUM extending from the UV to the visible (2000 - 5100 Å). The laboratory spectra indicate that the growth of the cation discrete absorption features and continuum under VUV irradiation is correlated to the depletion of the neutral precursor absorption. Because of its potential astrophysical importance, it was necessary to better determine the carrier of this continuous feature. A number of experiments [1,2] which were carried out to eliminate other possibilities (the anionic counterpart $C_{10}H_8^-$, or a dissociative transition within the neutral precursor) *showed that the discrete bands due to $C_{10}H_8^+$ and the continuum grow together as a function of VUV irradiation time* indicating that $C_{10}H_8^+$ *is most likely responsible for the continuum.* Such a broad, strong continuum has the potential to be very significant for astrophysics. If further laboratory experiments establish that this continuum is indeed a general characteristic of ionized PAHs, *this may be the channel through which interstellar UV-Visible radiation is converted to the discrete IR emission bands.*

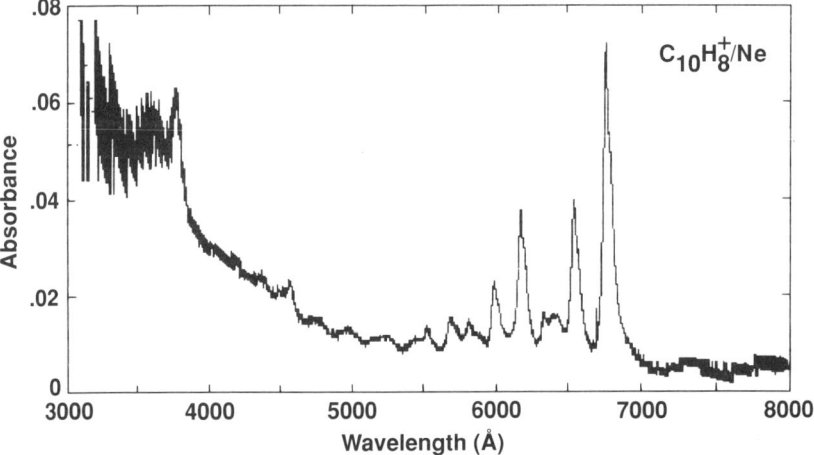

Figure 1: The absorption spectrum of $C_{10}H_8^+$ in a Ne matrix. Experimental conditions: $C_{10}H_8$/Ne=1:600, 4.2 K, 20 min VUV irradiation with 10.2 eV photons.

References:
1- Salama, F. and Allamandola, L. J. (1991) 'Electronic absorption spectroscopy of matrix-isolated polycyclic aromatic hydrocarbon cations. I. The naphthalene cation ($C_{10}H_8^+$)', J. Chem. Phys. 94 (11), 6964-6977.
2- Salama, F. and Allamandola, L. J. (1991) 'The role of matrix material and CCl_4 (electron acceptor) on the ionization mechanisms of matrix-isolated naphthalene', J. Chem. Phys. (in press, Oct. 15 edition).

INFRARED SPECTROSCOPY OF INTERSTELLAR AND SOLAR SYSTEM ICE ANALOGS: MEASUREMENT OF OPTICAL CONSTANTS

D.M. Hudgins, S.A. Sandford, A.G.G.M. Tielens, and L.J. Allamandola
NASA/Ames Research Center, MS 245-6, Moffett Field, CA, 94035

ABSTRACT: Laboratory spectra through the mid-infrared have been used to calculate the optical constants (n and k) for a variety of pure and mixed molecular ices. The ices studied were H_2O, CH_3OH, CO_2, OCS, CH_4, CO_2+CH_4, CO_2+OCS, $CO+CH_4$, $CO+OCS$, O_2+CH_4, O_2+OCS, N_2+CH_4, N_2+OCS, H_2O+CH_4, H_2O+OCS, and $H_2O+CH_3OH+CO+NH_3$.

1. Introduction

Band shapes, positions, and strengths in the absorption spectrum of a mixed molecular ice are diagnostic of the composition, temperature, and thermal history of that ice.[1] Particle properties such as size, shape, and morphology (e.g. pure versus core/mantle particles) may also induce observable variations.[2] Consequently, absorption spectra carry a great deal of information about the nature of ices in a variety of astronomical environments. Through detailed modelling using the pertinent optical constants it is often possible to deconvolve the factors affecting the observed spectra. In the laboratory it is necessary to study the mixtures of interest since the n's and k's for a mixed molecular ice are not given by the weighted sum of the n's and k's of each ice component in its pure form. Such an approach fails to account for the unique intermolecular interactions which occur between different molecules in a mixed solid.

2. Technique

The experiment involves an infrared window suspended from a cryogenically cooled cold finger in a high vacuum chamber. A thin layer of ice is layed down by vapor deposition onto the cooled window. Sample thickness is monitored by the interference fringes observed in a HeNe laser beam reflected off the front of the window during sample deposition. After deposition is complete, the sample is rotated under vacuum to face the infrared beam and spectra are collected as a function of ice temperature. The results were obtained through an iterative application of the Kramers-Kronig relation. A hypothetical set of n's and k's were used to calculate a theoretical transmission spectrum. This, in turn, was compared to the actual spectrum, and the set of n's and k's were corrected for any deviations between the two. This process was repeated until the maximum calculated deviation was less than 0.1%.

3. Results

We have derived optical constants for the following ices: Pure substances: H_2O (water), CH_3OH (methanol), CO_2 (carbon dioxide), OCS (carbonyl sulfide), and CH_4 (methane); Binary mixtures: $CO_2:CH_4=20:1$, $CO_2:OCS=20:1$, $CO:CH_4=20:1$, $CO:OCS=20:1$, $O_2:CH_4=20:1$, $O_2:OCS=20:1$, $N_2:CH_4=20:1$, $N_2:OCS=20:1$, $H_2O:CH_4=20:1$, $H_2O:OCS=20:1$, and $H_2O:OCS=2:1$; Interstellar mixtures: $H_2O:CH_3OH:CO:NH_3=100:50:1:1$ and $H_2O:CH_3OH:CO:NH_3=100:10:1:1$. An example is shown in figure 1. The results will be presented in a future paper.[3]

3. References

(1) Sandford, S.A., Allamandola, L.J., Tielens, A.G.G.M., and Valero, G.J., *Ap.J.*, **329**, 498 (1988)
(2) Tielens, A.G.G.M., Tokunaga, A.T., Geballe. T.R., and Baas, F., *Ap.J.*, (in press) 1 November, 1991
(3) Hudgins, D.M., Sandford, S.A., Tielens, A.G.G.M., and Allamandola, L.J., *Ap.J.Supp.Ser.*, (in preparation)

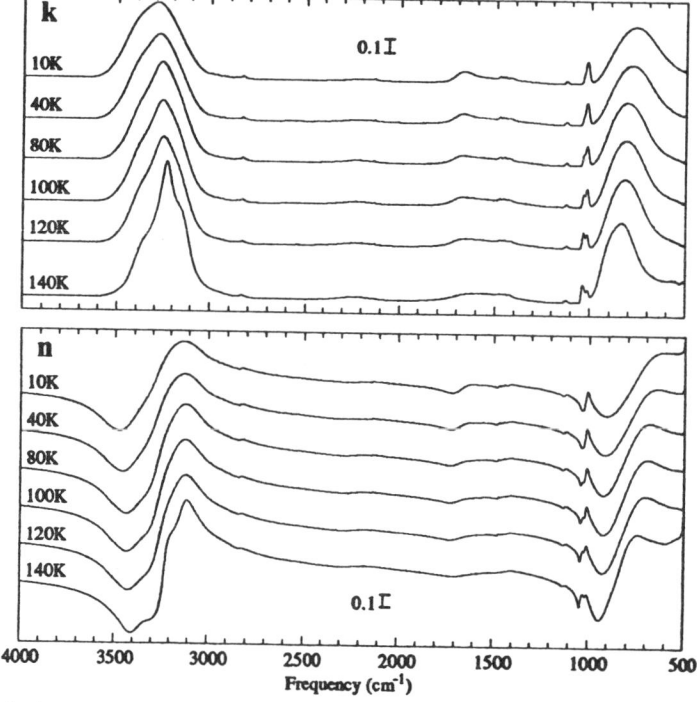

Figure 1. k and n values for the ice mixture $H_2O:CH_3OH:CO:NH_3=100:10:1:1$

FORMATION OF ORGANIC MOLECULES BY FORMALDEHYDE REACTIONS IN ASTROPHYSICAL ICES AT VERY LOW TEMPERATURES

W. A. SCHUTTE, L. J. ALLAMANDOLA, and S. A. SANDFORD
Space Science Division, MS:245-6, NASA/Ames Research Center, Moffett Field, CA 94035, U.S.A.

ABSTRACT. Warm-up of astrophysical ice analogues containing formaldehyde produced organic residues in large abundances. It is argued that formaldehyde reactions at very low temperatures could be an important source of interstellar and cometary organic molecules.

1. Introduction

The higly reactive molecule formaldehyde (H_2CO) is present at the level of a few percent in cometary and probably also in interstellar ices, making it one of the more abundant constituents after water (Mumma and Reuter 1989). This paper reports the results of experiments on analogues of astrophysical ice mixtures (H_2O, H_2CO, NH_3, CH_3OH) that were performed to investigate the possible role that formaldehyde may have in the formation of cometary and interstellar organic molecules.

2. Experimental Results

Ice mixtures containing water (H_2O), ammonia (NH_3) and H_2CO in various concentrations were prepared at 10 K. Upon warm-up, reactions involving formaldehyde took place between 40 - 180 K, producing organic residues in high yields. Remarkably, these residues seem to consist of only 3 products. The first product can be obtained from H_2CO ice with traces of NH_3. Its IR spectrum identifies it as polyoxymethylene (POM; $[-CH_2-O-CH_2-O]_n$), a well known H_2CO polymerization product. The second product, designated X, is produced by reactions between H_2CO and NH_3. Its IR spectrum reveals C-H, C-O, $-NH_2$ and possibly O-H groups. It evaporates below 230 K, implying a small structure with $< \approx$ 2 C atoms. The third product, designated Y, is obtained from reactions between H_2CO and H_2O. It contains O-H, C-H, and C-O groups and seems to have ether- as well as alcohol-type properties. Evaporating below 260 K, it should contain at most ~ 2 C atoms.

Figure 1 shows the yields of POM, X, and Y, i.e., the fraction of the carbon initially deposited as formaldehyde that ends up in these products, as well as the total residue yield as a function of the initial [NH_3/H_2CO] ratio at [H_2CO/H_2O] = 0.05. It can be seen that traces of ammonia ([NH_3/H_2CO] $> \approx$ 0.005) are required to initialize reactions involving H_2CO and the production of an organic residue. For [NH_3/H_2CO] $> \approx$ 0.4, the conversion of formaldehyde to organic molecules is almost 100 %. The produced relative amounts of POM, X and Y depend sensitively on the initial ice composition.

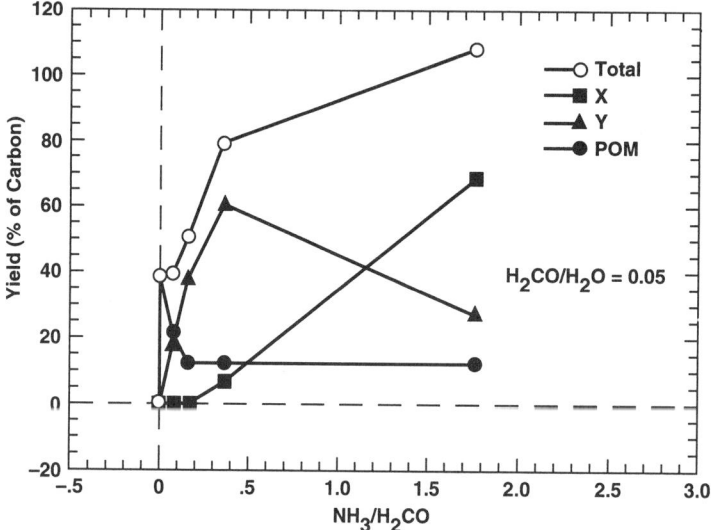

Figure 1. Yields of POM, X and Y obtained upon warm-up of ices with initial ratio $[H_2CO/H_2O] = 0.05$, as a function of NH_3 concentration.

Introduction of 1 - 10 % of methanol (CH_3OH) in water-rich ices ($[H_2CO/H_2O] \approx 0.05$) with $[NH_3/H_2CO] <\approx 0.01$ decreases the total yield of residue from 40 % to 10 - 20 %. Furthermore, the yield of POM drops to less than 3 % and the residue is dominated by new alcohol-type compounds. For water-rich ices with $[NH_3/H_2CO] >\approx 0.25$, adding CH_3OH does not significantly change the organic residues.

3. Astrophysical Implications

Our experimental results have a number of astrophysical implications. Low temperature formaldehyde reactions in astrophysical ices are probably an important source of interstellar and cometary organic compounds and may account for 1 - 10 % of the organic molecules detected in Comet Halley. Only a small number of compounds are produced (4 - 6). For astrophysically relevant ices the major products are small ($<\approx 2$ C atoms), highly O-rich (O/C $\approx$ 1) and have alcohol-, amine-, and ether-like properties, while POM is a minor product. The kind of molecules that are produced and their relative abundances trace the initial ice composition and concentrations.

References

Mumma, M. J., and Reuter, D. C. (1989) 'On the Identification of Formaldehyde in Halley's Comet', ApJ 344, 940-948.

COMPUTATIONAL CHEMISTRY APPROACH TO SPACE CHEMISTRY

Y. ELLINGER
Ecole Normale Supérieure et Observatoire de Paris
24 Rue Lhomond
75005 Paris
France

ABSTRACT. This review paper presents the results of state of the art Quantum Chemistry calculations in the field of Astrochemistry. It provides selected examples to illustrate the possible contribution of molecular orbital theories to solving a number of problems of astrophysical interest ranging from identification of new molecules to IR emission analysis and rate constants determinations.

Many of the roles played by laboratory work in the sequence going from the identification of a new molecule in space to the understanding of its chemistry, such as analysis of rotational spectra, evaluation of abundances, IR signature and possible mechanisms of formation and destruction, can be played by theoretical methods. Quantum Chemistry is ideally suited for this task as it refers directly to isolated gas phase systems and can be used to examine chemical species and processes that are difficult, if not impossible, to study in laboratories. Selected examples are presented below to illustrate how efficient a partner Quantum Chemistry can be for understanding chemistry in space.

1. Identifying Molecular Structures: the SiC_2 problem

The astrophysical importance of SiC_2 has been known for a long time. Though not identified, the electronic absorption spectrum was first observed in 1926 in the atmosphere of carbon stars. It was not until 1956 that laboratory experiments proved that the carrier of these bands was a molecule formed of C and Si atoms exclusively. On the basis of a partial vibrational analysis, the spectrum was assigned to SiC_2; by analogy with a cometary spectrum of C_3, SiC_2 was assumed to be linear, and the absorption bands interpreted in terms of a $^1\Pi \leftarrow {}^1\Sigma^+$ transition[1]. On the other side, theoretical arguments based on qualitative relations between the geometry of a molecule and the number of its valence electrons (Walsh's rules) pointed to a possibly non-linear structure.

This problem was definitively solved through a close collaboration between experiments and theory. The analysis of the visible spectra of jet cooled SiC_2 showed[2] that the rotational structure was inconsistent with a linear structure but, as demonstrated by ab-initio configuration interaction (CI) calculations[3], came from a triangular structure, more stable by 1.1 kcal/mol than the linear. The correct assignment of the spectrum to an $^1B_2 \leftarrow {}^1A_1$ was then possible.

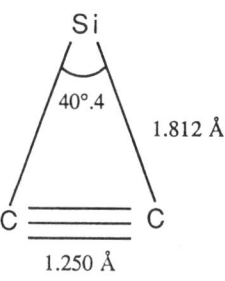
Experimental

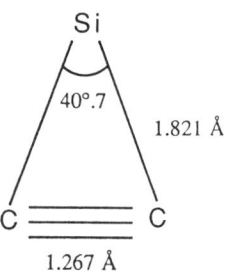
Ab-initio theory

It should be emphasized that a single configuration description of this system gives the reversed stability order, namely, linear more stable than cyclic by 1.5 kcal/mol, erroneously. Elaborate MCSCF calculations are necessary to provide orbitals of high enough quality to initiate the correlation energy calculations.

2. Probing Intuitive Spectroscopic Models: the structure of C_2H_3

One of the most important progenitors in the formation of hydrocarbons in ion-molecule reactions is protonated acetylene. The structure of this ion has been a puzzling problem for many years. Two geometries have been postulated on intuitive grounds, respectively called the linear and the bridge structures. Recently, high resolution infra-red studies have been possible[4]. Interpretation of the spectra using a simple internal rotor model in which the triangle formed by the three hydrogens rotates around the center of gravity of the C=C bond leads to an "experimental" determination of the energy difference of 6.0 kcal/mol between the two structures.

Exptl. intuitive models

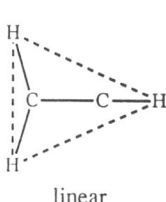
linear

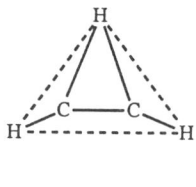
bridge

Ab-initio theory

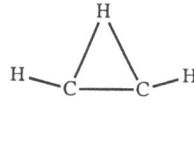
calculated

Ab-initio state of the art calculations[5] showed unambiguously that protonated acetylene was effectively more stable in a bridge structure, but with a geometry fundamentally different from that postulated in the spectroscopic model. The correct energy difference, i.e., 3.7 ± 1.3 kcal/mol has been established.

3. Assessing Electronic States : the $^2\Sigma$ - $^2\Pi$ alternative in linear radicals

The discovery in IRC+10216, by Guélin et al[6]. of a series of "doublets of roughly similar intensity, with center frequencies almost exactly in the ratio of half interger number and with doublet splitting regularly increasing with frequency" pointed, together with additional spectroscopic arguments and abondance considerations, to the conclusion that the carrier was a linear $^2\Pi$ radical. Comparing the observed effective rotational constant of 1386.2 MHz with an a-priori estimation using standard bond lengths suggested the C_6H or C_5N radicals. However, if the parent closed-shell molecules have an obvious $^1\Sigma$ ground state, radicals obtained by rupture of a terminal CH bond can be either $^2\Sigma$ or $^2\Pi$. The usual rule of thumb connecting the parity of the number of conjugated centers to the electronic state of the system (n odd $\rightarrow$ $^2\Pi$; n even $\rightarrow$ $^2\Sigma$) which was verified for all the already known radicals predicts a $^2\Sigma$ state for both C_6H or C_5N which cannot account for astronomical observations.

Exploratory calculations at the single configuration level of wave function showed a striking difference between the dipole moments of $^2\Sigma$ or $^2\Pi$ radicals in the two series. A large value (μ=4.3 D) was found for the $^2\Pi$ of C_6H, a small one (μ=0.9 D) for the nearby $^2\Sigma$ state; the opposite situation was found for C_5N, i.e., a very small dipole (μ=0.07 D) for the $^2\Pi$ state and a large one (μ=3.2 D) for the $^2\Sigma$. As the line strengths scale with μ^2, it was clear that the $^2\Pi$ state could not be that of C_5N. The attention was then focused on the C_nH series.

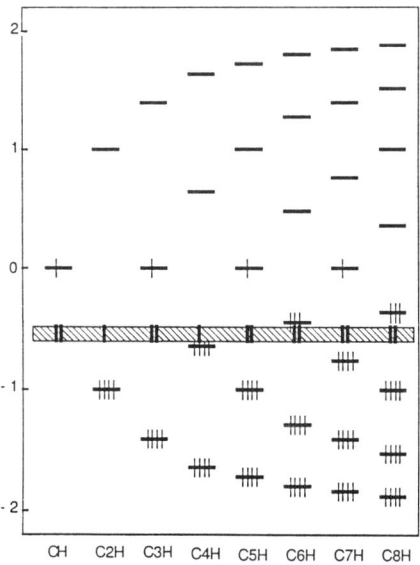

Molecular energy diagram for the π system and terminal σ orbital in the C_nH series. The stripped area gives the expected position of the σ orbital

The energy diagram reported here shows how the energies of the π orbitals are sensitive to the number of conjugated centers in the linear chain. By contrast, the σ orbital pointing opposite to the carbon backbone is not affected by the length of the chain and its energy is constant. It is clear from this qualitative picture that all systems with an odd number of centers are undoubtedly $^2\Pi$. For even numbers of centers, the symmetry of the electronic ground state depends on the relative position of the upper π shell with respect to the σ orbital on the terminal carbon. The intuitive chemical view which looks at the molecule as a series of conjugated triple bonds gives the correct answer only up to 2 acetylenic fragments. The energy of the upper π shell increasing regularly with the number of centers in the delocalized system, one can expect a $^2\Sigma$ to $^2\Pi$ crossing. A series of ab-initio large scale CI calculations designed to insure convergence in the basis set and in the n-particle space reached the conclusion that the state crossing occurs for 6 carbons with $^2\Pi$ more stable than $^2\Sigma$ by 0.25 eV; n-particle spaces up to 381566 ($^2\Sigma$) and 375217 ($^2\Pi$) configurations were considered in this study. Finally, it can be anticipated that all members of the C_nH series with n ≥ 6 (*n odd* or *even*) have a ground state with 1 or 3 electron in the π open shell according to the parity of the carbon chain[7].

4. Predicting Rotational Constants : the example of linear species

Ab-initio molecular orbital theory can also be used to estimate the rotational constants for systems that are candidates for discovery in interstellar space. Since the precision needed for that purpose is well beyond the present capability of computational chemistry for medium size systems, evaluation of the rotational constants is made by combining ab-initio quantum chemistry calculations and observed data on the starting members of the series under investigation. The principle of these estimations is to derive an analytic or numerical fitting relation between the optimized theoretical geometries giving Be(theory) and the observed Bo. Such an approach has been used to predict rotational constants of the "next" members in the cyanopolyyne family[8], the polycarbon monoxides[8] and in the radical series[9] C_nN and C_nH:

$B_0(HC_{13}N) = 0.1073 \pm 0.0002$ GHz ; $B_0(C_7H) = 0.873 \pm 0.002$ GHz
$B_0(C_5O) = 1.36 \pm 0.02$ GHz ; $B_0(C_4N) = 2.428 \pm 0.003$ GHz

with error bars tight enough to guide both laboratory and astronomical searches.

5. Interpreting IR Bands: the PAH hypothesis

The emission lines observed in many interstellar infrared sources at 3050, 1610, 1300, 1150 and 885 cm^{-1} (3.3, 6.2, 7.7, 8.6, and 11.3 microns) are hypothesized to originate from polycyclic aromatic hydrocarbon molecules (PAHs)[10,11]. These assignments are based on the analysis of laboratory infrared spectra of neutral PAHs. But, although the IR emission band spectrum resembles what one might expect from a mixture of PAHs, it does not match in details (frequency, band profile, or relative intensities) with the absorption spectra of any known PAH molecule.

Moreover, it is often assumed that PAHs, in the regions where they are observed, are mostly ionized, i.e., positively charged, and partly dehydrogenated[12]. If PAHs are considered as free flyers, the deductions made from the comparison of the interstellar emission with laboratory data are highly questionable. The spectra which are compared are spectra of different molecules: no a-priori reason can be quoted to get the same IR spectra, especially concerning intensities, for a molecule and its ionized or dehydrogenated derivatives. These systems which are generated from the parent PAHs are highly unstable in laboratory conditions. Experiments specifically designed for their study are extremely difficult to realize. Another source of information to test the PAH hypothesis is *ab initio* molecular orbital theory. It can be used to compute, from first principles, the geometries, vibrational frequencies, and vibrational intensities for model PAH compounds.

The IR spectra of naphtalene ($C_{10}H_8$), anthracene ($C_{14}H_{10}$), pyrene ($C_{16}H_{10}$) and the corresponding positive ions have been calculated as models for linear and compact PAHs. These calculations showed that, if frequencies are little displaced, intensities by contrast are strongly affected by ionization; their variations are related to the type of vibration but similar in all the molecules calculated. The main reult is the decline of the ratio of the CH(3.3 µ) to CC(7.7 µ) intensities with ionization. The ratios to the other bands are in much better agreement with the observed ones in the cations while out of range for the neutral systems[13]. Dehydrogenation as been studied on the simplest PAH, naphtalene, and results are summarized in the following table.

Naphtalene : integrated intensities normalized to the 7.7 µ band

	I(3.3µ)/I(7.7µ)	I(5.2µ)/I(7.7µ)	I(6.2µ)/I(7.7µ)	I(11.3µ)/I(7.7µ)	I(11.3µ)/ I(7.7+8.6µ)
$C_{10}H_8$	9.09		2.82	15.27	8.00
$C_{10}H_7$	4.50		1.56	10.37	6.83
	5.42		2.67	11.25	5.87
$C_{10}H_6$	2.30	0.04	1.70	5.81	3.75
	2.00	0.11	0.81	4.63	3.0
$C_{10}H_8^+$	0.02		0.58	0.27	0.14
$C_{10}H_7^+$	0.04		0.49	0.34	0.18
	0.04		0.43	0.22	0.12
$C_{10}H_6^+$	0.06	0.10	0.44	0.22	0.12
	0.04	0.19	0.37	0.22	0.10
Observed	0.084[a]	0.054[a]	0.54[a]	0.29[a]	0.125[b]

a) Ref. 14   b) Ref. 15

It is clear that the neutral PAH, dehydrogenated or not, cannot account for the observed intensity ratios, whereas the dehydrogenated positive ions compare favourably with the IR emission lines. The major result of this study[16] is that, if UIR bands originate from PAHs, then, these species are ionized and there is no longer the need for a massive dehydrogenation[17] to accomodate the IR emission.

6. Checking Molecular Abundances: the HCO^+/CO and HCS^+/CS ratios

Molecular abundances in interstellar clouds depend on the balance between a series of formation and destruction mechanisms whose relative importance cannot be easily established. Among those, dissociative recombination is an important process which is generally fast and governs the abundances of positive ions. The study of such a type of process, including the determination of associate reaction coefficients is an enormous task even at room temperature and the collision rates used in astrochemical models are low temperature extrapolated values obtained using more or less reliable procedures.

A typical challenge in this field was the difference of several orders of magnitude[18,19] between the abundance ratios HCO^+/CO and HCS^+/CS. The well admitted analogy between oxygen and sulphur chemistry being unable to give any rationalization of these facts, Millar suggested[20] that the abundance ratio HCS^+/CS could be higher, provided the dissociative recombination rate for HCS^+ was much lower than for HCO^+.

The potential energy curves of the systems ($HCO^+ + e^-$) and ($HCS^+ + e^-$) in Rydberg and dissociative valence states, as well as those of the positive ions, were computed taking into account correlation effects in appropriate CI calculations[21].

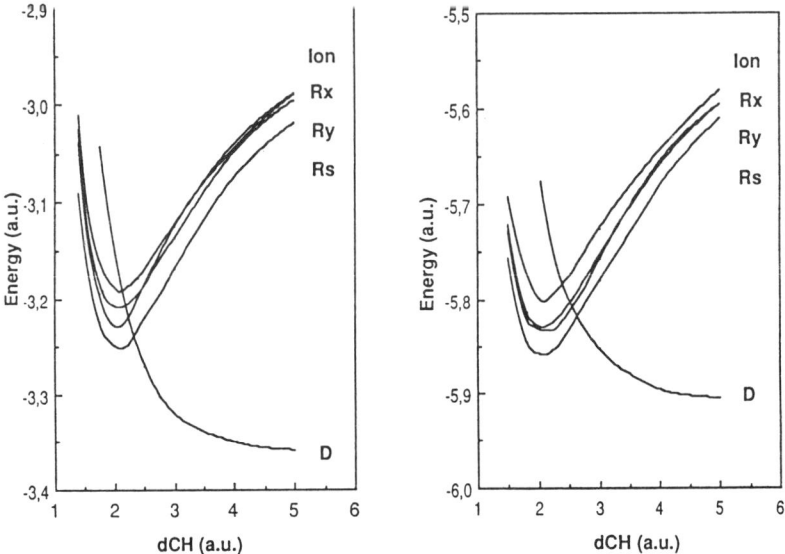

The quasi-diabatic representation of the potential surfaces shows striking differences:
- the crossing of the ground state surface of the ion with the dissociative curve of the neutral species is much closer to the energy minimum for HCO than for HCS.
- every term of the Rydberg series crosses the dissociative valence state at a position closer to the energy minimum for HCO than for HCS.

Numerical calculations of the rate constants show that
- HCO has a usual behavior with a preponderant direct mechanism.
- HCS is anomalous, showing quite large contributions from an indirect mechanism.

The recombination rate of electrons with HCO^+ is significantly larger, by one to two orders of magnitude, than with HCS^+. These ab-initio calculations proved that Millar's hypothesis is perfectly justified.

7. Explaining Anomalous Reaction Rates : the $NH_3^+ + H_2$ reaction

Experiments show that the ion-molecule reaction

$$NH_3^+ + H_2 \rightarrow NH_4^+ + H$$

is slow at room temperature, becomes slower as the temperature decreases, and then begins to increase in rate at still lower temperatures[22,23]. The authors hypothesized the mechanism for that reaction to be the initial formation of a complex, from which tunneling under a small transition state barrier could then occur. To understand the mechanism, both a quantum chemical study of the potential energy surface governing the reaction and a study of the reaction dynamics were undertaken[24]. The minimum energy pathway, taking vibrational energies into account, has been calculated using correlated wave functions and large basis sets.

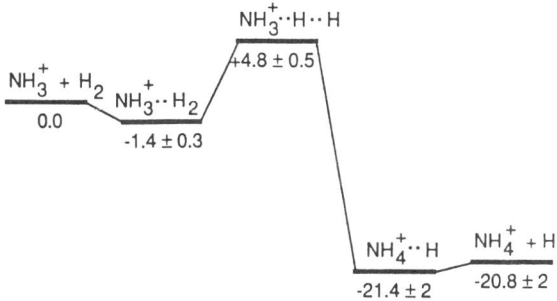

It proceeds through a weakly bound entrance channel complex $NH_3^+..H_2$, of binding energy 1.4±0.3 kcal/mol, then rises to a transition state $NH_3^+.H.H$, of energy 4.8±0.5 kcal/mol above that of the reactants, before going down to the products via a weakly bound exit channel complex with an overall exothermicity of 25.6±0.7 kcal/mol. This energy profile was then used in a theoretical treatment of the reaction dynamics by the phase space approach . The calculated rate coefficients are in quantitative agreement with experiment and reproduce the unusual temperature dependence observed in the laboratory.

8. Concluding remarks

The state of the art in Quantum Chemistry has now reached a point such that the wave functions determined by advanced computational techniques are accurate enough to provide reliable informations useful for a basic understanding of a number of chemical problems of astrophysical interest.Thus:
- Spectroscopic constants and rotational excitation of molecules that are difficult to study experimentally, can be calculated.
- Separations between spectroscopic states, as well as dipole moments necessary to determine column densities, can be obtained for exotic radicals.
- IR signatures can be calculated for neutral, ionized and dehydrogeenated PAHs, leading to a appealing modelisation of the UIR bands .
- Reaction mechanisms can be understood in greater detail than is generally possible from experiments alone; the role of tunneling can be analyzed easily.

- Reaction rates to be used in chemical models can be evaluated directly from chemical dynamics calculations instead of being estimated by extrapolating room-temperature rates to interstellar temperatures.

Time has come for a fruitful collaboration between Astrophysics and Quantum Chemistry.

References

1. Gausset, L., Herzberg, G., Lagerqvist, A., and Rosen, A., (1965) *Ap. J.*, **142**, 45
2. Michalopoulos, D.L., Geisic, M.E., Landridge-Smith, P.R.R., and Smalley, R.E., (1984) *J. Chem. Phys.*, **80**, 3556
3. Grev, R.S., and Schaeffer III, H.F., (1984) *J. Chem. Phys.*, **80**, 3552
4. Crofton, M.W., Jagod, M.F., Rehfuss, B.D., and Oka, T., (1989) *J. Chem. Phys.*, **91**, 5139
5. Lindh, R., Rice, J.E., and Lee, T.J., *J. Chem. Phys.*, (1991) **94**, 8008
6. Guélin, M., Cernicharo, J., Kahane, C., Gomez-Gonzalez, J., and Wamsley, C.M., (1987) *Astron. Astrophys.*, **175**, L5
7. Pauzat, F., and Ellinger, Y., (1989) *Astron. Astrophys.*, **216**, 305
8. DeFrees, D.J., and McLean, A.D., (1989) *Chem. Phys Lett.*, **158**, 540
9. F. Pauzat, F., Y. Ellinger, Y., and McLean, A.D., (1991) *Ap. J.*, **369**, L13
10. Leger A. and Puget J.L., (1984) *Astron. Astrophys.*, **137**, L5
11. Allamandola L.J., Tielens A.G. and Baker J.R., (1985) *Ap. J. Lett.*, **290**, L25
12. Puget J.L. and Leger A., (1989) *Ann. Rev. Astron. Astrophys.*, **27**,161
13. DeFrees, D.J., Miller, M.D., Talbi, D., Pauzat, F., and Ellinger, Y.,*Ap. J.*, **000**
14. Cohen, M., Tielens, A.G., Bregman, J., Witteborrn, F.C., Rank, D.M., Allamendola, L.J., Wooden D.H., and de Muizon, M., (1989)*Ap. J.*, **341**, 246
15. Zavagno A., Cox P., and Baluteau J.P., (1992) *Astron. Astrophys.*, **000**
16. Ellinger, Y., Talbi, D., and Pauzat, F., (to be published)
17. de Muizon M., d'Hendecourt L.B. and Geballe T.R., (1990) *Astron. Astrophys.*, **227**, 526
18. Mitchell, G.F., Ginsburg, J.L., and Kuntz, P.J., (1978) *Ap. J, Suppl.*, **38**, 39
19. Thaddeus, P., Guélin, M., and Linke, R.A., (1981) *Ap. J. Lett.*, **246**, L41
20. Millar, T.J., Adams, N.G., Smith, D., and Clary, D.C., (1985) *M.N.R.A.S.*, **216**, 1025
21. Talbi, D., Hickman, A.P., Pauzat, F., Ellinger, Y., and Berthier, G., (1989) *Ap. J.*, **339**, 231
22. Luine, J.A., and Dunn, G.H., (1985), *Ap. J. Letters*, **299**, L67
23. Bohringer, H., (1989) *Chem. Phys Lett.*, **122**, 185
24. Herbst, E., DeFrees, D.J., Talbi, D., Pauzat, F., Koch, W., and McLean, A.D., (1991) *J. Chem. Phys.*, **94**, 7842

METAMORPHISM OF COSMIC DUST: DIAGNOSTIC INFRARED SIGNATURES

Joseph A. Nuth III
Astrochemistry Branch, Code 691
NASA/Goddard Space Flight Center
Greenbelt, MD 20771

Introduction. Nucleation is a non-equilibrium process: the products of this process are seldom the most thermodynamically stable condensates but are instead those which form fastest. It should not be surprising that grains formed in a circumstellar outflow will undergo some degree of metamorphism if they are annealed or exposed to a chemically active reagent. As a consequence of this processing in the laboratory one observes a continuous increase in the strength of the silicate absorption band at 20 microns relative to the 10 micron feature. In Section 1 we show that this ratio can be used as an indicator of the relative age of silicate condensates. Metamorphism of refractory particles continues in the interstellar medium (ISM) where the driving forces are sputtering by cosmic ray particles, annealing by high energy photons and grain destruction in supernova generated shocks. Studies of the depletion of the elements from the gas phase of the ISM tell us that if grain destruction occurs with high efficiency, then there must be some mechanism by which grains can be formed in the ISM. Laboratory studies of such a process (Moore, Tanabe, and Nuth, Ap. J. (Lett) 373, L31-L34, 1990) have shown that the frequency of the -SiH stretch can be used as an indicator of the oxidation state of the silicon in such grains. Highly reduced grains exhibit an SiH absorption near 2100 cm^{-1} whereas highly oxidized silicates absorb near 2300 cm^{-1}: this point is discussed in Section 2.

1. The "Age" of Silicate Grains is Indicated by the Relative Strengths of the 10 and 20 Micron Bands

In the outflows around oxygen-rich stars the chemical environment is still quite reduced due to the high abundance of hydrogen. Consequently the vapors which eventually condense to form silicate grains consist of reduced species such as silicon monoxide (SiO), iron (Fe) and magnesium (Mg) metal: Excess oxygen is tied up as water or hydroxyl (OH) rather than in combination with either Fe or Mg. When refractory vapors become sufficiently supersaturated to condense, the processes of nucleation and grain growth occur quite rapidly - much too rapidly in most cases for equilibrium to be maintained. One result of rapid condensation is the production of "underoxidized" grains consisting primarily of SiO, Fe and Mg plus some additional oxygen incorporated as the grain reacted with water vapor and OH. Once the refractory vapors have all condensed onto the grain cores these cores will continue to oxidize until chemical equilibrium is achieved and stable, silicate minerals are formed. This sequence is outlined in the paper by Nuth and Hecht (Astrophys. Sp. Sci. 163, 79-94, 1990).

Such a definitive chemical sequence has very specific observational consequences. First, the initial condensate will consist mostly of isolated SiO units together with iron and magnesium metal. Very little of the SiO will have polymerized and only a small fraction of the metal will have oxidized. Because metals are not infrared active, metallic species in refractory grains will not be directly detectable. Isolated SiO units will absorb strongly near 10 microns due to the SiO vibrational stretch, while only the more polymerized SiO_x units absorb near 20 microns as a result of the O-Si-O bending vibration. Both MgO and FeO absorb near 20 microns, but since most Mg and Fe will have condensed as metal, these oxides will contribute very little to grain absorption in this bandpass. Freshly condensed grains will therefore absorb strongly near 10 microns and only weakly, if at all, near 20 microns.

As the grains "age", several chemical processes will occur which will affect the spectral properties of the dust. First, the grains will continue to react with oxygen in the ambient environment; this will produce more polymerized SiO_x and will also increase the oxidized fraction of Mg and Fe. Both of these processes will increase the grain opacity near 20 microns quite significantly, whereas the continued oxidation of the SiO to "SiO_2" will increase the opacity near 10 microns by at most a factor of two. The ratio of the grain absorption near 10 microns compared to that at 20 microns should therefore be a monotonically decreasing function with time even if the intrinsic strength of the 10 micron feature increases due to the oxidation of SiO. This point was used by Stencel et al. (Ap. J. (Lett.) 350, L45-8, 1990) to construct a dust chronology for Mira variables.

2. The "4.6 Micron" Feature is an Indicator of the Chemical State of Silicate Grains

Silicate grains have been produced in the laboratory by two separate and very different techniques, both of which allow the incorporation of hydrogen into the structure of the grains. In the first experimental setup, ices of $SiH_4 + CH_4$, $SiH_4 + NH_3$ or $SiH_4 + H_2O$ were irradiated by 1 MeV protons to a total dose of $\sim 10^{15}$ protons/cm^2 then slowly warmed to room temperature. In a carbon-rich grain, the SiH fundamental stretch occurs near 2110 cm^{-1} while in an oxygen-rich residue this feature appears as a triplet with components at 2140 cm^{-1}, 2187 cm^{-1} and 2244 cm^{-1}. The spectra of grains produced via the second experimental technique - the combustion of an $SiH_4 + O_2$ mixture in hydrogen at a variety of temperatures - follow a similar pattern. Combustion at higher temperature yields more complete oxidation of the silicate: grains produced at higher temperatures generally exhibit SiH absorption at higher energy since they are more oxygen-rich.

The SiH feature in an "SiC" grain gradually undergoes oxidation in air at room temperature: the position of the major peak shifts from 2110 cm^{-1} in the "reduced" grain to 2172 cm^{-1} in the more oxidized silicate while a shorter wavelength spectral component develops near 2230 cm^{-1}. Exposure of grains produced via combustion at intermediate temperatures to air and/or water (oxidizers) results in the loss of all long wavelength components of the SiH stretch, a small change in the position of the main peak, and some loss of hydrogen from the grains. Exposure to vacuum at elevated temperature for long time periods results in the loss of most of the hydrogen from the grains. Therefore for grains in circumstellar or interstellar environments where the grain temperature might exceed 300 K for considerable times, the elimination of the SiH feature is possible if the rate of replenishment by ion-implantation does not exceed the rate at which hydrogen is thermally driven from the grains. However, as long as some SiH exists in the grain, the energy of the SiH fundamental will be indicative of the oxidation state of the silicon in the grains.

POSITRONIUM IN ASTROPHYSICAL CONDITION

V.V. BURDYUZHA[1], KAUTS V.L.[2], YUDIN N.P.[3]

[1] Astro-Space Center Lebedev Physical Institute
Profsoyuznaya 84/32, Moscow, USSR

[2] Institute of Nuclear Research, 60 October Str.,7a,
Moscow, USSR

[3] Physical Department Moscow State University,
Leninskie Gory, Moscow, USSR

From a physical point of view we analyzed the possibility of observation of the L_α-line of the positronium (Ps). In a broad range of temperatures the processes of recombination to states of Ps with different nl, collisions of Ps with electrons and protons of medium are examined.

1. INTRODUCTION

Observations of e^+e^- annihilation line (511 keV) have put forward the question on the formation of lepton atoms Ps and therefore the question on the observation of L_α-line of positronium. However in a previous paper (Burdyuzha et al., 1987) it was pointed out that the observed details in UV spectra of SN 1987A can't be interpreted as a Lyman series of Ps. The attempt to detect the radiorecombination lines of Ps in the direction of the center of the Galaxy used VLA and IRAM was also unsuccessful (Anantharamaiah et al., 1989). On the other hand the annihilation processes in nonthermal sources take place everywhere (radiopulsars, X and γ ray sources, solar flares and so on) and production of lightest atom is evident. Undoubtedly future missions with more sensitive γ-spectrometers will display new kinds of sources of annihilation line. We intend to discuss the formation of Ps in excited states (mainly in 2S, 2p states) and therefore the possibility of detecting the L_α-line of Ps ($\lambda \approx 2431$Å). Note that this question was already considered in some degree in earlier literature (see for example, McClintock, 1984; Gould, 1989).

2. ATOM OF POSITRONIUM

Positrons in medium come into the sufficiently complex processes, the final result of which is, of course, the annihilation. The annihilation of Ps takes place from S-states (the nonrelativistic limit). This is because the process demands that the e^+ and the e^- - have to be at small distances but the wave function of Ps in states with nonzero angular momentum ($l \neq 0$) at small distances will become very small. The states of Ps have a spin equal to zero (para - Ps) or to one (orto Ps). The orto - Ps decays to three photons, the para Ps decays to two-photons. (see Fig. 1).

Let's also give also some additional information on the Ps. The average lifetime of Ps from singlet (para) and triple (orto) states are:

$$\tau_{para} \simeq 1.25 \times 10^{-10}\ n^3\ sec;\ \tau_{orto} \simeq 1.33 \times 10^{-7}\ n^3\ sec,$$

where: n is the main quantum number.
The energies E_n of Ps levels are given by the formula (in eV)

$$E_n = -6.8/n^2 \tag{1}$$

The probabilities A of radiative transitions in Ps are two times less than the probabilities of transitions between similar states of hydrogen $A_{Ps} = 1/2\ A_H$. The wave lengths of emitted photons in Ps are two times more than analogious wave lenghts of hydrogen $\lambda_{Ps} = 2\lambda_H$.

3. THE PROCESS OF RECOMBINATION

For radiative recombination of positron with electron the formulae are practically the same that for H with the trivial modification:

$$k = \sqrt{E_{e^+}/E_1^{Ps}},\quad E_1^{Ps} = 6.8\ eV,\ \sigma_{nl}^{Ps} = 4\sigma_{nl}^{H}$$

where E_{e^+} is $\mu v^2/2$, $\mu = \frac{1}{2} m_e$, v is relative velocity.
For $k \gg 1$ the cross section of radiation recombination is

$$\sigma_{nl} \approx \frac{1}{k} f_{nl} \tag{2}$$

where $f_{10} = 1/k^4$, $f_{20} = 1/8k^4$, $f_{21} = 3/32k^6$, $f_{30} = 1/27k^4$, $f_{31} = 8/243k^6$, $f_{32} = 1/246k^8$.

From the asymptotic formulas it's seen that in the process of recombination only low levels are populated (the factor of suppression is $1/n^3$). It's also seen that for $k \gg 1$ mainly the states with $l = 0$ are populated. The populations of orto and para states are defined by their statistical spin weights.

4. THE ROLE OF COLLISIONS

For Ps in quasiclassical approximation cross section takes the form

$$\sigma_{para}(2^1S_0 \to 2^1P_1) = 72\pi\left(\frac{\hbar}{m/2 \, v}\right)^2 \left[\ln\left(\frac{mv^2/2}{\Delta(2^1S_0 \to 2^1P_1)}\right) - 1.53\right]$$

$$\sigma_{orto}(2^3S_1 \to 2^3P_{0,1,2}) = 72\pi\left(\frac{\hbar}{m/2v}\right)^2 \left[\ln\frac{mv^2/2}{(\Delta'_0 \Delta^3_1 \Delta^5_2)^{1/9}} - 1.53\right], \quad (3)$$

where: $\Delta_{0,1,2}$ is the corresponding differences of energy (see Fig.1). Simple arithmetic of Maxwell averaging over electroneutral plasma ensures the following formula for the probability of $2S \to 2P$ transition under collisions of Ps with e^- and p:

$$\overline{\sigma v} = 1.88 \cdot 10^{-5} \frac{1}{T} \left\{2.73 \ln \frac{T}{2B} - 4.66\right\} \quad (4)$$

where: T is the temperature in eV, $B = (\Delta'_0 \Delta^3_1 \Delta^5_2)^{1/9}$ for the orto-Ps, $B = \Delta(2^1S_0 \to 2^1P_1)$ for the para-Ps.

Let's stress that the probability of intercombinational transitions for $n = 2$ is much less than the probability of basic processes.

5. DISCUSSION

As shown by Gould (1989) at temperature $T > 7 \cdot 10^5$ K the direct annihilation becomes dominant, i.e. this is the first limitation on production of Ps.

As seen from the results of our paper at small T (T < 10 eV) the recombination takes place, in general, to the 2P levels of Ps. The appearance in this case of the intensity of L_α-line is probable event with an effectiveness of 15% viewed from the annihilation line (in number photon), i.e. $L_{2431Å}/L_{511keV} \approx 1.5 \times 10^{-6}$.

At temperature $T > 10$ eV the recombination takes place, generally, to the 2S levels of Ps. The main result of collisions is the existence of critical densities of electrons N_{cr} for which the probability of annihilation is equal to the probability of collisional transition $2S \to 2P$. If the density of electrons is more than the $N_{cr}^{(2)}$ then the emission of L_α-line positronium must take place.

More detail about observations of P_s you can read in our total paper in Astron. and Astrophys. (1991).

REFERENCES

Anantharamaiah K.R., Radhakrishnan V., Morris D., Vivekanand M., Downes D., Shukle C.S., Preprint NRAO-88/178.
Burdyuzha V.V., Chechetkin V.M., Mickevich A.S., Shantarovich V.P., Yudin N.P., 1987, ESO workshop on the SN 1987 A. ed. Danziger, P. 113..
Burdyuzha V.V., Kauts V.L., Yudin N.P., 1991. Astron. Astrophys. (in press).
Gould R.J. 1989. Astrophys. J. 344, 232.
McClintock J.E. 1984. Astrophys. J., 282, 291.

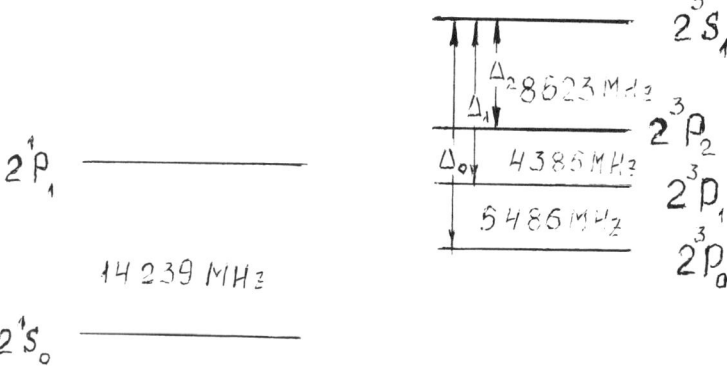

Fig.1   The low levels of Ps

Astrophysical Problems Involving Carbon Re-appraised

J P Hare and H W Kroto

*School of Chemistry and Molecular Sciences
University of Sussex, Brighton, BN1 9QJ UK*

ABSTRACT

The molecule C_{60}, Buckminsterfullerene, was discovered[1-3] during laboratory experiments motivated by problems associated with processes involving carbon in stars and space[4,5]. Astronomical puzzles also lay behind the experiments which led to the molecule's extraction and structure confirmation[6-8]. Although the resulting breakthrough has opened up exciting new avenues of chemistry, physics and materials science here on earth[9] the original astrophysical questions still remain and are even more tantalising now than they were before. Some of the puzzles are here re-addressed in the light of the new understanding which the fullerene discovery has brought. Indeed we shall look at the questions through magenta coloured spectacles and note that there are new and even more intriguing parallels between the behaviour of carbon on earth and space. The article contains a brief account of the processes responsible for the synthesis of carbon in stars and its dissemination throughout the Galaxy as this information is deemed necessary to gain an intrinsic understanding of the amazing role carbon plays in nature.

INTRODUCTION

There are several problems involving carbon where some links between C_{60}, chains and dust are possible[2,10,11], perhaps even likely, and it is useful to re-appraise some of the leading questions in the light of our new understanding. The aim here is to understand these questions better in the hope that answers will soon be forthcoming. Fullerene-60 analogues are only likely to be important in space if the geodesic cage structure (sigma electron network) remains essentially intact *and* that the integrity of the fullerene pi-electron system is disrupted as little as possible. Otherwise a vast range of analogues appears feasible and/or cage fragmentation would be expected to take place. Secondly, as fullerenes are very photostable in cold beams[12] and exhibit a peculiar instability in some condensed phases[13] they may be more stable in space than on earth. The detailed arguments which suggest that fullerene analogues may be the missing links in the Galactic Carbon Chain (of events) have been discussed by Kroto and Jura[14]

The Diffuse Interstellar Bands (DIBs).

The identity of the carrier of the DIBs (a set of some 80 electronic absorption features which are too broad to be atomic lines) has been a major puzzle ever since they were identified as interstellar features in 1936. Their properties have been summarised by Herbig[15,16]. Numerous suggestions have been made as to the nature of the carrier and at present the consensus lies between some moderately large "stable" molecule and grain-like material in which atoms are trapped giving rise to matrix-like spectra. The problem is that this material appears to be is ubiquitously abundant even in the hostile environment of space and yet, astoundingly, has never been identified on earth. Recently Sarre[17] and Fossey[18] have shown convincingly that five DIB features occur in *emission* from the "Red Rectangle" a carbon-rich star, HD44179, which exhibits a bipolar outflow with a roughly rectangular emission pattern. These features are sitting on top of a very broad intense red continuum emission[19] which Duley has suggested is from some form of H/C containing material[20]. C_{60} or an analogue was suggested as a possible carrier when the molecule was first found to form spontaneously[1]. There are intriguing parallels between fullerene-60 and the DIB carrier which have prompted the conjecture to be probed further[2,10,11,14,21]. The ubiquity of <13.6eV radiation indicates that if C_{60} is formed in space (and is as stable as beam experiments suggest) it will be ionised. The case for ionic analogues[10,11] has been discussed[21]. The discovery of endohedral fullerene complexes in which the atom is inside the cage[22] such as (La) has prompted interest in the astrophysical importance of such species and several calculations have been carried out on their spectra[23-26]. Perhaps the most interesting analogues are however *exo*hedral complexes and the possibility that charge transfer bands associated with such species as ( )M^+ where M is an abundant interstellar species such as Na, K, Ca, Fe, S, O etc has been discussed in detail[14]. These ideas are supported by the recent experimental results of Huang and Freiser[27].

The Unidentified 2170A Band.

Another feature which has also been the subject of numerous studies is a strong absorption centered at ca 2170A[28-30]. This feature is very strong and in general always observed at 2170A. Stecher and Donn suggested this was caused by carbon when they first detected the feature[29]. The line shape is also relatively constant although there are one or two objects in which the feature appears to shift slightly[30] Interestingly Day and Huffman[31] have studied the scattering from pure small carbon particle and finds a feature centered near 2400A. It has been noted[14] that in the highly H-depleted RCorBor stars the feature appears to shift considerably - to ca 2400A[30]. Protonated species are common in the ISM and thus the possibility that a protonated exohedral fullerene complex (+)H should be seriously considered[14]. Such a species is expected to have a very strong, very broad, charge-transfer electronic absorption near 6eV (ca 2100A)[14] which is quite close to 2170A, so a contribution to the interstellar extinction from such a transition may be important.

The Unidentified Infrared Bands (UIBs).

The development of sensitive IR telescopes has led to the detection of IR emission from circumstellar material[32]. Comparisons between some discrete circumstellar features with laboratory IR measurements of known polycyclic aromatic hydrocarbons (PAHs) are sufficiently good that the assignment to PAH-like carriers is quite convincing[33-35]. The C_{60} discovery suggests that hydrogenated curved/partly-closing structures might also occur and such objects have been shown to yield certain IR features consistent with observation[36].

There are of course a multitude of possible species which can form from some complex mix of small compounds to large aggregates of organic material such as soots and tars. It has been suggested[34] that the interstellar radiation flux which pumps the observed bands is such that the pattern of emission frequencies detected can be used to set a bound on the size of the emitting species as containing 20-40 atoms. This calculation however depends on the assumption that on absorption of a photon the energy is uniformly distributed and the species or grain emits IR radiation with a characteristic thermalised profile. It is not however clear that localised chromophore excitation cannot occur due to inefficient energy transfer in heterogeneous aggregates. Intramolecular vibrational relaxation and the relationship with radiative processes are poorly understood even for small molecules and even less so in complex, loosely bound aggregates.

Of course hydrogen is present in copious quantities in most astrophysical objects, however the observations of Gerhardt et al[37] and Howard et al[38] have shown polyynes are a prelude to soot formation and that C_{60} forms in a sooting flame and can be extracted. There is thus every prospect that fullerenes form wherever polyynes and dust form in space, even in the presence of hydrogen and oxygen. There is still some way to go before the question of how important the role of non-planar carbonaceous networks is during soot formation, however it should be investigated as soot formation appears to be so poorly understood[39].

Interstellar and Circumstellar Dust

The dust in space is a most fascinating material, playing key roles in many important astrophysical processes[28]. For instance it appears that grain surface catalysis is the only way to explain the formation of H_2 in space[40]. For the purposes of this discussion we note that the dust in the dark clouds protects molecules from dissociation by starlight and so enables the heat generated on gravitational collapse to be dissipated by low temperature rotational emission by CO and so allow later phases of cloud collapse to occur - ultimately to form new stars. Dust certainly forms in the shells of red giant stars and also around novae, supernovae, and planetary nebulae and the latter are proving to be rather interesting sources of carbon. The planetary AFGL 2688 is a hot star surrounded by a donut-shaped dense molecular gas and dust cloud[41,42]. The star sits in the hole in the middle of the donut and emanating from the star, along the axis of the donut, is a bipolar outflow of high velocity gas - H and CO at ca 50 km s^{-1}. The visible emission contour has caused it to be called the "Egg Nebula"[41]. Rieu et al[42] discovered a curious feature about this star: The donut contained the molecules that the ion-molecule theories[43,44] readily explain (ammonia etc) but far out from the center, the conical bipolar outflow is surrounded by hollow cone of gas, rich in polyynes[42]. It has been suggested that these chains are produced by grain fragmentation due to grain-grain collisions[45]. Carbon cluster beam photofragmentation studies have shown that macroscopic carbon clusters can break down to such chains and C_{60} also appears[12]. Thus laboratory experiments suggest that C_{60} analogues may also occur in such regions[2].

Studies of the polarisation of scattered light by interstellar dust suggests that it in some regions it may consist of elongated particles[28]. The work of Iijima[46] and Endo[47] indicates how such elongated carbon microfibres may grow under conditions in which C_{60} forms. Microfibres with diameters of 30-100A with graphite walls containing from 2-50 layers of graphite have now been detected. These structures are closely related to the structures observed earlier by Iijima[48]. On the basis of our new understanding of closed carbon giant networks[49,2] a possible growth mechanism for such microfibers has been proposed[50]. The possibility that such tiny tubular graphite structures (Zeppelenes?) are feasible adds a new dimension to the possible answers to a range of fascinating spectroscopic astronomical observations such as those by Sellgren who observed IR emission consistent with microscopic particles of these dimensions[51]. Wright has suggested that spheroidal graphite structure cannot account for the observed emission[52]. It should however be noted that attention has been drawn to the fact that the apparent epitaxial control of growth in general breaks down and in later stages of accretion from the gas phase, graphite layers become more chaotic and so-called "amorphous" texture appears to develop[53].

Meteoritic Carbon

Most interesting observations have been made by Lewis et al[54] who have detected diamond domains in meteorites. These inclusions are most interesting and how they might arise is food for thought. One suggestion is that they might form by metamorphosis, initiated perhaps by shock waves, of the internal structure of quasi icosahedral concentric shell graphite dust included in the meteorite[55].

One of the most intriguing aspects of the meteorites are their curious isotope anomalies. Clayton[56] has suggested that the existence of almost pure ^{22}Ne in some carbonaceous chondrites might be explained if it is a remanant of ^{22}Na formed in supernovae. He suggests that the ^{22}Na (which has a 2.5 yr half-life) is ejected from the star and co-condenses with carbon in the surrounding dust shell. The detection of endohedral metal complexes such as (M) where M=La etc[22,27] and the elegant recent observation of He encapsulation during collisions with fullerene-60 by Weiske et al[57] suggest that the endohedral fullerene complexes might well play a role in these well known isotope anomalies involving carbon. Soon after the discovery of C_{60}, Heymann[58] considered the (He) complex from an astrophysical viewpoint.

Conclusions

Several astrophysical problems have been described which involve carbon chains and dust and attention has been drawn to various chemical scenarios in the laboratory which yield very similar conditions. The cluster beam studies indicate that chains (or their monocyclic ring analogues) form first which then evolve into fullerenes and carbon microparticles in some yet to be fully understood way. The fascinating observation of Rubin et al[59] that C_{30} monocylic rings dimerise spontaneously to C_{60} can be rationalised by a scheme in which a concerted series of cyclo-addition steps can result in a graphitic/fullerene network[60]. Furthermore recent evidence suggests that fullerenes may even be the direct precursors of pure carbon microparticle accreting carbon directly from the vapor[50]. On the basis of a wealth of circumstantial evidence we have argued fullerene analogues must have an important role in the "Galactic Carbon Cycle". It has been pointed[14] out that fullerene analogues do have several properties that justify their carful consideration as carriers of some ubiquitous astrophysical features and that if they are not responsible there is some other as yet unidentified mystery involving carbon to be unraveled[14].

Acknowledgements We are grateful to Simon Balm, Laurence Dunne, Mike Jura, Sydney Leach, Peter Sarre and Jim Watson for valuable discussions. We thank the Royal Society, British Gas and SERC.

REFERENCES

1) Kroto, H. W.; Heath, J. R.; O'Brien, S. C.; C url, R. F.; Smalley, R. E., Nature (London) **1985**, *318*, 162-163.
2) Kroto, H. W. Science **1988**, *242*, 1139-1145.
3) Curl, R. F. and Smalley, R. E. Science **1988**, *242*, 1017-1022.
4) Heath, J. R.; Zhang, Q.; O'Brien, S. C.; Curl, R. F.; Kroto, H. W.; Smalley, R. E. J. Am. Chem. Soc. **1987**, *109*, 359-363.
5) Kroto, H. W.; Heath, J. R.; O'Brien, S. C; Curl, R. F.; Smalley, R. E. Astrophys. J. **1987**, *314*, 352-355.
6) Kraetschmer, W.; Lamb, L. D.; Fostiropoulos, K.; Huffman, D. R. Nature (London). **1990**, *347*, 354-358.
7) Taylor, R.; Hare, J. P.; Abdul-Sada, A. K.; Kroto, H. W. J. Chem. Soc. Chem. Commun. **1990**, 1423-1425.
8) Kroto, H. W., Angewandte Chemie **1990**, in press
9) Kroto, H. W., Allaf, A. W.; Balm, S. P., Chem. Revs. **1991**, *91*, 1213
10) Kroto, H. W. Polycyclic Aromatic Hyrocarbons and Astrophysics, Leger, A.; d'Hendecourt, L. B. eds.; Reidel: Dordrecht, 1987, pp. 197-206.
11) Kroto, H. W. Ann. Phys. Fr. **1989**, *14*, 169-179.
12) Heath, J. R.; O'Brien, S. C.; Curl, R. F.; Kroto, H. W.; Smalley, R.E. Comments. Condens. Matter Phys. **1987**, *13*, 119-141
13) Taylor, R.; Parsons, J. P.; Avent, A. G.; Rannard, S. P.; Dennis, T. J.; Hare, J. P.; Kroto, H. W.; Walton, D. R. M., Nature **1991**, *351*, 277.
14) Kroto, H. W.; Jura, M.; in press.
15) Herbig, G. H. Astrophys. J. **1975**, *196*, 129-160.
16) Herbig, G. H. Astrophys. J. **1988**, *331*, 999-1003.
17) Sarre, P. J. Nature, **1991**, *351*, 356.
18) Fossey, S. J. Nature, **1991**, *353*, 393.
19) Cohen, M. et al., Astrophys. J. **1975**, *197*, 179.
20) Duley, W. W., Mon. Not. R. astr. Soc. **1985**, *215*, 259-263.
21) Leger, A.; d'Hendecourt, L.; Verstraete, L.; Schmidt, W. Astron. Astrophys. **1988**, *203*, 145-148.
22) Heath, J. R.; O'Brien, S. C.; Zhang, Q.; Liu, Y.; Curl, R. F.; Kroto, H. W.; Smalley, R. E. J. Am. Chem. Soc. **1985**, *107*, 7779-7780.
23) Ballester, J. L.; Antoniewicz, P. R.; Smoluchowski, R. Astrophys. J. **1990**, *356*, 507-512.
24) Cioslowski, J.; Fleischmann, E. D., J. Chem. Phys. **1991**, *94*, 3730.
25) Chang, A. H. H.; Ermler, W. C.; Pitzer, R. M., J. Chem. Phys. **1991**, *94*, 5004.
26) Wastberg, B.;Rosen, A., Physica Scripta. **1991**, *44*, 276-288.
27) Huang, Y.; Freiser, B. S., Nature **1991**, *****

28) Savage, B. D.; Mathis, J. S., Ann. Rev. Astrophys. **1979**, *17*, 73-111.
29) Stecher, T. P.; Donn, B., Astrophys. J. **1965**, *142*, 1683.
30) Fitzpatrick, E. L.; Massa, D., Astrophys. J. **1986**, *307*, 286.
31) Day, K. L.; Huffman, D. R., Nature Physical Science, **1973**, *243*, 50-51.
32) Holm, A. V.; Wu, C.; Doherty, L. R., Astro. Soc. Pac. **1982**, *94*, 548-552.
33) Duley, W. W.; Williams, D. A. Mon. Not. R. Astron. Soc. **1988**, *231*, 969-975.
34) Leger, A.; Puget, L. J. Astron. Astrophys. **1984**, *137*, L5-L8.
35) Allamandola, L. J.; Tielens, A. G. G. M.; Barker, J. R. Astrophys. J. **1985**, *290*, L25-L28.
36) Balm, S.P.; Kroto, H. W. Mon. Not. R. Astron. Soc., **1990**, *245*, 193-197.
37) Gerhardt, P.; Homann, K. H. J. Phys. Chem. **1990**, *94*, 5381-5391.
38) Howard, J. B.; McKinnon, J. T.; Makarovsky, Y.; Lafleur, A. L.; Johnson, M. E., Nature **1991**, *352*, 139-141.
39) Harris, S. J.; Weiner, A. M., Ann. Rev. Phys. Chem. **1985**, *36*, 31-52.
40) McCrea, W. H.; McNally, D., Mon. Not. Roy. Astronom. Soc. **1960**, *121*, 238.
41) Ney, E. P., Sky and Telescope **1975**, Jan, 21.
42) Nguyen-Q-Rieu; Winnberg, A.; Bujarrabal, V., Astron. Astrophysics. **1986**, *165*, 204-210.
43) Herbst, E.; Klemperer, W. Astrophys. J. **1973**, *185*, 505-533.
44) Dalgarno, A.; Black, J. H. Rep. Prog. Phys. **1976**, *39*, 573-612.
45) Jura, M.; Kroto, H. W., Astrophys. J. **1990**, *351*, 222-229.
46) Iijima, S., Nature **1991**, *354*, 57.
47) Endo, M., personal communication.
48) Iijima, S. J. Cryst. Growth, **1980**, *5*, 675-683.
49) Kroto, H. W.; McKay, K. G. Nature (London), **1988**, *331*, 328-331.
50) Endo, M. and Kroto, H. W., to be published.
51) Sellgren, K., Astrophys. J. **1984**, *277*, 623-633.
52) Wright, E. L. Nature (London) **1988**, *336*, 227-228
53) Kroto, H. W., Iijima, S., to be published
54) Lewis, R. S.; Ming, T.; Wacker, J. F.; Anders, E.; Steel, E., Nature **1987**, *326*, 160.
55) McKay, K. G.; Dunne, L. J.; Kroto, H. W.; in preparation.
56) Clayton, D. D. Nature (London) **1975**, *257*, 36-37.
57) Weiske, T.; Bohme, D. K.; Hrusak, J; Kratschmer, W; Schwarz, Angew. Chem. Int. Ed. Engl. **1991**, *30*, 884.
58) Heymann, D. J. Geophys. Res. B **1986**, *91*, E135-138.
59) Rubin, Y.; Kahr, M.; Knobler, C. B.; Diederich, F.; Wilkins, C. L., J. Am. Chem. Soc. **1991**, *113*, 495-500.
60) Kroto, H. W.; Walton, D. R. M.; *Post-Fullerene Organic Chemistry* in *Chemistry of Three-Dimensional Polycyclic Molecules,* Editors; Osawa, E. and Yonemitsu, O., VCH International: New York.

GREENBERG

a) What is the visual absorption by C_{60} as a function of wavelength.
b) Could it contribute to the observed extinction.
c) How much C is needed to produce visual extniction if in C_{60}

KROTO

a+b) The C_{60} spectrum is now known and it is not ovious that it contributes to any optical features. The visible absorption is very weak. The bands at 2100 and 2500A are very strong. However in the regions where it is likely to be seen (visibly) it is also likely to be ionised so C_{60}^+ is probably the dominant analogue. Also in any region where HCO^+ is abundant the most likely species is $C_{60}H^+$.

c) Visible extinction of the neutral C_{60} molecule is fairly weak as there are only forbidden transitions in this region

GREENBERG

If DIBs are due to K then there should be an anticorrelation between K absorption and the DIBs. (Incidentally, I once - 30 years ago - questioned the Ca line anticorrelation with the 4430A band to indicate Ca as a possible source of DIB carrier). Maybe Ca is a better choice for your mechansim.

KROTO

Our paper (Kroto and Jura) suggests that the DIB carriers might be $C_{60}M$ ionised complexes where M is an alkali metal or any abundant atom. Ca, Na K and Mg as well as O and S etc are all excellent candidates for the attached species. The Ca and K anticorrelation are possibly interesting support for the proposal and in fact we have discussed Ca in our paper. In my verbal presentation I discussed K as an example of the type of atom likely to give rise to charge transfer bands. (doubly charged species such as $C_{60}^-Ca^{++}$ are also very interesting species).

GREENBERG

If DIBs are due to K in/on C_{60} do you need to give the DIB strength. Could we be running into a cosmic abundance problem since N_C/N_{Ca} is large per DIB molecule.

KROTO

Under certain conditions in the laboratory the amount of carbon in C_{60} and other fullerenes to carbon in soot varies from 10-20%. In benzene combustion flames it can be as high as ca 7% of the soot collected. The charge transfer bands associated with exohedral coplexes of the kind discussed are associated with electron jumps of the order of 6-7A which are more than an order of magnitude larger than for standard atomic and molecular electronic transitions resulting in bands some 50-100 times stronger than usual fully allowed transitions.

WILLIAMS

What is the C:H ratio in the experiments you do in the lab?

KROTO

This is a difficult question to answer as the process we observe is highly inhomogeneous. A pure carbon plasma is vaporised from graphite surface by a laser into a He/H_2 atmosphere. As the hot plasma expands the contact surface causes complex reactions between carbon atoms (which are themselves clustering) at the interface and reacting with C and C/H radicals. Even in pulses with He/H_2 ratio of ca 1:1 (by volume) we still see C_{60} formation. Even more remarkable is the observation by Ulf Sassenberg and Bosse Lindgren in Stockholm who have seen C_{60} formation in pure oxygen gas pulses! Also, Howard and his co-workers at MIT have seen some 7% of fullerene material in soot from a benzene flame. To sum up, C_{60} is detected whenever carbon particles are produced in the gas phase.

L d'HENDECOURT

The production of C_{60} by Krätschmer (as well as others) yields large amounts of C_{60} in a very specific way (plasma of carbon in He at "high" pressure). Are such experiments applicable to the atmospheres of (carbon) stars where hydrogen dominates and the expected pressures are orders of magnitude smaller?

KROTO

I feel that the key factor is whether the conditions favor carbonaceous dust production. Whenever we see carbon particles In the laboratory we also invariably now see C_{60}. Intuition certainly suggests that the higher the C/H ratio is, the higher the C_{60}/soot ratio is likely to be. When H/C = 0 we know C_{60} can make up some 15±5% of the soot and RCorBor and similar stars are clearly ideal objects. The recent <u>combustion</u> data suggest that even in high (H+O)/C regions the C_{60} to dust ratio may be between 1-10%

ADVANCEMENT OF PHOTOIONIZATION AND PHOTODISSOCIATION RATES RELEVANT TO ASTROCHEMISTRY

S P TARAFDAR
Tata Institute of Fundamental Research
Homi Bhabha Road
Bombay 400 005, INDIA

ABSTRACT: The advancement of photoionization and photodissociation rates relevant for interstellar diffuse and dense clouds, circumstellar medium, cometary coma and planetary atmosphere has been reviewed with a mention of uncertainties involved in these rates.

1. INTRODUCTION

One of the most importance processes governing the molecular abundances in astronomical scenario is the photodissociation and photoionization. The role played by these processes in determining the molecular abundances in cometary coma and in planetary atmosphere were realised as early as 1930 (cf. Chapman 1930, Wurm 1932) bringing the subject of photochemistry in the forefront of astronomical studies. In interstellar medium the importance of photodissociation and photoionization were appreciated as early as 1941 by Swings (1941) when he attempted to understand the relative abundances of newly discovered (Swing and Rosenfeld 1937, McKeller 1940, 1941, Douglas and Herzburg 1941) CH, CN and CH^+ molecules though the possibility of photodissociation of molecules in interstellar space was mentioned earlier by Eddington (1926). The importance of photodissociation and photoionization in circumstellar medium has been appreciated only recently (cf. Goldreich and Scoville 1976, Scalo and Slavsky 1980, Clegg et al 1983), though the importance of ionization of atoms and ions were realised very early (cf. Stromgren 1948). The progress on the determination of photoionization and photodissociation rates were, however, hindered due to the lack of our knowledge about radiation field in ultraviolet (i.e. $\lambda < 3000 Å$) where most of the photodissociating and photoinoizing thresholds lie. With the advent of space astronomy, the radiation fields in the UV spectral region for different astronomical scenario starting with solar values become available and the determination of photoionization and photodissociation rates was possible. Some of these early determinations have been reviewed by Van Dishoeck (1987) in the first Astrochemistry Symposium at Goa (cf. Vardya and Tarafdar 1987). Van Dishoeck (1987) has also discussed various processes which lead to photodissociation. A discussion on processes leading to photoionization and photodissociation of molecules can also be found in Duley and Williams (1984). Here I would like to summarize the advancement made since meeting at Goa on the determination of photorates relevant to different astronomical objects which are distinguished from the flux of unattenuated radiation $F(\lambda)(phcm^1 s^{-1})$ used in the photorate expression:

$$\Gamma = \int_0^{\lambda_{th}} \sigma(\lambda)F(\lambda)e^{\tau}d\lambda, s^{-1}, \tag{1}$$

where $\Gamma(s^{-1})$ is the photorate, $\sigma(\lambda)$ the wavelength (λ) dependent cross section, λ_{th} the threshold wavelength and τ is the optical depth of the medium at wavelength λ.

2. INTERSTELLAR MEDIUM

2.1 Rates with interstellar Diffuse radiation:

As reported by Van Dishoech (1987), the first comprehensive photorates for interstellar diffuse clouds were given by Black and Dalgarno (1977). Prasad and Huntress (1980) included these photorates in their compilation of reaction rates adding a number of photorates on their own. The rates given by Black and Dalgarno (1977) and Prasad and Huntress(1980) used interstellar radiation field which is in agreement with the interstellar field at 1000Å given by Habing (1968) and has a λ^3 dependence for $\lambda \leq 3000$Å. For $\lambda \geq 3000$Å, the used radiation field was the diluted blackbody field proposed by Spitzer (1968). The radiation field assumed to be zero for $\lambda < 912$Å, as absorption by hydrogen atom will inhibit the presence of radiation field at wavelengths smaller than its threshold wavelength of 912Å. Van Dishoeck (1987, 1988) determined photorates for a number of species using more reliable cross sections and a radiation field, the wavelength dependence of which determined from impirical fit to the interstellar radiation field (Witt and Johnson 1973, Jura 1974) and is given by the relation (Draine 1978):

$$F(\lambda) = \frac{3.210 \times 10^{15}}{\lambda^3} - \frac{5.17 \times 10^{17}}{\lambda^4} + \frac{2.064 \times 10^{20}}{\lambda^2} photons\, cm^{-2} s^{-1} Å^{-1} \tag{2}$$

for $\lambda \leq 2000$Å and

$$F(\lambda) = 6.973 \times 10^2 \lambda^{0.7} photons\, cm^{-2} s^{-1} Å^{-1} \tag{3}$$

for $\lambda \geq 2000$Å. Fig. 1 compares the radiation field used by Black and Dalgarno (1977), that given by equations (2) and (3) and the observed and computed radiation field by different authors. Fig. 1 shows that the flux given by equations (2) and (3) passes through maximum available values of radiation field and gives photorates a factor 1.5 to 2 larger depending on the threshold wavelength of dissociation or ionization than that obtained with other values of the field. Another uncertain factor related to interstellar radiation field is its attenuation by dust and molecular hydrogen. Recently Van Dishoeck and Black (1988) have shown that the H_2 and HI absorptions could reduce CO dissociation by a factor of 1.5. The grain absorption, however, remains very uncertain, as it depends on unknown grain properties which may vary from cloud to cloud. Different authors have given prescription for the inclusion of the effect of dust attenuation.

The calculation of photoionization and photodissociation rates was extended to more and more complex molecules (cf. Herbst and Leung 1986) resulting a comprehensible table (Miller et al 1991) of photorates for 92 species with 134 channels. The accuracy of these rates, however, is not very good. Besides, the uncertainty associated with the interstellar

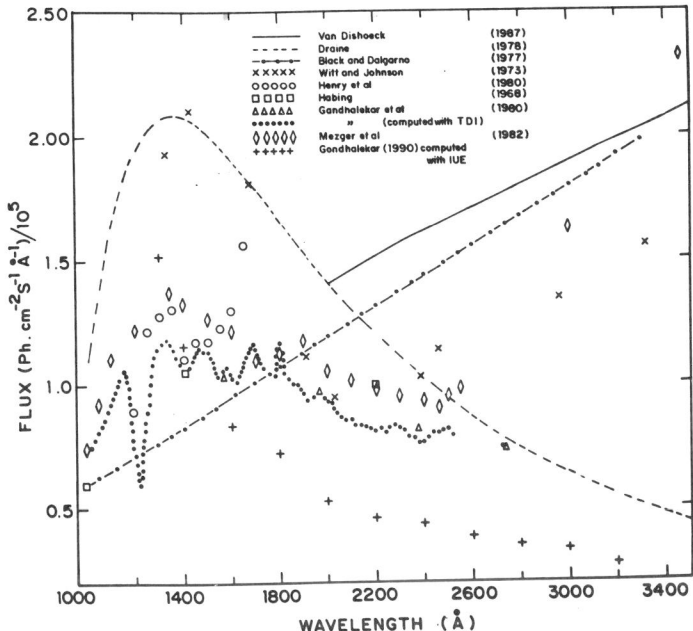

Fig 1: Comparison of observed and theoretically determined interstellar radiation field with that used in photorate determination.

radiation field and its attenuation, the photodissociation and photoionization cross sections used for the determination of rates are very uncertain. To give an idea of uncertainty present, the photodissociation cross section are infered from molar extinction coefficient for CH_3OH, C_2H_5OH (Harrison et al 1959), C_2H_4O (Lake and Harrison 1959) and for CH_3OCH_3 (Harrison and Price 1959) and from low resolution absorption coefficients for C_2H_5OH (Ogawa and Cook 1958). It is needless to mention that better cross sections for many molecules are needed.

2.2 Photorates with cosmic ray induced emission from H_2:

The interior of dense clouds is not completely devoid of ultraviolet photons as it was assumed conventionally. Prasad and Tarafdar (1983) have shown that the electrons produced by cosmic ray interaction with H_2, HI and He can excite H_2 molecules to its upper electronic states, the spontenuous decay of which act as source of UV-photons in the dense clouds. Prasad and Tarafdar (1983) estimated the flux of these photons in the interior of dense clouds and determined CO dissociation rates there. They came to the conclusion that dissociation rate of CO is large enough to give observed CI/CO ratio in dense cloud in the steady state. Note that observed CI/CO ratio could be achieved in evolving models (Tarafdar et al 1985) and in time dependent models, if the life time of the cloud is not long enough to reach the steady state. Sternberg, Dalgarno and Lepp (1987) determined the UV-flux from cosmic ray induced H_2 emission using Franck-Condon distribution for electron excitation to different vibrational levels of $B^1\Sigma_u^+$ and $C^1\Pi_u$ upper electronic states and used the flux to determine the photodissociation rates for seven molecules $-CH_4, C_2H_2, OCS, NH_3, HCN, H_2O$ and OH. Examining the effect of cosmic ray induced photons in dense clouds, they concluded that its presence will inhibit the formation of complex molecules. Gredel, Lepp and Dalgarno (1987) have determined the photodestruction rate of CO by cosmic ray induced photons as a function grain albedo, temperature and CO/H_2 ratio, as cosmic ray induced photons are line photons and photodestruction of CO is also by absorption of line photons. They concluded from their chemistry that CI/CO ratio depends on the abundance, the rational population and the velocity distribution of CO. For a uniform cloud with $CO/H_2 = 1.5 \times 10^{-4}$ and CO rotational temperature of $30K$, $CI/CO = 5 \times 10^{-3}$ which is much larger than its value of 5×10^{-6} without the presence of cosmic ray induced photons but smaller than the observed value of 0.1. An agreement between theory and observation can be obtained, however, by postulating, a severe depletion of C/O ratio and a differential velocity structure within the cloud. Gredel et al (1989) have improved the cosmic ray induced photon flux by taking the actual vibrational excitation distribution by electrons instead of Franck-Condon distribution and used this flux to determine photoionization and photodissociation rates for 37 species with 65 channels. The list of available photodestruction rates has been extended to 88 species with 193 channels by Tarafdar et al (1992) after determining the cosmic ray induced H_2-emission flux independently including the effect of H_2 and CO absorptions together with the absorption by grains. The photodestruction rates for common species in Gredel et al (1989) and Tarafdar et al (1992) agree farely well except for species having threshold near 1100Å or at shorter wavelength. For these species rates given by Tarafdar et al (1992) is smaller as H_2 absorption is more effective in the calculation of Tarafdar et al (1992) (Table 1). This discrepancy, however, is smaller than the uncertainty associated with the uncertain values of cross section mentioned before.

TABLE 1: comparison of photorates (s^{-1}) given by Tarafdar et al (1992) and Gredel et al (1989) for species with threshhold wavelength near 1100Å.

Species	Product	With H_2-absorption		Without H_2-absorption	
		Gradel et al	Present	Gradel et al	Present
CN	$C + N$	1.8-13	1.9-14	2.0-13	6.0-11
O_2	$O_2^+ + e$	1.8-15	4.7-16	2.3-15	1.7-15
Hcl	$H + Cl$	9.0-15	3.0-14	1.2-14	3.4-14
NH_2	$NH + H$	1.6-15	5.8-15	1.6-15	6.3-15
CH_3OH	$H_2CO + H_2$	6.3-14	4.0-13	6.0-14	4.0-13
C_2H_2	$C_2H_2^+ + e$	2.6-14	6.3-15	2.4-14	1.2-14
	$C_2H + H$	1.0-13	2.3-14	1.0-13	2.4-14
C_2H_3N	$C_2H_3N^+ + e$	4.5-14	4.6-15	3.8-14	1.9-14
C_2H_4	$C_2H_4^+ + e$	1.5-14	3.1-15	1.4-14	6.3-15
C_2H_4O	$CH_3 + CHO$	1.0-14	2.8-15	1.1-14	3.0-15
	$CH_4 + CO$	1.0-14	3.2-15	1.1-14	3.4-15

$*a - n$ imply $a \times 10^{-n}$.

2.3 Photodissociation by photons from decay of Dark Matters:

Interior of dense clouds could have another source of photons - the decay of all pervading dark matters. Sciama (1990a) has shown that the source function φ of photons arising from decay of dark matter (neutrinos) of density $(N_d) 5 \times 10^7 cm^{-3}$ (Salucci and Sciama 1990) and life time τ of $1.5 \times 10^{23} s$ (Sciama 1990b) is given by

$$\phi \equiv \frac{n_d}{\tau} = 3.3 \times 10^{-16} ph\, cm^{-3} s^{-1}. \tag{4}$$

Sciama (1990b) has also argued that the photons arising from the decay of dark matters (DDM-photons) can have energy of $13.8 ev$. The observations were, however, not able to support the existence of such photons in cluster of galaxies (cf. Dadidson et al 1991, Fabian et al 1991). Certain loopholes in the observations need to be plugged before their existence is overruled (cf. Hogan 1991, Davidson et al 1991). Keeping the need of further proof in support or in against the possibility of the existence of DDM-photons Tarafdar (1991) determined the photon flux inside dense clouds considering absorption by HI and grains. The photodissociation rate of CO was calculated and the effect of such photons on CI/CO ratio were examined. It was shown that if such DDM-photons exists, they can photodissociate CO to give the observed CI/CO ratio.

3. PHOTODISSOCIATION RATES FOR CIRCUMSTELLAR MEDIUM:

The photoionization and photodissociation in circumstellar medium could take place due to stellar or interstellar radiation fields. In the circumstellar medium of cool stars like evolved high mass or newly formed low mass objects, the stellar radiation field in UV has low value so that the photoionization and photodissociation in circumstellar medium is by the interstellar radiation field with proper attenuation (cf. Glassgold, Huggins and Langer 1985; Omont 1986; Glassgold, Lucas and Omont 1986; Nejad and Miller 1987; Glassgold et al 1987; Mamon, Glassgold and Omont 1987; Mamon, Glassgold and Huggins 1988; Nejad and Miller 1988; Beiging and Nguyen-Quang-Rieu 1989; Nercession et al 1989; Truong-Bach et al 1990 and Howe and Miller 1990). In some cool stars, the stellar radiation field could also be important (cf. Clegg et al 1983). The dissociation of molecules by cosmic ray induced photons may not be important as its time scale of $10^{13} - 10^{15} s$ is much longer than the expansion time scale ($\sim 10^{11} s$) of the circumstellar medium. However, in a static circumstellar medium the cosmic ray induced photons could play some limited role. The photodissociation and photoionization in circumstellar medium around hot stars (O and B stars) are mainly by stellar photons, as the ultraviolet flux from these stars is large. However, the interstellar photoionization and photodissociation rates scaled by appropriate factor (Cf. Tielens and Hullenbach 1985) can be used, as the stellar emission spectra (Fig. 2) has similar spectral shape in UV as interstellar radiation field. The error introduced by this assumption may be as large as an order or more (See Fig. 2) depending on the molecules (i.e. its threshold wavelength) and the spectral type (the temperature) of the central star.

4. PHOTOIONIZATION AND PHOTODISSOCIATION IN COMETARY COMA:

For cometary coma, the photoionizaation and photodissociation are governed by solar radiation field in UV. The earlier available rates (cf. Jackson 1976a, b and Huebner and Carpenter 1979) have been improved by Swift and Mitchell (1981) and more recently by Huebner, Keady and Lyon (1991). Note that photorates could be different from comets to comets and also in the same comet depending on its solar approach velocity which can

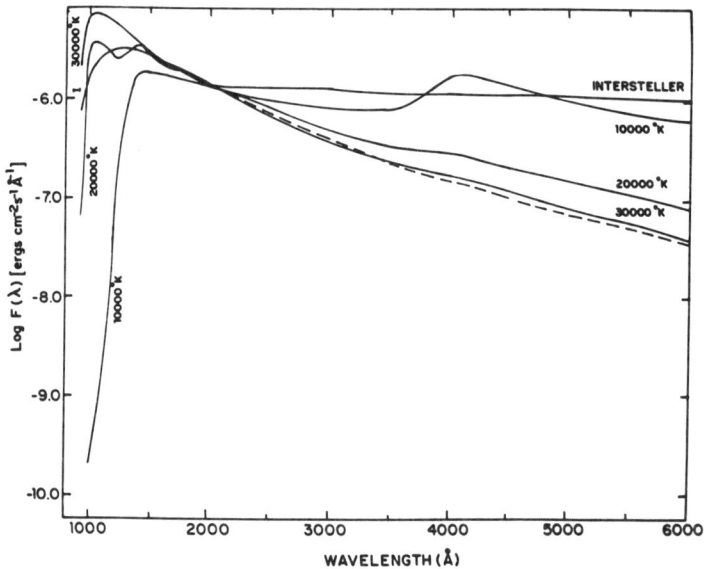

Fig. 2: Comparison of Interstellar radiation field used in photorate determination with stellar radiation fields (Kurucz 1779). Fields are normalised at 2000Å.

shift the threshold wavelength of photoprocess resulting a variation of radiation field and the photorates, as solar radiation in UV varies significantly over small wavelength interval.

5. PHOTOIONIZATION AND PHOTODISSOCIATION RATES FOR PLANETARY AND EARTHS ATMOSPHERE:

The photodissociation and photoionization in planetary atmosphere is governed by solar ultraviolet radiation field as in cometary coma without the problem associated with velocity, as planetary orbits are circular. However, the photodissociation rates in planetary atmosphere has the inherent difficulty of accounting for the attenuation by atmospheric dust particles and the absorption by major constituent molecules. The photochemistry in different layers of Earth's atmosphere and in the atmospheres of planets has been discussed in a book edited by Levine (1985) who has also compiled the photodissociation rates in the appendix.

Acknowledgements: I would like to thank P. D. Singh and A. Dalgarno for enabling my participation in the symposium.

6. REFERENCES:

Black, J. and Dalgarno, A.: 1977, Astrophys. J. Suppl. 34, 405.
Bieging J. H. and Nguyen-Quang-Rieu: 1989, Astrophy. J. 343, L25.

Chapman, S.: 1930, Mem. R. Meteorol. Soc. 3, 103.
Clegg R. E. S., Van Ijzendoorn, L. J. and Allamandola, L. J.: 1983, Mon. Not. R. Astr. Soc. 203, 125.
Davidson, A. F. et al: 1991, Nature, 351, 128.
Douglas, A. E. and Herzberg, G.: 1941, Astrophys. J. 94, 381.
Douglas, A. E. and Herzberg, G.: 1942, Can. J. Res. A20, 71.
Draine, B. T.: 1978, Astrophys. J. Suppl. 38, 595.
Duley, W. W. and Williams, D. A.: 1984, *Interstellar Chemistry*, Academic Press, London.
Edington, A. S.: 1926, Proc. Roy. Soc. A111, 424.
Fabian, A. C., Naylor, T. and Sciama, D. W.: 1991, Mon. Not, R. Astr. Soc. 350, 461.
Glassgold, A. E., Huggins, P. J. and Langer, W. D.: Astrophys. J. 290, 615.
Glassgold, A. E., Lucas, R. and Omont, A.: 1986, Astron. Astrophys. 154, L12.
Glassgold, A. E., Mamon, G. A., Omont, A. and Lucas, R.: 1987, Astron. Astrophys. 180, 183.
Goldreich, P. and Scoville, J. M.: 1976, Astrophys. J. 205, 144.
Gondhalekar, P. M.: 1990, In *The Galactic and Extragalactic Background Radiation*. IAU Symp. No. 139, P49, eds. S. Bowyer and C. Leinert, Kluwer, Dordrecht.
Gondhalekar, P. M., Phillips, A. P. and Wilson R.: 1980, Astron. Astrophys. 85, 272.
Gredel, R., Lepp, S. and Dalgarno, A.: 1987, Astrophys. J. 323, L137.
Gredel, R., Lepp, S. and Dalgarno, A. and Herbst, E.: 1989, Astrophys. J. 347, 289.
Habing, H. J.: 1968, Bull. Astr. Inst. Neth. 19, 421.
Harrison, A. J. and Price, D. R. W.: 1959, J. Chem. Phys. 30, 357.
Harrison, A. J., Cederholm, B. J. and Terwilliger, M. A.: J. Chem. Phys. 30, 355.
Henry, R. C., Anderson, R. C. and Fastie, W. G.: 1980, Astrophys. J. 239, 859.
Herbst, E. and Leung, C. M.: 1986, Mon. Not. R. Astr. Soc. 226, 689.
Hogan, C. J.: 1991, Nature, 351, 96.
Howe, D. A. and Miller, T. J.: 1990, Mon. Not. R. Astr. Soc. 244, 444.
Huebner, W. F. and Carpenter, C. W.: 1979, Los Alamos Report LA8085-MS.
Huebner, W. F., Keady, J. and Lyon, S.: 1991, Astrophys. Space Sci. 1991 (in press).
Jackson, W. M.: 1976, J. Photochem. 5, 107.
Jackson, W. M.: 1976, In *Study of Comets*, IAU Coll. No. 25, p679, ed. B. Don.
Jura, M.: 1974, Astrophys. J. 191, 375.
Kurucz, R. L.: 1979, Astrophys. J. suppl. 40, 1.
Lake, J. S. and Harrison, A.: 1959, J. Chem. Phys. 28, 747.
Levine, J. S.: 1985, *The Photochemistry of Atmospheres*, Academic Press, New York.
Mamon, G. A., Glassgold, A. E. and Omont, A.: 1987, Astrophys. J. 323, 306.
Mamon, G. A., Glassgold, A. E. and Huggins, P. J.: 1988, Astrophys. J 328, 797.
McKellar, A.: 1940, Pub. Astr. Soc. Pac. 52, 307.
McKellar, A.: 1941, Pub. Dominion Astrophys. Observ. 7, 251.
Mezger, P. G., Mathis, J. S. and Panagia, N.: 1982, Astron. Astrophys. 105, 372.
Miller, T. J., Rawlings, J. M. C., Bennett, A., Brown, P. D. and Charnley, S. B.: 1991, Astron. Astrophys. Suppl. 87, 585.
Nejad, L. A. M. and Miller, T. J.: 1987, Astron. Astrophys. 183, 279.
Nejad, L. A. M. and Miller, T. J.: 1988, Mon. Not. R. Astr. Soc. 230, 79.
Nercession, E., Guillotean, S., Omont, A. and Benayoun, J. J.: 1989, Astron. Astrophys. 210, 225.

Ogawa, M. and Cook, G. R.: 1958, J. Chem. Phys. 28, 747.
Omont, A.: 1986, in *Circumstellar Matter*, IAU Symp. No. 122, eds. I Appenzeller and C. Jordan, P511, Reidel, Dordrecht.
Prasad, S. S. and Huntress, W. T. Jr.: 1980, Astrophys. J. Suppl. 43, 1.
Prasad, S. S. and Tarafdar, S. P.: 1983, Astrophys. J. 267, 603.
Scalo, J. M. and Slavsky, D. B.: 1980, Astrophys. J. 239, L73.
Sciama, D. W.: 1990a, Astrophys. J. 364, 549.
Sciama, D. W.: 1990b, Phys. Rev. Lett. 65, 2839.
Spitzer, L.: 1968, *Diffuse Matter in Space*, Interscience, New York.
Stromgren, B.: 1948, Astrophys. J. 108, 242.
Sternberg, A., Dalgarno, A. and Lepp, S.: 1987, Astrophys. J. 320, 676.
Swing, P.: 1942, Astrophys. J. 95, 270.
Swing, P. and Rosenfeld, L.: 1937, Astrophys. J. 86, 483.
Swift, M. B. and Mitchell, G. F.: 1981, ICARUS 47, 412.
Tarafdar, S. P.: 1991, Mon. Not. R. Astr. Soc. 252, P55.
Tarafdar, S. P., Prasad, S. S., Huebner, W. F., Keady, J. and Lyon, S.: 1991, Astrophys. J. (to be published).
Tielens, A. G. G. M. and Hollenbach, D.: 1985, Astrophys. J. 291, 722.
Truong-Bach, Morris, D., Nguyen-Quang-Rieu and Deguchi, S.: 1990, Astron. Astrophys. 230, 431.
Van Dishoeck, E. F.: 1987, in *Astrochemistry*, IAU Symp. No. 120, P51, eds. M. S. Vardya and S. P. Tarafdar, D'Reidel, Dordrecht.
Van Dishoeck, E. F.: 1988, in *Rate Coefficients in Astrochemistry*, eds T. J. Miller and D. A. Williams, P49, Kluwer, Dordrecht.
Van Dishoeck, E. F. and Black, J.: 1988, Astrophys. J. 334, 771.
Vardya, M. S. and Tarafdar, S. P.: 1987, *Astrochemistry*, IAU Symp. No. 120, D'Reidel, Dordrecht.
Witt, A. N. and Johnson, M. W.: 1973, Astrophys. J. 181, 363.

QUESTIONS AND ANSWERS

P.Shapiro: Recently, space UV measurements have placed an upper limit on the flux near 13.6 eV from a cluster of galaxies which is 10 times less than the prediction of Sciama's decaying neutrinos. Will this result not affect your conclusions?

S.P.Tarafdar: Recent space measurements (Davidson et al. 1991, Fabian et al. 1991) do put a limit on 13.7 eV photon flux and if it is correct, then there should not be enough DDM-photons in ISM to affect its chemistry. However, there are some low photons in observations (e.g. Davidson et al. 1991 and Hogan et al. 1991) which puts some doubts on the observational interpretation.

V.Burdyuzha: Why you used the model hot dark matter?

S.P.Tarafdar: In hot dark matter model, neutrinos could be a possible constituent and Sciama (1990a) used this model to put values on its density, mass and lifetime.

P.Craine: (comment) If ther exist regions of the ISM which are convincing shown to be less than predicted by Sciama, then this can be used as an argument against the 27.8 eV.

S.P.Tarafdar: Yes, if we can show that DDM-photons arising from neutrinos of mass 27.8 eV do some thing contrary to observation in ISM, it will argue against such neutrinos. Initially my aim was to try this. But I found that neither ionization and heating rates nor the chemical effects of such photons are large enough to contradict any observation of ISM.

R.Gredel: Are there enough "Dark Matter Photons" to ionize atomic carbon and produce the observed abundances of [CII]?

S.P.Tarafdar: No, I do not think that DDM-photons can give observed C^+, if it is from dense part of the cloud. The flux of DDM-photons inside dense clouds is $\sim 65/(1 + 2 \times 10^{-4} n_H)$ (Tarafdar, 1991) which will give C^+/CO ratio of $0.16/n(H_2)$ for $n_H < 5 \times 10^3 cm^{-3}$ assuming formation of C^+ by photoionization of C with a cross section of $10^{-17} cm^2$ and destruction of C^+ by reaction with H_2 with a rate of $4 \times 10^{-16} cm^3/s$. The above C^+/CO ratio is less than the observed value for any reasonable value of $n(H_2)$.

GRAIN SPECTROSCOPY

L.J. ALLAMANDOLA
NASA-Ames Research Center
Mountain View, CA 94035
USA

I. Introduction

Our fundamental knowledge of interstellar grain composition has grown substantially during the past two decades thanks to significant advances in two areas: astronomical infrared spectroscopy and laboratory astrophysics. The opening of the mid-infrared, the spectral range from 4000-400 cm^{-1} (2.5-25 µm), to spectroscopic study has been critical to this progress because spectroscopy in this region reveals more about a material's molecular composition and structure than any other physical property.

Infrared spectra which are diagnostic of interstellar grain composition fall into two categories: absorption spectra of the dense and diffuse interstellar media, and emission spectra from UV-Vis rich dusty regions. The former will be presented in some detail, with the later only very briefly mentioned. This paper summarizes what we have learned from these spectra and presents "doorway" references into the literature. Detailed reviews of many aspects of interstellar dust are given in [1]. Reviews of the IR spectra of dense clouds, emphasizing observations, can be found in [2,3] and those emphasizing laboratory and theoretical considerations in [4-9].

2. Absorption Spectroscopy

2.1 DENSE INTERSTELLAR MEDIUM

IR spectroscopy shows that dust in dense molecular clouds is comprised principally of mixed molecular ices and silicates. This conclusion is based on the <u>direct</u> comparison of astronomical spectra with the spectra of laboratory analogs as illustrated in Figure 1. The ice compositions and abundances derived from these spectra for several molecular clouds are summarized in Table 1. The major conclusions drawn from each spectral region comprising the mid-IR are discussed below.

2.1.1. *The 4000-2500 cm^{-1} (2.5-4 µm) Region.* The deep 3250 cm^{-1} (3.08µm) band, a common characteristic of molecular cloud spectra, is attributed to the OH stretch in frozen H_2O, the most abundant interstellar ice constituent known. Small contributions from O-H (alcoholic) and N-H (amine) stretches are also likely. In dense clouds this band has a low frequency wing, [10-14] in contrast to the more symmetric profile associated with OH IR stars [15,16]. The wing, which extends from approximately 3000 to 2700 cm^{-1} (3.3 to 3.7 µm), has been attributed to H_2O-base interactions, hydrocarbons, and scattering [4, 17,18]. Spectral observations tend to support the first two [13,10] while polarization studies seem marginally inconsistent with the latter [19]. Theoretical discussions of this band are in references [17, 4, 20].

Spectral features associated with aliphatic hydrocarbon C-H stretches were long expected on the wing [4] and have recently been found. The importance of methanol [5] has been confirmed by detection of a band at 2825 cm^{-1} (3.5 µm) [21,14]. When present, CH_3OH is the second most abundant ice constituent (~10-50% wrt H_2O). A broader feature peaking near 2880 cm^{-1} (3.4 µm) has also been found. This frequency is

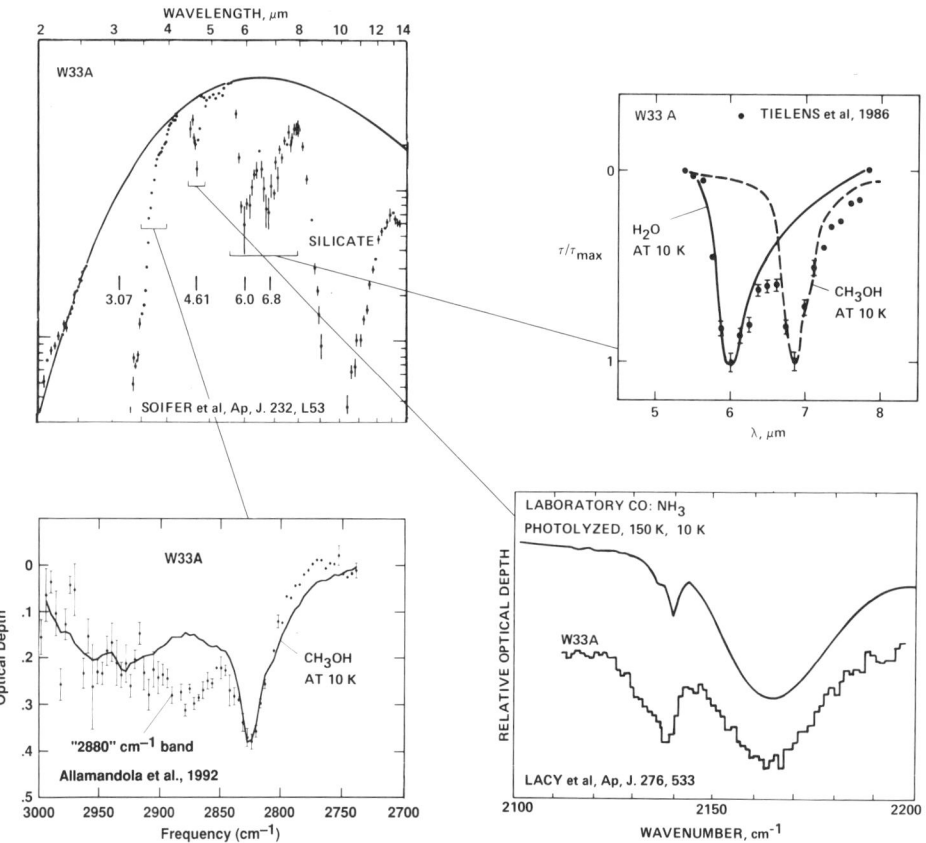

Figure 1. The mid-infrared spectrum of W33A compared with spectra of laboratory analogs.

characteristic of C-H stretching vibrations for H on "tertiary carbon", i.e. aliphatic carbon atoms which are bound to three other carbon atoms and one hydrogen atom. The dominance of this feature over primary and secondary CH stretching features suggests that a material with a diamond-like carbon skeleton is ubiquitous in dense clouds [14].

2.1.2 *The 2500-2000 cm^{-1} (4-5μm) Region.*   Frozen CO is responsible for the band near 2140 cm^{-1} found in many, but not all, clouds [22-24]. The peak position of this band shows that the solid CO is distributed between two forms of interstellar ice, one dominated by H_2O, the other by some non-polar material [25,26]. When CO is present, the CO/H_2O concentration of interstellar ices is typically about a few percent.

The broad feature near 2165 cm^{-1} (4.6μm) in Figure 1 has also been detected in one or two other objects (Tielens, Teglar-private communications). Laboratory experiments show that the peak position and profile indicate a CN stretch [22]. The initial assignment to a nitrile (-C≡N) or iso-nitrile (C≡N-) has recently been questioned. The CN stretch in the OCN^- ion [27], and the Si-H stretch [28,29] have been suggested in its place. In the author's opinion, the initial assignment remains the most attractive. Further experiments are needed to settle this.

A weak feature at 2040 cm^{-1} (4.9 μm) in the spectrum of W33A suggests that OCS is also an interstellar ice constituent [30].

2.1.3 *The 2000-1000 cm^{-1} (5-10μm) Region.* The main ice bands here peak near 1670 and 1470 cm^{-1} (6.0 and 6.8 μm). These are evident in several clouds and are fit quite well by H_2O (1670 cm^{-1}) and CH_3OH (1670 cm^{-1}) [5].

The column densities of both H_2O and CH_3OH derived from these bands exceeds those derived from their corresponding OH and CH stretches near 3000 cm^{-1} (Table 1). This inconsistency can be rationalized in the following ways: 1 - Other species may contribute significantly to the 1470 cm^{-1} (6.8 μm) band [21]. NH_4^+ has been suggested [31]. However, if this were the case, clearly resolved features which correspond to the NH stretching vibrations should be evident in the 3500-3000 cm^{-1} (2.8-3.3 μm) region ; 2- The 2825 cm^{-1} CH stretch blends with other features and its optical depth is a lower limit [14]; and 3- Higher frequency (~3000 cm^{-1}, 3 μm) radiation reaching the telescope includes scattered light which fills in the absorptions in this region whereas lower frequency radiation (<2000 cm^{-1}; >5 μm) does not [18,32].

Recently, weak features in the 1400-1250 cm^{-1} (7-8 μm) region have been detected [33]. That at 1301 cm^{-1} (7.685 μm) is attributed to methane, implying rather low solid state abundances for this molecule. Frozen CH_4 appears to be approximately 100 times less abundant than solid CO [33] which in turn is typically a few percent the abundance of H_2O ice.

2.1.4 *The 1000-500 cm^{-1} (10-20 μm) Region.* This region is dominated by the broad "SiO" stretch near 1000 cm^{-1} (10 μm) [2,3,34]. This interstellar material had been thought to be amorphous due to the lack of spectral structure on the band. However, recent higher resolution spectroscopic and polarization measurements imply that crystalline material is also present [35,36].

As with the ices, laboratory studies give important insight into the molecular nature of this interstellar refractory material [37-43]. Laboratory experiments show that the band profile reveals much about the material's formation conditions and subsequent processing [40, 42,43]. Unfortunately, although this feature was extensively studied in the 70's , high resolution spectroscopic studies utilyzing modern astronomical instruments have not been carried out. In view of the wealth of information one can extract from such studies, it is hoped that a significant amount of observing time with the new generation of astronomical, high resolution, IR spectrometers is devoted to measuring this feature, looking specifically for substructure. In particular, crystalline silicates can produce resolvable structure near 890 cm^{-1} (11.2 μm). Furthermore, space-borne spectrometers will provide the frequency, strength, and profile of the bending modes near 400 cm^{-1} (20 μm). Such complete coverage can provide important insight into the interstellar material's history [36,43] and degree of crystallinity.

In addition to these features which have long been known to be associated with refractories, absorptions due to ice constituents have recently been reported in this region. There is good evidence for the librational band of H_2O (and probably some alcohols) at 830 cm^{-1} (12 μm) [44] and the 667 cm^{-1} (15 μm) band of CO_2 [45] in IRAS-LRS spectra of a few clouds. For the objects studied, the solid CO_2 abundance seems comparable to that of solid CO.

Table 1. Column densities of some molecular cloud dust components.

MATERIAL	OBJECT			
	NGC 7538 IRS 9	W33A	W3 IRS 5	S140 IRS 1
Hydrogen [a]	1.6×10^{23}	2.8×10^{23}	2.7×10^{23}	1.4×10^{23}
H_2O				
3250 cm^{-1} [d]	$>5.3 \times 10^{18}$	$>8.6 \times 10^{18}$	3.7×10^{18}	2.1×10^{18}
1650 cm^{-1} [d]	1.1×10^{19}	4.2×10^{19}	4.3×10^{18}	8.8×10^{18}
CH_3OH				
2825 cm^{-1} [b]	9.1×10^{17}	3.9×10^{18}	$<5.3 \times 10^{17}$	$<3.8 \times 10^{17}$
2600+2540 cm^{-1} [c]	...	2.5×10^{18}	...	...
1470 cm^{-1} [d]	7.1×10^{18}	2.3×10^{19}	3.5×10^{18}	...
CO (non-polar ice) [e]	6.4×10^{17}	1.1×10^{17}	1.1×10^{17}	...
CO (polar ice) [e]	3.2×10^{17}	2.8×10^{17}	5.4×10^{16}	...
CO (gas) [e]	1.4×10^{19}	2.0×10^{19}	2.2×10^{19}	...
"Diamonds" [b]	3.8×10^{18}	1.5×10^{18}	2.8×10^{18}	2.8×10^{18}

a) From $N_H = 1.9 \times 10^{21} A_v$; b), c) [14]; d) [5], except for S140 IRS 1 from [2]; e) [26].

2.2 DIFFUSE INTERSTELLAR MEDIUM

Mid-IR spectra of the diffuse ISM are dominated by the SiO stretch. As mentioned above, high resolution spectroscopic studies of this feature, searching for substructure, are very important to carry out.

Most conclusions regarding other diffuse medium dust components derived from IR spectroscopy are based on studies in the OH/NH/CH stretch region (3500-2700 cm^{-1}; 2.9-3.7 µm). These studies have focused on Galactic Center sources, particularly infrared source number 7 (GC IRS7) [46-49]. This work has recently been expanded to include VI Cyg 12 and several other objects [50,51,52].

Figure 2 shows that there is a reasonable match between the spectrum of GC IRS7 and a laboratory residue produced by the vacuum-UV irradiation of an interstellar ice analog [51]. The match between the interstellar feature and the C-H stretch band in the spectrum of hydrocarbons sublimed from the Murchison meteorite is remarkable [52]. Based on these observations it appears that the aliphatic component of diffuse medium dust has a -CH_2-/-CH_3 ratio of about 2, with chains containing 3-4 carbon atoms that have one end attached to an electronegative group such as -OH, -C≡N, or perhaps an aromatic network. About 8-20% of cosmic carbon is tied up in this component.

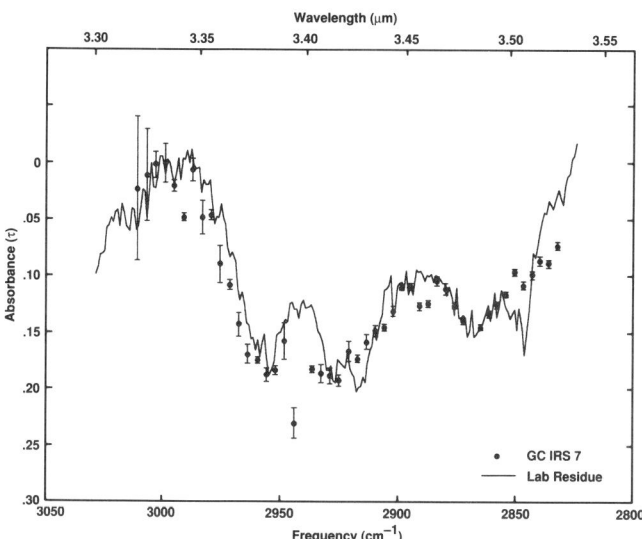

Figure 2. The spectrum, in the CH stretch region, of Galactic Center source IRS 7 compared to that of a residue produced by the irradiation of an interstellar ice analog.

3. Emission Spectroscopy

A family of infrared emission bands from 3050-870 cm^{-1} (3.28-11.5 μm) carry a significant fraction of the energy radiated by many different objects in which an intense UV-Vis radiation field is attenuated by dust. An example of a typical spectrum is given in Figure 3. The "resemblance" of this spectrum to spectra of polycyclic aromatic hydrocarbons (PAHs) and related materials such as amorphous carbon (HAC) has been taken as compelling evidence that these materials are responsible. Analysis of these mid-infrared emission spectra indicates that a few percent of the cosmic carbon is tied up in the small (20-40 carbon atom) PAHs which are responsible for the sharp IR features, and a similar amount is tied up in the larger (200-500 carbon atom) PAHs, PAH clusters, and amorphous carbon particles responsible for the underlying broad spectral components.

The aromatic nature of the carrier seems secure. However, there is still some debate as to whether or not the emitters of the discrete features are free molecular PAHs, or PAH units which make up HAC or other carbonaceous material. All this and more can be found in the following extensive reviews and papers: Molecular PAHs: [53,54] HAC: [55,56,57]; QCC: [58,59].

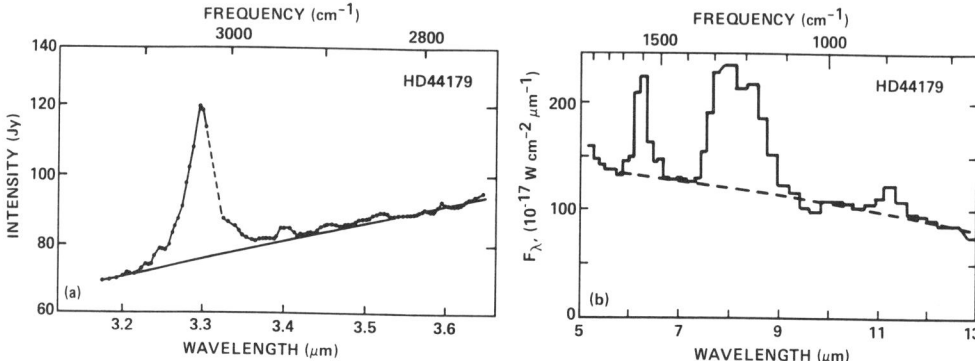

Figure 3. The spectrum of HD 44179 showing the family of emission bands associated with PAHs. These features are emitted from many different astronomical objects.

4. References

1. Allamandola, L.J. and Tielens, A.G.G.M.(1989) eds.Interstellar Dust: IAU Symposium 135, (Kluwer, Dordrecht).
2. Willner, S.P., Gillett, F.C., Herter, T.L., Jones, B., Krassner, J., Merrill, K.M., Pipher, J.L., Puetter, R.C., Rudy, R.J., Russell, R.W., and Soifer, B.T.(1982) Ap.J., 253, 174.
3. Whittet, D.C.B.(1988) in Bailey, M.E. and Williams, D.A. (eds.), Dust in the Universe, Cambridge Univ. Press, Cambridge, 25-54.
4. Allamandola, L.J.(1984) in Kessler, M.F. and Phillips, J.P. (eds.), Galactic and Extragalactic Infrared Spectroscopy; Reidel, Dordrecht, 5-35.
5. Tielens, A.G.G.M., and Allamandola, L.J. (1987) in Morfill, G.E. and Scholer, M., (eds.), Physical Processes in Interstellar Clouds, D. Reidel, Dordrecht, 333-376.
6. Tielens, A.G.G.M., and Allamandola, L.J. (1987) in Hollenbach, D.J. and Thronson, H. A. (eds.) Interstellar Processes, D. Reidel, Dordrecht, 397-469.
7. Allamandola, L.J. and Sandford, S.A. (1988) in Bailey, M.E. and Williams, D.A. (eds.), Dust in the Universe, Cambridge Univ. Press, Cambridge, 229-264.
8. Tielens, A.G.G.M. (1989) in Allamandola, L.J., and Tielens, A.G.G.M. (eds.) Interstellar Dust:, Kluwer, Dordrecht, 239-262.
9. Bussoletti, E. and Colangeli, L. (1990) Rivista del Nuovo Cimento 13, 1-70
10. Smith, R.G., Sellgren, K., and Tokunaga, A.T., (1989) Ap.J. 344, 413.
11. Nagata T., Tokunaga, A. T., Sellgren, K., Smith, R.G., Onaka, T., Nakada, Y., and Sakata, A. (1988) Ap. J. (Letters) 287, L51.
12. Whittet, D.C.B., Bode, M.F., Longmore, A.J., Baines, D.W.T., and Evans, A. (1983) Nature 303, 218-221.

13. Eiroa, C. and Hodapp, K.W. (1989) A&A, 210, 345-350
14. Allamandola, L.J., Sandford, S.A., Tielens, A.G.G.M., and Herbst, T. (1992) Ap.J. in press
15. Soifer, B.T., Willner, S.P., Capps, R.W., and Rudy, R.J. (1981) Ap. J. 250,631.
16. Smith, R.G., Sellgren, K., and Tokunaga, A.T.(1987) Ap.J. 334, 209-219.
17. Tielens, A.G.G.M., and Hagen, W.(1982) A&A 114, 245.
18. Leger, A., Gauthier, S., Defourneau, D., and Rouan, D. (1983) A&A 117, 164-169.
19. Hough, J.H. et al., (1988) MNRAS 230, 107.
20. Greenberg, J.M., van de Bult, C.E.P.M., and Allamandola, L.J. (1983) J. Phys. Chem. 87, 4243.
21. Grim, R.J.A., Baas, F., Geballe, T.R., Greenberg, J.M., and Schutte, W. (1991) A&A, 243, 473-477
22. Lacy, J.H., Baas, F., Allamandola, L.J., Persson, S.E., McGregor, P.J., Lonsdale, C.J., Geballe, T.R., and van de Bult, C.E.P.M. (1984) Ap. J. 276, 533-543.
23. Geballe, T.R. (1986) A&A 162, 248-252.
24. Whittet, D.C.B., Adamson, A.J., Duley, W.W., Geballe, T.R., and McFadzean, A.D. (1989) MNRAS 241, 707-720.
25. Sandford, S.A., Allamandola, L.J., Tielens, A.G.G.M., and Valero, G.J. (1988) Ap.J. 329, 498-510.
26. Tielens, A.G.G.M., Tokunaga, A.T., Geballe, T.R., and Baas, F. (1991) Ap. J. 381, 181-199.
27. Grim, R., and Greenberg, J.M. (1987) Ap. J. (Letters) 321, L91.
28. Nuth, J.A.,and Moore, M.H. (1988)Ap. J. (Letters) 329, L113.
29. Moore, M.H., Tanabe, T., and Nuth, J.A. (1991) Ap. J. (Letters) 373, L31-L34
30. Geballe, T.R., Baas, F., Greenberg, J.M., and Schutte, W. (1985) A&A 146 L6-L8.
31. Grim, R.J.A., Greenberg, J.M., Schutte, W.A., and Schmitt, B.(1989) Ap. J. (Letters) 241, L87.
32 Pendleton, Y., Tielens, A.G.G.M., and Werner, M.W. (1990) Ap.J. 349, 107.
33. Lacy, J.H., Carr, J.S., Evans II, N.J., Baas, F., Achtermann, J.M., and Arens, J.F.(1991) Ap.J. 376, 556-560.
34. Merrill, K.M. and Stein, W.A. (1976) PASP 88, 874-887.
35. Aitken, D.K., Roche, P., Smith, C.H., James, S.D., and Hough, J.H. (1988) MNRAS 230, 629-638.
36 Aitken, D.K., Smith, C.H., and Roche, P. (1989) MNRAS 236, 919-927.
37. Day, K.L.(1979) Ap.J. 234, 158-161.
38. Knacke, R.F. and Kratschmer, W.(1980) A&A 92, 281-288.
39. Koike, C., Hasegawa, H., Asada, N., and Hattori, T. (1981) Ast.&Sp. Sci. Rev 79, 77-85
40. Nuth, J.A. and Donn, B. (1983) J. Geophys. Res. 88, Supplement, A847-A852.
41. Nuth, J.A., and Hecht, J.H. (1990) Ast.&Sp. Sci. 163, 79-94.
42. Mukai, T. and Koike, C. (1990) Icarus 87, 180-187.
43. Nuth, J.A. (1992) in Singh, P.D. and Almeida, (eds.) Astrochemistry of Cosmic Phenomena: IAU Symposium 150, Kluwer, Dordrecht, in press
44. Cox, P. (1990) A&A 225, L1.
45. d'Hendecourt, L.B., and Jourdain de Muizon, M. (1990) A&A 223, L5.
46. Wickramasinghe, D.T., and Allen, D.A. (1980) Nature 287, 518.
47. Jones, T.J., Hyland, A.R., and Allen, D.A. (1983) Mon. Not. R. Astr. Soc. **205**, 187.
48. Butchart, I., McFadzean, A.D., Whittet, D.C.B., Geballe, T.R., and Greenberg, J.M. (1986) Astron. Astrophys. 154, L5.
49. McFadzean, A.D., Whittet, D.C.B., Longmore, A.J., Bode, M.F., and Adamson, A.J. (1989) Mon. Not. R. Astr. Soc. 241, 873.
50. Adamson, A.J., Whittet, D.C.B., and Duley, W.W. (1990) Mon. Not. R. Astr. Soc. 243, 400.

51. Sandford, S.A., Allamandola, L.J., Tielens, A.G.G.M., Sellgren, K., Tapia, M., and Pendleton, Y.(1991) Ap.J. 371, 607-620.
52. Pendleton et al (1992) in preparation
53. Allamandola, L.J., Tielens, A.G.G.M., and Barker, J.R.(1989) Ap.J. Suppl., 71, 733-775.
54. Puget, J.L. and Leger, A. (1989) ARA&A, 27, 161.
55. Duley, W.W.and Williams, D.A. (1988) MNRAS 231, 969-975.
56. Duley, W.W.and Williams, D.A. (1990) MNRAS 247, 647-650.
57. Blanco, A., Bussoletti, E., and Colangeli, L. (1988) Ap.J. 334, 875-882.
58. Sakata, A., Wada, S., Onaka, T., and Tokunaga, A.T. (1987) Ap. J. (Letters) 320, L63-L67.
59. Sakata, A., Wada, S., Onaka, T., and Tokunaga, A.T. (1990) Ap. J.353, 543.

QUESTIONS AND ANSWERS

B. Khare: Could you comment on the particle size distribution from the laboratory spectra?

L. Allamandola: Apart from the interstellar 3.09 µm H_2O band, these bands don't have wings which would arise from scattering. Thus, these spectra don't give size information.

A. Leger: Comment- It is not correct to say that there is no scattering in the IR. Very nice polarization measurements by Capps et al. in Orion have shown unambiguously scattering at two different wavelengths and allow a size determination. Actually the size found is about 0.5 µm, definitely larger than in the diffuse medium (Rouan et al. ~1987)

L. d'Hendecourt: Comment- The identification of the 3.4 µm band towards IRS 7 is far from unique. Carbon extract from the Orguel meteorite shows, in this region, exactly the same spectral fit to IRS 7 as was shown during this talk to a hydrocarbon produced by ice irradiation (Ehrenfreund Contributed paper, this meeting). Yet, this Orguel material is mostly made up of aromatic material while your material is aliphatic. This may shed some light on the origin of this material which might originate directly from carbon stars (eg CRL 618), and not from the UV photolysis of dirty ices which may not produce aromatic structure.

L. Allamandola: I agree. It is important to stress that this absorption only probes the aliphatic component and gives no insight into what other types of hydrocarbons may be present. In this regard it is also important to remember that olefinic hydrocarbons absorb in a similar region to aromatics, and olefins *are* produced in irradiation experiments. Thus even if there were a match of the 3.3 µm band, aromatics are not necessarily implied.

D. Williams: The new feature at 2880 cm^{-1} is attributed by you to "diamond'-like carbon. Please describe this structure.

L. Allamandola: The smallest structure possible that meets the spectral requirements must contain more than 30, mainly sp^3 hybridized, carbon atoms which are bound to each other and hydrogenated on the surface. The spectra allow for a few -CH_3 and -CH_2- edge groups.

CHEMISTRY IN THE EARLY UNIVERSE

PAUL R. SHAPIRO
Department of Astronomy
The University of Texas at Austin
Austin, TX 78712 USA

ABSTRACT. Galaxies and the first stars in the universe formed billions of years ago as a result of the cooperative effects of gravitational collapse and nonequilibrium chemistry. Gravity drew the primordial gas together into lumps; the formation of the first molecules in the universe, simple diatomic molecules like H_2, H_2^+, HD, HeH^+, LiH, and LiH^+, may then have ensured that the heat generated by gravitational collapse and shock waves was radiated away rapidly enough to allow the gravitational collapse and fragmentation of these gaseous lumps to proceed to the point of forming stars and galaxies. We briefly mention a few of the latest studies of this primordial chemistry, including that in the evolving intergalactic medium (IGM) in a Cold Dark Matter (CDM) model cosmology and that in radiative shocks in the early universe.

1. Introduction

Molecule formation in the early universe has been studied under three related circumstances: (1) the uniform pregalactic IGM at high redshift z in the postrecombination epoch ($z \lesssim 10^3$)[1-4]; (2) the gravitational collapse of primordial gas clouds [5-16]; and (3) the radiative shocks which occur in a wide range of galaxy and primordial star formation theories [17-29]. In view of the space limitation here, I will just mention a few highlights of work completed since the last reviews of this subject [25,30] and refer the reader to those reviews for the full background and references.

When the universe recombined at $z \sim 10^3$, the gas was composed primarily of H, D, ^{4}He, ^{3}He, and ^{7}Li, with nuclear abundances by number relative to H of order $1, 5 \times 10^{-5}, 10^{-1}, 10^{-5}$, and 10^{-10}, respectively [31]. In this essentially metal-free gas, H_2 was likely to have been the most abundant molecule and was potentially important as a source of radiative cooling in collapsing pregalactic clouds and in shock-heated pregalactic gas. Since H_2 has no dipole moment, formation by direct radiative association of two H atoms is very slow, and H_2 formation is thought to have been dominated, instead, by the creation of H^- by radiative attachment, followed by associative detachment [5,6]:

$$H + e^- \to H^- + \gamma$$
$$H + H^- \to H_2 + e^-$$

and by the creation of H_2^+ by radiative association, followed by charge transfer [32]:

$$H + H^+ \to H_2^+ + \gamma$$
$$H_2^+ + H \to H_2 + H^+.$$

2. H_2 in the IGM in a Postrecombination CDM Universe

Previous studies of molecule formation in the pregalactic IGM beginning with the recombination epoch at $z \cong 10^3$ considered a uniform expanding IGM of fixed comoving density Ω_{IGM} (in units of the critical density for an Einstein-deSitter universe) and focused on the epoch prior to nonlinear structure formation in the universe [1,2]. (An exception is [3], which considered the effect of radiation from postulated primordial stars at high z.) As part of our general study of the IGM, we have reconsidered this problem within the context of the recently popular CDM model for galaxy and large-scale structure formation. We have coupled our detailed, numerical calculations of the thermal and ionization balance, molecule formation, and radiative transfer in a uniform IGM of H and He to the linearized equations for the growth of density fluctuations in both the gaseous and dark matter components [4,33,34]. The mean IGM density parameter in this case is $\Omega_{IGM}(z) = \Omega_b[1-f_c(z)]$, where Ω_b is the total baryon mass fraction in the universe and $f_c(z)$ is the time-varying, collapsed baryon fraction. We consider a standard CDM model with $\Omega_{tot} = 1$, $\Omega_b = 0.1$, and Hubble constant $h = H_o/(100\ km\ s^{-1}\ Mpc^{-1}) = 0.5$, with a power spectrum normalized so that the rms density fluctuation evaluated using a "top-hat" filter is $\sigma_o = 1/b$ at $R = 8h^{-1}$ Mpc at present, where b is the so-called bias parameter. In order to account for the observed absence of a detectable H Ly α absorption trough in the spectra of high z quasars due to a smoothly distributed IGM (so-called Gunn-Peterson effect; "GP"), we assume that the collapsed fraction releases just enough ionizing radiation at a constant rate per collapsed baryon, with an AGN-like spectrum ($F_\nu \propto \nu^{-1.5}$), to reduce the GP optical depth to less than 0.1 by $z = 4.1$. This minimal emissivity corresponds to 1200 ($b = 1$) or 24,400 ($b = 2.6$) ionizing photons per collapsed baryon per present Hubble time.

Shown in Figure 1 are some results of our calculations, which solve rate equations for the concentrations of H^o, H^+, He^o, He^+, He^{++}, H_2, H_2^* (vibrationally excited), H_2^+, H^-, and e^-, the energy conservation equation, including the effects of cosmological expansion, radiative cooling, and cooling (or heating) by Compton scattering of the cosmic microwave background (CMB), and the equation of radiative transfer, including the opacity of the IGM and its own diffuse emission, as well as the mean opacity of an evolving distribution of gas clumps (i.e. quasar absorption-line clouds) embedded in the smoothly distributed IGM. As in previous studies, H_2 formation shortly after recombination was limited by the photodestruction of H^- and H_2^+ by the CMB. It is important in this case to take account of the population of excited states of H_2^+ with an excitation temperature equal to the CMB radiation temperature. Otherwise, photodissociation of H_2^+ at high z is underestimated, and H_2 formation is spuriously accelerated and enhanced. For $200 \lesssim z \lesssim 600$, the H_2^+ process dominated the formation of H_2, yielding a concentration $\lesssim 10^{-6}$ by $z \sim 300$. By $z \sim 100$, the H^- process boosted this to $\sim 2 \times 10^{-6}$, before the collapsed fraction finally released enough radiation by $z \lesssim 25$ ($b = 2.6$) or $z \lesssim 50$ ($b = 1$) to heat and ionize the IGM, destroying the H_2. Figure 1 shows that once IGM reionization began, unshielded photodissociation would have destroyed the H_2 more quickly than collisional dissociation, but even in the fully self-shielded limit (no photodissociation), collisions would have been more than enough to reduce the H_2 concentration by many orders of magnitude to undetectable levels even before the GP constraint was satisfied.

3. Radiative Shocks and Nonequilibrium H_2 Chemistry in the Early Universe

Shock waves in a metal-free gas are predicted to occur in the IGM and inside protogalaxies under a wide range of circumstances in the theory of galaxy and primordial star formation, including gravitational collapse of density fluctuations, cloud-cloud collisions, and blast waves from the explosive heating of the IGM. In all cases, the radiative cooling of the postshock gas is essential in order to dissipate enough gravitational or explosion energy to make gravitational instability and fragmentation possible in the shock-heated gas, a prerequisite for galaxy or star

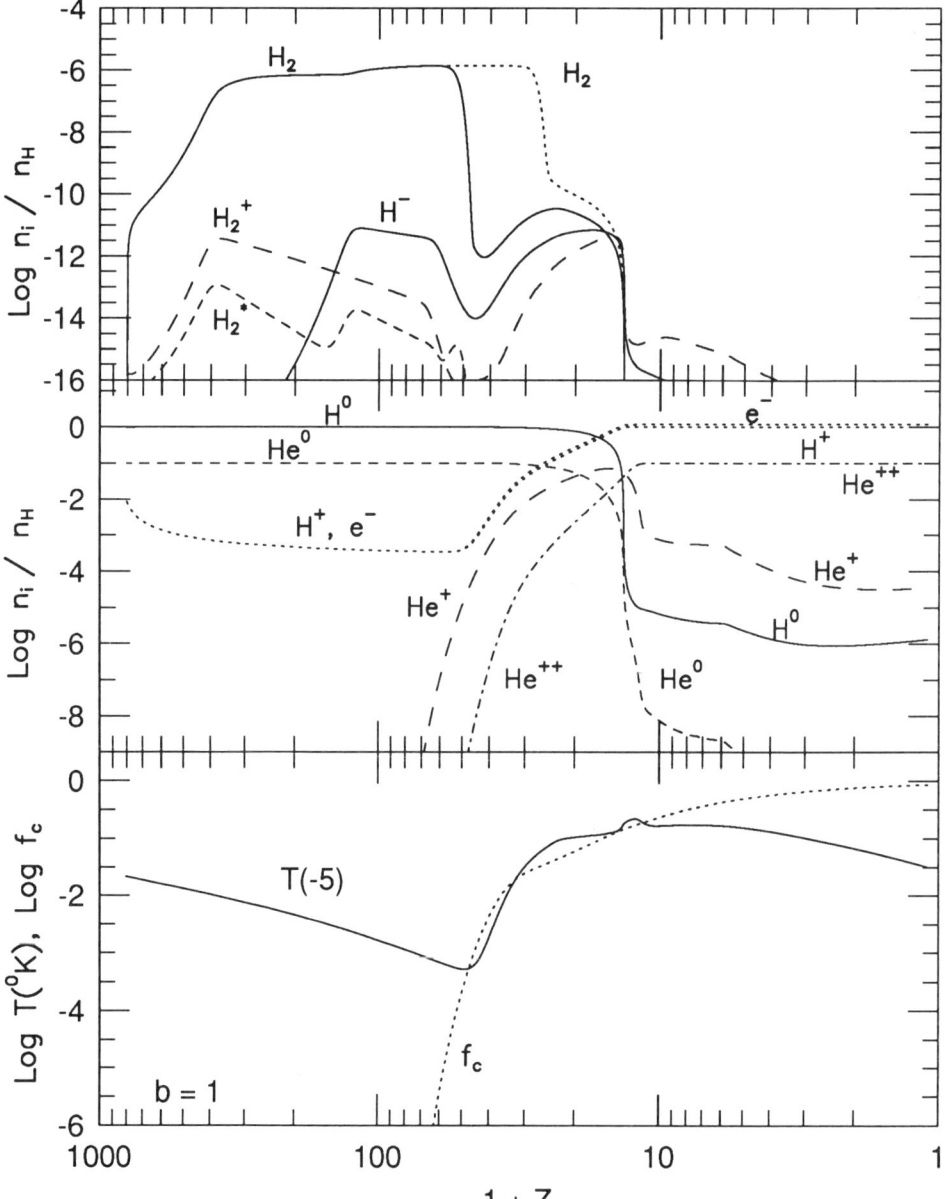

Figure 1. IGM temperature $T(°K)(\times 10^{-5})$, collapsed baryon fraction f_c, and species concentration by number relative to H, n_i/n_H, as labeled for bias $b = 1$. Solid curve for H_2 assumes unshielded photodissociation; dashed H_2 curve neglects photodissociation.

formation. We have numerically solved the hydrodynamical conservation equations, along with the rate equations for nonequilibrium ionization, recombination, molecule formation and dissociation, and the equation of radiative transfer, in detail for steady-state, planar shocks of velocity in the range $20 \leq v_s \leq 400$ km s^{-1} for a range of conditions appropriate for shocks in the IGM and inside protogalaxies [22-29]. (Calculations for shocks in the IGM were also reported by [20].) We include the same species as in the calculations for Figure 1. Figure 2 shows results for shocks inside protogalaxies, with $v_s = 300$ km s^{-1} such as would be produced by the typical gravitationally-induced motions within a protogalaxy, in a gas of preshock density $n_{H,1} = 0.1$. The shocked gas generally cools faster than it can recombine and, as a result, is able to form an H_2 concentration as high as 10^{-3} or higher via the H^- and H_2^+ processes, thanks to the enhanced nonequilibrium ionization at $10^4 K$. With such an H_2 concentration, the gas cools rapidly by H_2 rotational-vibrational line excitation to $T \sim 10^2 - 10^3 K$, well below the canonical final temperature of $10^4 K$ for a molecule-free gas without metals. This cooling below $10^4 K$ significantly lowers the characteristic gravitationally unstable fragment mass for such shocks, relative to the value if the gas cooling stops at $10^4 K$.

This presents a problem for the suggestion that globular clusters formed within protogalaxies by gravitational instability in the compressed gas resulting from radiative shocks with velocities of the order of the virial velocity of the protogalaxy or from thermal instability in gas at the virial temperature [35,36]. Under such circumstances, gas which cools to $10^4 K$ and, thereafter, remains at $10^4 K$ for a time longer than its internal free-fall time will lead naturally to gravitational instability with a characteristic mass comparable to those of globular clusters. If, instead, the gas forms enough H_2 to cool too rapidly below $10^4 K$, however, the model fails. Our shock calculations have led to a new version of the model, in which a strong enough UV or soft X-ray source ($L \gtrsim 10^{45}$ erg s^{-1}), such as a quasar or early-type stars, is required to be present within the protogalaxy during the globular cluster formation epoch to suppress H_2 formation in the shocks [26].

Figure 3 shows the effect of adding to the shock in Figure 2 a preshock magnetic field of strength $B_1 = 1$ μG, oriented parallel to the shock front [29]. Flux-freezing and the nearly isobaric compression of the radiatively cooling postshock gas together cause magnetic pressure to dominate downstream before the gas reaches $\sim 10^4 K$. This halts the compression and makes the photon-to-atom ratio higher in the temperature plateau at $10^4 K$ than in the nonmagnetic case. The latter encourages the photodestruction of H_2 and makes it easier to suppress H_2 formation and cooling with a given radiation flux level than in the nonmagnetic case, thereby lowering the threshold level required for the globular cluster formation model (e.g. $L \sim 10^{44}$ erg s^{-1} or less, for $B_1 \gtrsim 1.6$ $n_{H,1}^{1/2}$ μG).

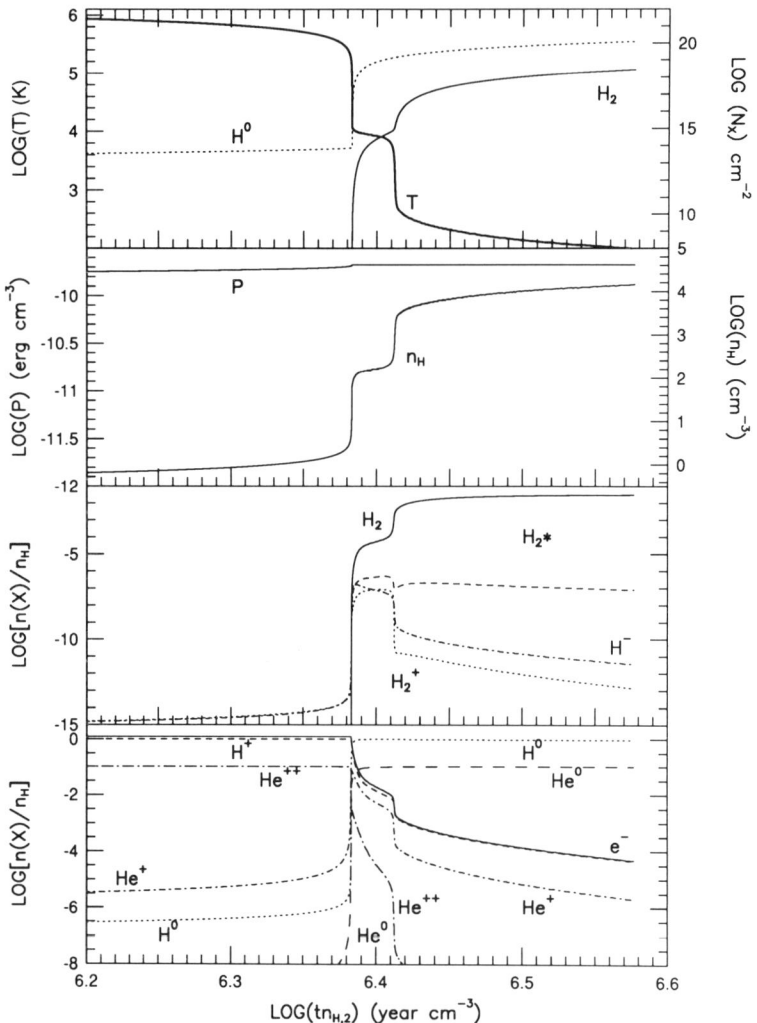

Figure 2. Postshock flow quantities versus $tn_{H,2}$ (years cm^{-3}) for $v_s = 300$ km s^{-1} and $n_{H,1} = 0.1$ cm^{-3}, where $n_{H,2}$ is postshock H atom density and t is time since fluid element was shocked. Horizontal coordinate also corresponds to N_H cm^{-2}, total H column density between fluid element and shock, at time t, where $N_H = n_{H,1}v_s t = (n_{H,1}/n_{H,2})\, v_s tn_{H,2}$, as long as we take $log_{10}N_H = log_{10}(tn_{H,2}) + 14.382$. Top panel shows temperature (scale at left) and total H^0 and H$_2$ column densities between fluid element and shock (scale at right) as labeled. Second panel shows gas pressure (scale at left) and H atom density (scale at right) as labeled. Third and fourth panels show concentrations n_i/n_H as labeled.

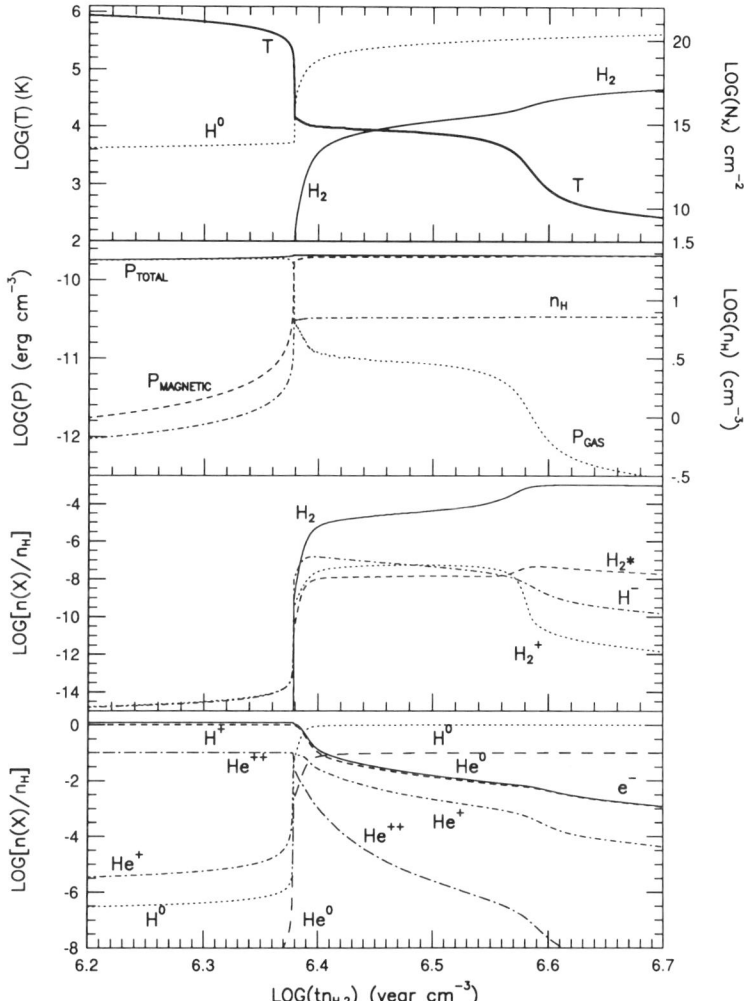

Figure 3. Same as Figure 2, but magnetized case with $B_1 = 1\mu G$. All panels are the same as in Figure 2 except the second, where gas, magnetic and total pressure are plotted, along with H atom density.

Acknowledgments. I am especially grateful to my collaborators M. Giroux, H. Kang, and A. Clocchiatti. This work was supported in part by Robert A. Welch Foundation Grant F-1115, Texas Advanced Research Program Grant 4132, NASA Grants NAGW-2399, NGT-50316, and NGT-50519. Our numerical computations were performed on the University of Texas Center for High Performance Computing Cray Y/MP.

References

1. Lepp, S. and Shull, J. M. (1984) 'Molecules in the Early Universe', Ap. J., 280, 465-469.
2. Latter, W. B. and Black, J. H. (1991) 'Molecular Hydrogen Formation By Excited Atom Radiative Association', Ap. J., 372, 161-166
3. Couchman, H.M.P. (1985) 'Reheating of the Intergalactic Medium at $z > 10$', M.N.R.A.S., 214, 137-159.
4. Giroux, M. L. and Shapiro, P. R. (1992) 'The Reionization of the Intergalactic Medium', Ap. J., to be submitted.
5. Peebles, P.J.E. and Dicke, R. H. (1968) 'Origin of the Globular Star Clusters', Ap. J., 154, 891-908.
6. Hirasawa, T., Aizu, K., and Taketani, M. (1969) 'Formation of Galaxies from Hydrogen Gas', Progr. Theor. Phys., 41, 835-838.
7. Hirasawa, T. (1969) 'Formation of Protogalaxies and Molecular Processes in Hydrogen Gas', Progr. Theor. Phys., 42, 523-543.
8. Matsuda, T., Sato, H., and Takeda, H. (1969) 'Cooling of Pre-Galactic Gas Clouds by Hydrogen Molecule', Progr. Theor. Phys., 42, 219-232.
9. Yoneyama, T. (1972) 'On the Fragmentation of a Contracting Hydrogen Cloud in an Expanding Universe', Pub. Astr. Soc. Japan, 24, 87-98.
10. Hutchins, J. B. (1976) 'The Thermal Effects of H_2 Molecules in Rotating and Collapsing Spheroidal Gas Clouds', Ap. J., 205, 103-121.
11. Silk, J. (1977) 'On the Fragmentation of Cosmic Gas Clouds. I. The Formation of Galaxies and the First Generation of Stars', Ap. J., 211, 638-648.
12. Carlberg, R. G. (1981) 'An Estimate of the Mass of Zero Metal Stars', M.N.R.A.S., 197, 1021-1029.
13. Palla, F., Salpeter, E. E., and Stahler, S. W. (1983) 'Primordial Star Formation: The Role of Molecular Hydrogen', Ap. J., 271, 632-641.
14. Izotov, Yu. I and Kolesnik, I. G. (1984) 'Kinetics of H_2 Formation in the Primordial Gas', Soviet Astr., 28, No. 1, 15-21.

15. Murray, S. D. and Lin, D.N.C. (1989) 'The Fragmentation of Proto-Globular Clusters. I. Thermal Instabilities', Ap. J., 339, 933-942.
16. Murray, S. D. and Lin, D.N.C. (1990) 'On the Fragmentation of Protogalactic Clouds', Ap. J., 363, 50-56.
17. Struck-Marcell, C. (1982) 'Gas Cloud Collisions in Protogalaxies: I. Numerical Simulations', Ap. J., 259, 116-126.
18. Struck-Marcell, C. (1982) 'Star Formation in Protogalactic Gas Cloud Collisions', Ap. J., 259, 127-132.
19. Shapiro, P. R. (1986) 'Extragalactic Gas at High Redshift: A Chronograph of Nonlinear Departures from Hubble Flow', in B. F. Madore and R. B. Tully (eds.), Galaxy Distances and Deviation from Universal Expansion, Reidel, Dordrecht, pp. 203-213.
20. MacLow, M. M. and Shull, J. M. (1986) 'Molecular Processes and Gravitational Collapse in Intergalactic Shocks', Ap. J., 302, 585-589.
21. Shapiro, P. R. and Kang, H. (1987), 'Hydrogen Molecules and the Radiative Cooling of Pregalactic Shocks', in G. R. Knapp and J. Kormendy (eds.), IAU Symposium 117: Dark Matter in the Universe, Reidel, Dordrecht, p. 365.
22. Shapiro, P. R. and Kang, H. (1987) 'Hydrogen Molecules and the Radiative Cooling of Pregalactic Shocks', Ap. J., 318, 32-65.
23. Shapiro, P. R. and Kang, H. (1987) 'Hydrogen Molecules and the Radiative Cooling of Pregalactic Shocks II: Low Velocity Shocks at High Redshift', Rev. Mexicana Astron. Astrof., 24, 58-65.
24. Shapiro, P. R., Giroux, M. L., and Kang, H. (1987) 'New Results in the Theory of the Intergalactic Medium at High Redshift', in J. Bergeron, D. Kunth, B. Rocca-Volmerange, and J. Tran Thanh Van (eds.), High Redshift and Primeval Galaxies, Editions Frontières, pp. 501-515.
25. Shapiro, P. R. and Kang, H. (1990) 'Radiative Shocks and Nonequilibrium Chemistry in the Early Universe: Galaxy and Primordial Star Formation', in R. Capuzzo-Dolcetta *et al.* (eds.), Physical Processes in Fragmentation and Star Formation, Kluwer Academic, Dordrecht, pp. 55-70.
26. Kang, H., Shapiro, P. R., Fall, S. M., and Rees, M. J. (1990) 'Radiative Shocks Inside Protogalaxies and the Origin of Globular Clusters', Ap. J., 363, 488-498.
27. Kang, H. and Shapiro, P. R. (1992) 'Radiative Shocks and Hydrogen Molecules in Pregalactic Gas: The Effects of Postshock Radiation', Ap. J., 386, 432-451.
28. Shapiro, P. R., Clocchiatti, A., and Kang, H. (1991) 'Magnetic Fields and Radiative Shocks in Protogalaxies and the Origin of Globular Clusters', in K. Janes (ed.), The Formation and Evolution of Star Clusters, (Astronomical Society of the Pacific Conference Series, Vol. 13), pp. 176-179.

29. Shapiro, P. R., Clocchiatti, A., and Kang, H. (1992), 'Magnetic Fields and Radiative Shocks in Protogalaxies and the Origin of Globular Clusters', Ap. J., 387, in press.
30. Dalgarno, A. and Lepp, S. (1987) 'Chemistry in the Early Universe', in M. S. Vardya and S. P. Tarafdar (eds.), I.A.U. Symposium 120: Astrochemistry, Kluwer Academic, Dordrecht, pp. 109-120.
31. Walker, T. P., Steigman, G., Schramm, D. N., Olive, K. A., and Kang, H. S. (1991) 'Primordial Nucleosynthesis Redux', Ap. J., 376, 51-69.
32. Saslaw, W. C. and Zipoy, D. (1967) 'Molecular Hydrogen in Pre-galactic Gas Clouds', Nature, 216, 976-978.
33. Shapiro, P. R., Giroux, M. L., and Babul, A. (1991), 'The Evolving Intergalatic Medium: The Uncollapsed Baryon Fraction in a Cold Dark Matter Universe', in S. S. Holt, C. L. Bennett, and V. Trimble (eds.), After The First Three Minutes (American Institute of Physics Conference Proceedings No. 222), pp. 347-351.
34. Shapiro, P. R., Giroux, M. L., and Babul, A. (1992) 'The Evolving Intergalatic Medium: The Uncollapsed Baryon Fraction in a Cold Dark Matter Universe', Ap. J., to be submitted.
35. Fall, S. M. and Rees, M. J. (1985) 'A Theory for the Origin of Globular Clusters', Ap. J., 298, 18-26.
36. Fall, S. M. and Rees, M. J. (1988) 'The Origin of Globular Clusters', in J. E. Grindlay and A.G.D. Philip (eds.), I.A.U. Symposium 126: Globular Cluster Systems in Galaxies, Reidel, Dordecht, pp. 323-332.

QUESTIONS AND ANSWERS

L.Blitz: I thought that the lowest temperature you could cool pure H_2 gas was ~500 K which is the temperature of the lowest quadrupole H_2 rotational transition. Your curves level off at temperatures a little more than 100 K. Where does the other factor of 4 - 5 in the temperature come from?

P.R.Shapiro: Although it is true that the lowest quadrupole rotational transition for H_2 corresponds to an excitation energy of 510 K (in temperature units), this only means that the collisional excitation rate for this transition is exponentially cut-off at temperatures below 510 K. Hence, while the rate is reduced by this exponential T dependence, it is not zero. Another words, the high energy tail of the H atom Maxwellian velocity distribution still has enough energy per atom to excite the H_2 transition. In addition, in our post shock gas, as the temperature drops below 500 K, the H_2 concentration continues to increase, so this offsets the exponential fall-off of the T dependence of the excitation cooling rate per H_2 molecule.

PRIMORDIAL MOLECULAR HYDROGEN PRODUCTION AND THE FORMATION OF POPULATION III OBJECTS

J. C. N. de Araujo and R. Opher
Instituto Astronômico e Geofísico - U.S.P.
C.P. 9638, 01065 São Paulo, S.P. - Brasil

ABSTRACT. We study the formation of Population III objects (and stars) taking into account the expansion of the Universe and a series of physical processes relevant to the primordial plasma. In particular, a very important way to cool the first condensations in the Universe is with molecular hydrogen. The minimum mass of the Population III objects formed is $\sim 10^4 M_\odot$. We show also that stars of mass $\gtrsim 50 M_\odot$ can be formed from the fragmentation of Population III objects of mass $\sim 10^6 M_\odot$.

INTRODUCTION

The isothermal density perturbations can be nonlinear for subgalactic scales (e.g. Hogan 1978, Lahav 1986, de Araujo & Opher 1988, 1989) and lead to the formation of Population III objects.

It is possible to produce nonlinear perturbations for scales $M < 10^8 M_\odot$ by isocurvature or adiabatic cold dark matter density perturbations (see de Araujo and Opher 1991).

These Population III objects can collapse directly or fragment, forming in this way the first stars (Population III).The cooling caused by the H_2 molecules have a very important role in the evolution and formation of Population III objects, as well as in the masses of the fragments (Population III stars).

CALCULATIONS AND DISCUSSION

In the study of the formation of Population III objects and stars, we performed calculations take into account a series of processes: the photon-drag and cooling due to the background radiation, recombination, photoionization, collisional ionization, Lyman-α cooling and the formation and cooling of H_2 molecules.

The H_2 molecules have an important contribution to the thermal evolution of the clouds that we study. Several authors have studied the possible ways in which molecular hydrogen can be formed. In our calculations, we take into account the most efficient processes for the production and the destruction of the H_2 molecules:

$H + e^- \to H^- + h\nu$ (Hirasawa 1969); $H + H^- \to H_2 + e^-$ (Browne & Dalgarno); $H^- + h\nu \to H + e^-$ (Matsuda et al 1969); $H^+ + H^- \to 2H$ (Peterson et al 1971); $H + H^+ \to H_2^+ + h\nu$ (Ramaker & Peek 1976); $H_2^+ + H \to H_2 + H^+$ (Karpas, Anicich & Huntress 1979); $H_2^+ + h\nu \to 2H^+$ (Matsuda et al 1969); $H_2^+ + e^- \to 2H$ (Giusti-Suzor, Bardsley & Derkits 1983); $H_2 + H \to 3H$ (Lepp & Shull 1983).

For the cooling function of the H_2 molecules we adopted Lepp & Shull (1983) which gives good results for $100K < T < 10^5 K$ and $n > 0.1 cm^{-3}$.

We begin the calculations at the recombination era $z_{rec} \sim 1500$ and consider the expansion of the Universe and take $\Omega = 0.1$ (density parameter) and h=1.0 (Hubble constant in units of $100 km.s^{-1}.Mpc^{-1}$).

We take a spectrum of perturbations of the form

$$\delta = \frac{\delta\rho}{\rho} = (M/M_o)^\alpha/(1 + z_{rec}) \tag{1}$$

where M_o is the mass scale and z_{rec} is the redshift at recombination. We take $M_o = 10^{15} M_\odot$ and $\alpha=1/3$.

Our calculations show that the minimum mass that can collapse has $\sim 10^4 M_\odot$.

We study the formation of Population III stars that can be produced by the fragmentation of clouds of mass $M \sim M_j$ (Jeans mass at the beginning of the recombination era, $\sim 10^{5-6} M_\odot$). We use the fact that a perturbation for $M \ll M_j$ that can survive with some residual amplitude, can produce the fragmentation of clouds of mass $M \sim M_j$ in the first free fall time scale of M (see de Araujo & Opher 1989).

We obtain that the minimum mass that can fragment from the M_c cloud is $\sim 50 M_\odot$ at $z \sim 181$.

REFERENCES

Browne, J. C. & Dalgarno, A. 1969, J. Phys. B, 2, 885
de Araujo, J. C. N. & Opher, R. 1988 , MNRAS, 231, 923
de Araujo, J. C. N. & Opher, R. 1989 , MNRAS, 239, 271
de Araujo, J. C. N. & Opher, R. 1991, ApJ (in press)
Giusti-Suzor, A., Bardsley, J. N. & Derkits, C. 1983, Phys. Rev. A, 28, 682
Hirasawa, T. 1969, Prog. theor. Phys., 42, 523
Hogan, C. 1978, MNRAS, 185, 889
Karpas, Z., Anicich, V. & Huntress, W. T., 1979, J. Chem. Phys., 70, 2877
Lahav, O. 1986, MNRAS, 220, 259
Lepp, S. & Shull, J. M., 1983, Ap. J., 270, 578
Matsuda, T., Sato, H. & Takeda, H., 1969, Prog. theo. Phys., 42, 219
Peterson, J. R., Abertz, W. H., Moseley, J. T. & Sheridan, J. R., 1971, Phys. Rev. A, 3, 1651
Ramaker, D. & Peek, J. 1976, Phys. Rev. A, 13, 58

DEUTERIUM IN THE DIFFUSE INTERSTELLAR MEDIUM

R. FERLET
Institut d'Astrophysique de Paris, CNRS
98 bis Boulevard Arago, 75014 Paris, France

ABSTRACT. We review the observational status in evaluating the interstellar deuterium abundance and show that the most reasonable value is of the order of 10^{-5}. Although in general agreement with the standard Big Bang nucleosynthesis, the situation is still unclear and deserves much more observations.

1. Introduction

It is largely accepted that the main site of formation of the light elements ^{2}H, ^{3}He, ^{4}He, ^{7}Li is during the first few minutes of the Universe, in the frame of the so-called standard Big Bang nucleosynthesis. Among these, deuterium is the most sensitive to primordial conditions, and determining its primordial abundance is therefore of great cosmological significance (see e.g. [1]).

Since the formation of the Galaxy, the evolution of the deuterium abundance X_D is relatively straightforward because it is simply burned when passing through stars. It is predicted that this Galactic astration should decrease X_D by no more than a factor of ~ 2, along with a negative gradient from the Galactic center to the edge small enough to expect a constant D/H ratio in the solar neighbourhood [2]. Although this simplest approach has not yet been proved to be wrong, one should keep in mind that several poorly known parameters and/or unknown pregalactic events could strongly affect these predictions.

2. Deuterium Observations

Prior to 1972, deuterium was only measured on the Earth, either in the ocean or inside meteorites, at a $\sim 1.5\ 10^{-4}$ level. Then, it was observed indirectly through ^{3}He in the solar wind or directly in planets, in the interstellar medium through deuterated molecules (see respectively Owen and Tielens in these proceedings) or through its atomic form, and in the atmosphere of massive stars.

Concerning stars, it has been shown by [3] that in Canopus, D/H $\lesssim 5.5\ 10^{-7}$, while in the main sequence star α Pav, D/H $\lesssim 10^{-5}$ [4]. This latter result implies that deuterium was either also destroyed in a main sequence star, in contradiction with existing model predictions, or that the present value of D/H is at most 10^{-5} with the important consequences which will be discussed later.

Although extremely difficult, attempts to observe directly interstellar deuterium atoms through their hyperfine transition at 91.6 cm are still underway. Following the original approach of [5], [6] and [7] did not confirm the deuterium line in absorption in the direction of the Galactic center used as a background source and set $D/H < 5\ 10^{-4}$. More recently, [8] searched for deuterium emission in the anti-center direction and found a marginal detection yielding a present $D/H < 6\ 10^{-5}$.

The more precise evaluations of the deuterium abundance are made in the diffuse interstellar medium through the observation of the Lyman lines of both H and D seen in absorption in the spectra of early B and O type fast rotating stars used as background sources. This was one of the major accomplishments of the Copernicus satellite which operated in the far UV at high spectral resolution (15 km s^{-1}) from 1972 to 1980 (see fig.1a, ref. [9] to [16]).

From about a dozen of lines of sight observed, a scatter of more than a factor of 4 clearly shows up, which it is extremely difficult to attribute to instrumental and/or analysing errors only in view of the quality of the Copernicus data and the number of independent spectral lines used in the Lyman series (up to 5). However, these variations may not be due to deuterium, since H atoms blue shifted by about 80 km s^{-1} will produce the same spectral signature.

At least two sources of high velocity HI can provide such a contamination : shocks in the interstellar medium and structures in the stellar winds themselves of the hot target stars. By eliminating the lines of sight which seems to be polluted in that way [17], as well as those with lower quality data or for which only one Lyman line was observed, we finally end up with an average D/H value toward ϵ Ori, δ Ori, γ Cas and λ Sco (see fig. 1a) of $8.6\ 10^{-6}$. Even more carefully, γ Cas is well known to exhibit a shell-like activity with a X-ray emission and should be also rejected. Thus, the present day D/H ratio in the interstellar medium is $7\ 10^{-6}$ in number or $X_D \sim (1 \pm 0.5)\ 10^{-5}$ by mass.

The interstellar deuterium can also be observed toward cool nearby stars, in absorption against the Lyman α stellar emission (see fig.1b, ref. [18] to [26], which includes results both from the Copernicus and the IUE satellites; IUE is still in operation since 1978 but with a lower spectral resolution of 30 km s^{-1}). In spite of the necessary modelling of the stellar emission for defining the continuum for the interstellar absorption and of the eventually multiple velocity components on the line of sight not resolved with IUE, the more recent and detailed study by [26] toward α Aur and λ And (fig.1b) confirms the high D/H ratios found in the local interstellar medium, indicating that this ratio might change by possibly a factor of four over scales as small as few parsecs. If these variations are confirmed to be a local phenomenon, they should not affect the "average" D/H value derived above over larger scales.

In conclusion, following the standard Galactic evolution model, the pre-Galactic deuterium abundance is :

$$X_{Dp} \sim 2 \pm 1\ 10^{-5} \qquad \text{by mass.}$$

3. Discussion

The comparison of this result, along with other light elements abundance evaluations, with primordial nucleosynthesis predictions leads to a reasonably good agreement showing that the baryons contribute between 0.02 and 0.11 to the critical energy density of the Universe, in the frame of the standard Big Bang model calculations [27].

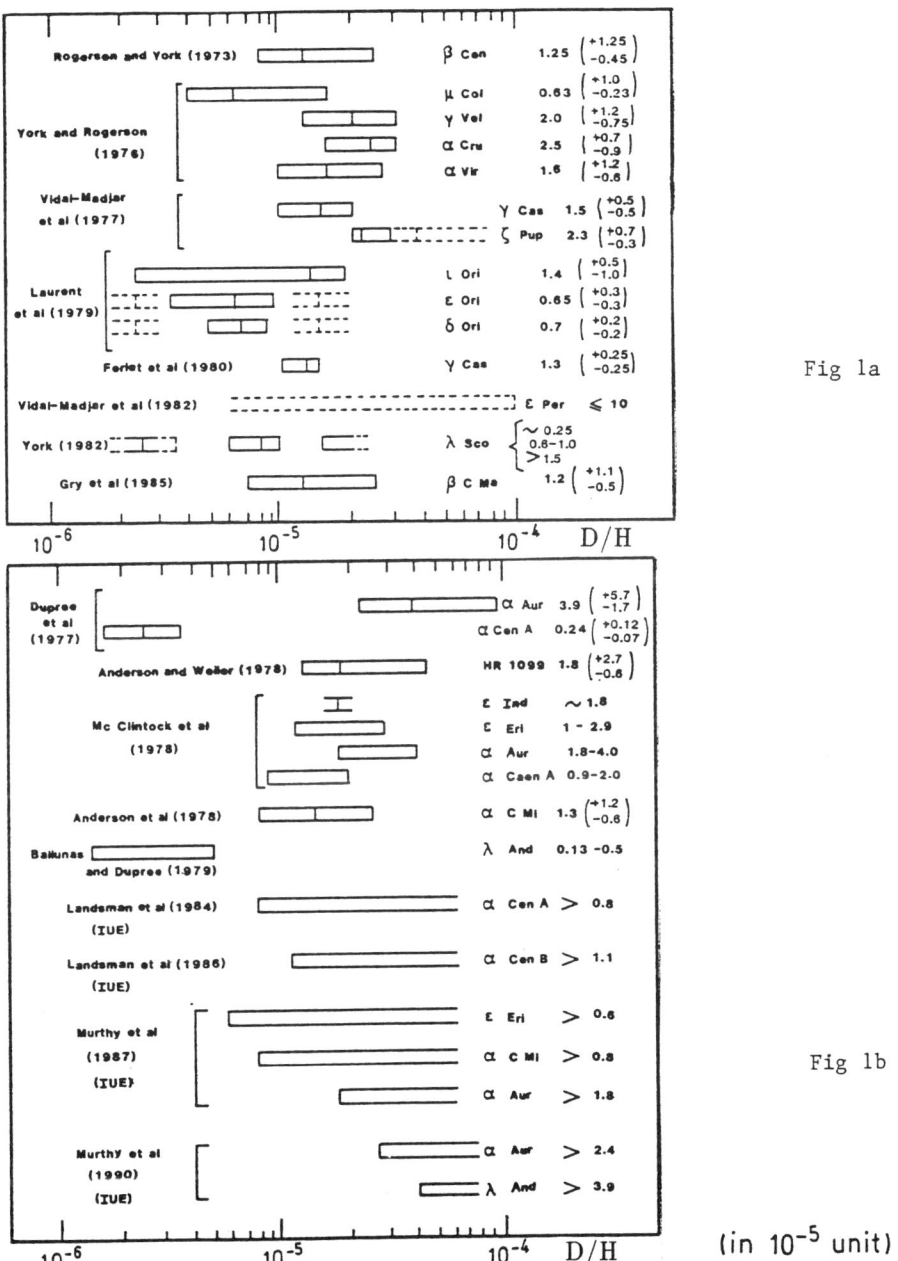

Figure 1. Observations of the D/H ratio in the diffuse interstellar medium through Lyman lines in absorption toward hot stars in the range 100–1000 pc (Fig.1a, ref. [9] to [16]), or toward very nearby cool stars (Fig.1b, ref. [18] to [26]). The average interstellar D/H ratio is around 10^{-5}, while local variations seem to be present (see text).

It has to be stressed that this model has now reach a real maturity, being able to produce correct predictions in the field of particle physics. In effect, it was easily shown by [28] that probably only three families of neutrinos were compatible with the observed helium abundance, a fourth one being barely acceptable. LEP experiments at CERN beautifully confirmed that prediction [29].

This should now lead to a more detailed comparison between the predicted and the observed abundances in order to detect possible fine tuning disagreement which might be the signature of still unclear issues. Among them, the Galactic evolution is certainly poorly known, as it is shown by [30] through the history of the light element abundance measurements : in spite of their large variations with instrumental progress, it was invariably concluded to a good agreement with the predictions.

The D constraint can now be replaced by the ^{3}He + D one since the sum is measured in the solar wind [31]. This is the unique site of measurement, and it seems to fit the model. We have shown here that more numerous deuterium observations have brought difficulties. Much more are needed to rise interesting constraints, as is already the case for ^{4}He and ^{7}Li, either at 92 cm or in the UV and visible wavelength ranges. Thus, Lyman lines will be seen at different redshifts with the Hubble Space Telescope and the FUSE-LYMAN explorer, and will perhaps be detected in "primordial" clouds in absorption toward quasars [32] accessible to large ground-based telescopes. We may at the end reach the solution.

4. References

[1] Vidal-Madjar, A., 1982, in "Diffuse Matter in Galaxies", Audouze et al. (eds), NATO ASI Series, N° 110, 57
[2] Audouze, J. and Tinsley, B.M., 1976, Ann. Rev. Astron. Ap., **14**, 43
[3] Ferlet, R., Dennefeld, M. and Spite, M., 1983, Astron. Astrophys., **124**, 172
[4] Vidal-Madjar, A., Ferlet, R., Spite, M. and Coupry, M.F., 1988, Astron. Astrophys., **201**, 273
[5] Césarsky, D.A., Moffet, A.T. and Pasachoff, J.M., 1973, Astrophys. J. Letters, **180**, L1
[6] Sarma, N.V.G. and Mohanty, D.K., 1978, M.N.R.A.S., **184**, 181
[7] Anantharamaiah, K.R. and Radhakrishnan, V., 1979, Astron. Astrophys. Letters, **79**, L9
[8] Blitz, L. and Heiles, C., 1987, Astrophys. J. Letters, **313**, L95
[9] Rogerson, J.B. and York, D.G., 1973, Astrophys. J. Letters, **186**, L95
[10] York, D.G. and Rogerson, J.B., 1976, Astrophys. J., **203**, 378
[11] Vidal-Madjar, A., Laurent, C., Bonnet, R.M. and York, D.G., 1977, Astrophys. J., **211**, 91
[12] Laurent, C., Vidal-Madjar, A. and York, D.G., 1979, Astrophys. J., **229**, 923
[13] Ferlet, R., Vidal-Madjar, A., Laurent, C. and York, D.G., 1980, Astrophys. J., **242**, 576
[14] Vidal-Madjar, A., Laurent, C., Gry, C., Bruston, P., Ferlet, R. and York, D.G., 1983, Astron. Astrophys., **120**, 58
[15] York, D.G., 1983, Astrophys. J., **264**, 172
[16] Gry, C., York, D.G. and Vidal-Madjar, A., 1985, Astrophys. J., **296**, 593
[17] Gry, C., Lamers, H.J.G.L.M. and Vidal-Madjar, A., 1984, Astron. Astrophys., **137**, 29
[18] Dupree, A.K., Baliunas, S.L. and Shipman, H.L., 1977, Astrophys. J., **218**, 361
[19] Anderson, R.C. and Weiler, E.J., 1978, Astrophys. J., **224**, 143

[20] McClintock, W., Henry, R.C., Linsky, J.L. and Moos, H.W., 1978, Astrophys. J., **225**, 465
[21] Anderson, R.C., Henry, R.C., Moos, H.W. and Linsky, J.L., 1978, Astrophys. J., **226**, 883
[22] Baliunas, S.L. and Dupree, A.K., 1979, Astrophys. J., **227**, 870
[23] Landsman, W.B., Henry, R.C., Moos, H.W. and Linsky, J.L., 1984, Astrophys. J., **285**, 801
[24] Landsman, W.B., Murthy, J., Henry, R.C., Moos, H.W. Linsky, J.L. and Russel, J.L., 1986, Astrophys. J., **303**, 791
[25] Murthy, J., Henry, R.C., Moos, H.W., Landsman, W., Linsky, J.L., Vidal-Madjar, A. and Gry, C., 1987, Astrophys. J., **315**, 675
[26] Murthy, J., Henry, R.C., Moos, H.W., Vidal-Madjar, A., Linsky, J.L. and Gry, C., 1990, Astrophys. J., **356**, 223
[27] Olive, K.A., Schramm, D.N., Steigman, G. and Walker, T.P., 1990, Phys. Letters, **B236**, 454
[28] Yang, J., Schramm, D.N., Steigman, G. and Rood, R.T., 1979, Astrophys. J., **227**, 697
[29] L3 Collab., Adeva, B., et al., 1989, Phys. Letters, B**231**, 509
ALEPH Collab., Decamp, D., et al., 1989, Phys. Letters, B**231**, 519
OPAL Collab., Akrawy, M.Z., et al., 1989, Phys. Letters, B**231**, 530
DELPHI Collab., Aarnio, P., et al., 1989, Phys. Letters, B**231**, 539
[30] Vidal-Madjar, A., 1990, in COSPAR XXVIII, The Hague
[31] Bochsler, P., Geiss, J. and Maeder, A., 1990, Solar Physics, **128**, 203
[32] Webb, J.K., Carswell, R.F., Irwin, M.J. and Penston, M.V., 1991, "On measuring the deuterium abundance in QSO absorption systems", preprint.

QUESTIONS AND ANSWERS

S.P.Tarafdar: What are the equivalent width of Hydrogen Lyman-α forest lines? What will be the expected equivalent width of corresponding deuterium Ly-α forest lines and, can we observe them with our present day capability?

R.Ferlet: Using existing spectrographs at a resolution of ~ 10 km/s on a ground-based $4m$ telescope, with an integration time of about 10 hours on a $m_r \sim 17.5$ quasar of redshift > 2.6 in order to observe the first 5 lines of the Lyman series with a continuum S/N ratio ~ 15, the lowest HI column density in a "Ly α forest" absorption line required for detecting DI at Ly α if $D/H = 10^{-4}$ is $\sim 10^{17}$ atoms cm^{-2} (assuming a 5σ detection limit or a limiting equivalent width of about 70 $m\mathring{A}$). In that case, the DI Ly α equivalent width is predicted to be ~ 110 $m\mathring{A}$. The highest HI column densities seen in QSO absorption clouds are $\sim 10^{21} cm^{-2}$; in that case, deuterium (still if $D/H = 10^{-4}$) will be detectable in Ly γ, δ and ϵ with equivalent widths ~ 300 $m\mathring{A}$. Note that in all cases, the DI line is blended with the HI one. These simulations by Webb et al. (1991; preprint) are for $b = 25$ km/s. For $b = 35$ km/s, or if $D/H < 10^{-4}$, the detection constraints are more stringent.

J.C.Pecker: You have insisted quite convincingly, in my opinion, upon the "pollution " which is endeavored by the D/H ratio from place to place, and time to time, in the evolution of a galaxy. On the other side, Shapiro, before you, has stressed, as you did also, the bad knowledge of the early stages of the galactic evolution, which may imply highly intense UV sources so far and unknown evolutive processes. I would like therefore to insist upon the use (and abuse!) of the adjective "primordial" or "primaeval", when refering to varius determinations of abundances, be it D/H or, say, He^4/H or else. It seems to me that one should, each time, to be clearer: some determinations lead to "primordial" abundances at the time scale of the solar system; some others aim at the (different) "primordial" abundances at the time scale of the lifetime of our Galaxy; whether some refer really to the scale time of the universe is very much in doubt, to my eyes. On line with this comment, I would add the following question: what do predict the models differing from the standard Big Bang, and I am thinking of $\Lambda \neq 1$, of inflationary models,etc. which constitute a very different "new Big Bang" from what is known as the standard one? What abundances can the "new Big(s) Bang(s)" predict?

R.Ferlet: Amongst non-standard Big Bang nucleosynthesis models, inhomogeneous ones inspired from the quark-hadron phase transitions have been the subject of recent and active research. However, they involve several free parameters and are still controversial. Nevertheless, they could become especially interesting if the observed "primordial" helium abundance is derived to be smaller than currently thought, if the "primordial" ^7Li abundance is found to be higher than the abundance measured in very metal-poor Population II stars (i.e. if it is demonstrated that surface lithium has been destroyed in these stars), if there is a significant cosmological yield to elements as ^9Be, and/or if it exists a neutrino with a mass of 17 keV.

L.Blitz: Two comments on the 92-cm line work: (1) In the work I published with Heiles, done with the help of a number of people in the room, we were careful not to claim a detection, in spite of the suggestiveness of the spectrum. (2) More important, the DI line has now been detected by Heiles and McCollough toward Cas A in a completely convincing spectrum that has now been confirmed. The problem is that it is in a difficult region to interpret since an atomic line has been observed in a clump of both atomic and molecular gas, where there is likely to have been some stellar processing, but a preliminary estimate gives $D/H \sim 1.5 \times 10^{-5}$.

THE D/H RATIO IN MOLECULAR CLOUDS

A.G.G.M. Tielens
MS 245-3, NASA Ames Research Center, Moffett Field, CA 94035, USA.

1. ABSTRACT

This paper summarizes our understanding of the dominant D-reservoirs in molecular clouds and suggests possible direct determinations of the D-abundance. It is concluded that rotational HD lines from shock regions provide the best way to determine the gas phase D-abundance. In cold dense cores, the dominant gas phase D-reservoir is likely to be atomic D, because of expected inefficient HD formation on grain surfaces. The gas phase D-abundance derived from observations of dense cores is $\approx 5 \times 10^{-6}/n_4$ (with n_4 the total density in units of 10^4 cm^{-3}). A large fraction of the D ($\approx 50\%$) may be locked up in deuterated molecules in grain mantles. A small fraction ($\approx 2\%$) may be locked up in the photolyzed residues of such grain mantles. PAHs will also be deuterated (PADs), containing $\approx 1\%$ of the D. Finally, it is likely that all of these processes have contributed to the D-enrichment observed in solar system materials.

2. INTRODUCTION

The elemental D-abundance and its gradient in the galaxy has important implications for our understanding of the origin and early evolution of the universe; ie., is the universe closed by exotic dark matter or by white dwarfs resulting from an early epoch of star formation[1]. Furthermore, various components enriched in D have been isolated from carbonaceous meteorites and interplanetary dust particles[2]. This D-fractionation probably reflects an ancestry dating back to molecular clouds. Past studies on the D-abundance in molecular clouds have relied on measuring trace species and relating their abundance back to the dominant D-reservoir. However, because of the zero-point energy difference, abundances of deuterated species are very susceptible to fractionation effects and thus sensitive to the local physical conditions. A reliable determination of the D-abundance requires a direct study of the dominant D-reservoirs.

3. DEUTERIUM IN WARM DENSE CLOUDS

In warm dense media, such as PDRs and shocks, the deuterium chemistry is dominated by neutral-neutral reactions, in particular, $D + H_2 \rightarrow H + HD$. The conversion of D into HD by this reaction is much faster than the shock cooling time ($\tau \approx N(H_2)/[n_0 v_s]$) and all D should be in the form of HD in this shock (Fig. 1). Likewise, although previously neglected, this neutral reaction is much faster than grain surface formation in PDRs (Fig. 1). However, in PDRs, photodestruction quickly reverts the HD back into D and, unlike shocks, D is the dominant reservoir.

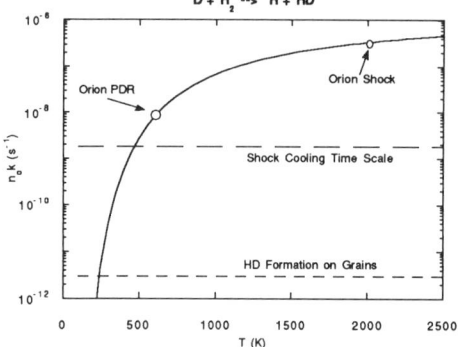 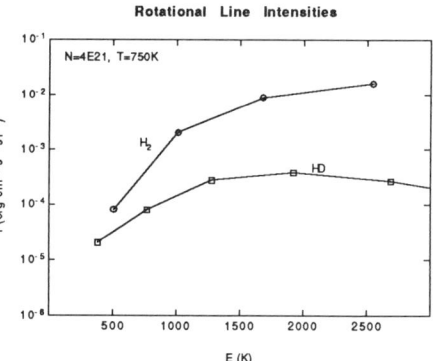

Fig. 1: The rate of D-HD conversion in warm gas for an H_2 density of $n_0=10^5$ cm^{-3}. Also indicated are the conditions characteristic for the Orion shock (derived from the 2μm vibrotational H_2 lines) and the Orion Bar PDR (derived from the rotational H_2 lines).

Fig. 2: Predicted intensities of rotational H_2 and HD lines for the Orion shock as a function of their excitation energy. The assumed column density and temperature are characteristic for the observed CO and pure rotational H_2 lines, gas downstream from that observed in the 2μm H_2 lines.

Since all D is in the form of HD and the gas is warm, shocks present an excelent probe of the total gas phase deuterium abundance (Fig. 2). Although line-to-continuum forms somewhat a problem for the HD lines, these lines are measurable with present day techniques. Careful selection of HD and H_2 lines from similar excitation energies will allow a direct estimate of the HD/H_2 ratio, circumventing the rather uncertain aspects of shock theory. Although the figures are geared towards the Orion shock, any strong shock (≈35 km/s) will do and this may present the best opperunity to measure the elemental D gradient in the galaxy.

4. DEUTERIUM IN COLD DENSE CLOUDS

In recent years it has been realized that D-fractionation is more complex than previously thought since both D and HD may be important reservoirs[3,4]. The atomic D abundance can be directly and reliably determined from observations. The atomic H and D abundances are governed by very similar reactions; ie., formation by dissociative recombination of molecular ions (HCO+ and DCO+), and destruction by accretion and reaction on grain surfaces (Fig. 3). The atomic D/H abundance ratio in reflects therefore directly that of these gas phase ions. The DCO+/HCO+ ratio is observed to be about 0.05 in dark clouds[5,6]. Thus, the atomic D/H ratio is ≈0.05, much larger than the deuterium cosmic abundance. Now, the H abundance is essentially a balance between cosmic ray ionization of H_2 and accretion on grain surfaces and is ≈1 cm^{-3}. Thus, the fraction of the deuterium in the form of atomic D is about $0.5/n_4$, assuming an elemental D abundance of 10^{-5}.

The HD abundance is much more difficult to estimate. Generally, it has been assumed that HD is efficiently formed on grain surfaces, which implies then: HD/D≈$4n_4$ (Fig. 3). In that case, most of the deuterium would be in HD in dense cores. However, the atomic H abundance is very low in dense dark clouds (≈1 cm^{-3}) and the H accretion rate on grains is less than that of the heavier species (ie., O, CO). Almost all accreted H and D will then

THE GAS PHASE D/HD RATIO

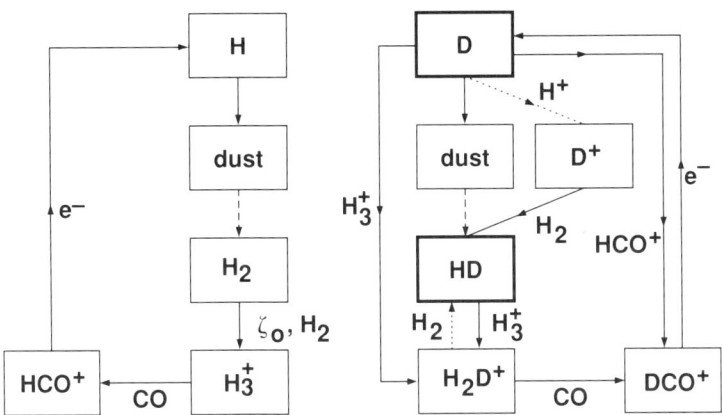

Fig. 3: A schematic view of the important reactions involved in the exchange between the two important gas phase reservoirs of deuterium, D and HD.

react to form H_2O, HDO, etc (see below) and HD formation is inefficient[7]. Gas phase formation of HD is also inefficient (Fig. 3) and leads to HD/D$\approx 10^{-2}$ H^+/H_3^+. Although somewhat depending on the details of the chemistry, the HD/D ratio in that case is $\approx 10^{-1}$-10^{-2} and D is the dominant reservoir. In warm gas (T>25K), HD is the important reservoir since H_2D^+ will be channeled to HD rather than DCO^+ (Fig. 3).

Thus, in summary, HD formation on grain surfaces is probably inefficient and D is the main gas phase reservoir in dense cores. The total D-abundance in the gas phase is then estimated to be $\approx 4 \times 10^{-6}$. This is less than the generally accepted elemental deuterium abundance ($\approx 10^{-5}$) and, thus, some of the D may well be in the solid state.

5. DEUTERATED GRAIN MANTLES

The atomic D/H ratio in the gas phase is much larger than the elemental D abundance, leading to large D-fractionation in grain mantles[7]. Moreover, since the activation barrier for abstraction of a D-atom is larger than for H abstraction by the zero-point energy difference, D abstraction from a deuterated molecule has very low probability. As a result, the abundance of some deuterated molecules such as HDCO and D_2CO can be considerably enhanced over the gas phase D/H ratio. The recent observations of the high D_2CO/HDCO relative to the HDCO/H_2CO ratio in Orion lends some support for this scheme[8].

Due to the mass difference, the IR active modes involving deuterium will be shifted by $\approx \sqrt{2}$ from those of the hydrogenated species - typically to the 4.5-4.8μm and 13-15μm region, both completely opaque due to telluric CO_2. Direct detection of deuterated grain mantle molecules has therefore to await space borne missions (ISO). Such observations will be crucial to determine the total D-abundance in molecular clouds. Of course, they will also be important in determining the composition of interstellar grain mantles, testing grain chemistry schemes, as well as testing the importance of grain mantles for the gas phase composition of hot molecular cloud cores. In the absence of direct measurements, the fraction of D in grain mantles can be guestimated from observations of hot cores. If the molecules in hot cores result from evaporation of grain mantles[9] and if we assume that accretion was complete prior to the formation of the protostar, the observed D-fractionation (HDO/H_2O≈1%) in hot cores implies that ≈50% of all the D was in grain mantles. Finally, this D-fractionation is expected to be passed on to the organic residue formed by UV photolysis of these grain mantles. Since ≈3x10^{-5} of the H is observed to be in these organic grain mantles (via the 3.4μm band) in the diffuse interstellar medium, ≈2% of the D is expected to be locked up in this dust component. Although this estimate may be somewhat on the low side due to preferential loss of H over D by photolysis, organic grain mantles seem to be a rather unimportant sink for D.

6. DEUTERATED PAHS

Two different chemical pathways may play a role in the deuteration of PAHS: 1) Small PAHs (<25 C-atoms) highly excited by an UV photon will preferentially "cool" by loss of a peripheral H-atom rather than IR emission[10]. The resulting radical may react with atomic D rather than atomic H. Due to the zero-point energy difference, the ratio of the fission rates of H to D is ≈3 and fractionation can occur this way. In steady state, the fraction, f_D, of peripheral D relative to H is then given by f_D≈3 [n_D/n_H]. In PDRs - the prime emission zones of the IR emission features - the atomic D/H abundance ranges from ≈10^{-5} at the surface to ≈10^{-3} at an A_V of 5 (H_2 goes molecular before HD). Thus, fractionation due to this process, although substantial, is too small to be observable. Inside dense clouds, the UV field due to cosmic ray excitation of H_2 is still large enough to drive the PAD/PAH abundance to steady state. Now, of course, n_D/n_H is much larger (≈0.05, §4) and every small PAH is expected to have 1 or 2 of its peripheral H's replaced by a D.

2) Proton transfer or isotope exchange reactions (PAH + H_2D^+ --> PAD + H_3^+ and PAH^+ + HD (D) --> PAD^+ + H_2 (H)) can also lead to D-fractionation of PAHs inside dense cores. Assuming typical abundances for the coreactants and realizing that PAH^+/PAH≈10^{-4}, the D-fractionation timescale ≈3x$10^6/n_4$ yrs for both reactions. Thus, over the timescale of a dense core every PAH would acquire ≈1 peripheral D-atom. Consequently, in a dense core, about 1% of the elemental D could be locked up in PADs. Note that, while the e-sticking coefficient for small PAHs (<50 C-atoms) is small (≈10^{-4}-10^{-2}), larger PAHs will predominantly (90%) be negatively charged. These timescales are then 10 times longer and PADs are correspondigly less important. Finally, once the small PADs find themselves in a UV bright region (ie., IR emission zone), they will quickly equilibrate with the local atomic D/H ratio and only PAHs in the 25-50 C-atom range may retain the D-signature of dense cores.

References

[1] Kurucz, R.L., 1991, APJL, preprint.
[2] Kerridge, J.F., 1989, in *Interstellar Dust*, eds. Allamandola and Tielens, (Reidel, Dordrecht), p.383.
[3] Dalgarno, A. & Lepp, S., 1984, ApJL,287,L47.
[4] Millar, T.J., Bennett, A. & Herbst,E., 1989, ApJ,340,906
[5] Wootten, A., Loren, R., & Snell, R.,1982, ApJ,255,160.
[6] Butner, H., 1990, PhD Thesis, Univ. Texas; see also these proceedings.
[7] Tielens, A.G.G.M., 1983, A&A, 119, 117
[8] Turner, B., 1990,ApJL, 362, L29.
[9] Walmsley, C., 1989, in *Interstellar Dust*, eds. Allamandola and Tielens, (Reidel, Dordrecht), p.263.
[10] Allamandola, L.J., Tielens, A.G.G.M., & Barker, J.R., 1989, ApJS, 71,733.

QUESTIONS AND ANSWERS

C.M.Walmsley: Why do you think HD is not formed efficiently on grains?

A.G.G.M.Tielens: When the gas density is higher than $\sim 10^4 cm^{-3}$, the accretion rate of heavy molecules (such as CO, O_2 and O) is larger than of atomic H. In that case, all accreted H atoms will go into hydrogenating these heavy molecules rather than form H_2, forming an icy grain mantle (ie., H_2O, H_2CO, CH_3OH).

M.Gray: Do you think molecules of Buckminster-Fullerene type could hold a significant of D?

A.G.G.M.Tielens: Buckminster-Fullerene (C_{60}) itself does not have any H and thus will not (directly) fractionate. However, partially closed fullerenes are probably hydrogenated in the ISM. They are expected to behave like PAHs and could thus D_- fractionate as well.

DEUTERIUM IN THE SOLAR SYSTEM

T. OWEN
Institute for Astronomy
2680 Woodlawn Drive
Honolulu, HI 96822
USA

ABSTRACT. Values of D/H measured in the methane on the giant planets and Titan indicate the presence of two distinct reservoirs of deuterium in the outer solar system. The dominant reservoir is in hydrogen gas, the second, multi-component reservoir is found in the hydrogen that is bound in condensed compounds. Both reservoirs appear to have originated in the interstellar medium. In contrast, the values of D/H in water vapor on Mars and Venus (especially) exhibit a large enrichment from the "condensed matter" starting value. Interpretation of this enrichment may illuminate the history of water on these two planets.

1. Introduction

During the past fifteen years, Drs. Catherine de Bergh, Barry Lutz, Jean Pierre Maillard and I have been engaged in a systematic study of the ratio of D/H in planetary and satellite atmospheres. We decided from the outset to determine this isotopic ratio in the outer solar system by studying CH_3D and CH_4, rather than using HD and H_2. This decision was based on our perception that it is extremely difficult to measure equivalent widths of the very weak lines of HD and H_2 in the presence of ubiquitous weak lines of CH_4 in the spectra of all of the giant planets and Titan. Smith et al. (1989) have recently described this difficulty in detail.

To study deuterium on the inner planets, we elected to use HDO and H_2O, as these are the most abundant carriers of hydrogen in the atmospheres of Mars and Venus. We considered using DCl on Venus, but had to abandon this idea because of strong interference from absorptions of CO_2 and CO.

2. THE OUTER SOLAR SYSTEM

The first step was to make the necessary quantity of CH_3D and examine the spectrum of this molecule to identify and analyze absorption bands that could be used for our purpose (Danehy et al. 1977, Lutz et al. 1978, 1983). This exercise led to the conclusion that the second overtone of the vibrational frequency v_2 at 6425 cm^{-1} was ideal, in that it falls in the 1.6 µm window of the CH_4 spectrum. Having made this identification, it was possible to find suggestive evidence for the presence of this band in existing spectra of Titan and Uranus that had been recorded by Fink and Larson (1979) at a resolution of 3.6 cm^{-1} (Lutz et al. 1981).

To progress, we had to obtain our own spectra at a higher resolution of 1.2 cm^{-1}, first of Uranus (de Bergh et al. 1986) and Titan (de Bergh et al. 1988). To these we finally added Neptune at a resolution of 4.0 cm^{-1} (de Bergh et al. 1990). Along the way, we had studied Saturn as a control, since a value of D/H = 2.2 +1.8/-1.7 x 10^{-5} had been deduced for this planet by Courtin et al. (1984). These authors used Voyager observations of the 8.6 µm methane band complex in emission to derive this result, a technique which had been employed successfully on Jupiter by Kunde et al. (1982). We obtained D/H = 1.7 = +1.7/-0.8 x 10^{-5} for Saturn, suggesting that our method did not introduce any large systematic errors. We could not use our approach on Jupiter because absorption bands of ammonia block the window at 1.6 μm where our CH_3D band occurs.

The results of this investigation are summarized in Figure 1, which includes values of D/H determined in CH_4 on Jupiter (discussed by Gautier and Owen 1989), and an updated value of D/H = 1.7 ± 1.1 x 10^{-5} for Saturn (Noll and Larson 1991). The value of D/H = 1 ± 0.5 x 10^{-5} for the Interstellar Medium is based on the discussion by R. Ferlet at this conference. The protosolar value of 3.0 ± 1.0 x 10^{-5} is derived from solar $^4He/H$ = 0.097 and meteoritic $^3He/^4He$ = 1.4 ± 0.2 x 10^{-4} given by Anders and Grevesse (1989) and solar wind $^3He/^4He$ = 4.4 ± 0.5 x 10^{-4} from Bochsler et al. (1990).

We conclude from inspection of Figure 1 that there are two principal reservoirs of deuterium in the outer solar system (Owen et al. 1986). The largest reservoir consists of the hydrogen gas that originally dominated the solar nebula and is now the main constituent of Jupiter and Saturn. This should be the protosolar value of D/H and indeed, the Jupiter and Saturn average of D/H = 2.0 ± 1.0 x 10^{-5} overlaps the value of D/H = 3.0 ± 1.0 10^{-5} derived from meteorites and the solar wind. The second reservoir must actually consist of several different species of condensed hydrogen-containing compounds, presumably dominated by H_2O. Thus the value of D/H that we measure in gas from this reservoir must be an average of several different values set by ion-molecule and grain-catalyzed reactions in the interstellar medium. We find evidence for this reservoir in the methane on Titan, Uranus and Neptune. A high value of D/H resulting from the mixing of material from the icy cores of these planets was predicted long ago by Hubbard and MacFarlane (1980). These predictions are illustrated in Figure 1 by an "x".

3. THE INNER SOLAR SYSTEM

The condensed matter reservoir is also manifested in the high value of D/H seen in H_2O in meteoritic hydrated silicates (Yang and Epstein 1983), and in Halley's Comet (Eberhardt et al., 1987) as illustrated in Figure 1. Still higher values are found in some of the organic compounds identified in carbonaceous chondrites (Yang and Epstein 1983, Epstein et al. 1987) and in interplanetary dust particles (Zinner et al. 1983) (see Figure 1). Given the meteoritic and cometary bombardment of the early inner planets, we may reasonably expect that the condensed matter reservoir furnished the starting value of D/H on all of these bodies. On Earth, we find that the present value of D/H in the oceans is overlapped by the various determinations of this ratio in objects representing the condensed matter reservoir (Figure 1).

Some slight enrichment of D/H on Earth must have occurred as a result of the dissociation of H_2O and resulting hydrogen escape (Yung et al. 1989). Studying H_2O and HDO on Mars and Venus, we found values of D/H that are respectively 6 ± 3 and 120 ± 40 times the telluric values (Owen et al. 1988, de Bergh et al. 1991). The Martian value has since been confirmed by Bjoraker et al. (1989) who obtained an enrichment of 5.2 ± 0.2 times telluric, while the 100-fold enrichment of D/H on Venus was originally discovered by McElroy et al. (1982) and Donahue et al. (1982) with mass spectrometers on the Pioneer Venus Spacecraft.

The interpretation of these enrichments is complicated by the variety of processes that are involved: the addition of water by cometary impact, chemical reactions with surface materials, changes in the luminosity of the sun, etc. The original interpretation of the Venus enrichment by

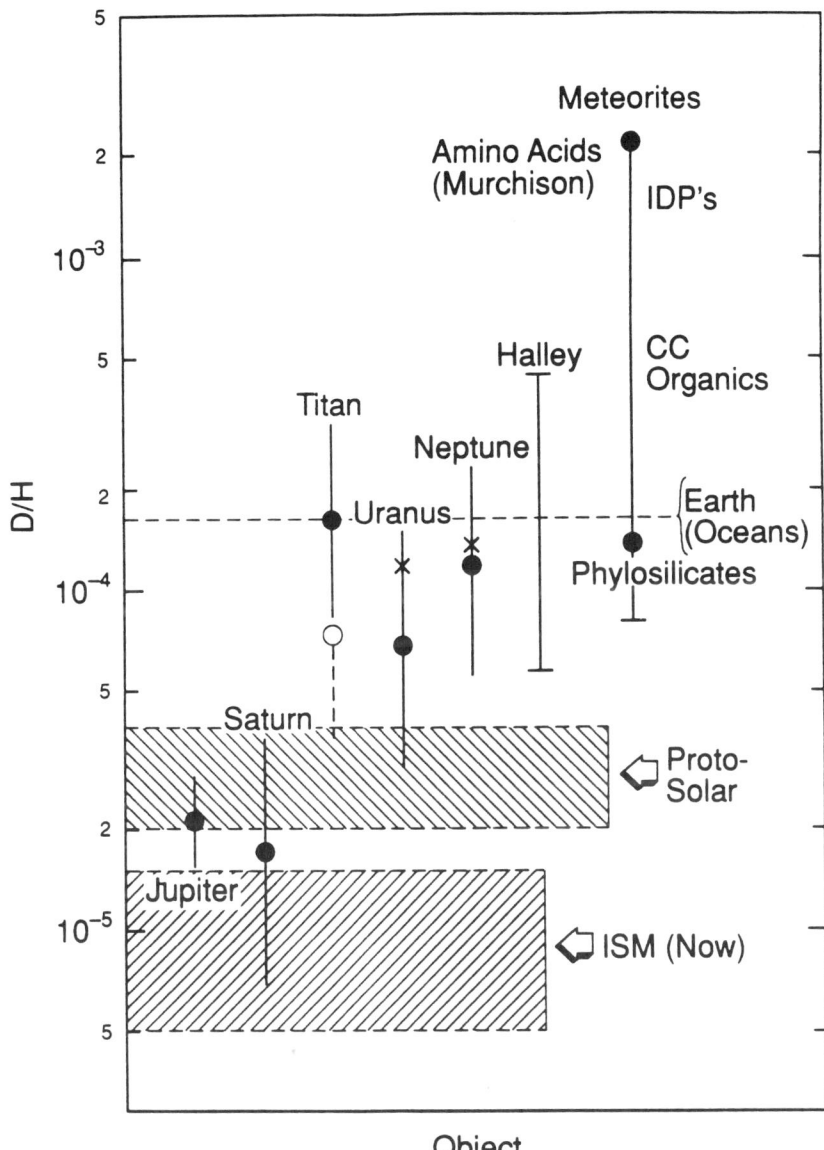

Figure 1: Values of D/H as measured in various bodies in the solar system compared with the present interstellar value and the protosolar value deduced from meteorites and the solar wind. The open circle shown for Titan represents a correction for enrichment of D/H through atmospheric escape.

Donahue et al. (1982) in terms of a runaway greenhouse that depleted a starting inventory of ≥ 0.3% of the Earth's oceans still seems the most reasonable scenario (de Bergh et al. 1991—see Carlson et al. 1991 for a new value of the global H_2O abundance that reinforces this conclusion). The Martian atmosphere is so thin that a single cometary impact has the potential of re-setting the hydrogen isotopes (Owen 1992). Thus it is difficult to know how significant the present D/H value is for the long-term history of the planet.

4. CONCLUSIONS

The results for the outer planets suggest that the hydrogen gas in the atmospheres of Jupiter and Saturn can be used to obtain an accurate measurements of the solar nebula value of D/H. In effect, this will be a measurement of D/H in an interstellar cloud as it existed 4.5 billion years ago. Compared with measurements of the present value of D/H in the ISM, this value can be used in principle to constrain models for big-bang nucleosynthesis and galactic evolution, particularly when combined with an accurate value of $^3He/^4He$. Both ratios can be determined by the mass spectrometer on the Galileo Probe, now set for delivery into the Jovian atmosphere in December 1995 (Niemann et al. 1992).

The relatively high value of D/H measured in Titan's methane excludes models for the origin of this satellite's atmosphere that involve reduction of interstellar CO to CH_4 in the Saturn sub-nebula (Owen and Gautier 1989). It seems more likely at this point that the methane comes from the decomposition of organic compounds (grains) during the accretion of the satellite.

Finally, there is much more to be done before the interpretation of the high values of D/H on Venus and Mars will be completed. The outstanding problem on Venus is to obtain a reliable value for the global atmospheric water abundance. This is possible from Earth-based observations, and may even be achieved this year. On Mars, we need additional information on oxygen isotopic ratios in the atmosphere and in the rocks as well as in situ measurements of D/H in sub-surface and polar ice. Only then will we be able to unravel the time-history of the atmosphere with any confidence.

5. References

Anders, E., and Grevesse, N. 1989, *Geochim. et Cosmochim. Acta,* **53**, 197-214.
Bjoraker, G. L., Mumma, M. J., and Larson, H. P. 1989, *Bull. Amer. Astron. Soc.*, **21**, 990 (abstract).
Bochsler, P., Geiss, J., and Maeder, A. 1990, *Solar Physics* **128**, 203-215.
Carlson, R. W., et al. 1991, *(Science* in press).
Courtin, R., Gautier, D., Marten, A., and Bèzard, B. 1984., *Astrophys. J.*, **287**, 899-916.
de Bergh, C., Bèzard, B., Owen, T., Crisp, D., Maillard, J. P., and Lutz, B. L. 1991, *Science,* **251**, 547-579.
de Bergh, C., Lutz, B. L., Owen, T., Brault, J., and Chauville, J. 1986, *Astrophys. J.*, **311**, 501-510.
de Bergh, C., Lutz, B. L., Owen, T., and Chauville, J. 1988, *Astrophys. J.*, **329**, 951-955.
de Bergh, C., Lutz, B. L., Owen, T., and Maillard, J. P. 1990, *Astrophys. J.*, **355**, 661-666.

Danehy, R. G., Lutz, B. L., Owen, T., Scattergood, T. W. and Goetz, W. 1977, *Astrophys. J. Lett.*, **273**, 397-409.
Donahue, T. M., Hoffman, J. H., Hodges, R. R. Jr., and Watson, A. J. 1982, *Science* **216**, 630-633.
Eberhardt, P., Dolder, U., Schulte, W., Krankowsky, D., Lämmerzahl, P., Hoffmann, J. H., Hodges, R. R., Berthelier, J. J., and Illiano, J. M. 1987, *Astron. Astrophys.*, **187**, 435-437.
Epstein, S., Krishnamurthy, R. V., Cronin, J. R., Pizzarello, S., and Yuen, G. U. 1987, *Nature*, **326**, 477-479.
Fink, U., and Larson, H. P. 1979. *Astrophys. J.*, **233**, 1021-1040.
Gautier, D., and Owen, T. 1989, in *Origin and Evolution of Planetary and Satellite Atmospheres*, ed. S. K. Atreya, J. B. Pollack, and M. S. Matthews (Tucson: U. of Arizona press) 487-512.
Hubbard, W. B. and MacFarlane, J. J. 1980, *Icarus*, **44**, 676-682.
Kunde, V. G., Hanel, R. A., Maguire, W. C., Gautier, D., Baluteau, J. P., Marten, A., Chedin, A., Husson, N., and Scott, N. 1982, *Astrophys. J.*, **263**, 443-467.
Lutz, B. L., de Bergh, C. and Maillard, J. P. 1983, *Astrophys. J.*, **273**. 397-409.
Lutz, B. L., Danehy, R. G., and Ramsay, D. A. 1978, *J. Molec. Spectrosc.* **72**, 128.
McElroy, M. B., Prather, M. J., and Rodriquez, J. 1982, *Science*, **215**, 1614-1615.
Niemann, H. B., Harpold, D. N., Atreya, S. K., Carignan, G. R., Hunten, D. M., and Owen, T. C. 1992, *Space Sci. Rev.* (in press).
Noll, K. S., and Larson, H. P. 1991, *Icarus* **89**, 168-189.
Owen, T. 1992, in MARS, ed. H. Kieffer, B. Jakosky, and M. Matthews (Tucson: U. of Arizona Press) (in press).
Owen, T. and Gautier, D. 1989, *Adv. Space Res.* **9**, (2)73-(2)78.
Owen, T., Lutz, B. L., and de Bergh, C. 1986, *Nature* **320**, 244-246.
Owen, T., Maillard, J. P., de Bergh, C., and Lutz, B. L. 1988, *Science*, **240**, 1767-1770.
Smith, W. H., Schempp, W. V., Simon, J., and Baines, K. H. 1989, *Astrophys. J.* **336**, 962.
Yang, J. and Epstein, S. 1983, *Geochim. et Cosmochim. Acta,* **47**, 2199-2216.
Yung, Y. L., Wen, J.-S., Moses, J. I., Landry, B. M., Allen, M. and Hsu, K.-J. 1989, *J. Geophys. Res.*, **94**, 14971-14989.
Zinner, E., McKeegan, K. D., and Walker, R. M. 1983, *Nature* **305**, 119-121.

H₂ Emission from External Galaxies

Kimiaki KAWARA
National Astronomical Observatory
Mitaka, Tokyo 181, Japan

ABSTRACT. 2 μm spectroscopic observations by many authors have revealed significant rotation-vibrational H_2 emission is widespread from starburst to bare nucleus galaxies. Near-IR H_2 emission lines can arise from various excitation sources: UV radiation by hot stars, shock excitation by supernova remnants or AGN driven winds, and UV/X-ray radiation by an AGN. In this review recent data will be compared with such H_2 excitation models.

1. H_2 line ratio; Is H_2 emission is thermal?

Rotation/vibrational H_2 emission around 2μm has been observed in all types of emission line galaxies from HII region galaxies to bare nucleus objects (e.g., Kawara, Nishida, and Gregory 1987 & 1990; Fisher et al. 1987; Moorwood and Oliva 1988). There are three plausible H_2 sources to excite H_2 gas; shocks, far-UV radiation and X-ray radiation. The shock excitation thermalizes the level population at temperatures of ~ 2000 K, resulting in the line ratio of 2-1S(1)/1-0S(1) ~ 0.1. In the H_2 excitation due to far-UV radiation (912 - 1108 Å), radiative fluorescent H_2 emission is produced in cold gas with density below ~ 10^4 cm^{-3} where relative line intensities are characterized by the non-thermal population, such as 2-1S(1)/1-0S(1) ~ 0.5, while thermal H_2 emission is produced in warm (T > 1000 K) dense gas heated by intense UV radiation (Sternberg and Dalgarno 1989). In the H_2 excitation due to X-ray radiation modeled by Lepp and McCray (1983), H_2 emission can also consist of the thermal and non-thermal components. Thermal H_2 emission is produced in warm clouds near a X-ray source, while non-thermal H_2 emission is produced in cool clouds by non-thermal electrons. The non-thermal component becomes significant and the ratio of the thermal to non-thermal component is 0.25 in the 1-0S(1) line for clouds with column densities of 10^{24} cm^{-3} located near the X-ray source.

In the sample galaxies exhibiting Seyfert and/or starburst activity taken by Moorwood and Oliva (1990), the data [2-1S(1)/1-0S(1) vs 1-0S(0)/1-0S(1)] falls between the Orion cloud (pure shock excitation) and the prediction by the pure UV fluorescent excitation (or pure non-thermal excitation). This suggests that extragalactic H_2 emission is a mixture of thermal and non-thermal components.

Another example is the spectrum of NGC 6240 taken by Lester, Harvey, and Carr (1988), although this remarkable interacting/merging galaxy is not a typical example. Analyzing this spectrum, Tanaka, Hasegawa, and Gatley (1991) demonstrated that H_2 emission in NGC 6240 consists of the thermal and non-thermal components; 70% of 1-0S(1) originates from gas thermalized at 1600 K, and 70% of the total H_2 luminosity is non-thermal. They suggested UV radiation by B stars as the major source for the excitation. Drain and Wood (1990) also decomposed the H_2 emission lines, and found the similar contribution from the thermal and non-thermal contribution. They explained the results in terms of X-ray radiation as the source of the thermal component and of UV radiation as that of the non-thermal component. It should be noted that 1-0S(1)/3.28μm is 10 times greater in NGC 6240 than in other starburst galaxies (Mouri et al. 1990a). This suggests that the observed 1-0S(1) intensity has an excess of 10 over the predicted value from the UV radiation in NGC 6240, since the 3.28μm dust emission is powered by the similar UV photons as used for H_2 emission.

2. Comparison with other emission lines

The 1-0S(1) line luminosity relative to the far-infrared luminosity is enhanced in AGNs than in starburst galaxies (Kawara, Nishida, and Gregory 1987 & 1990). Finding the linear correlation between the 1-0S(1)/Brγ and [OI]6300/Hα, Mouri et al. (1989) demonstrated that 1-0S(1)/Brγ is a powerful tool to distinguish between AGN and starburst galaxies as in the case of [OI]6300; 1-0S(1)/Brγ is clearly greater in AGNs than in starburst galaxies. The linear correlation suggests H_2- and [OI]-emitting regions are closely related to each other. The ionization potential of O^0 (13.62 eV) exactly matches that of H^0 (13.60 eV), and so [OI]-emitting regions must be located outside fully ionized regions. It is generally considered the [OI] emitting region is due to electron collisions in the partly ionized regions. In the case of AGNs, extended partly ionized regions are formed by intense X-ray radiation from the central source. In starburst galaxies, the [OI] emission comes from the narrow transition zone at the boundary of the Stromgen sphere (e.g., Veilleux and Osterbrock 1987) or from shock-heated gas (Cambell 1988) probably associated with supernova remnants. This linear correlation thus favors X-ray excitation in AGNs and shock excitation in starburst galaxies. However, Mouri and Taniguchi (1991) recently found that 1-0S(1)/Brγ of starburst galaxies can also be explained by UV florescent models only if [OI]6300 is powered by supernova remnants.

Comparison with [FeII]1.644μm is also of great interest, because [FeII]/Brγ is greater in AGNs than in starburst galaxies, being linearly correlated with 1-0S(1) (Kawara, Nishida, and Taniguchi 1988; Moorwood and Oliva 1988; Mouri et al. 1990b). In starburst galaxies, the [FeII]1.644μm emission is probably powered by shocks associated with supernova remnants (Joseph et al. 1987). Recently, the [FeII]1.644μm and 1-0S(1) lines were detected by Graham, Wright, and Longmore (1990) in the Crab nebula where a UV-X-ray power-low continuum excites line emission. They concluded, "Given the many similarity between the NLR (narrow line region) in NGC 4151 and the Crab Nebula filaments, ... the IR [FeII] 1.644μm emission from NGC 4151 can be entirely accounted for by the standard NLR model. There is no evidence for any excess [FeII] 1.644μm emission and no need to invoke shocks, either from supernova remnants or from galactic winds."

Hence if a single excitation mechanism controls [OI]6300, [FeII]1.644, and 1-0 S(1), H_2 emission in AGNs results from X-ray radiation and that in starburst galaxies is excited through shocks by supernova remnants. However, it is more likely that UV radiation by young stars and shocks by supernova remnants share H_2 emission in starburst galaxies because the H_2 emission appears to be a combination of the thermal and non-thermal components.

3 H_2 excitation source in AGNs

Detecting broad emission lines in polarized light from NGC 1068 (type 2 Seyfert), Antonucci and Miller (1985) hypothesized that the molecular torus completely hides the inner BLR (broad line region) from our view. The classification of Seyfert types is determined by the angle of the line of sight relative to the torus: type 2 Seyferts for the edge-on view and type 1 Seyferts for the face-on view. Recent X-ray observations by the Ginga satellite are finding X-ray emission from type 2 Seyferts and LINERs (e.g., Koyama 1989). It is natural to consider that all AGNs are X-ray sources which are surrounded by molecular clouds. In fact, the H_2 1-0S(1) was detected in NGC 4151 (Fisher et al. 1987) and NGC 3783 (Kawara, Nishida and Gregory 1989) which are classified into bare nucleus objects that has no excess emission in the far-infrared and no reddening in the optical. In the torus model developed by Krolik and Lepp (1989), dense clouds located near the central source (torus's inner edge is ~ 1 pc from the central source) are heated up to ~ 1,000 K by hard X-ray radiation. Hence, H_2 spacial distribution of type 1 Seyferts would have a point source at the center surrounded by a diffuse extended envelope which is caused by the heating of circumnuclear clouds. In type 2 Seyferts, we would not see a bouble-peaked H_2 emission because the inner part of the torus is obscured by outer clouds. Although there is a correlation between 1-OS(1) and X-ray (2-10Kev) intensities (Kawara, Nishida, and Gregory 1990), it is not clear that the observed H_2 emission in type 1 Seyferts is dominated by the X-ray powered H_2 emission, because the H_2 luminosity of the type 2 Seyfert NGC 1068 is comparable to those of type 1 Seyferts.

Rotaciuc et al. (1991) published the extended H_2 emission in a type 2 Seyfert NGC 1068. The H_2 emission is more extended than the Bry region. The H_2 emission is minimized at the compact 2 μm continuum source with two unequal emission peaks ~ 1.3" (100 pc) on either side. The lowest contours extended about 4" (300 pc) from the compact continuum source. We should keep in mind that 2-1S(1)/1-0S(1) is $\leq$ 0.14 in a 6"x6" slit, thus the H_2 emission is dominated by the thermal component (Moorwood and Oliva 1990). The H_2 emission appears to be a torus-like structure at the center of which the 5GHz jet start to outflow 30° away from the pole of the molecular torus. The radio jet appears to be interacting with the molecular torus just above the major H_2 peak. If this is the case, the gas outflow from the center would be responsible for the extended H_2 emission. Observing [NII]6548, 6583 and using a mass outflow rate of ~ 0.4 M☉ yr^{-1} predicted by Krolik and Begelman 1986, Cecil, Bland, and Tully (1991) estimate ~ 10^{44}

ergs s^{-1} for the wind kinetic luminosity, which is much greater than 3×10^{41} ergs s^{-1} that is required to be dissipated to produce the observed H_2 1-0 S(1) luminosity (Rotaciuc 1991). The UV to X-ray continuum is another obvious source for the H_2 excitation. In fact, Rotaciuc et al. (1991) discussed that the observed surface brightness of the H_2 emission is consistent with that in the photodissociation model with gas densities $\geq 5 \times 10^5$ cm^{-3} (Sternberg and Dalgarno 1989), in which case the level population is thermalized. The X-ray luminosity (2-10KeV) of NGC 1068 is possibly $\sim 10^{44}$ ergs s^{-1} (Koyama 1989), which can produce the 1-0S(1) H_2 luminosity of $\sim 10^{41}$ ergs s^{-1} if the X-ray source is spherically surrounded by molecular clouds (Lepp and McCray 1983). This value is also consistent with the observed 1-0S(1) luminosity 7×10^{39} ergs s^{-1}. If this is the case and if the column density of clouds is 10^{23} cm^{-2}, the non-thermal component contributes only 10% of the observed 1-0S(1) luminosity, resulting in 2-1S(1)/1-0S(1) = 0.06 (Lepp and McCray 1983).

References

Antonucci, R.R.J. and Miller, J.S. 1985, Ap.J., **297**, 621.
Cambell, A. 1988, Ap.J., **335**, 644.
Cecil, G., Bland, J., and Tully, R.B. 1990, Ap.J., **355**, 70.
Drain, B.T., and Woods, D.T. 1990, Ap.J., **363**, 464.
Fisher, J., Geballe, T.R., Smith, H.A., Simon, M., and Storey, J.W.V. 1987, Ap.J.,**320**, 667.
Graham, J.R., Wright, G.S., and Longmore, A.J. 352, 172.
Joseph, R.D., Wright, G.S., Wade, R., Graham, J.R., Gatley, I.,
    and Preswich, A.H. 1987, in Star Formation in Galaxies, ed. C.J. Lonsdale (Washington, D.C.: US Government Printing Office), p421.
Kawara, K., Nishida, M., and Gregory, B. 1987, Ap.J. (Letters) **321**, L35.
Kawara, K., Nishida, M., and Gregory, B. 1989, Ap.J. (Letters) **342**, L55.
Kawara, K., Nishida, M., and Gregory, B. 1990, Ap.J. **352**, 433.
Kawara, K., Nishida, M., and Taniguchi, Y. 1988, Ap.J. (Letters) **328**, L41.
Koyama, K. 1989, in IAU symposium 134, Active Galactic Nuclei, ed. D.E. Osterbrock and J.S. Miller (Dordrecht: Kluwer), p. 167.
Krolik, J.H. and Begelman, M.C. 1986, Ap.J. (Letters) **308**, L55.
Krolik, J.H. and Lepp, S. 1989, Ap.J. **347**, 179.
Lester, D.F., Harvey, P.M., and Carr, J. 1988, Ap.J., **329**, 641.
Lepp, S., and McCray, R. 1983, Ap.J. **269**, 560.
Moorwood, A.F.M, and Oliva, E. 1988, Astr. Ap., **203**, 278.
Moorwood, A.F.M, and Oliva, E. 1990, Astr. Ap., **239**, 78.
Mouri, H., Taniguchi, Y., Kawara, K., and Nishida, M. 1989, Ap.J. (Letters) **346**, L73.
Mouri, H., Kawara, K., Taniguchi, Y., and Nishida, M. 1990a, Ap.J. (Letters) **356**, L39.
Mouri, H., Nishida, M., Taniguchi, Y., and Kawara, K. 1990b, Ap.J. **360**, 55.
Mouri, H, and Taniguchi,Y. 1991, Ap.J. Private communication.
Rotacius, V., Krabbe, A., Cameron, M., Drapatz, S., Genzel, R., Sternberg, A., and Storey, J.W.V. 1991, Ap.J. (Letters) **370**, L23.
Sternberg, A., and Dalgarno, A. 1989, Ap.J., **338**, 197.
Tanaka, M., Hasegawa, T., and Gatley, I. 1991, Ap.J., **374**, 516.
Veilleux, S., and Osterbrock, D.E. 1987, Ap.J. Suppl. **63**, 295.

QUESTIONS AND ANSWERS

C.Henkel: What additional insight in the pumping mechanism can be gained by measuring higher excited H_2 lines?

K.Kawara: I am not a right person to answer this question. What I can say is that in the UV-pumping, for example, pumped levels cascade down to lower level, resulting in more chances to emit higher level emission than in the thermal process. So, the higher level lines can provide a diagnostics to determine the dominant excitation mechanism.

A.Sternberg: It is unlikely that the H_2 emission in Seyfert 1's are produced in an observing torus of 1 pc size since the luminosities of Sy 1's and Sy 2's are comparable and in Sy 2's the H_2 emission is extended over $\sim$ 100 pc scales.

K.Kawara: I agree with your argument. However, it seems to me that the torus model is one of plausible models to explain the double-peaked H_2 map of NGC 1068 and the more enhanced H_2 emission in Sy 1's than in Sy 2's. Otherwise, we have to figure out an alternative cloud configuration which can minimize the H_2 emission at the galactic center of NGC 1068. The best estimate of the x-ray luminosity (2 - 10 keV) of NGC 1068 is $\sim 10^{44} erg/s$ (Koyama,1989: IAU Symp.,#134) based on the hard-x-ray observation, which can produce $\sim 10^{41} erg/s$ for the S(1) luminosity. This value is 10 times larger than the observed S(1)' luminosity of $\sim 10^{40} erg/s$. If you look at NGC 1068 on face-on-view, you may observe more H_2 emission?

INTERSTELLAR POLARIZATION IN THE SMALL MAGELLANIC CLOUD

C. RODRIGUES[1], A.M. MAGALHÃES[1,2], G. COYNE[3], V. PIIROLA[4]
[1] *Inst. Astron. e Geof., Univ. São Paulo - BRAZIL*
[2] *Space Astronomy Lab., Univ. Wisconsin - USA*
[3] *Vatican Obs. Research Group, Univ. Arizona - USA*
[4] *Obs. and Astrophysics Lab., Univ. Helsinki - FINLAND*

ABSTRACT. Data of our on-going program on the interstellar polarization in the Small Magellanic Cloud (SMC) seem to show a correlation between the UV extinction and the parameters of the Serkowski law which describe the polarization. Fits of popular interstellar dust models to the polarization data are not always satisfactory, being better for stars with small wavelength of maximum polarization, λ_{max}.

Introduction

Studies of dust in the SMC, mostly based on extinction data, show in the infrared and visible no strong differences to the galactic law and indicate an R = $A_V/E(B-V)$ = 2.7 ± 0.2(Bouchet *et al*, 1985), slightly smaller than the galactic value. The UV extinction is generally characterized by the absence of the 2175Å bump and shows an extremely steep FUV rise (Prévot *et al*, 1984). However, one object, AzV456, possesses an extinction similar to the typical galactic law and also its gas-to-dust ratio is typical of the Milky Way(Lequeux *et al*, 1984). Our aim is to use optical polarization in order to further constrain grain models.

Results

Magalhães *et al* (1989) have presented the first wavelength dependent polarization data for the SMC. From that sample, we consider here only the stars whose polarization is large compared to the foreground (galactic) polarization correction. They are AzV211, AzV215, AzV221, AzV398 and AzV456.

Fits to the Serkowski law show that these stars can be divided in two groups. AzV211, AzV221 and AzV398 show maximum polarization at wavelengths bluer than the galactic average, while AzV215 and AzV456 possess λ_{max} in the V filter (Table 1). Also, the growth of K (i.e, the progressive narrowing of the curve) with λ_{max} seems to be faster in the SMC than in the Galaxy. AzV456 presents, as mentioned, a Galactic extinction law while AzV398 shows a 'typical' SMC law.

Preliminary IUE results for AzV211 (Magalhães et al, 1991) indicate that it does not show the bump as well. Therefore, it seems that the distinct SMC extinctions are related to distinct polarization curves as well.

Conclusion

The SMC polarization data show typically small λ_{max} values. The absence (or presence) of the 2175Å extinction bump in the SMC may be correlated with smaller (or larger) λ_{max} values found in our polarization data. It would also seem that the dust models able to reproduce the Galactic interstellar polarization may face difficulties in reproducing some of the polarization curve in the SMC.

Table 1: Fitted Parameters of the Serkowski Law

Star	$\lambda_{max}(\mu)$	K	$P_{max}(\%)$	UV bump
211	0.37 ± 0.22	0.48 ± 0.67	0.956 ± 0.089	no
215	0.537 ± 0.047	2.5 ± 1.9	0.706 ± 0.071	
221	0.42 ± 0.12	1.2 ± 1.3	0.892 ± 0.073	
398	0.459 ± 0.039	1.26 ± 0.55	1.870 ± 0.067	no
456	0.572 ± 0.017	2.06 ± 0.53	1.189 ± 0.051	yes

References

- Bouchet, P., Lequeux, J., Maurice, E., Prévot, L. and Prévot-Burnichon, M.L. 1985, A&A, **149**, 330.
- Lequeux, J., Maurice, E., Prévot, L., Prévot-Burnichon, M.L., Rocca-Volmerange B. 1982, A&A, **113**, L5.
- Magalhães, A.M., Piirola, V., Coyne, G.V. and Rodrigues, C.V. 1989, in *Interstellar Dust Contributed Papers*, eds. A.G.G.M. Tielens and L.J. Allamandola, NASA CP-3036, 347
- Magalhães, A.M., Coyne, G.V., Piirola, V. and Rodrigues, C.V 1991, in preparation.
- Prévot, M.L., Lequeux, J., Maurice, E., Prévot, L. and Rocca-Volmerange, B. 1984, A&A, **132**, 389.

Molecular abundances in extragalactic sources

C. HENKEL and R. MAUERSBERGER

Max-Planck-Institut für Radioastronomie, Auf dem Hügel 69, D-5300 Bonn 1, F.R. Germany

August 30, 1991

Abstract. Chemical properties of extragalactic molecular clouds are summarized. Column densities, relative abundances, and chemical implications are briefly discussed and references for more detailed studies are given. The need for model calculations is emphasized.

Key words: galaxies - molecules - abundances

1. Introduction

Until recently, molecular spectroscopy of extragalactic sources was confined to the observation of a few molecular species only. Since then, however, the situation has dramatically improved and 21 molecules, aside of CO and H_2, have been detected to date. Previous articles summarizing these achievements are given by Henkel and Mauersberger (1991) and Henkel et al. (1991; see e.g. their Table 1). Here, we present a brief overview emphasizing measured column densities and abundances of the 21 species to provide an appropriate input for chemical models.

Most of the molecules are observed at mm-wavelengths and exhibit quasithermal emission. Their electric dipole moment is usually much larger than that of the commonly observed CO molecule, thus allowing the study of regions with typical densities $n(H_2) \geq 10^4 \, cm^{-3}$. One may distinguish four classes of sources which have been measured: (1) irregulars of the Local Group, (2) nearby spirals or interacting irregulars of nearby groups of galaxies, (3) a nearby elliptical, and (4) distant mergers at the upper end of the far infrared luminosity range of galaxies.

2. Results and discussion

2.1. THE LMC AND SMC

Johansson (1991) report the detection of CN, CS, SO, HCN, HCO^+, C_2H, and H_2CO toward the HII region N 159 of the LMC. CS and HCO^+ are also observed in the SMC, toward N 19. A very preliminary analysis (assuming optically thin line emission, excitation temperatures between 5 and 50 K, and a spatial extent of the emission relative to the CO cloud size similar to that in Galactic sources) suggests that abundances relative to H_2 may be similar to those in 'local' molecular clouds. The exception is SO, which may have a very large relative abundance ($X(SO) \sim 3 \, 10^{-4} ... 10^{-2}$), but additional data are required to support the estimate.

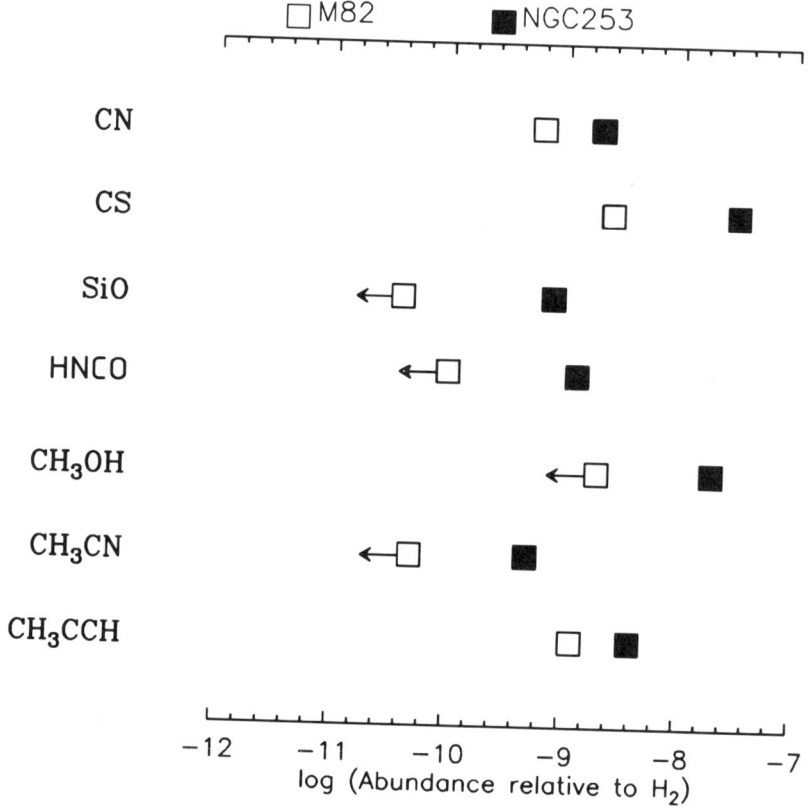

Fig. 1. Relative abundances for NGC 253 and M 82. Adopted H_2 column densities are $10^{23.2}$ and $10^{23.6}$ cm^{-2}, respectively. For HNCO in the SW-hotspot of M 82 we assumed an excitation temperature of 10 K, similar to that found in IC 342 and Maffei 2 (see Nguyen-Q-Rieu et al. 1991). For the other molecules, compare with Mauersberger and Henkel (1991; CN has been corrected).

2.2. NGC 253 versus M 82

Among the nearby spirals and interacting irregulars the outstanding sources are NGC 253, Maffei 2, IC 342, M 82, and NGC 4945. The dense gas, emitting a large variety of molecular lines, is confined to the inner few 100 pc. NGC 253 and M 82 are the sources observed in greatest detail (e.g. Henkel and Mauersberger 1991; Henkel et al. 1991; Mauersberger and Henkel 1991; Mauersberger et al. 1991). Both have approximately the same distance. While NGC 253 contains a bar or ring (or both) with an angular distance between the two major hotspots of 9″ (∼ 150 pc), the main lobes of the rotating ring in M 82 are separated by ∼30″. Because the source size of M 82 is larger than most telescope beams, Fig. 1 compares relative abundances between NGC 253 and the south-western lobe of M 82. Note that the H_2 col-

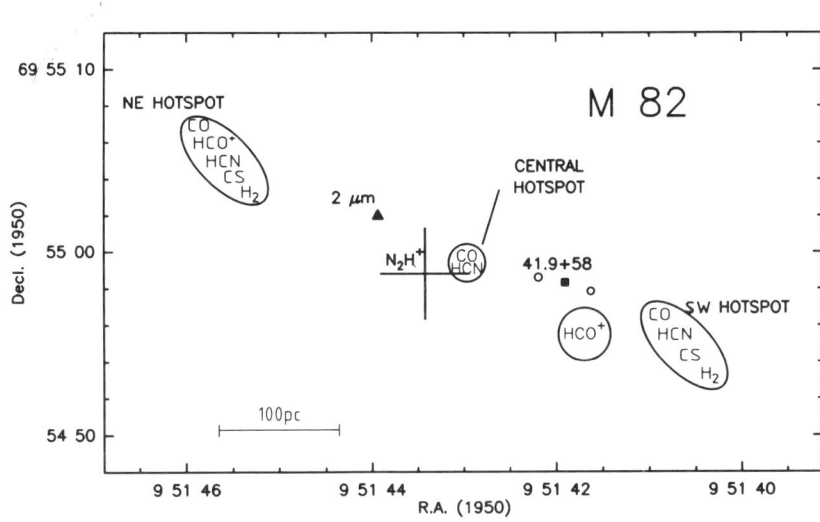

Fig. 2. A finding chart for molecular and continuum features in the nuclear region of M 82 (Mauersberger and Henkel 1991). Besides the molecular hotspots discussed in the text, the filled triangle marks the 2μm peak close to the kinematical center. The filled square marks the strongest radio continuum source, 41.9+58, the open circles the location of the OH masers.

umn densities are poorly determined for such small regions and that relative abundances may be uncertain by at least an order of magnitude. Assuming, however, that the same H_2 column density can be applied for all molecular species within a single source, a *comparison of relative abundances* can be made with high accuracy. There are at least two groups of molecules: CN and CH_3C_2H have similar relative abundances. CS is less abundant by an order of magnitude in M 82. The other molecules in M 82 may be even more 'underabundant' relative to NGC 253. All species have been observed in more than one transition so that non-LTE effects in a single transition are not responsible for the values given. In M 82, the CO/HCN line temperature ratio is extremely large (Nguyen-Q-Rieu et al. 1989), so that we have reason to suspect that HCN is another molecule which is 'underabundant' in M 82. Possible explanations in terms of different kinetic temperatures, CH_3^+ abundances, and non-steady state chemistry are outlined by Henkel et al. (1991) and Mauersberger et al. (1991). None of these interpretations is, however, satisfactory. Each of them only addresses a few of the observed species and the difference in the HNCO abundances (see Nguyen-Q-Rieu et al. 1991) remains entirely unexplained.

2.3. SPATIAL STRUCTURE WITHIN NUCLEAR REGIONS

Besides studying differences between various galaxies, there are a few sources where we can trace spatial abundance variations within a single nucleus. A possible example are the HCN and HNC profiles toward the central region of NGC 4945 (see Henkel et al. 1990) which show different lineshapes and may arise from different nuclear rings. This interpretation is, however, based on indirect arguments and is not yet supported by data which directly resolve the line emission.

A much more impressive case is hence the nuclear region of M 82 (see Fig. 2). We distinguish the main lobes (NE- and SW-hotspot), the central hotspot, and the SW-HCO^+-hotspot which is located closer to the nucleus than the 'associated' main lobe. Neutral molecules peak at the NE- and SW-hotspots of the hypothesized rotating ring. Vibrationally excited H_2 may arise from the inner molecular shell directly exposed to the active nuclear region (Lester et al. 1990). The density of the NE-hotspot, $n(H_2) \sim 10^5$ cm^{-3} (Mauersberger and Henkel 1989), is large enough to cause a low N_2H^+ abundance. In the SW-hotspot where densities are $> 10^5$ cm^{-3} (Baan et al. 1990), both N_2H^+ and HCO^+ do not exhibit noticeable peaks (Carlstrom 1988; Mauersberger and Henkel 1991). The central hotspot, however, exhibits most of the N_2H^+ emission so that by analogy with Galactic clouds, the density should be at order 10^4 cm^{-3} (Mauersberger and Henkel 1991). The $HCO^+(1-0)/HCN(1-0)$ or $HCO^+(1-0)/CO(1-0)$ intensity ratios drop considerably from the inner to the outer edges of the molecular lobes (Carlstrom 1988). One may speculate that this is due to shocks or a high cosmic ray flux at the inner edges of the molecular structure. The HCO^+ clump between the central and the SW-hotspot might represent dense molecular gas entrained by the the outflowing plasma from the nucleus.

2.4. CEN A

Toward this 'active' elliptical, H_2, CO, OH, HCN, HNC, HCO^+, NH_3, H_2CO, and C_3H_2 have been detected (e.g. Bell and Seaquist 1988; Seaquist and Bell 1990; Eckart et al. 1990; Israel et al. 1990, 1991). There are at least four major components: (1) emission from the optically visible dust lane, only seen in CO; (2) broad emission from the circumstellar gas observed in CO and C_3H_2; (3) narrow absorption lines at the systemic velocity, ~ 550 km s^{-1}, present in all detected molecular species except H_2, and (4) narrow absorption lines at ~ 600 km s^{-1} observed in CO, HCO^+, OH, and perhaps H_2CO. The absorption lines sample gas along the line-of-sight toward the nuclear continuum source. The column densities are small (Seaquist and Bell 1990) so that either the clouds are small or the line-of-sight is intersecting the edges of the clouds only. A comparison of the column densities derived from the $\sim 550\,\mathrm{km\,s^{-1}}$ data (see Table 2 of Seaquist and Bell 1990) indicates 'normal' abundances when compared with Galactic dark cloud material. Note that the NH_3 column density is as low as expected when only observing the edge of a clump. The absorption component at $\sim 606\,\mathrm{km\,s^{-1}}$ may represent gas falling toward the nucleus and arises from warm gas (Israel et al. 1991). Its HCO^+/CO optical depth ratio appears to be large, indicating either enhanced cosmic ray ionization or shock induced processes. This component and the SW-HCO^+-hotspot in M 82 (Fig. 2) might be chemically related.

2.5. Megamaser galaxies

It is also possible to estimate OH column densities in megamaser galaxies, where the 18 cm main lines amplify the nuclear radio continuum. Assuming that the main lines have an optical depth ratio of 1.8:1 (this is the LTE intensity ratio in the optically thin case) and that excitation temperatures and continuum source covering factors are equal, the relative intensity of the 1665 and 1667 MHz lines allows an estimate of the optical depth (Henkel and Wilson 1990; Henkel et al. 1991). In the likely case that the absolute values of of excitation temperatures are similar to the rotation temperature, $\sim 60\,\mathrm{K}$ (see Henkel et al. 1987), we obtain, irrespective of the detailed pumping mechanism, OH column densities of order $10^{18...19}\,\mathrm{cm^{-2}}$; $X(OH)\sim 10^{-6}$. These large values might be caused by shocks, which should be common near the nuclei of merging systems, giving rise to megamaser emission.

The radiative 2 cm pumping mechanism suggested by Baan et al. (1986) to explain the H_2CO maser in Arp 220 requires $X(H_2CO)\sim 10^{-8\pm 1}$. This relative abundance is 'normal' by Galactic standards.

3. Outlook

We have shown that there exists already a large amount of data relevant for an analysis of chemical processes in extragalactic sources. While further drastic observational improvements are expected, it is remarkable that chemical models for the sources discussed above do not exist. It is yet too early for a thorough analysis of the Magellanic clouds. However, our knowledge of the nuclear regions of the other galaxies (see Sects. 2.2...2.5) is good enough to justify model calculations involving shock chemistry and/or a greatly enhanced flux of cosmic ray particles.

References

Baan. W.A.. Güsten, R., Haschick, A.D. 1986, *Ap. J.* **305**, 830
Baan, W.A., Henkel, C., Schilke, P., Mauersberger, R., Güsten, R. 1990, *Ap. J.* **353**, 132
Bell, M.B., Seaquist, E.R. 1988, *Ap. J.* **329**, L17
Carlstrom, J.E. 1988, in 'Galactic and Extragalactic Star Formation', eds. R.E. Pudritz, M. Fich, Kluwer, p571

Eckart, A., Cameron, H., Rothermel, H., Wild, W., Zinnecker, H., Rydbeck, G., Olberg, M., Wiklind, T. 1990, *Ap. J.* **363**, 451
Henkel, C., Baan, W.A., Mauersberger, R. 1991, *Astr. Ap. Rev.* **3**, 47
Henkel, C., Güsten, R., Baan, W.A. 1987, *Astr. Ap.* **185**, 14
Henkel, C., Mauersberger, R. 1991, in 'Dynamics of Galaxies and Their Molecular Cloud Distributions', eds. F. Combes and F. Casoli, Kluwer, *IAU Symp.* **146**, p195
Henkel, C., Whiteoak, J.B., L.-Å. Nyman, J. Harju 1990, *Astr. Ap.* **230**, L5
Henkel, C., Wilson, T.L. 1990, *Astr. Ap.* **229**, 431
Israel, F.P., van Dishoeck, E.F., Baas, F., Koorneef, J., Black, J.H., de Graauw, T. 1990, *Astr. Ap.* **227**, 342
Israel, F.P., van Dishoeck, E.F., Baas, F., de Graauw, T., Phillips, T.G. 1991, *Astr. Ap.* **245**, L13
Johansson, L.E.B. 1991, in 'Dynamics of Galaxies and Their Molecular Cloud Distributions', eds. F. Combes and F. Casoli, Kluwer, *IAU Symp.* **146**, p1
Mauersberger, R., Henkel, C. 1989, *Astr. Ap.* **223**, 79
Mauersberger, R., Henkel, C. 1991, *Astr. Ap.* **245**, 457
Mauersberger, R., Henkel, C., Walmsley, C.M., Sage, L.J., Wiklind, T. 1991, *Astr. Ap.* **247**, 307
Nguyen-Q-Rieu, Henkel, C., Jackson, J.M., Mauersberger. R. 1991, *Astr. Ap.* **241**, L33
Nguyen-Q-Rieu, Nakai, N., Jackson, J.M. 1989, *Astr. Ap.* **220**, 57
Lester, D.F., Carr, J.S., Joy, M., Gaffney, N. 1990, *Ap. J.* **352**, 544
Seaquist, E.R., Bell, M.B. 1990, *Ap. J.* **364**, 94

QUESTIONS AND ANSWERS

M.Guelin: In view of the small number of galaxies observed (and the way they were selected e.g. NGC 253 and M 82), I think it is preliminary to speak of "underabundances" of species such as SiO and HNCO. A pair of excitation problems, these species are detected only in few clouds in our own Galaxy; their non-detection in an external galaxy seems thus "normal".

C.Henkel: Since the extragalactic H_2 column densities may be uncertain and since it is difficult to detect some of the 'rare' species in galactic sources, it would indeed be preliminary to speak of 'underabundances' relative to galactic clouds. The term was used here exclusively in the context of comparing abundances between NGC 253 and M 82.

R.Opher: If we would make the exact same analysis that you made for the 'other' molecules, to CO abundances, wouldn't we obtain an error in the Hydrogen abundance of more than 30%?

C.Henkel: The H_2 column densities are based on the integrated intensity of the CO $J = 1 - 0$ line, using a standard conversion ratio derived from the galactic disk. This analysis differs from that used for the 'other molecules'. However, even if the applied conversion ratio is correct and does not vary, say with galactic radius, uncertainties may be >30%. This is perhaps most obvious in the case of Arp 220, where the CO based H_2 column density refers to a region 300 pc in diameter, while the OH maser arise from ~ 1 pc sized hotspots.

A 158μm [CII] MAP OF NGC 6946: DETECTION IN EXTRAGALACTIC
ATOMIC AND IONIZED GAS

S. C. Madden, N. Geis, R. Genzel, F. Herrmann, A. Poglitsch
Max-Planck Inst. für extraterrestrische Physik, Garching, FRG

J. Jackson
Dept. of Astronomy, Boston U., Boston, MA

G. J. Stacey and C. H. Townes
Dept. of Physics, U. C. Berkeley, Berkeley, CA

1. Introduction

The first observations of the [CII] line toward the nuclei of gas-rich external galaxies, showed that the far-infrared line emission contributes up to 1% of the total luminosity and most likely originates from dense photon-dominated regions (PDRs) associated with the surfaces of molecular clouds exposed to FUV from external or embedded OB stars (Crawford et al. 1985, Lugten et al. 1986, Stacey et al. 1991). We have mapped the [CII] emission toward NGC 6946 over an 8'x 6' (23 x 17 kpc) (Madden et al. 1991) using the Max-Planck Instutute/U.C.Berkeley Far-Infrared Imaging Fabry-Perot Interferometer (FIFI) on the Kuiper Airborne Observatory (KAO).

2. Origin of the [CII] Emission

Figure 1 suggests a close resemblance between the morphology of the nuclear and spiral arm [CII] components and that of the optical continuum/Hα that traces the current massive star formation activity. Three spatial regions can be distinguished in the [CII] map of NGC 6946: 1) The nucleus associated with the peak intensity (7×10^{-5} erg $s^{-1} cm^{-2} sr^{-1}$; 2) a component tracing the spiral arms, and 3) an extended component existing out to 11 kpc which contains most of the [CII] luminosity. We consider possible origins of the [CII] emisssion in the 3 spatial components of NGC 6946 from neutral atomic gas, extended low density gas (ELD) and PDRs by comparison of the various tracers [CII] (Madden et al. 1991), FIR (Engargiola 1991), CO (Tacconi and Young 1989), HI (Tacconi and Young 1986), and 2.8 cm (Klein et al. 1982).

2.1 Extended Component. Since the [CII] in the extended component (~3.5'west of the nucleus) is associated with a relatively large mass (10.7×10^7 M☉), and the HI emission is prominent even well beyond the spiral arms, unlike the distribution of the molecular gas, we estimate the contribution of the [CII] from the diffuse atomic hydrogen ISM.

Assuming standard atomic clouds with $n(H) \sim 30$ cm^{-3} and kinetic temperature of 100 K (Kulkarni and Heiles 1988), we estimate the [CII] intensity arising from the observed 21cm HI gas to be $\sim 40\%$ of our observed [CII] emission.

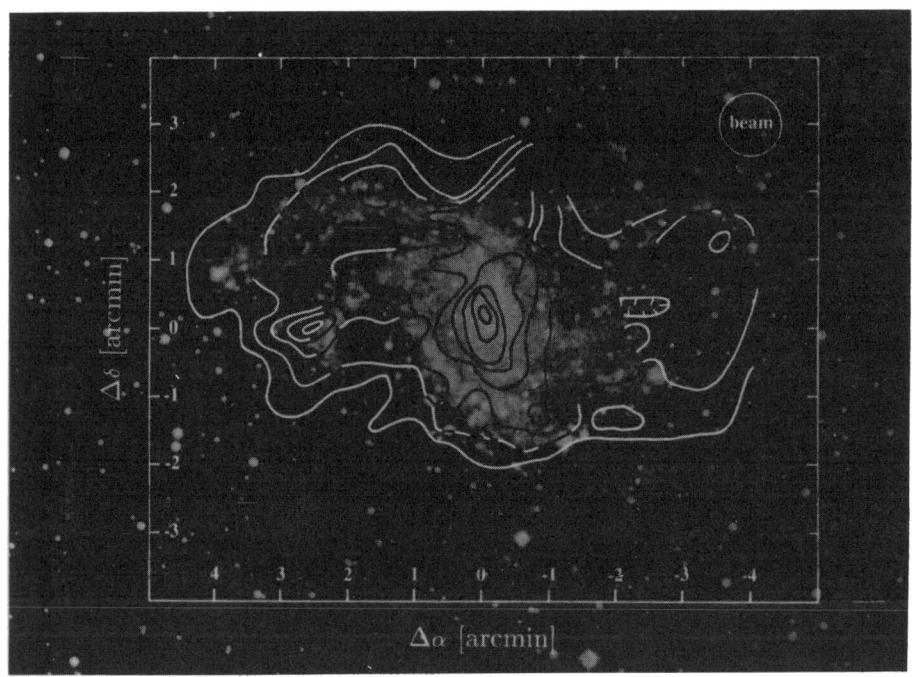

Figure 1. *Integrated [CII] line intensity contours superposed on an optical image of NGC 6946. The contour interval is 1×10^{-5} erg s^{-1} cm^{-2} sr^{-1} and the peak value is 7×10^{-5} erg s^{-1} cm^{-2} sr^{-1}. The beam size is 55". The center position is: RA(1950) $20^h 33^m 48.8^s$, DEC(1950) $+59^\circ 58' 50$.*

[CII] emission may also arise from the fully ionized interstellar medium that has been modeled as extended low density ($n(e) \sim 3$ cm^{-3}) (ELD) HII regions with electron temperatures $\sim 10^4$K (Mezger 1978). We derive an expected [CII] line intensity from the ionized gas to be $\sim 34(\pm 11\%)$ of our measured [CII] value in the extended component. This supports measurements of the [CII] in the ELD in the Galaxy (see references in Madden et al. 1991).

If we assume that [CII] and CO line emission as well as the far-infrared continuum all originate at the surfaces of the UV exposed molecular clouds, then PDR models can be used to derive the physical conditions of the emitting cloud. After subtracting the contribution of the [CII] emission from the ELD HII regions and the HI medium, we determine a PDR solution to the remaining CII emission (5×10^{-6} erg

$s^{-1} cm^{-2} sr^{-1}$). The regions require UV field intensities, $\chi(UV)$, of 700 times $\chi(o)$, the UV field in the solar neighborhood (2.0×10^{-4} erg $s^{-1} cm^{-2} sr^{-1}$), very dense hydrogen densities ($\sim 7 \times 10^5$ cm^{-3}) and a FIR filling factor, $\Phi(FIR)$, of $\sim 1\%$.

2.2 Spiral Arms. We calculate the contribution of the diffuse HI gas to be $\sim 10\%$ of our total [CII] emission in the prominent spiral arm 2.5' east of the nucleus. The ELD HII regions can contribute $\sim 20\%$ of the total [CII] emission observed toward this spiral arm region assuming the standard conditions for the HI clouds and the ionized gas described above.

The remaining observed [CII] emission in the spiral arm after the ELD and HI contributions are removed, is 3.5×10^{-5} erg $s^{-1} cm^{-2} sr^{-1}$ giving a PDR solution of $n(H) \sim 600$ cm^{-3} with $\chi(UV)/\chi(o) \sim 120$. The filling factor for the FIR sources is then 20% ($\sim 25"$ sources in our 55" beam), consistent with high resolution Hα and CO data (Bonnarel *et al.* 1986; Weliachew *et al.* 1987, Ball *et al.* 1989).

2.3. Nucleus. The [CII], FIR and CO all peak in the nucleus, substantiating an interpretation of the origin of [CII] from PDRs. The PDR solution implies very dense (n (H)$\sim 8 \times 10^5$ cm^{-3}) molecular clouds exposed to UV fields of $500\chi(o)$. The beam area filling factor is 12% corresponding to 19" source sizes, in fairly good agreement with the diameter of the nuclear source when observed at high spatial resolution in Hα or CO.

3. References

Ball, R., Sargent, A. I., Scoville, N. Z., Lo, K. Y., and Scott, S. L. 1985, *Ap. J. (Letters)*, 298, L21.
Bonneral, F., Boulesteix, J. and Marcclin, M. 1986, *Astr. Ap. Suppl.*, 66, 149.
Crawford, M. K., Genzel, R., Townes, C. H., and Watson, D. M. 1985, *Ap. J.*, 291, 755.
Engargiola, G. 1991, *Ap. J. Suppl.*, 76, 875.
Klein, U., Beck, R., Buczilowski, U. R., and Wielebinski, R. 1982, *Astr. Ap.*, 108, 176.
Kulkarni, S. and Heiles, C. 1988, *Galactic and Extragalactic Radio Astronomy* (Springer-Verlag).
Lugten, J. B., Watson, D. M., Crawford, M. K., and Townes, C. H. 1986 *Ap. J.*, 306, 691.
Madden, S. C., Geis, N., Genzel, R., Herrrmann, F., Jackson, J., Poglitsch, A., Stacey, G. J., and Townes, C. H. 1991, in prep.
Mezger, P. G. 1978, *Astr. Ap.*, 70, 565.
Stacey, G. J., Geis, N., Genzel, R., Lugten, J. B., Poglitsch, A., Sternberg, A., Townes, C. H. 1991, *Ap. J.*, 373, 423.
Tacconi, L. J. and Young, J. S. 1986, *Ap. J.*, 308, 600.
Tacconi, L. J. and Young, J. S. 1989, *Ap. J.*, 71, 455.
Weliachew, L., Casoli, F., and Combes, F. 1988, *Astr. Ap.*, 199, 29.

Molecule Formation in External Galaxies

T J MILLAR

Department of Mathematics, UMIST, PO Box 88, Manchester M60 1QD, UK

and

E HERBST

Physics Department, Ohio State University, 174 W. 18th Ave, Columbus, Ohio 43210, USA

September 9, 1991

Abstract. We discuss the parameters needed to model chemistry in extragalactic clouds. While density and temperature can be constrained by multiline observations, molecular abundances may be severely affected by the adopted elemental abundances. While the observations of the Magelllanic Clouds can be reasonably interpreted in terms of dark cloud models, molecular gas in starburst galaxies could well be dominated by photoeffects. The detection of deuterium in extragalactic molecules would provide a valuable diagnostic.

1. Introduction

At the time of writing, August 1991, 23 molecules have been detected in external galaxies (see Henkel, this volume). Most molecules have been found in the starburst galaxies NGC 253 and M82, although a significant number have been detected in the metal-poor Large Magellanic Cloud (LMC). Derived molecular abundances from distant galaxies are beam averages over a very large area. From observations of molecular clouds in our own Galaxy, we know that structure exists on all accessible scales and that certain of these structures, for example hot molecular cores, have distinct chemical compositions and abundances from those typical of quiescent molecular clouds. For this reason, the few chemical models applied to external galaxies have been fairly simplistic, although they may be useful in delineating the main synthetic routes to certain species. On the other hand, models of the nearby Magellanic Clouds may be tested by observations with reasonable spatial resolution. Another advantage of studying the Magellanic Clouds is that we have a certain amount of information on stellar properties and on the dust particles which have been studied from ultraviolet to infrared wavelengths.

2. Chemical Parameters

In any calculation of molecular abundances, a number of parameters must be specified. We discuss these in turn.

Input densities are usually derived from multi-transition observational studies of polar species such as CS and HC_3N. Density variations are obvious. Henkel et al. (1991) have derived densities in the central region of M82 varying from $\sim 10^5$ cm^{-3} in the NE molecular 'hotspot' to $\sim 10^4$ cm^{-3} in the central hotspot, to $> 10^6$ cm^{-3} in the SW hotspot, where 2 cm H$_2$CO emission is observed. Similar variations in density have also been derived for NGC 253 (Mauersberger et al. 1990).

Temperatures are derived in several ways. Vibrationally excited H$_2$ which traces hot molecular gas at $\sim$ 1000-2000 K is observed in many galaxies, particularly in starbursts, and may trace the interaction of a flow from the nucleus with surrounding molecular clouds. The molecular observations give a variety of temperatures, typically 20-40 K, although HC$_3$N in NGC 253 indicates that the highest density gas, $\sim 10^5$ cm^{-3}, has a kinetic temperature greater than 60 K. The detection of large abundances of SiO and CH$_3$OH in NGC 253 (Mauersberger and Henkel, 1991) may indicate even higher temperatures since Galactic observations indicate that the abundances of these species are enhanced at high temperature, probably due to the evaporation of mantle material.

There is no reason to believe that the abundances of the elements C, N, O, etc. are similar in external galaxies. Indeed in the LMC and SMC, the abundances of C, N and O are known to be much less in HII regions than in Orion. For example in the SMC, carbon and nitrogen are reduced by factors of 16 and 14 respectively (see Bel et al., 1986; Barbuy et al., 1991). As we shall discuss below, molecular abundances are, in some cases, extremely sensitive to the adopted elemental abundances.

The dust-to-gas ratio is important because it is related to the dust cross-section which is directly related to the extinction caused by the dust and because the grain surface area affects the rate at which H$_2$ may form.

Molecular clouds can experience UV fluxes much different from those in the solar neighbourhood. The UV flux has a direct effect on photodissociation and photoionisation rates as well as on, through grain photoelectric emission and photoionisation, gas temperatures. The UV flux may vary dramatically on a small scale in external galaxies and may be very large in starbursts. Maloney and Black (1988) find that in the central 450 pc of M82, the mean UV intensity is at least 250 times the local value. This implies that photon-dominated regions (Sternberg, this volume) may play an important role in determining the chemistry of starburst galaxies.

Cosmic rays provide the source of ionisation which drives the chemistry in dark clouds as well as a low-level flux of UV photons. The resulting ionisation rate is poorly known in our own Galaxy (Lepp, this volume) and even more so in external galaxies. If cosmic rays originate in supernovae, the flux may be large in starbursts. X-ray radiation can also provide a source of ionisation in these galaxies.

3. Molecular Abundances in External Galaxies

3.1. CARBON MONOXIDE

Maloney and Black (1988) studied the $C^+/C/CO$ transition for model clouds with elemental abundances, dust-to-gas ratios and UV fields typical of those in the LMC and SMC. Because H_2 self-shields efficiently against photodissociation, they found that its distribution with depth into the cloud was relatively unaffected by changing the parameters. On the other hand, they found the CO distribution to differ dramatically because CO self-shields less efficiently than H_2. At the centre of a Galactic cloud, 99% of the carbon was in CO, for the LMC 5% and for the SMC 1%. This argues that for CO depth-dependent calculations need always to be carried out. For other molecules, this may not be so essential, at least in the cases of the LMC and SMC, since more complex molecules will probably form only in regions in which CO is abundant.

3.2. MOLECULES IN THE LMC DARK CLOUDS

Millar and Herbst (1990) calculated pseudo-time-dependent models for dark clouds in the LMC and SMC and discussed the variation of abundances with adopted parameters. Since then, Johansson (1991) has published molecular abundances derived from the SEST spectral scan of Cloud 2 associated with the HII region N159, just south of 30 Dor in the LMC. We have updated our earlier calculations to include sulphur chemistry and to include ion-dipolar rate coefficients. A comparison of molecular abundances, relative to CO, is given in Table I. Note that the uncertainties in the derived abundances are large (Johansson, 1991) while the time-dependent behaviour of the calculated abundances for some species, particularly SO, HCN and HNC, is severe. The calculations were performed for $n(H_2) = 10^4$ cm^{-3}, T = 10 K and O, C and N depleted by factors of 2.8, 4.8 and 12.5 below Galactic dark cloud values, respectively (Millar and Herbst, 1990). While one should not take the agreement between theory and observation too seriously, the agreement may indicate that such crude modelling is worthwhile. If so, this can be tested by observations of the SMC. Millar and Herbst (1990) show that certain molecular abundances should depend not only on the C/O ratio but also on the <u>difference</u> O-C. They find that in their SMC model, which has a carbon abundance only four percent of the Galactic value, the C_2H fractional abundance is similar to that in Galactic dark clouds, since the abundance of atomic oxygen, which destroys C_2H, is smallest in the SMC model.

Millar and Herbst (1990) found that nitrogen-bearing molecules have a complex behaviour in their various model calculations, in the main because N_2 and NH_3 formation depend on the abundance of OH, which itself depends on the amount of oxygen not tied up in CO, as well as on the atomic oxygen abundance.

Pineau des Forêts et al. (1991) have calculated molecular abundances in dark cloud models in which they systematically vary the depletions of carbon, nitrogen and oxygen, δ_{CNO}, from 1 - 1000 times less than solar. They found, as did Millar and Herbst (1990), that the CO abundance follows that of elemental carbon. In addition, the fractional abundances of species destroyed by atoms, thus principally radicals, tend to increase as depletion increases because of the decreasing abundances of atoms, so that at the largest depletions, the abundance of OH is larger than that of CO. The abundance of NH_3 is remarkably constant in these models with a value which varies by less than an order of magnitude and is close to 10^{-8} at both $\delta = 1$ and $\delta = 1000$. Thus, although NH_3 contains only $\sim 10^{-4}$ of the total nitrogen abundance for solar abundances it contains ten percent at $\delta_{CNO} = 1000$.

4. Conclusions

The chemical models presently developed for interpreting extragalactic observations are rather crude, as indeed is the spatial resolution available in most observations. Abundances in metal poor, dwarf galaxies such as the Magellanic Clouds, appear to be in reasonable agreement with the results of dark cloud models. No models have, as yet, been developed for starburst galaxies. The molecular gas in these galaxies appears to exist in warm, dense clouds - denser than those at the Galactic Centre - bathed by intense radiation fields. Although the molecules detected in NGC 253 and M82, with the exception of SiO in NGC 253, are also found in Galactic dark clouds, the evidence for a significant dark cloud chemistry has yet to be found. Photon-dominated regions may dominate the chemical evolution of these galaxies.

Finally it is worth stating that a prime observation would be the detection of deuterium in external galaxies. This would give not only important cosmological information but also help delineate the molecule formation process.

References

BARBUY, B, SPITE, M, SPITE, F and MILONE, A: 1991, *Astron. Astrophys.* **247**, 15
BEL, N, VIALA, Y P and GUIDI, I: 1986, *Astron. Astrophys.* **160**, 301
HENKEL, C, BAAN, W A and MAUERSBERGER, R: 1991, *Astron. Astrophys. Rev.* **3**, 47
JOHANSSON, L E B: 1991, in F COMBES and F CASOLI, ed(s)., *Dynamics of Galaxies and Their Molecular Cloud Distributions*, Kluwer: Dordrecht, p.1
MALONEY, P and BLACK, J H: 1988, *Ap. J.* **325**, 389
MAUERSBERGER, R and HENKEL, C: 1991, *Astron. Astrophys.* **245**, 457

MAUERSBERGER, R, HENKEL, C and SAGE, L J: 1990, *Astron. Astrophys.* **236**, 63
MILLAR, T J and HERBST, E: 1990, *Mon. Not. R. astr. Soc.* **242**, 92
PINEAU DES FORÊTS, G, FLOWER, D R and MILLAR, T J: 1991, *Mon. Not. R. astr. Soc.*, in press

TABLE I

A comparison of molecular abundances, log N_{CO}/N_X, in the LMC (Johansson, 1991) with theory. The model parameters are taken from Millar and Herbst (1990). Ion-dipole rate coefficients are adopted and the calculated results are given for the period $(1\text{-}3.16)\ 10^5$ yr.

Species	Observed	Calculated
CS	1.7-4.5	3.2-3.3
CN	2.9-4.5	3.6-4.8
SO	2.1-3.4	4.5-3.1
HCO^+	2.7-4.2	3.9
HCN	3.0-4.5	4.3-5.4
HNC	3.4-4.9	4.2-5.4
C_2H	2.1-3.6	2.7-2.9
H_2CO	2.1-4.3	2.2-3.0
C_3H_2	2.6-4.6	3.5-3.8

QUESTIONS AND ANSWERS

M.Guelin: It seems that in your comparisons you should rather speak of molecular column densities (or even integrated line intensities) rather than molecular abundances. Those are far from being known in external galaxies.

T.J.Millar: The determination of column densities from model calculations can be performed if a cloud model is adopted. The problem is that model clouds (spheres, plane-parallel slabs) are not at all like real interstellar clouds. Indeed, the poor spatial resolution of observations in external galaxies implies that the observations are averaging over many types of clouds. One thing which might force the theoreticians to calculate line intensities and/or line profiles would be if the observers stopped tabulating fractional abundances!

J.M.Greenberg: I could not understand how 99% of carbon could be in CO in the center of galactic molecular clouds (from Maloney & Black). What about the dust which accounts for *at least* 50% of the carbon or probably more?

T.J.Millar: Maloney & Black refer to the *gas phase* distribution of carbon, not to that in the solid state.

J.M.Greenberg: Getting back to the presence of 99% of C in CO is **gas**: How about the accretion of CO on the dust? If $n_{H_2} \simeq 10^4$ and we use a small dust/gas ratio then, since τ_{ac} (accretion time scale) is $\sim 2 \times 10^9/(2n_{H_2}) \approx 10^5$ years, I would expect about $1/2\, n_{CO}$ on dust and not in gas; i.e. only 50% of **free** C would be in gas CO, **unless** dust is pretty warm in the cloud certainly which seems exceptional.

T.J.Millar: Maloney and Black performed calculations for $n_H = 10^3 cm^{-3}$. Also, since the dust-to-gas ratios in the LMC and the SMC are known to be less than the Galactic value, the accretion time-scales in their model clouds are much longer than 10^6 yr, and the gas-grain interaction is therefore negligible.

E.F.van Dishoeck: (comment) The calculations by Maloney & Black on the $C^+ \to CO$ conversion in metal-poor systems stress the importance of depth-dependent models and the comparison of column densities rather than local densities with observations. The assumed size of the cloud is crucial in this respect.

T.J.Millar: This is indeed true in a system such as the SMC, with very low metallicity. It is perhaps not so critical for the LMC - unless one is modelling CO - because the many molecules seen in this galaxy, almost certainty arise from well-shielded regions and not from extended clouds edges.

Observations of Diffuse and Translucent Clouds

PHILIPPE CRANE

European Southern Observatory
Garching, Germany

October 11, 1991

Abstract. Observations of interstellar CN, CH, and CH^+ in diffuse and translucent clouds obtained with a spectral resolution of more the 500,000 are reported. First results from a survey of CH and CH^+ are discussed. The CH^+ isotopic ratio has been determined to be 68 ± 5 toward ζ Oph and a component of $^{12}CH^+$ has been found with a velocity width of 15 km s^{-1}. A review of the rotational excitation temperature of interstellar CN and its implications is given.

1. Introduction

Advances in optical techniques in the last decade have provided a new dimension for the study of molecular species in diffuse and translucent clouds. Improved detectors have provided the means to detect weak lines, or to study stronger lines with vastly improved dynamic range. Advances in spectrograph design and implementation have combined with the improved detectors to permit studies at vastly increased spectral resolution.

These advances have begun to yield new insights into the physical conditions and dynamics of diffuse and interstellar clouds. For the first time(1), interstellar absorption lines have been observed with roughly the same velocity resolution as the millimeter radio observations. These result are discussed in section 2, where the first results of a survey at very high spectral resolution are mentioned.

Section 3 presents results for the CH^+ isotopic ratio. Section 4 presents some data on the excitation temperature of interstellar CN.

The visible interstellar absorption features discussed here are those of CN R(0) at 3874.607Å, of CH^+ R(0) at 4232.548Å and of CH $R_2(1)$ at 4300.313Å. For CN, in addition to the R(0) line, the R(1) and P(1) lines in this system are also considered. The $^{13}CH^+$ isotopic analog of $^{12}CH^+$ is at 4232.267Å.

2. Ultra High Resolution Observations

Optical observations with a spectral resolution of 500,000 or more are difficult not only because the spectrometers are not often capable of this resolution, but even when they are, such high resolutions require relatively bright stars. The work reported here has been performed at the Coudé feed of the 2.7 m telescope of the MacDonald Observatory of the University of Texas.

A first set of observations toward the star ζ Oph have been reported(1) for the CN, CH and CH$^+$. Following the success of these observations, a survey about 20 sight lines was started to study the CH and CH$^+$ lines.

2.1. ζ Oph Results

Figure 1 shows a summary of the data obtained toward ζ Oph for the CN, CH, and CH$^+$ lines. These data have an effective resolution of $\lambda/\Delta\lambda = 600,000$. The CN line can clearly be seen to have two velocity components with a velocity separation of 1.18 km s^{-1}. This coincides very closely with the velocity separation seen in the CO line profiles(2; 3) The CH$^+$ line in these data consists of a single broad component with a velocity width of b = 2.1 km s^{-1}. A major surprise in these data is that the CH line shows elements of both the narrow features seen in the CN data and of the broad feature seen in the CH$^+$ line. This clearly demonstrates that CH exists in two different regions and under quite different physical conditions.

2.2. Survey of CH and CH$^+$

Following the observations toward ζ Oph, a more comprehensive survey of other sight lines was undertaken. However, partly due to instrumental limitations, this survey, which is still in progress, was restricted to the CH and CH$^+$ species.

Table 1 provides a list of the sight lines which are being studied, and a very preliminary estimate of the line widths(FWHM) and relative velocities. The relative velocities given are the velocities of the sub-components relative to the strongest component, and not the relative velocity of CH relative to CH$^+$. A separate observing program is underway to determine this.

One interesting result emerging from this survey is the relative frequency of multiple features in the CH lines but not in the corresponding CH$^+$ lines.

3. High Signal-to-Noise Observations

In order to further investigate a reported discrepancy(4; 5) in the ^{12}C/^{13}C isotope ratio toward ζ Oph as determined from CH$^+$ observations, several spectra were obtained of the CH$^+$ line at 4232.548Å. The resulting spectrum is shown in Figure 2. In addition to finally providing close to a definitive result on the isotope ratio(^{12}C/^{13}C = 68±5), this spectrum also revealed a very broad feature in the ^{12}CH$^+$ feature. This broad feature has a width of about 15 Km s^{-1} FWHM and contributes 1.9 mÅ of equivalent width to the line whose total equivalent width is 23.25 mÅ.

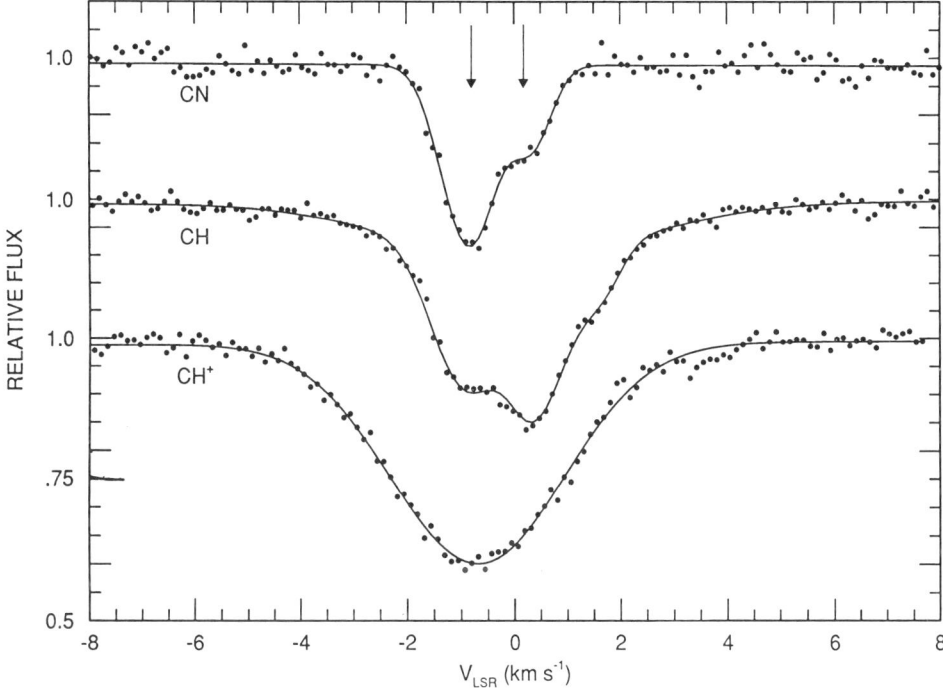

Fig. 1. Plot of CN lines toward ζ Oph. The top spectrum shows the CN line, and indicates the positions of the two velocity components. The middle spectrum shows the CH line, where the effects of the lambda doubling are evident, as well as the effects of a broad component underlying the narrower components. The bottom spectrum shows the CH^+ line which consists of a single Gaussian.

4. CN Excitation Temperature

The excitation temperature of the lowest rotational levels of interstellar CN has proved to be a good means for determining the temperature of the Cosmic Background Radiation(6). However, after the launch of the COBE satellite(7) and the successful rocket experiment of Gush, Halpern, and Wishnow(8), there appears to be a possible discrepancy between the CBR temperature determined by the space experiments, and the best determinations of the CN excitation temperature(9). It has been proposed(10) that the origin of this difference could be local processes in the interstellar clouds which could excite the CN rotational levels a few milliKelvin above the CBR.

In an attempt to test the local excitation hypothesis, Palazzi, Mandolesi, and Crane(9) have collected data from several new sight lines as well as from sources in the literature. By using the primary data — equivalent widths, and

TABLE I
Preliminary Survey Results

	CH Data			CH$^+$ Data		
Star	v$_{rel}$	W$_\lambda$ (mÅ)	σ (mÅ)	v$_{rel}$	W$_\lambda$ (mÅ)	σ (mÅ)
23 Ori				0.00	3.90	26.14
67 Oph	0.00	0.88	15.43	0.00	1.33	21.60
	1.44	0.176	12.11			
κ Aql	0.00	1.38	10.57			
68 Cyg	0.00	1.27	10.94			
	-0.254	1.25	31.67			
λ Cep	0.00	14.57	6.85			
	0.418	3.88	24.95			
μ Sgr	0.00	1.54	14.38			
	-1.422	0.28	18.51			
ρOphA	0.00	2.23	6.51			
	0.00	2.06	18.80			
ψ Per	0.00	1.68	11.21	0.00	3.51	17.27
	1.19	1.41	12.04	1.14	4.25	16.72
	2.09	0.91	54.69	2.22	0.74	30.83
ζ Per	0.00	2.92	8.80	0.00	0.86	18.37
	-0.28	3.01	20.22			
o Per				0.00	1.99	21.38
χ Oph	0.00	7.53	14.49	0.00	4.14	18.61
δ Sco	0.00	0.00	7.00	0.00	0.40	12.03
	0.48	0.26	10.81			
β^1 Sco	0.00	0.81	24.06			
20 Tau				0.00	8.96	18.47
23 Tau				0.00	5.81	20.09

b values, they have produced a homogeneous set of CN excitation temperatures. They find a weighted mean CN excitation temperature of 2.817±0.022 K which is about 82 mK above the space values. They have searched their data set for correlations of the excitation temperatures with a wide range of parameters. There is no correlation with electron density, distance from the Sun, or with any other plausible parameter. More observations will be necessary to confirm the excitation temperature discrepancy, or to understand its origin.

Acknowledgements

Much of the work presented here has been done in collaboration with David Lambert, Reno Mandolesi, and Eliana Palazzi. It is a pleasure to acknowledge their contributions to this work.

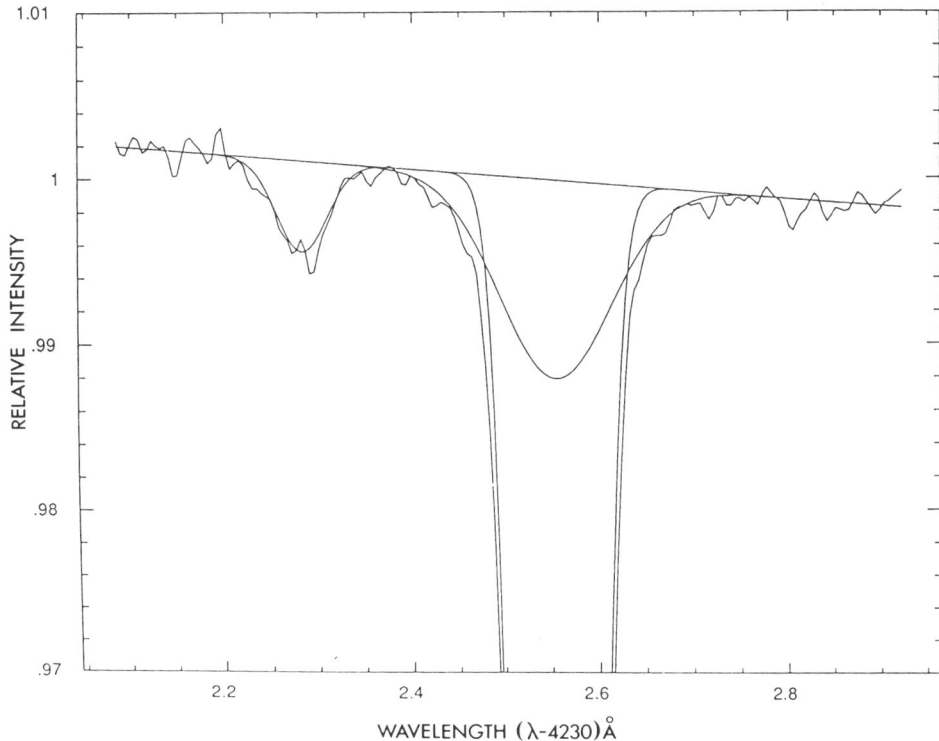

Fig. 2. High signal-to-noise data for CH^+ toward ζ Oph. The depth of the stronger feature reaches a relative intensity of 0.67. The $^{13}CH^+$ line is clearly visible on the left of the strong $^{12}CH^+$ line. The solid line indicates the subcomponents used to fit the data, and illustrates the presence of the very broad component.

References

1) Lambert,D.L., Sheffer,Y., and Crane,P., *CN, CH, and CH^+ Toward ζ Ophiuchi*, Astrophysical Journal (Letters),**359**,L19, 1990.
2) Langer,W.D., Glassgold,A.E., and Wilson,R.W., 1987, *Radio Observations of Carbon Monoxide toward Zeta Ophiuchi: Velocity Structure, Isotopic Abundances, and Physical Properties* Astrophysical Journal,**322**,450.
3) Crutcher,R.M., and Federman,S.R.,1987, *The CO J=2-1 Emission from the Interstellar Gas toward Zeta Ophiuchi*, Astrophysical Journal (Letters),**316**,L71.
4) Hawkins,I., and Jura,M., 1987, *The $^{12}C/^{13}C$ Ratio of the Interstellar Medium in the Neighborhood of the Sun*, Astrophysical Journal,**317**,926.
5) Stahl,O., Wilson,T.L., Henkel,C., and Appenzeller,I., 1989 *CH^+ toward Zeta Oph* Astronomy and Astrophysics,**221**, 321.
6) Crane,P., Hegyi,D.J., Kutner,M.L., and Mandolesi,N., *Cosmic Background Radiation Temperature at 2.64 Millimeters* Astrophysical Journal,**346**,136, 1989
7) Mather,J., et al., *A Preliminary Measurement of the Cosmic Microwave Background Spectrum by the COBE Satellite*, Astrophysical Journal (Letters),**354**,L37, 1990.
8) Gush,H.P., Halpern,M., and Wishnow,E.H., *Rocket Measurement of the Cosmic Background Radiation mm Wave Spectrum*, Physical Review Letters,**65**,537, 1990.

9) Palazzi,E., Mandolesi,N., and Crane,P., *CN Rotational Excitation*, Astrophysical Journal, submitted, 1992.
10) Black,J.H., and van Dishoeck,E.F., *Electron Densities and the Excitation of CN in Molecular Clouds*, Astrophysical Journal (Letters),**369**,L9, 1991.

DISCUSSION

- **N Wright:**
  = Comment 1: The largest signal from the galaxy in the COBE 3 mm maps is ≤ 1 mK, so dust emission is unlikely to raise the $T_{exc}(CN)$ above T_{CMB}
  = Comment 2: The equivalent width of the ^{12}CN R(0) line in your high spectral resolution data is less than in your high SNR data, suggesting that CN also has a weak broad component.
  = Question: The line widths you quote are too large to be velocity dispersions. Do you mean FWHM?
- **P. Crane:** I agree with both your comments! The line widths quoted are indeed FWHMs and they are in mÅ.
- **M. Guelin:** Is there in your CN excitation temperature a correlation of the excitation temperature with the error in determining the excitation temperature?
- **P. Crane:** There does not appear to be such a correlation.
- **J.C. Pecker:** This difficulty about the excitation temperature of CN reminds me of a very different case, that of the solar atmosphere. There, in spite of large temperatures and densities, we found (in the fifties!) a ΔT_{exc} very large between the rotational levels and the electronic levels. We could not explain it because of lack of reliable collisional cross sections involving all relevant levels. In spite of the highly different conditions, this lack of physical data, could it be a reason for the unexplained ΔT_{exc} discrepancy? Or perhaps an underestimate of the number of transitions involved?
- **P. Crane:** I have never thought of this, although it seems at first glance unlikely, I will look more into your suggestion.

THE INTERSTELLAR C-H STRETCHING BAND NEAR 3.4 μm: CONSTRAINTS ON THE COMPOSITION OF ORGANIC MATERIAL IN THE DIFFUSE ISM

S. A. Sandford[1], L. J. Allamandola[1], A. G. G. M. Tielens[1], K. Sellgren[2], M. Tapia[3], and Y. Pendleton[1]

[1] NASA/Ames Research Center, MS 245-6, Moffett Field, CA 94035
[2] Dept. of Astronomy, Ohio State Univ., Columbus, OH 43210
[3] Instituto de Astronomia, UNAM, Ensenada, Mexico

ABSTRACT. The spectra of objects suffering extinction by diffuse interstellar dust contain a broad feature centered at ~3300 cm^{-1} (~3.0 μm), attributed to O-H stretching vibrations, and/or a feature near 2950 cm^{-1} (3.4 μm) attributed to C-H stretching vibrations. The 2950 cm^{-1} feature can be attributed to C-H stretching vibrations in the -CH$_2$- and -CH$_3$ groups of a fairly complex carbonaceous material containing aliphatic functional groups.

1. The Observations and Data

To constrain and quantify the composition of material in the diffuse interstellar medium (ISM), we took 3600-2700 cm^{-1} (2.8-3.7 μm) absorption spectra of objects which have widely varying amounts of visual extinction due to diffuse medium dust along different lines-of-sight. The data were obtained using NASA's Infrared Telescope Facility (IRTF) on Mauna Kea. The spectra of these objects contain a feature centered near 3300 cm^{-1} (~3.0 μm), attributed to O-H stretching vibrations, and/or a feature near 2950 cm^{-1} (3.4 μm) attributed to C-H stretching vibrations. The lack of correlation between the strengths of these two bands indicates that they do not arise from the same molecular carrier.

2. Discussion

We attribute the feature near 2950 cm^{-1} (3.4 μm) to material in the diffuse ISM on the basis of the similarity between the band profiles along the very different lines-of-sight to Galactic Center source IRS7 and VI Cygni #12 (Figure 1). Similar features were also found in the spectra of Galactic Center source IRS3, Ve 2-45, and AFGL 2179.

The interstellar C-H stretching feature for Galactic Center source IRS7 has sub-peaks near 2955, 2925, and 2870 cm^{-1} (±5 cm^{-1}), which we attribute to C-H stretching vibrations in the -CH$_2$- and -CH$_3$ groups of aliphatic hydrocarbons. These band positions fall within 5 cm^{-1} of the values normal for saturated aliphatics. The absence of a distinct band near 2855 cm^{-1} suggests that the material contains small amounts of electronegative groups like O-H or -C≡N. The relative strengths and profiles of the 2955 and 2925 cm^{-1} features towards five objects suggests an average diffuse ISM line-of-sight $-CH_2-/-CH_3$ ratio of about 2.5, indicating the presence of relatively complex organic materials. The strengths of the subpeaks at 2925 and 2955 cm^{-1}, due to $-CH_2-$ and $-CH_3$ groups, respectively, correlate with visual extinction, strongly suggesting that the C-H stretching band is a general feature of the material along different lines-of-sight in the diffuse ISM. We find average ratios of $A_v/\tau(2925$ cm$^{-1}) = 240 \pm 40$ and $A_v/\tau(2955$ cm$^{-1}) = 310 \pm 90$ for the objects we have observed. We deduce that 2.6-35% of the cosmic carbon in the ISM is tied up in the carrier of this band with the most likely value falling near 10%.

The interstellar C-H band is remarkably similar to the feature in lab residues produced by irradiating analogs of dense molecular cloud ices. This is consistent with a model in which the hydrocarbon component in the diffuse interstellar medium consists of complex hydrocarbons containing aliphatic side chains and bridges which are produced in dense molecular clouds and subsequently modified in the diffuse medium. A more complete discussion of these results can be found elsewhere (Sandford et al. 1991).

3. References

Sandford, S.A., Allamandola, L.J., Tielens, A.G.G.M., Sellgren, K., Tapia, M., and Pendleton, Y. (1991) 'The interstellar C-H stretching band near 3.4 microns: Constraints on the composition of organic material in the diffuse interstellar medium', Astrophys. J. 371, 607-620.

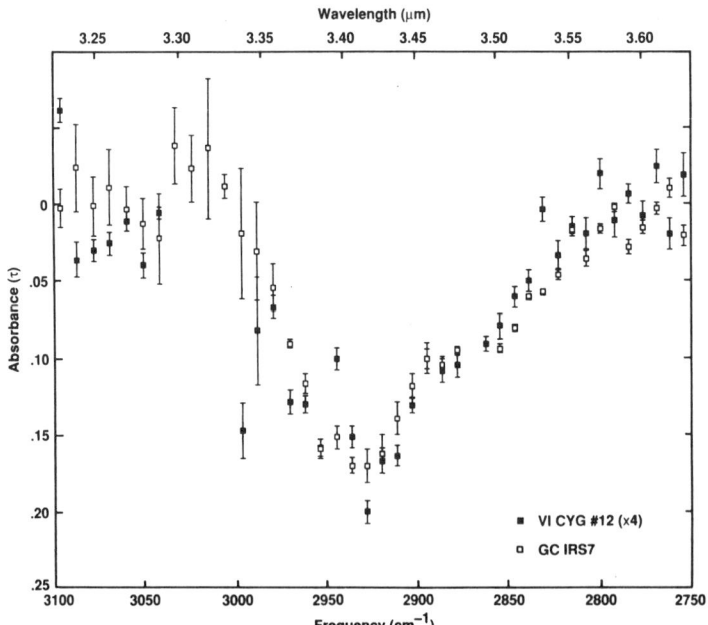

Figure 1 - Comparison between the C-H stretch feature (in absorbance) of Galactic Center source IRS7 and VI Cygni #12. The VI Cygni #12 absorbance spectrum has been scaled up by a factor of 4.

SEARCH FOR THE 4430 Å DIB IN THE SPECTRUM OF CORONENE

P. EHRENFREUND [1,2], L. d'HENDECOURT[1], L. VERSTRAETE[1],
A. LEGER[1], W. SCHMIDT[3]
[1] *Groupe de Physique des Solides, Univ. Paris 7, Paris, France*
[2] *Huygens Laboratory Astrophysics, Leiden, The Netherlands*
[3] *Institut fuer PAH-Forschung, Greifenberg, Germany*

ABSTRACT. Polycyclic Aromatic Hydrocarbon (PAH) molecules have been proposed as candidates to explain the Diffuse Interstellar Bands (DIBs). We have performed laboratory measurements of coronene, using rare gas matrix isolation techniques and UV photolysis. Our aim was to search for a possible identification of the 4430 Å DIB, but also to provide data almost free from environmental band shifts and broadening, which can be used for astronomical identification of the species.

1. Introduction

In 1985, independently, van der Zwet & Allamandola, and Léger and d'Hendecourt have proposed that Polycyclic Aromatic Hydrocarbons (PAHs) could be attractive candidates of the Diffuse Interstellar Bands (DIBs), seen in many galactic and extragalactic objects. The main arguments in favour of PAHs are their stability against photodissociation and their high abundance (Léger and d'Hendecourt, 1985). Furthermore, large neutral PAHs and PAH$^+$ ions have strong electronic transitions in the visual (Crawford et al., 1985) and the expected rotational structures of these transitions are narrow enough to be compatible with the observed widths of DIBs (van der Zwet & Allamandola, 1985).

In the ISM a selection mechanism can originate from the physical conditions that favors a few species, whose absorptions give rise to the DIBs.

The visible spectrum of the coronene cation, as measured in different solvents, shows the presence of a strong band at a wavelength close to the 4430 Å DIB (Shida & Iwata, 1973, Khan, 1988). It is therefore quite attractive to measure the intrinsic spectrum for this species. We present the visible spectra of partially ionized coronene, isolated in a neon matrix and compare the spectrum with those in different solvents, to determine the perturbation they introduce.

2. Results

In Fig. 1 the spectrum of partially ionized coronene, isolated in neon matrix is compared with ionized coronene in solid boric acid, obtained by Khan, 1988. In our experiment we used photoionization by irradiation with UV light from a H_2 (0.1 - 1 mbar) microwave excited discharge lamp. Comparing the spectra in both environments shows, as expected, that the interaction in boric acid produces important broadening and red shifts of the bands with respect to that in neon matrices. For the 9465 Å band vibrational replica can be seen at shorter wavelengths in neon.

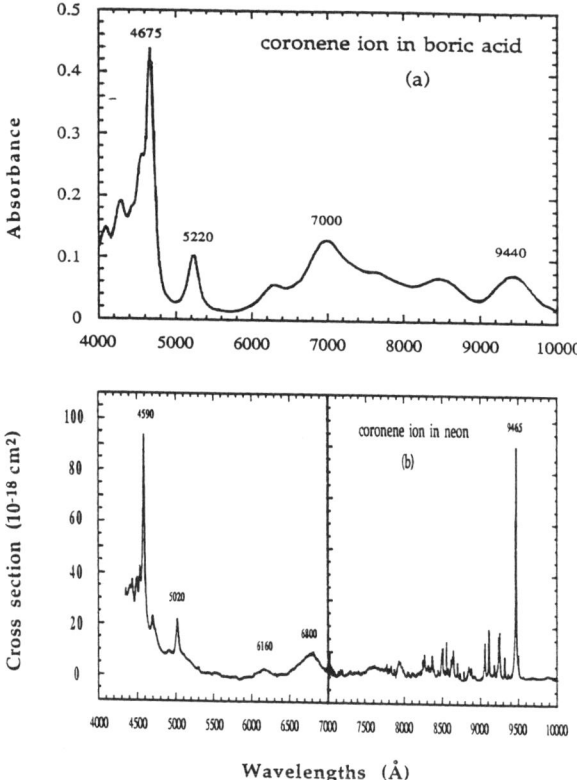

Figure 1. (a) Electronic absorption spectrum of the coronene cation, obtained in solid boric acid (adapted by Khan, 1988) (b) Coronene in a neon matrix

3. Discussion

The coronene cation was a potential candidate for the carrier of the 4430 Å DIB and has been measured in a neon matrix. The 4590 Å band does not coincide with the observed DIB at 4430 Å, rejecting coronene as a possible carrier. Conversely, the reported data on the spectrum of the coronene cation can be used to search for the presence of this species in the interstellar medium. Finally, the measurement of band shifts for large PAHs, caused by organic and polar matrices can be useful to estimate the intrinsic band position for similar molecules from the knowledge of their spectra in solvents.

References
Crawford M.K., Tielens A.G., Allamandola L.J. (1985), Ap. J. Let. 293 L 45
Khan Z.H. (1988), Spectrochimica Acta., 44A, 3, 313
Léger A., d'Hendecourt L. (1985), A & A 146, 81
Shida T., Iwata S. (1973), J. of the American Chemical Society 95, 3473
van der Zwet G.P., Allamandola L.J. (1985), A & A 146, 76

THEORETICAL MODELLING OF THE INFRARED FLUORESCENCE BY INTERSTELLAR POLYCYCLIC AROMATIC HYDROCARBONS

W. A. SCHUTTE, A. G. G. M. TIELENS, and L. J. ALLAMANDOLA
Space Science Division, MS:245-6, NASA/Ames Research Center, Moffett Field, CA 94035, U.S.A.

ABSTRACT. We modelled the IR emission of interstellar PAHs. Substantial differences between the IR properties of interstellar and laboratory PAHs are found, possibly resulting from ionization. The various IR features being dominated by distinctly different size PAHs, their observed relative intensities are sensitive indicators of the size distribution. A number of applications of our model related to future ISO and SIRTF IR data are pointed out.

1. Introduction

The interstellar emission features at 3.3, 6.2, 7.7, 8.6, 11.3, and 12.7 μm have been ascribed to fluorescence by interstellar Polycyclic Aromatic Hydrocarbons (PAHs). We investigated the relation between the PAH size distribution and the observed emission spectrum by theoretical modelling. We furthermore studied to what extend interstellar PAHs may be dehydrogenated.

2. The Model

We defined the PAH size distribution by extrapolating the MRN power law distribution for visual grains down to a minimum size of 24 C atoms. To calculate the fluorescence we used the thermal approximation (Léger and Puget 1984). The results differ by less than 10 % from exact quantum-mechanical calculations.

3. Results

Figure 1 shows how various sizes of PAHs contribute to the emission in the mid-IR features and the far-IR 25 μm IRAS band. The various interstellar emission bands are dominated by PAHs of distinctly different sizes, from ≤ 80 C atoms for the 3.3 μm feature, to ~ 10^3 C atoms for the far-IR emission. The relative intensities of the various mid-IR bands and the far-IR emission are therefore sensitive indicators of the size distribution.

Using cross sections of PAHs measured in the laboratory for the mid-IR features to model the observed emission results in a considerable underestimate of the 6.2 and 7.7 μm C-C stretching and 8.6 μm C-H bending features relative to the 11.3 and 12.7 μm C-H out-of-plane bending modes. Although dehydrogenating the PAHs by ~ 95 % reproduces the correct 7.7/11.3 μm ratio (de Muizon et al. 1990), this solution faces a number of problems. First, it results in a much too weak 8.6 μm feature. Second, the observed 12.7 μm feature which is likely due to the coupled out-of-plane modes of 2 adjacent C-H groups, is not reproduced, since only isolated C-H groups can survive at this severe dehydrogenation. Alternatively, we can solve the discrepancy using fully hydrogenated PAHs by increasing the intrinsic strengths of the 7.7 and 8.6 μm features by a factor of 6 and the 6.2 μm feature by a factor of 2.4. Such an increase could possibly be related to the interstellar PAHs being ionized (deFrees et al. 1991).

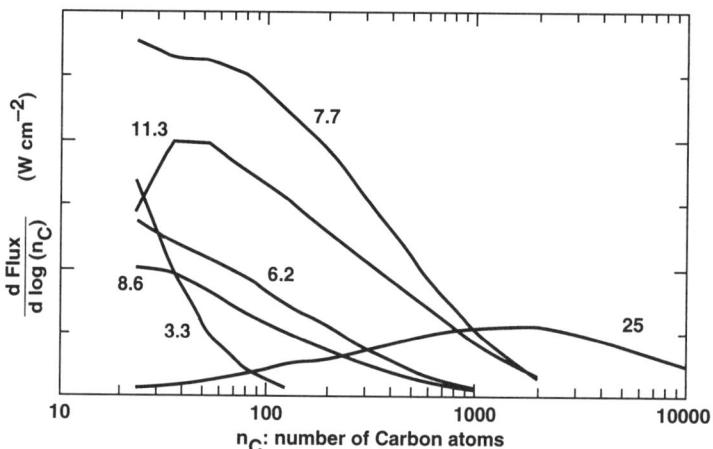

Figure 1. Distribution over the PAH size distribution of the emission in the mid-IR features and the IRAS 25 μm band.

4. Future Apllications

High sensitivity, single aperture observations of the entire emission spectrum from 3 μm to far-IR wavelengths as will likely be obtained by the ISO and SIRTF satellites, together with theoretical modelling of the PAH emission may answer a number of important questions. First, how does the size distribution vary from object to object ? Second, are there characteristic emission features in the far-IR ($\lambda > 15$ μm) that could be used to determine the abundance of individual PAHs ? Next, is there a feature around 14 μm due to the out-of-plane bending mode of 3 adjacent C-H groups and, if so, what do the relative intensities of the 3 out-of-plane modes reveal about the molecular structure and degree of hydrogenation ? And, finally, how do the properties change for "young" PAHs in planetary nebula and "old" PAHs in reflection nebulae and cirrus clouds ?

References

deFrees, D. J., Miller, M. D., Talbi, D., Pauzat, F., and Ellinger, E. (1991) 'Theoretical IR spectra of PAHs and the effect of ionization', preprint.

Jourdain de Muizon, M. J., d'Hendecourt, L. B., and Geballe, T. R. (1990) 'Polycyclic aromatic hydrocarbons in the near-infrared spectra of 24 IRAS sources', A&A 227, 526-541.

Léger, A., and Puget, J. L. (1984) 'Identification of the "unidentified IR emission features of interstellar dust ?', A&A 137, L5.

ARE CARBON CLUSTERS THE CAUSE OF INTERSTELLAR DIFFUSE BANDS?

S P TARAFDAR, K S KRISHNA SWAMY, C BADRINATHAN and
D MATHUR
Tata Institute of Fundamental Research
Homi Bhabha Road
Bombay 400 005, INDIA

ABSTRACT: Theoretically determined vertical excitation energies of C_5 and C_7 molecules have been shown to agree with the strongest diffuse interstellar band at 4430Å. Several other weak diffuse bands can be identified with vibrational transitions, if 4430Å band is taken as 0-3 transition and vibrational constant have a value of $2190 cm^{-1}$. The vertical excitation energy of a second electronic transition of C_7 molecule agrees with diffuse band at 6177Å. This electronic band system may account for diffuse bands at 5778Å, 6660Å and other bands near them.

1. INTRODUCTION

Since the discovery by Merril (1936), advancement in observations lead to the identification of a large number (about 50) of interstellar diffuse bands, a comprehensive study of which are given by Herbig (1975, 1988). The bands, however, still elude identification. Here we compare the wavelengths of theoretically determined vertical transitions of C_5 and C_7 molecules with the interstellar diffuse bands.

2. CALCULATION OF EXCITATION ENERGIES

The calculation was done in two steps. Firstly, geometrical information regarding equilibrium bond lengths was obtained in all-electron ab initio molecular orbital calculations using unrestricted self-consistent-field Hartree-Fock techniques and optimizing by means of Murtagh-Sargent algorithm (Murtagh and Sargent 1970). The excitation energies and oscillator strengths to low-lying electronic states from ground level were carried out in the second step by means of the semi-empirical HAM/3 (cf. Lindholm and Asbrink 1985) method using the sum of the kinetic energy, electron-electron potential energy, exchange energy and correlation energy as electronic energy. The transition energies and oscillator strengths for C_3, C_5 and C_7 are given in Table 1. For each transition energies of C_5 and C_7 two values

— one for the determined geometry with different inner and outer bond lengths and the other for the assumed equal bond lengths – are given. The fare agreement between theoretical and experimental (2.992ev) transition energies for C_3 suggests that transition energies of C_5 and C_7 may be reasonably accurate. However, the same confidence cannot be put on the oscillator strengths which should, therefore, be used for relative comparison.

TABLE 1. Transition Energies and relative oscillator strengths of lowest transitions for C_3, C_5 and C_7. Two entries for C_5 and C_7 are for two different geometries.

C_3		C_5		C_7	
Trans En(ev)	Rel Osc St	Trans En(ev)	Rel Osc St	Trans En(ev)	Rel Osc St
2.992	0.026	2.793-2.764	0.032-0.029	1.962-1.994	0.010-0.008
–	–	–	–	2.799-2.816	0.002-0.013
7.395	2.018	7.008-6.981	5.99-6.09	6.386-6.357	8.539-9.151
9.836	0.054	7.677-7563	0.051-0.003	6.720-6.678	2.231-1.001

3. COMPARISON OF THEORETICAL WAVELENGTHS WITH INTERSTELLAR DIFFUSE BANDS

Table 1 shows that both C_5 and C_7 have electronic transition near 2.8ev. The wavelengths corresponding to these transitions match well with strongest interstellar diffuse bands at 4428Å. Our calculation does now allow the determination of upper (v') and lower (v'') vibrational levels of transition nor the vibrational constants w'_e and $w'_e x'_e$. However, if we assume $v'' = 0, v' = 3, w'_e = 2190 cm^{-1}$ and $w'_e x'_e = 15 cm^{-1}$, we can determine the wavelengths for different vibrational transitions and compare them with the observed diffuse bands (Table 2). The observed band wavelengths are put in two columns - one column giving diffuse band wavelengths falling within the theoretical range defined by the geometry and the other outside the theoretical range. Table 2 shows that a large number of diffuse bands fall in the theoretical wavelength range for C_5 and C_7. The diffuse bands falling outside but near the theoretical range could arise from other vibrational mode of transition together with the vibrational modes given in the table and if the vibrational constants of the new mode are in the range of $300 - 500 cm^{-1}$ Such values of w'_e are plausible (cf. Herzberg 1966). Besides the transition near 2.8ev, C_7 has a strong transition near 2.0 ev. The corresponding wavelength range overlaps a number of diffuse bands (Table 2). The wavelengths of theoretical vibrational transitions are determined assuming 2.0 ev transition to be 0-1 transition, $w'_e = 1170 cm^{-1}$ and $w'_e x'_e = 25 cm^{-1}$. Table 2 shows that almost all observed interstellar diffuse bands can arise from C_7 molecule. The observed strengths of 6177Å and 4428Å bands are also consistent with their relative theoretical oscillator strengths. Table 1 shows that C_5 has transitions in the range $1615 - 1639$Å and $1769 - 1776$Å and C_7 in the ranges $1941 - 1946$Å and $1847 - 1851$Å besides those discussed above. The IUE

TABLE 2. Comparison of diffuse bands with transition wavelengths of C_5 and C_7 arising from transition near 2.8 ev and of C_7 from that near 2ev (2nd rows).

Vib trans $v'' - v'$	C_5-trans	C_7-trans	Observed diffuse bands	
			Within*	Outside*
0–3	4439-4485	4403-4429	4428	4501
0–2	4895-4952	4851-4883	4882	4726,4754,4763
0–2†		5789-5829	5778,5780,5797	5705,5844,5849
0–1	5465-5536	5410-5450	5420,5449,5487	5361,5404,5544
0–1†		6171-6217	6177,6195,6203	6314,6353,6362
0–0	6196-6287	6126-6178	6234,6264,6283	6010,6042,6113
0–0†		6621-6682	6613,6660,6666	6507,6742,6770

*Within and outside the theoretical range of wavelengths.
† From C_7 transition near 2.0ev.

spectra shows that there are no diffuse bands in above wavelength ranges. It is, however, possible, that these bands in UV have disappeared as a result of merger of upper electronic levels of these bands with the continuum in the event of C_5 and C_7 residing on a grain surface or in the form small solids. Note that the C_5 and C_7 molecules have to reside on a grain surface or in the small solid form for their survival in interstellar diffuse clouds.

4. REFERENCES

Herbig G. H. 1975, Astrophys. J. 196, 129.
Herbig G. H. 1988, Astrophys. J. 331, 999.
Herzberg 1966, Molecular Spectra and Molecular Structure, Vol. 3, Van Nostrand Reinhold, New York.
Lindholm E. and Asbrink, L. 1985, Molecular orbitals and their Energies, studies by the semi-empirical HAM Method, Springer-Verlag, Berlin, p. 107.
Merril P. W. 1936, Astrophys. J. 83, 166.
Murtagh B. A. and Sargent R. W. 1970, Comput J. 13, 185.

DIFFUSE, TRANSLUCENT & HIGH–LATITUDE CLOUDS: THEORETICAL CONSIDERATIONS

E.F. van Dishoeck
Leiden Observatory
P.O. Box 9513
2300 RA Leiden, The Netherlands

ABSTRACT. The relative importance of gas–phase, shock and grain–surface chemistry is discussed with reference to recent observational data. The relation between diffuse, translucent and high–latitude clouds is illustrated with models in which carbon is transformed from atomic (C^+, C) to molecular (CO) form with increasing extinction.

1. Introduction

The study of diffuse molecular clouds has traditionally attracted a lot of theoretical attention, since these clouds are thought to be the simplest clouds to model and thus to form the most severe test of the basic chemical networks. Compared with the astrochemistry meeting in Goa (see Black 1987), it appears at first sight that little progress has been made in the last six years. For example, apart from the recent discovery of interstellar NH (Meyer & Roth 1991), there have been no new detections of molecules at optical wavelengths since that of C_2 in 1977. The diffuse interstellar bands are still unidentified; there still is no satisfactory solution to the CH^+ problem; and the rate coefficient for the $C^+ + H_2$ radiative association reaction – the reaction that is thought to initiate the carbon chemistry – is still uncertain by at least a factor of five.

Our understanding of diffuse clouds has increased, however, in other areas since 1985. First, we have expanded our view to different classes of diffuse clouds. In addition to the "classical" diffuse clouds with $A_V \lesssim 1$ mag observed with the *Copernicus* satellite, a significant body of optical data is now available for translucent (A_V=1–5 mag) and high–latitude clouds (van Dishoeck & Black 1989). There has also been increased spectral resolution at optical wavelengths (see Crane, this volume) and extension of the observations to millimeter wavelengths (see Blitz, this volume). Finally, in terms of the rate coefficients, there has been enormous progress in our understanding of the photodissociation of the CO molecule. With this perspective in mind, a number of aspects of diffuse cloud chemistry will be discussed. The reader is referred to several recent reviews (van Dishoeck & Black 1988a; Dalgarno 1988; van Dishoeck 1990) for more details.

2. Physical structure

In order to test the chemistry, an estimate of the physical parameters in the clouds is needed, since the reaction rates depend sensitively on them. The strength of the incident ultraviolet radiation field is particularly important since photodissociation and photoionization play a major role in the chemistry. These parameters are usually derived from the observed excitation of the various species, as summarized in Table 1. The primary diagnostic in diffuse clouds is H_2, for which lines arising from levels up to J=7 have been observed with

TABLE 1. Diagnostics of Physical Conditions

Species	Phys. Parameter Probed[a]	Diffuse Cloud	Transl. Cloud	High–lat. Cloud	Method	Ref.
H_2 low J	T	+	-	-	UV abs	1,2
H_2 high J	I_{UV}, formation	+	-	-	UV abs	1,2
C_2 low J	T	+	+	+	VIS abs	3
C_2 high J	I_R/n_H	+	+	+	VIS abs	3
C, C^+, O J	n_H,T	+	-	-	UV abs	2,4
CO low J	n_H,T	+	-	-	UV abs	2,5,6
		+	+	+	mm em	7,8
CN low J	n(e),n_H	+	+	+	VIS abs	9
		+	+	+	mm em	9

[a] I_{UV}=scaling factor for radiation at ultraviolet wavelengths λ=912–1100 Å; I_R=scaling factor in far–red $\lambda \approx 1$ μm.

References: 1. Jura 1975; 2. van Dishoeck & Black 1986; 3. van Dishoeck & Black 1982; 4. Jenkins & Shaya 1979; 5. Smith et al. 1978; 6. Crutcher & Watson 1981; 7. van Dishoeck et al. 1991; 8. Falgarone et al. 1991; 9. Black & van Dishoeck 1991a.

the *Copernicus* satellite. The lower J levels are populated by collisions, and are therefore sensitive to temperature; however, the higher J levels lie much too high in energy to be collisionally excited in gas with T <100 K. It is usually assumed that ultraviolet pumping is responsible for this excitation, so that the high–J level population provides a direct measure of the strength of the ultraviolet radiation field between 912 and 1100 Å. However, it can also result from the H_2 formation process on grains (Black & Dalgarno 1977; van Dishoeck & Black 1986; Wagenblast 1991), or from collisional excitation in a (shock–heated) layer of gas at 2000 K. The H_2 example thus demonstrates a common problem with diagnostics, namely that it is very rare that a single excitation mechanism dominates. Often a mixture of processes plays a role, and it is difficult to disentangle them.

Most of these diagnostics have been applied to the classical diffuse clouds studied with the *Copernicus* satellite. Unfortunately, for the more recently discovered translucent and high–latitude clouds, many of them are not available, in particular at ultraviolet wavelengths. Although the first observations of C, C^+, O and CO with the *Hubble Space Telescope* (HST) have just been published for the diffuse cloud toward ζ Per with $A_V \approx 1$ mag (Savage et al. 1991; Cardelli et al. 1991; Smith et al. 1991), data for more reddened lines of sight have not yet been taken. Moreover, *HST* will not provide any information on the H_2 rotational excitation, so that it is not possible to perform exactly the same analysis for these clouds as for the classical diffuse clouds. In particular, there are currently no constraints on the strength of the incident radiation field.

For the diffuse clouds such as the ζ Oph cloud, the various diagnostics give consistent results for the temperature and density structure within factors of two. The best fit to the data is for models in which the temperature decreases from 100 K or more at the edge to about 30 K in the center, and the density from about 100 cm^{-3} at the edge to several hundred cm^{-3} in the center. The ultraviolet radiation field is usually somewhat enhanced over the average field by a factor $I_{UV} \approx$2–4, which is not unreasonable since most of the clouds lie in regions of active star formation such as Ophiuchus, in which there are plenty of young bright stars. This estimate of I_{UV} refers to the field at the edge of the cloud. Deeper inside, the radiation is reduced because of absorption and scattering by grains and because of processes such as self–shielding (van Dishoeck 1988; Roberge et al. 1991). The shape of the radiation field at the shortest wavelengths also plays an important role. For the translucent and high–latitude clouds, the C_2 and CN excitations give somewhat higher central densities, $n_H \approx 500 - 5000$ cm^{-3}, and lower temperatures, $T \approx 15 - 30$ K. The CO rotational excitation can be interpreted either with a low density, higher temperature solution (van Dishoeck et al. 1991), or with a very cold ($T=10$ K) and very dense ($n_H > 10^4$ cm^{-3}) model (Falgarone et al. 1991). The former solution is consistent within the observational errors with the parameters derived from C_2 and CN.

3. Steady–state models

Given the set of physical conditions $n(z)$, $T(z)$ and $I(z)$, the basic network of gas–phase ion–molecule reactions can be solved at each depth z into the cloud. The availability of accurate reaction rate coefficients is a prerequisite for testing the chemistry. Other ingredients to the models are the elemental gas–phase abundances, which are constrained to fit the observations of the atomic lines where available, and the cosmic ray ionization rate ζ_o, which is chosen such as to reproduce the observed OH abundance. The inferred value of ζ_o depends sensitively on the H_3^+ dissociative recombination rate coefficient at low temperatures (Amano 1990).

Consider as an example one of the simplest molecules, CH. It is thought to be formed by the sequence starting with the radiative association of C^+ with H_2 and is destroyed primarily by photodissociation. Thus, only few reactions are involved and roughly $n(CH) \propto k_{ra} n(C^+) n(H_2)/k_{pd}$ (Federman 1982). Models of diffuse clouds with the physical structure described in §2 can reproduce the observed CH column densities very well, so that CH has always been considered to be one of the strongest cases of support for the steady–state gas–phase chemistry. Yet there are several major uncertainties in this analysis. First, k_{ra} is not known to better than a factor of five, in spite of several decades of theoretical and experimental work. The (indirect) experiments by Gerlich (1989) suggest $k_{ra} \approx 7 \times 10^{-16}$ cm^3 s^{-1}, whereas the theory by Smith (1989) gives $k_{ra} \approx (1-2) \times 10^{-15}$ cm^3 s^{-1} at interstellar temperatures. Second, $n(C^+)$ depends on the adopted gas–phase carbon abundance, which is not known to better than 50%. Also, the decrease of $n(C^+)$ with depth into the cloud depends on the physical structure (§4). Future observations of the semi–forbidden C II] line at 2325 Å with *HST* may provide better constraints on the carbon abundance (Cardelli et al. 1991). Third, the model CH abundance is proportional to the density $n(H_2)$; at higher densities, significantly more CH is produced than is observed. Fourth, the CH abundance is inversely proportional to the photodissociation rate, and thus I_{UV}, for which few constraints are available.

Given these uncertainties, it is not surprising that it is possible to fit the observational CH data with some combination of parameters. However, it would be impossible to establish whether *all* observed CH were produced by this mechanism; it is certainly possible that some fraction (up to 30%) could by formed by another process, as seems implied by the high–resolution data of Lambert et al. (1990) (see also Crane, this volume). The same conclusion holds for each of the other molecules individually. The real test of the models is whether a *consistent* model can be built that reproduces the whole array of observational data for a cloud with the *same* set of parameters. Such comprehensive models have been developed by e.g. Black & Dalgarno (1977), van Dishoeck & Black (1986), Viala et al. (1988a) and Wagenblast & Hartquist (1989), and are consistent with the observed abundances (and upper limits) of most species within factors of two. These same models can also reproduce the measured abundances in translucent and high–latitude clouds with only slightly different parameters, such as a lower incident radiation field and/or lower gas–phase carbon abundance (van Dishoeck & Black 1989). However, there are a number of species for which the steady–state models fail. The best-known case is that of CH$^+$, which the steady–state models underproduce by 1–2 orders of magnitude. Second, with the new CO photodissociation rate, the computed CO column densities are too low by factors of a few in some (but not all!) diffuse clouds. Third, the models fail to account for the observed NH abundance by an order of magnitude. Finally, they cannot reproduce the observed abundances of more complex species such as H_2CO and C_3H_2 in the thicker translucent and high–latitude clouds. The CO problem is discussed in more detail in van Dishoeck & Black (1988b) and van Dishoeck (1990). The other problems will be addressed in §5-7.

4. Relation between diffuse, translucent & high–latitude clouds

Because the diffuse, translucent and high–latitude clouds are usually studied by different observational techniques (optical absorption, millimeter emission, IRAS 100 μm), it is often thought that they are of a different nature. The diagnostics discussed in §2 already suggest that the physical parameters in the translucent and high–latitude clouds do not differ significantly from those in the classical diffuse clouds: they may be somewhat denser, colder and/or exposed to less radiation, but probably not by large factors. In this section, it is illustrated that the observational division results mostly from the chemistry of CO in the transition zone between diffuse and dense interstellar clouds.

The crucial parameter in the CO chemistry is its destruction by photodissociation. Owing to the detailed laboratory work by Letzelter et al. (1987), Eidelsberg & Rostas (1990) and Stark et al. (1991), the photoprocesses of CO are now fairly well characterized. The CO photodissociation occurs by line absorptions in the 912–1118 Å range, where overlap with lines of H_2, H and CO isotopes can occur. Detailed radiative transfer methods are therefore needed to compute the CO photodissociation rate as a function of depth into the cloud (van Dishoeck & Black 1988b; Viala et al. 1988b). At the edge, all carbon is in atomic form as C^+, but with increasing depth carbon is gradually transformed into C and CO. Thus, the total column density of CO and its abundance with respect to H_2 depend sensitively on the total extent of the cloud. In Figure 1, the CO/H_2 column density ratio is presented as a function of the total H_2 column density or visual extinction. For diffuse clouds with $A_V \lesssim 1$ mag, most of the carbon is still in the form of C^+ and $CO/H_2 \approx 10^{-6}$. For thick, dense clouds with $A_V \approx 10$ mag, all gas–phase carbon is transformed into CO and $CO/H_2 \rightarrow 2[C]_{gas}/[H]$, which is about 3×10^{-4} in these models in which 40% of the solar carbon abundance is assumed to be in the gas phase. It is just in the region of the translucent clouds of $A_V \approx 1-5$ mag that the CO/H_2 column density ratio increases very rapidly by two orders of magnitude. Where exactly the transition occurs depends on the physical parameters of the cloud. For higher density and/or lower radiation field, it is shifted to lower A_V.

Figure 1 forms a convenient framework in which to characterize the various clouds. The observed CO/H_2 column density ratios of a number of diffuse and translucent clouds are included, and are found to follow the general trend. Since the detection limit for CO emission at millimeter wavelengths is $N(CO) \approx$ a few $\times 10^{14}$ cm^{-2} (or $CO/H_2 \approx 10^{-6}$), it can be understood that Lada & Blitz (1985) divided these same clouds into "CO poor" and "CO rich". Similarly, the high–latitude clouds can be fitted into this scheme. These clouds do not form a homogeneous sample, but actually cover the full range of A_V from the very diffuse to dense interstellar clouds. Even within a single cloud, there may be significant structure. For example, there are the high–latitude clouds recognizable through their IRAS 100 μm emission in which no CO millimeter emission has been detected (Blitz et al. 1990), which are indicated with "BBD90" in Figure 1. These clouds have $A_V < 1$ mag and are probably similar to the diffuse clouds studied with *Copernicus*. On the other hand, high–latitude clouds such as MBM 12 show CO antenna temperatures similar to those found in thick, dense clouds such as TMC–1 (Pound et al. 1990). The majority of the high–latitude clouds detected by Magnani et al. (1985) probably fall in the middle region of the translucent clouds, where carbon is being transformed from atomic to molecular form

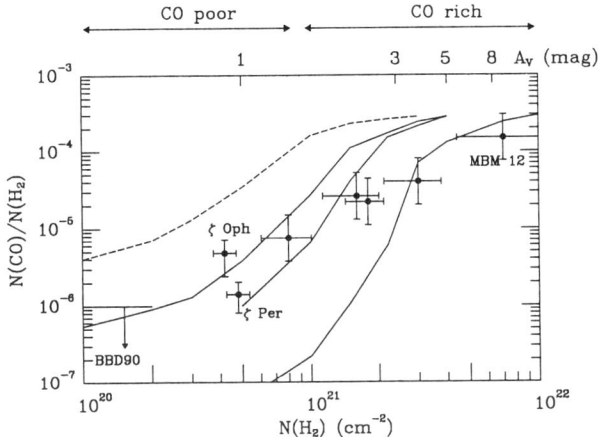

Figure 1. Computed and observed $^{12}CO/H_2$ column density ratios as functions of total H_2 column density (adapted from van Dishoeck & Black 1988b). The full lines from left to right are model results for I_{UV} =0.5, 1.0 and 10, respectively, and $n_H \approx 500$–2000 cm^{-3}. The dashed line is for models with I_{UV}=1, $n_H = 10^4$ cm^{-3} and $T=10$ K. In such models, ^{13}CO is heavily fractionated, with $N(^{12}CO)/N(^{13}CO)$=3–6. Observed values for a number of diffuse, translucent and high–latitude clouds are included. The unlabelled points are for HD 210121, HD 169454, HD 154368 and HD 29647, respectively, from left to right. The limit marked BBD90 refers to Blitz et al. (1990).

In this regime, the column densities are very sensitive to slight changes in the physical parameters, so that the interpretation of the small–scale structure seen in the maps is not obvious, as discussed for the high–latitude cloud toward HD 210121 by Gredel et al. (1991).

5. Evidence for shock chemistry?

Ever since its identification in 1941, the large abundance of CH$^+$ in diffuse clouds has remained a puzzle. The problem stems from the fact that the reaction C$^+$ + H$_2$ → CH$^+$ + H is endoergic by 0.4 eV and therefore does not proceed at low T. At the same time, rapid destruction of CH$^+$ is ensured in all environments through reactions with H, H$_2$ or e, or by photodissociation. Shocks which heat a narrow layer of the cloud to about 2000 K therefore seem a promising explanation (Elitzur & Watson 1978). These shocks should not overproduce molecules such as CH and OH, however, which are already accounted for in the steady–state models. The most successful shock models have therefore been of the magnetohydrodynamic type, in which part of the energy necessary for overcoming the barrier stems from the differential streaming of the ions and the neutrals (Draine & Katz 1986; Pineau des Forêts et al. 1986). One interesting feature of the MHD models is that they predict CH column densities and line widths consistent with those found in the broad components of Lambert et al. (1990). However, there are several recent observational and theoretical developments which cast severe doubts on their validity.

First, the predicted velocity shift between CH$^+$ and quiescent CH or CN is not observed in recent high–quality data, which indicate ΔV <1 km s^{-1} (Crawford 1989; Jenkins et al. 1989; Hawkins & Craig 1991). Second, the CH$^+$ column density continues to increase with A_V in translucent clouds (Souza 1979; Gredel et al. 1992). Also, the abundance of the ion tends to be inversely correlated with density, as if its formation is "quenched" at higher densities (Cardelli et al. 1990; Gredel et al. 1992). Third, there are examples of regions where the evidence for shocks is very strong but where the CH^+ column density is very low, such as toward HD 62542 (Cardelli et al. 1990) and in some high–latitude clouds (de Vries & van Dishoeck 1988). Finally, the latest shock models of Hartquist et al. (1990) suggest that no consistent model can be found that accounts for both CH^+, the rotationally–excited H_2 and OH. On the other hand, no satisfactory alternative solution to the CH^+ problem has yet been found. Its broad (but symmetric) line profiles indicate a different formation environment from that of the quiescent CH and CN, but the absence of velocity shifts and the increase in column density with A_V suggest that these two environments are well mixed. A possible solution, originally proposed by Dalgarno (1976) and recently explored by Duley et al. (1992), is that the ion is formed at the turbulent interfaces between the warm atomic and molecular gas (see also Falgarone, this volume). Detailed models for such regions must await a physical description of turbulence.

Large abundances of OH have also been claimed as evidence for shock chemistry. However, the OH velocities derived from 18 cm emission lines also agree within 1 km s^{-1} with those of the quiescent CO (Magnani & Siskind 1990). Moreover, the abundances found by Magnani & Siskind do not indicate large enhancements in translucent and high–latitude clouds compared with classical diffuse clouds. The large OH abundance found in one particular high–latitude cloud (Grossman et al. 1991) most likely results from an underestimate of the H_2 column density. OH column densities up to an order of magnitude larger than found in diffuse clouds can easily be produced in steady–state models in which the radiation field is reduced. In summary, neither the CH^+ nor the OH observations appear to support the current shock models.

6. Evidence for grain–surface chemistry?

One of the more exciting recent developments in diffuse cloud chemistry is the detection of NH by Meyer & Roth (1991) toward ζ Per. Its abundance is $NH/H_2 \approx 2 \times 10^{-9}$, compared with $CH/H_2 \approx OH/H_2 \approx 5 \times 10^{-8}$. The fact that the CH and OH abundances are much larger than that of NH has been one of the strongest arguments in favor of gas–phase chemistry (Crutcher & Watson 1976). However, grain–surface chemistry can contribute to the formation of CH and OH at the 1% level, and can dominate the formation of NH. NH is not easily formed in the gas phase because N cannot be photoionized, the N + H_3^+ reaction is slow, and the $N^+ + H_2$ reaction is slightly endothermic. The steady-state gas–phase models of van Dishoeck & Black, using the latest information on the NH photodissociation (Kirby and Goldfield 1991) and the $N^+ + H_2$ reaction (Galloway & Herbst 1989; Le Bourlot 1991), give for the ζ Per cloud a NH column density a factor 20 below the observed value. Mann & Williams (1985) considered the formation of NH resulting from grain surface reactions, either directly or indirectly through formation of NH_3 which then photodissociates to NH. The latter model predicts NH column densities similar to those observed. Possible alternative formation schemes of NH in e.g. shocked or turbulent interface layers still need to be investigated.

Another molecule whose abundance is difficult to reproduce by gas–phase or shock chemistry is H_2CO. Although large column densities for this molecule have been observed in translucent and high–latitude clouds (Magnani et al. 1988; Heithausen et al. 1987; Turner et al. 1989), it must be stressed that its actual abundance is not well determined, since it is not clear observationally how "diffuse" the positions at which H_2CO is seen really are. Indeed, it is likely that the molecule is not detected by its 2 cm or 6 cm absorption lines until $A_V > 3$ mag. Still, even models with $A_V \approx 3-5$ mag fail by several orders of magnitude. The possible formation of H_2CO on grains in translucent clouds through photoreactions of solid H_2O and CO has most recently been investigated by Breukers (1991) and Federman & Allen (1991). Such models appear capable of reproducing the observed H_2CO within an order of magnitude, although detailed depth–dependent models have not yet been made.

Finally, grains affect the chemistry in several indirect ways. The abundances of molecules like CO and CN are sensitive to the shape of the ultraviolet extinction curve at the shortest wavelengths (Cardelli 1988; van Dishoeck & Black 1989). Also, the presence of very small grains or PAHs can affect the ionization balance (Lepp et al. 1988), and the thermal balance (d'Hendecourt & Léger 1987; Lepp & Dalgarno 1988; van Dishoeck 1990).

7. Small–scale structure and evolution

There is growing evidence that interstellar clouds, including the more diffuse ones, have a very "clumpy" or "fractal" structure (see Falgarone, this volume; Black & van Dishoeck 1991b). How can this affect the models? Since the ultraviolet radiation can penetrate deeper in a clumpy cloud, it becomes more difficult to build up column densities of molecules such as CO. The effects can be explored in first approximation by using different effective grain scattering parameters (Boissé 1990). Also, large density fluctuations can significantly affect the chemistry of simple molecules such as C_2, CN and CO, as illustrated by van Dishoeck et al. (1991). The complicated velocity structure, however, may not have much effect, since the damped ultraviolet lines through which H_2 and CO photodissociate are so broad that the various clumps shield each other. This is demonstrated for the case of π Sco by models of Jenkins et al. (1989).

A related question is the evolutionary state of diffuse clouds and their relation to dense clouds. It should be realized that at the edges of diffuse and translucent clouds, the chemical time scales are so short, of order 10^3-10^4 yr, that chemical equilibrium is readily attained. The same holds for most time scales in the center of the clouds. The major exception is formed by the H_2/H equilibrium, and possibly by the $CO/C/C^+$ case in translucent clouds. Time dependent models of the H_2/H chemistry in diffuse clouds have been presented by Wagenblast & Hartquist (1989), and indeed, the results depend on whether the hydrogen is initially atomic or molecular. Further study of such evolutionary effects may provide important clues to the origin of diffuse, translucent and high–latitude clouds.

Acknowledgment. The author is grateful to J.H. Black for useful discussions and to the Netherlands Organization for Scientific Research (NWO) for financial support.

Amano, T. 1990, J. Chem. Phys., 92, 6492.
Black, J.H. 1987, in IAU Symposium **120**, *Astrochemistry*, eds. M.S. Vardya and S.P. Tarafdar (Reidel, Dordrecht), p. 217.
Black, J.H. & Dalgarno, A. 1977, ApJS, 34, 405.
Black, J.H. & van Dishoeck, E.F. 1991a, ApJ, 369, L9.
Black, J.H. & van Dishoeck, E.F. 1991b, in *Fragmentation of Molecular Clouds and Star Formation*, IAU Symposium 147, eds. E. Falgarone et al. (Kluwer, Dordrecht), p. 139.
Blitz, L., Bazell, D., & Désert, F.X. 1990, ApJ, 352, L13.
Boissé, P., 1990, A&A, 228, 483.
Breukers, R. 1991, PhD Thesis, University of Leiden.
Cardelli, J.A. 1988, ApJ, 335, 177.
Cardelli, J.A., Suntzeff, N.B., Edgar, R.J. & Savage, B.D. 1990, ApJ, 362, 551.
Cardelli, J.A. et al. 1991, ApJ, 377, L57.
Crawford, I.A. 1989, MNRAS, 241, 575.
Crutcher, R.M. & Watson, W.D. 1976, ApJ, 209, 778.
Crutcher, R.M. & Watson, W.D. 1981, ApJ, 244, 855.
Dalgarno, A. 1976, in *Atomic Processes and applications*, eds. P.G. Burke and B.L. Moiseiwitsch (North–Holland), p. 110.
Dalgarno, A. 1988, Astro. Lett. and Communications, 26, 153.
de Vries, C.P. & van Dishoeck, E.F. 1988, A&A, 203, L23.

d'Hendecourt, L.B. & Léger, A. 1987, A&A, 180, L9.
Draine, B.T. & Katz, N.S. 1986, ApJ, 306, 655; 310, 392.
Duley, W.W. et al. 1992, MNRAS, in press.
Eidelsberg, M. & Rostas, F. 1990, A&A, 235, 472.
Elitzur, M. & Watson, W.D. 1978, ApJ, 222, L141.
Falgarone, E., Phillips, T.G., & Walker, C.K. 1991, ApJ, 378, 186.
Federman, S.R. 1982, ApJ, 257, 125.
Federman, S.R. & Allen, M. 1991, ApJ, 375, 157.
Galloway, E.T. & Herbst, E. 1989, A&A, 211, 413.
Gerlich, D. 1989, private communication.
Gredel, R., van Dishoeck, E.F., de Vries, C.P., and Black, J.H. 1991, A&A, in press.
Gredel, R., van Dishoeck, E.F. & Black, J.H. 1992, A&A, submitted.
Grossman, V., Heithausen, A., Meyerdierks, H., & Mebold, U. 1990, A&A, 240, 400.
Hartquist, T.W., Flower, D.R., & Pineau des Forêts, G. 1990, in *Molecular Astrophysics —A volume honoring Alexander Dalgarno*, ed. T.W. Hartquist (Cambridge University Press).
Hawkins, I. & Craig, N. 1991, ApJ, 375, 642.
Heithausen, A., Mebold, U. & de Vries, H.W. 1987, A&A, 179, 263.
Jenkins, E.B. & Shaya, E.J. 1979, ApJ, 231, 55.
Jenkins, E.B., Lees, J.F., van Dishoeck, E.F., & Wilcots, E.M. 1989, ApJ, 343, 785.
Jura, M. 1975, ApJ, 197, 581.
Kirby, K.P. & Goldfield, E.M. 1991, J. Chem. Phys., 94, 1271.
Lada, E.A. & Blitz, L. 1988, ApJ, 326, L69.
Lambert, D.L., Sheffer, V., & Crane, P. 1990, ApJ, 359, L19.
Le Bourlot, J. 1991, A&A, 242, 235.
Lepp, S. & Dalgarno, A. 1988, ApJ, 335, 769.
Lepp, S., Dalgarno, A., van Dishoeck, E.F., & Black, J.H. 1988, ApJ, 329, 418.
Letzelter, C. et al. 1987, Chem. Phys., 114, 273.
Magnani, L., Blitz, L., & Mundy, L. 1985, ApJ, 295, 402.
Magnani, L., Blitz, L., & Wouterloot, J.G.A. 1988, ApJ, 326, 909.
Magnani, L. & Siskind, L. 1990, ApJ, 359, 355.
Mann, A.P.C. & Williams, D.A. 1984, MNRAS, 209, 33.
Meyer, D.M. & Roth, K. 1991, ApJ, 376, L49.
Pineau des Forêts, G. et al. 1986, MNRAS, 220, 801.
Pound, M.W., Bania, T.M., & Wilson, R.W. 1990, ApJ, 351, 165.
Roberge, W.G., Jones, D., Lepp, S., & Dalgarno, A. 1991, ApJS, 77, 287.
Savage, B.D. et al. 1991, ApJ, 377, L53.
Smith, A.M., Krishna Swamy, K.S. & Stecher, T.P. 1978, ApJ, 220, 138.
Smith, A.M. et al. 1991, ApJ, 377, L61.
Smith, I.W.M. 1989, ApJ, 347, 282.
Souza, S.P. 1979, PhD thesis, State Univ. of New York at Stony Brook.
Stark, G. et al. 1991, ApJ, 369, 574.
Turner, B.E., Rickard, L.J, & Xi, L.P. 1989, ApJ, 344, 292.
van Dishoeck, E.F. 1988, in *Rate Coefficients in Astrochemistry*, eds. T.J. Millar and D.A. Williams (Kluwer, Dordrecht), p. 49.
van Dishoeck, E.F. 1990, in *The Evolution of the Interstellar Medium*, A.S.P. Conference series nr. 12, ed. L. Blitz (Astronomical Society of the Pacific, San Francisco), p. 207.
van Dishoeck, E.F. & Black, J.H. 1982, ApJ, 258, 533.
van Dishoeck, E.F. & Black, J.H. 1986, ApJS, 62, 109.
van Dishoeck, E.F. & Black, J.H. 1988a, in *Rate Coefficients in Astrochemistry*, eds. T.J. Millar and D.A. Williams (Kluwer, Dordrecht), p. 209.
van Dishoeck, E.F. & Black, J.H. 1988b, ApJ, 334, 771.
van Dishoeck, E.F. & Black, J.H. 1989, ApJ, 340, 273.
van Dishoeck, E.F., Black, J.H., Phillips, T.G., & Gredel, R. 1991, ApJ, 366, 141.
Viala, Y.P., Roueff, E., & Abgrall, H. 1988a, A&A, 190, 215.
Viala, Y.P., Letzelter, C., Eidelsberg, M., & Rostas, F. 1988b, A&A, 193, 265.
Wagenblast, R. 1991, preprint.
Wagenblast, R. & Hartquist, T.W. 1989, MNRAS, 237, 1019.

QUESTIONS AND ANSWERS

D.A.Williams: (comments) The failure of diffuse clouds models to account for CO is particularly disturbing. There is a new work (see Wagenblast's poster) which allows H_2 (high J) formation on dust, so that the UV field can be relatively low. This work leads to a consistent model of $\zeta\, Oph$. It suggests that parameter space may not yet be thoroughly explored. A model of $\zeta\, Per$ with NH formation on dust has been made by Wagenblast & Williams. It gives results entirely consistent with all atomic and molecular observations on this line of sight. The failure of shocks to account satisfactorily for CH^+ seems to be established. An alternative (suggested by A.Dalgarno) is that CH^+ may be formed in warm interface region. An exploratory calculation (see Poster: Duley et al.) indicates that such a solution is possible without contravening the constraints of $H_2(J)$, OH and CH.

E.F.van Dishoeck: I agree that the H_2 formation model may be important for the population of the higher J levels, an effect which was already explored to some extent by Black and Dalgarno (1976) and van Dishoeck and Black (1986). See also van Dishoeck 1990 (in "The evolution of the ISM", ed. L.Blitz) for a model for the $\zeta\, Oph$ cloud with $I_{uv} = 1$. The problem is how to distinguish between the models; the UV pumping models make specific predictions about the vibrationally excited H_2. Note also that the high UV field models only fail for $\zeta\, Oph$, but not for $\zeta\, Per$ etc if a steep rising UV extinction curve is taken into account. Regarding CH^+, I fully agree that warm interfaces are a likely site for the formation of the ion. However, until there is a physical model for turbulence, we cannot prove anything quantitatively.

ABUNDANCE OF CH⁺ IN TRANSLUCENT MOLECULAR CLOUDS: PROBLEMS FOR SHOCK MODELS?

R. Gredel[1], E.F. van Dishoeck[2] and J.H. Black[3]

[1] *European Southern Observatory, Casilla 19001, Santiago 19, Chile*
[2] *Leiden Observatory, P.O. Box 9513, 2300 RA Leiden, The Netherlands*
[3] *Steward Observatory, Univ. of Arizona, Tucson, AZ 85721, USA*

1. Introduction

The large abundance of CH⁺ in diffuse clouds has been a mystery for more than 50 years (Dalgarno 1976). Many different explanations have been proposed, but only (shock) models with a substantial column of warm gas appear capable of approaching the observed column densities (Elitzur and Watson 1978). In these models, CH⁺ is formed in the warm postshock gas through the reaction $C^+ + H_2 \to CH^+ + H$, which is endoergic by 4650 K. Although the most sophisticated MHD shock models are consistent with various observational aspects of CH⁺, they require substantial "fine-tuning" of the parameters (Draine and Katz 1986; Hartquist et al. 1990). In addition, they predict a shift in velocity between CH⁺ and other molecules found in the cold quiescent gas, which is not observed in recent data (e.g. Crawford 1989). In order to test further the shock models, we have searched for CH⁺ in a number of translucent clouds ($A_V \approx$ 1–5 mag), which have H_2 column densities that are up to an order of magnitude larger than the clouds studied so far (cf. Souza 1979). Observations of the chemically related molecules CH, C_2, CN and CO have been obtained as well.

2. Observations and Analysis

Spectra of CH⁺, CH, CN and C_2 at $\lambda/\Delta\lambda$=60,000–100,000 were observed with the ESO 1.4m Coudé Auxiliary Telescope equipped with the Coudé Echelle Spectrometer and CCD detector. For CH⁺, we obtained spectra of both the A–X (0,0) and (1,0) bands. For the other molecules, we generally also measured at least two different transitions (Gredel et al. 1991). Millimeter emission lines of ^{12}CO and ^{13}CO 1-0 were observed with the SEST.
Most of the optical absorption lines are saturated, so that the Doppler parameter b along the line of sight needs to be accurately known in order to derive column densities. For CH⁺, b can be determined from the ratio of the strengths of the A–X (0,0) and (1,0) bands. For lines of sight with high enough S/N, this suggests $b = 2 - 3$ km s⁻¹. We assume that similar b–values apply to the CH⁺ observations for other lines of sight. For CN, the saturation corrections have been determined from comparison of the violet B–X and red A–X systems, resulting in accurate b values of 0.4–1.0 km s⁻¹ which agree well with those derived from the widths of ^{13}CO millimeter lines. For CH, we assume b–values in the range $(1.0-1.4) \times b(CN)$.

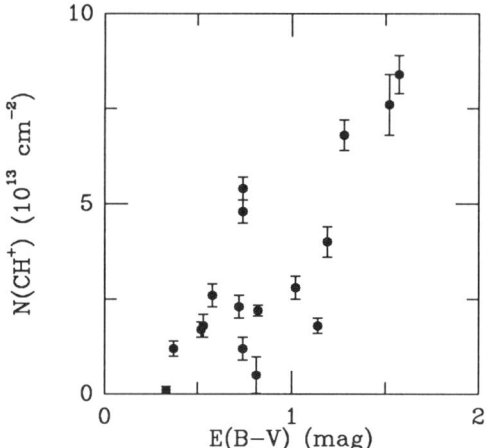

Figure 1. Observed CH^+ column densities as functions of reddening $E(B-V)$.

3. Results and Discussion

Figure 1 shows the CH^+ column densities obtained in this work as functions of $E(B-V)$. The most significant result is that they continue to increase with reddening. Moreover, the CH^+ velocities agree with those of CH, CN and C_2 within the errors. The largest CH^+ column densities are found for lines of sight for which the CO emission profiles are broader or more complex. The C_2 and CN column densities are strongly correlated with that of CH, which is probably a good tracer of the H_2 column density. CH^+ shows a larger scatter with CH, but still an overall increase. Finally, there is a tendency for the column density of CH^+ to decrease with increasing density n as derived from the C_2 excitation. In contrast, those of CN and C_2 increase strongly with n.

The fact that the CH^+ column density continues to increase with total column density is difficult to reconcile with the shock models, unless the number of shocks also increases. The absence of significant velocity shifts between CH^+ and other species further argues against the shock models. However, the larger b-values inferred for CH^+ together with the inverse correlation with density and possible association with more complex CO line profiles suggest that some energetic mechanism is responsible for the formation of the ion. Possibilities include translationally "hot" C^+ ions or H_2 molecules, since only a fraction 10^{-4} is needed to explain the observed CH^+ abundances. The warm turbulent interfaces of the clouds or clumps are a possible formation site (Duley et al. 1991), but quantitative calculations must await a physical description of turbulence.

Crawford, I.A. 1989, MNRAS 241, 575.
Dalgarno, A. 1976, in *Atomic Processes and Applications*, eds. P.G. Burke and B.L. Moiseiwitsch (North-Holland), p. 110.
Draine, B.T. and Katz, N. 1986, ApJ 310, 392.
Duley, W.W., Hartquist, T.W., Sternberg, A., Wagenblast, R., and Williams, D.A. 1991, this conference.
Elitzur, M. and Watson, W.D. 1978, ApJ 222, L141.
Gredel, R., van Dishoeck, E.F., and Black, J.H. 1991, A&A in press.
Hartquist, T.W., Flower, D.R., and Pineau des Forêts, G. 1990, in *Molecular Astrophysics— a volume honoring A. Dalgarno*, ed. T.W. Hartquist (Cambridge University), p. 99.
Souza, S.P. 1979, PhD Thesis, State University of New York at Stony Brook.

CHEMICAL EFFECTS OF SHEAR ALFVÉN WAVES IN MOLECULAR CLOUDS

S.B. CHARNLEY[1,†] and W.G. ROBERGE[2]
1. *Space Science Division, NASA Ames Research Center, California 94035.*
2. *Physics Department, Rensselaer Polytechnic Institute, New York 12180.*

ABSTRACT. We consider the propagation of low-amplitude MHD waves in partially-ionised plasmas. Ion-neutral drift (ambipolar diffusion) can lead to significant variations in the deuterium fractionation ratios of several molecules (e.g. HCO^+ and N_2H^+) on spatial scales of between a few hundredths and a few tenths of a parsec, depending upon the fractional ionisation of the plasma. It is possible that interstellar Alfvén waves could be detected by molecular spectroscopy and that these waves may produce other small-scale abundance gradients in molecular clouds.

The dense interstellar medium consists of neutral molecular gas, ions, electrons, and charged grains. This partially-ionised plasma is threaded by a weak magnetic field ($\sim$ 1-40μG) to which the ions, electrons and the negatively charged dust grains are attached; the plasma and dust are collisionally coupled to the neutral component. Perturbations of this system can generate a spectrum of magnetohydrodynamic (MHD) waves of which, for molecular clouds, the two most important types are compressional (magnetosonic) waves and shear Alfvén waves (Zweibel & Josafatsson 1983).

We have explored the chemistry of molecular gas in which shear Alfvén waves propagate. The magnetohydrodynamical structure and evolution of the wave packet are governed by Maxwell's equations, plus dynamical equations for the coupled fluids of neutral particles, ions, electrons, and dust grains (e.g. Elmegreen 1979; Draine 1986). Solution of the *linearised* problem (Roberge & Hanany 1991) indicates that appreciable ion-neutral drift speeds, v_{ni}, can occur ($v_{ni} <$ 1 km s^{-1}). Charnley & Roberge (1991a) have pointed out that this may have important consequences for interstellar chemistry: several important endothermic ion-molecule reactions may proceed rapidly in such a wave. Enhanced rates for reaction of H_2 with H_2D^+, CH_2D^+ and C_2HD^+, suggest that the high degree of deuterium fractionation observed in cold molecular clouds may offer a test for the chemical signature of Alfvén waves. An important quantity for the time-dependent ion-molecule chemistry is the time-scale of the coupling of the neutrals to the wave by elastic collisions with the streaming ions, τ_{ni}, which is inversely proportional to the fractional ionisation, x_e. A related quantity is the slowing-down length, $L_{ni} = \tau_{ni} v_A$, where v_A is the Alfvén speed. If the characteristic time-scale for a given chemical reaction to occur is τ_R, it will be affected by the presence of a wave if the ratio τ_{ni}/τ_R is (much) greater than unity. The maximum value of v_{ni} determines the most endothermic reaction that can be driven efficiently in the wave.

For isothermal shear Alfvén waves propagating in cold gas the only effect which can alter the molecular chemistry is the kinetic energy contributed to ion-neutral processes by ambipolar diffusion. We have incorporated this effect in a similar manner to that used to study the destruction of deuterated molecules in C-type MHD shocks (Pineau des Forêts *et al.* 1989). The deuterium chemistry model evolves in a 'pseudo-time-dependent' manner (Millar *et al.* 1989) until some time, t_{wave}, when the parcel of gas is swept by an Alfvén wave; the chemical evolution is then followed for times $\sim 5\tau_{ni}$. Figure 1 shows the chemistry in one representative model for $t_{wave} = 10^5$ years, at which point the chemistry is well-developed and a significant degree of D fractionation has occurred.

†*NAS/NRC Resident Research Associate*

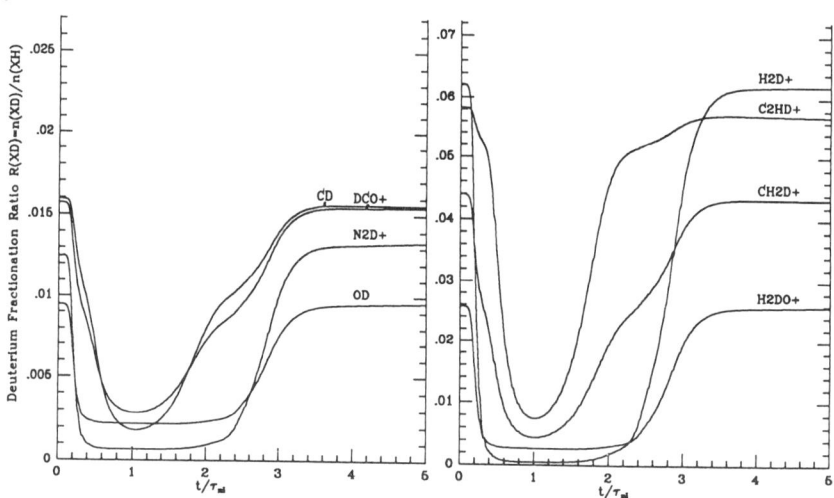

Figure 1. Evolution of D fractionation ratios in an Alfvén wave propagating in gas with $n_H = 2 \times 10^4 \text{cm}^{-3}$, $T_{kin}=12\text{K}$, and $x_e = 1.7 \times 10^{-7}$. For this model $\tau_{ni} = 5.2 \times 10^3$ years, $L_{ni} = 0.01$ pc, and the peak value of v_{ni} is 1.05 km s^{-1}.

As H$_2$D$^+$ and H$_3^+$ are difficult to detect in the ISM, the species which appear to be the best candidates for diagnostics of MHD wave propagation are DCO$^+$ and N$_2$D$^+$. Depending upon the ion-neutral drift speed ($v_{ni} \sim 0.5 - 1$ km s^{-1}) and the fractional ionisation of the gas, $R(\text{DCO}^+)$ and $R(\text{N}_2\text{D}^+)$ can easily vary by factors of between 3 and 10 over spatial scales of $\sim 0.01 - 0.1$ pc (Charnley & Roberge 1991b). High resolution mapping of dark clouds may discover the predicted variation in deuterium fractionation ratios and could allow the detection of hydromagnetic waves by molecular spectroscopy. We note that published maps of TMC-1 do show spatial variations in the emission of DCO$^+$ and HCO$^+$ (Guélin et al. 1982). The theory predicts that the molecular gas with the narrowest linewidths should contain the highest values of $R(\text{XD})$; observations by Bell et al. (1988) of $R(\text{C}_3\text{HD})$ in several dark clouds clearly show this trend.

We have developed this simple model to study the deuteration effects of *one* shear Alfvén wave, however, other ion-molecule reactions with small activation barriers may also be driven by these waves, and this work suggests that the presence of small-scale abundance gradients in dark clouds is related to the local level of MHD turbulence. MHD turbulence will generally act to produce spatial magnetic field gradients within interstellar clouds and the resulting ambipolar diffusion can provide a comprehensive mechanism to drive endothermic ion-molecule reactions.

References

Bell, M.B. et al. 1988, *Ap.J.*, **326**, 924.
Charnley, S.B. & Roberge, W.G. 1991a, in *Molecular Clouds*,
    eds. R.A. James & T.J. Millar, Cambridge, p.247
Charnley, S.B. & Roberge, W.G. 1991b, *in preparation*.
Draine, B.T. 1986, *M.N.R.A.S.*, **220**, 133.
Elmegreen, B.G. 1979, *Ap.J.*, **232**, 729.
Guélin, M., Langer, W.D. & Wilson, R.W. 1982, *Astr.Ap.*, **107**, 107.
Millar, T.J., Bennett, A. and Herbst, E. 1989, *Ap.J.*, **340**, 906.
Pineau des Forêts, G., Roueff, E. & Flower, D. 1989, *M.N.R.A.S.*, **240**, 167.
Roberge, W.G. & Hanany, S. 1991, *Ap.J.*, , submitted.
Zweibel, E.G. & Josafatsson, K. 1983, *Ap.J.*, **270**, 511.

Interpretation of the level population distribution of highly rotationally excited H_2 molecules in diffuse clouds

R WAGENBLAST

Department of Mathematics, UMIST, PO Box 88, Manchester M60 1QD, UK

September 13, 1991

The observed H_2 level population distribution of the rotational levels $J = 5, 6$ and 7 in the ground electronic and vibrational state $X(v = 0)$ is analysed using available Copernicus data (Spitzer and Morton, 1976) for the diffuse clouds toward the stars ζ Oph, δ Per and ξ Per. The abundances of H_2 in these high purely rotationally excited energy levels are due to the influence of one or more of three possible excitation mechanisms: collisions with abundant particles (e.g. H atoms and H_2 molecules), UV-photoexcitation and H_2 formation.

The UV-photoexcitation process on its own produces an occupation of the $J = 5$ level which is about 40 times bigger than that of the $J = 7$ level; the population of the $J = 6$ and $J = 7$ level of H_2 in diffuse clouds is dependent on the para-ortho ratio of molecular hydrogen. The observations show that the column density of H_2 in the $J = 7$ level is about 3 times higher than expected if UV pumping was the main excitation mechanism. Furthermore, the observed column density ratio of H_2 in the $J = 6$ and 7 level is found to be independent of the measured para-ortho of H_2. Therefore the UV pumping process cannot be the primary excitation mechanism for H_2 in these high J levels.

In contrast to UV pumping, collisional excitation of H_2 molecules could be the predominant action responsible for the observed column density distribution between the high J levels. However, the physical conditions needed to generate this distribution are quite special; the kinetic gas temperature would need to be about 2000K and the excitation of the H_2 molecules would have to be caused by H atoms at a density of about 300 cm^{-3} in quite efficient collisions. Collisions with other H_2 molecules are much less effective and so the density required would be much higher. While such a solution is possible, it is quite inconsistent with other determinations of these parameters.

We have investigated the alternative explanation of the J level population distribution which is based on the H_2 formation process. Assuming that the internal energy distribution of newly formed H_2 molecules is narrow, the observed column density distribution of H_2 in the $J = 5, 6$ and 7 level can be reproduced if H_2 molecules are initially in one of two neighbouring rovibrational levels (one for ortho-H_2, one for para-H_2) following: $(v = 1; J = 7, 8)$, $(v = 4,...11; J = 9, 10)$ or $(v = 10, 11; J = 11, 12)$; this choice is consistent with the H$_2$ formation model of Hunter and Watson (1978).

If the population of high J levels is mainly due to H$_2$ formation, a simple estimate for the evolution time t of a diffuse cloud can be made supposing the cloud contained at the beginning of its development virtually only atomic hydrogen and there is still plenty of it available

$$t = \frac{N(\mathrm{H}_2)}{N(\mathrm{H}_2 | J = 5)} \times 1.5 \mathrm{yr}. \tag{1}$$

The typical 'age' of diffuse clouds is within 5×10^5 to 1×10^6 yr for this kind of estimate. This is another reason why the formation of H$_2$ on grains is the most favourable process which can explain the observed J level population distribution of H$_2$.

Chemical models of diffuse clouds are strongly affected by the assumption that H$_2$ formation rather than UV pumping is the vital process for the population of high J levels in H$_2$ molecules, because there is no need any more for a strong UV field to explain why these levels are so heavily populated. A (non-equilibrium) model for the diffuse cloud toward ζ Oph was made which reproduces not only the H to H$_2$ ratio and the measured H$_2$ (J) column density distribution but also very satisfactorily the abundances of all observed chemical species (except CH$^+$), especially the high amount of CO. The model assumes that the cloud has a plane-parallel structure and a uniform density ($n_H = 240$ cm^{-3}). The cloud is subdivided in two parts with different kinetic temperatures (30 K and 140 K). The UV field which irradiates the cloud from the warm side has an intensity which is only a fifth of the standard value for the interstellar UV field. The cosmic ray ionization rate is 2×10^{-17} s^{-1}. The evolution time for the ζ Oph cloud is 1×10^6 yr.

Additional observations of H$_2$ in high rotational levels are needed to find out more about their main excitation mechanism. A far-UV spectrometer with high sensitivity and resolution is required for these observations (Lyman-FUSE in 1997). The detection of water in ζ Oph which is predicted by the model could give additional support for the theory that H$_2$ formation (or collisional excitation) rather than UV pumping is the vital excitation process for high J levels in H$_2$.

References

HUNTER, D A and WATSON, W D: 1978, *Astrophys. J.* **226**, 477
SPITZER, L and MORTON, W A: 1976, *Astrophys. J.* **204**, 731

THE STRUCTURE OF QUIESCENT CLOUDS

E. FALGARONE
*Radioastronomie Millimétrique, Ecole Normale Supérieure,
24 rue Lhomond, 75005 Paris, France*

Abstract. Several tracers of the cold interstellar medium in its quiescent phase reveal common unexpected properties: the fractal nature of the emission contour levels, the self-similar brightness distribution, the existence of small unresolved fragments of dense gas in various environments, and a highly turbulent velocity field. In spite of the difficulties met in interpreting these data, they are important bearings in our understanding of the structure, the physics and the chemistry of this medium.

This paper is a review of the observational grounds of our present knowledge of the structure of quiescent gas. Quiescent gas is defined as the component of the cold interstellar medium which it has not yet formed stars at a level large enough to significantly perturb the gas dynamics and chemistry. Quiescent gas is either mostly atomic or mostly molecular, depending on its shielding from the ambient radiation field. Its hydrogen average column density *at the parsec scale* does not exceed a few 10^{21} cm^{-2} and its degree of gravitational binding depends on the linear scale considered. It is a general terminology based on the physics of the gas. It clearly includes diffuse, translucent, and high latitude clouds which have been distinguished in the past on observational grounds (see van Dishoeck, this volume).

1. New observations: general trends

With the recent advent of extremely sensitive detectors in the millimetre, submillimetre and infra-red ranges, a new generation of maps characterized by large dynamical ranges even over regions of weak emission, has brought into light hidden properties of the cold interstellar medium. Eventhough their interpretation is not straightforward, it is encouraging to recognize that most tracers of the column density of quiescent gas agree on several general characteristics which are given below. Those tracers are: the IRAS 100μm emission which traces the warm dust column density in areas far from star forming regions and the line integrated CO and HI emissions which trace the H$_2$ and HI column density.

(i) A large connectivity of the emission makes the concept of "cloud" more and more elusive. This is spectacularly illustrated by the most recent IRAS 100μm maps of the sky produced by the Image Processing and Analysis Center at Caltech (Boulanger, private communication). One of them is shown on the front cover of the proceedings of the IAU Symposium 147.

New CO observations of large areas on the sky, down to unprecedented sensitivity limits, reveal that the previously believed isolated clouds are in fact connected by widespread emission at a very low level. Lee et al. (1990) for instance note that it is not possible to define cloud boundaries at a level of 1.2K in the spatial velocity maps of the inner Galatic plane, because, at this level, the emission from all clouds in the map merge together. Heithausen and Thaddeus (1990) have extended their survey of the Polaris Flare to the North Galactic Pole and find that the CO emission extends over $\sim 40\,\mathrm{pc}$ at a level $W(CO) = 0.4\,\mathrm{K\,km\,s^{-1}}$. It corresponds, for the standard CO/H_2 conversion factor, to quite a low column density, $N_{H_2} \sim 10^{20}\,\mathrm{cm^{-2}}$. In these new maps, CO peaks clearly exist but are connected in projection by large areas of low intensity emission.

Recent observations of an HI nearby cloud in Ursa Majoris with the Penticton interferometer also reveal long filamentary structures, the morphology of which is rapidly changing with velocity (Joncas, Boulanger and Dewdney, 1992).

(ii) Unresolved structure exists in all the maps (the smallest observed structures in quiescent gas are $\sim 0.01\,\mathrm{pc}$, a limit provided by single dish observations of rotational transitions of CO in the nearest clouds). This structure is usually more visible in channel maps and is found even in the most transparent parts (Falgarone and Pérault 1988; Falgarone, Phillips and Walker 1991; Falgarone, Puget and Pérault 1992). In the Taurus-Auriga-Perseus complex, a region of average H_2 column density $\sim 8 \times 10^{20}\,\mathrm{cm^{-2}}$ was mapped at high angular resolution in the $^{12}CO(J=2-1)$ and $(J=3-2)$ transitions. Holes appear between bright regions of all possible sizes. As in all other maps (infra-red and HI), no characteristic scale is visible.

(iii) There is a clear scale invariance of the maps of integrated emission. The maximal variations of $W(CO)$ in a map, for example, scale with the separation over which they are observed. This scale invariance is at the origin of the fractal geometry of the emission contours discussed below.

Contrasting with the agreement shared by the tracers of column density, large variations are found among "spectral" tracers. Illustrations are provided by Boulanger et al. (1990) who found small scale infrared color fluctuations in the IRAS maps of nearby molecular clouds. They cannot be explained by variations in the ambient radiation field because the UV shielding in these clouds is low everywhere. They likely reveal the existence of complex and rapid exchanges of constituents between the gas phase and the dust grain surfaces.

Another puzzle has been raised by a by-product of the deep CCD survey of faint field galaxies of Guhathakurta and Tyson (1990). They find that, in several high latitude clouds, the visible emission is correlated at large scale with the IRAS 100 μm emission although at small scale fluctuations of the ratio $I_B/I_{100\mu m}$ as large as 10 appear over all sizescales. Alike the IRAS color variations described above, these fluctuations are not easily understood in terms of extinction variations because they occur in gas of extremely low column density.

2. Fractal structure

The first point over which, surprisingly, many authors agree is that the complex structure of the maps can be described with the tools of fractal geometry.

Is fractal a structure in which the number of elements necessary to cover it increases as a power law of either the size of the elements (in a fractal of finite size) or the size of the structure (in a growing fractal). The exponent may be an integer. The fractal behaviour appears as soon as this exponent is smaller than the dimension of the embedding space. The essential property of fractals is their scale invariance.

All the attempts to measure the fractal dimension of the emission contour levels of several tracers of quiescent gas have lead to similar values of D_B. The method used is the same. It is an estimate of the area and perimeter of a given contour, which scale as $P \propto A^{D_B/2}$ if the contour is fractal of dimension D_B. Another independent determination of D_B is provided by the scaling of the perimeter with the resolution ϵ at which it is measured, $P \propto \epsilon^{1-D_B}$ (Lovejoy, 1982).

In the Taurus complex, Falgarone, Phillips and Walker (1990) find that the dimension is the same, $D_B \sim 1.4$, not only at three different linear scales but for three different rotational transitions of CO which are sensitive to gas of different densities. This dimension is comparable to that found on IRAS $100\mu m$ emission maps of the Taurus complex (Scalo, 1990), of a high latitude cloud (Bazell and Désert, 1988) and of other nearby clouds (Dickman, Horvarth and Margulis, 1990) as well as for the HI emission in a high velocity cloud (Wakker 1990).

The surprising fact, namely whichever tracer is used the fractal dimension is the same, suggests that lower density molecular gas and even cold atomic gas are distributed on sets which have the same fractal topology as that of denser gas, for instance that seen in $^{12}CO(J=3-2)$.

This value is indicative of a possible link between the topology of the cold interstellar medium and the role of turbulence in structuring it. Sreenivasan and Méneveau (1986) measure the same fractal dimension for a variety of interfaces in turbulent flows and more specifically for the surfaces of isodissipation of the kinetic energy.

The knowledge of the fractal dimension of a projected quantity, such as the integrated brightness, does not provide the actual spatial distribution of the gas nor the physics underlying this structure, but it may be used as an indicator to be compared with the dimensions found in other systems in Physics, or for other tracers of matter in Astrophysics. We describe below the methods followed to derive gas densities from the complex maps obtained in the CO rotational lines, for example.

3. Density structure

In addressing the issue of the density of (molecular) clouds, one has to distinguish between the average density over a given sizescale and the local density which may be very different from the former.

The average density in molecular clouds is usually derived from an excess above a local background of integrated CO (or isotopes) emission found in space-velocity maps. This excess is then converted into a column density by using either empirical relations between observed line integrated intensities and the H_2 column density or conversion factors (see the pannel discussion, this conference). The last step is the conversion of the column density N_H into an average density over l, $\bar{n}_H(l) = N_H/l$ where l is the projected size of the observed excess.

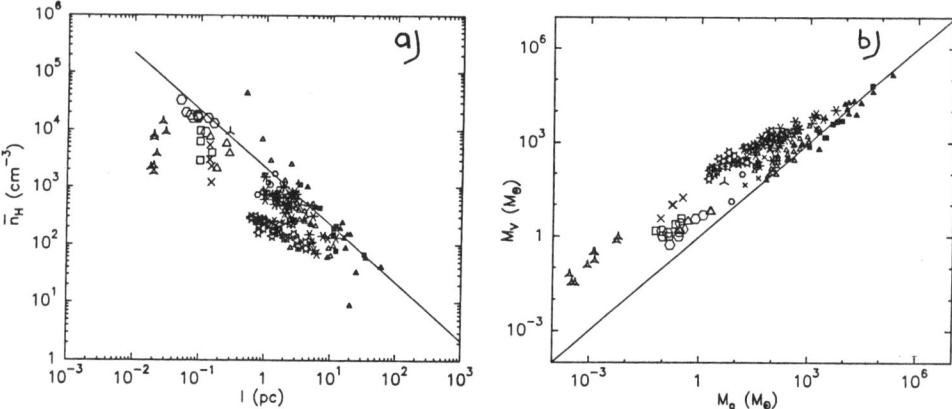

Fig. 1. *(a)* Average hydrogen density versus size in an ensemble of structures from different samples of quiescent gas (large symbols: Falgarone et al. 1992; Herbertz et al. 1991; Carr 1988; small symbols: Falgarone, Pérault and Puget 1985; Falgarone and Pérault 1987). The line $\bar{n}_H \propto l^{-1}$ is that given in Falgarone and Puget (1986) for an ensemble of self-gravitating structures. *(b)* Virial mass versus total gas (hydrogen plus helium) mass for the same structures. The line shows $M_v = M_g$. Note that the sample includes structures up to $M_v \sim 100\ M_g$.

When the works of different groups following comparable methods of analysis over samples of quiescent gas are put together, the net result, illustrated in Fig. 1, is a gross dependence of the maximum average density at a given scale as the inverse of the size down to 0.02 pc with a large spread in each range of size toward low average densities (Falgarone, Puget and Pérault, 1992). As a result, the sample includes structures which span the whole range from far below up to close to virial equilibrium (Fig.1b). The results shown in Fig. 1 have been derived from ^{12}CO, ^{13}CO and C^{18}O data. It is remarkable that no segregation appears among the various density determinations.

The local density in turn is derived from our knowledge of the excitation mechanisms of the rotational levels of molecules. But the derivation is far from straightforward. It implicitly assumes that quiescent gas has general properties, whichever individual cloud it belongs to. It also heavily relies on the scale invariance properties described above and the conclusions are derived from a body of elements rather than from a single set of observations.

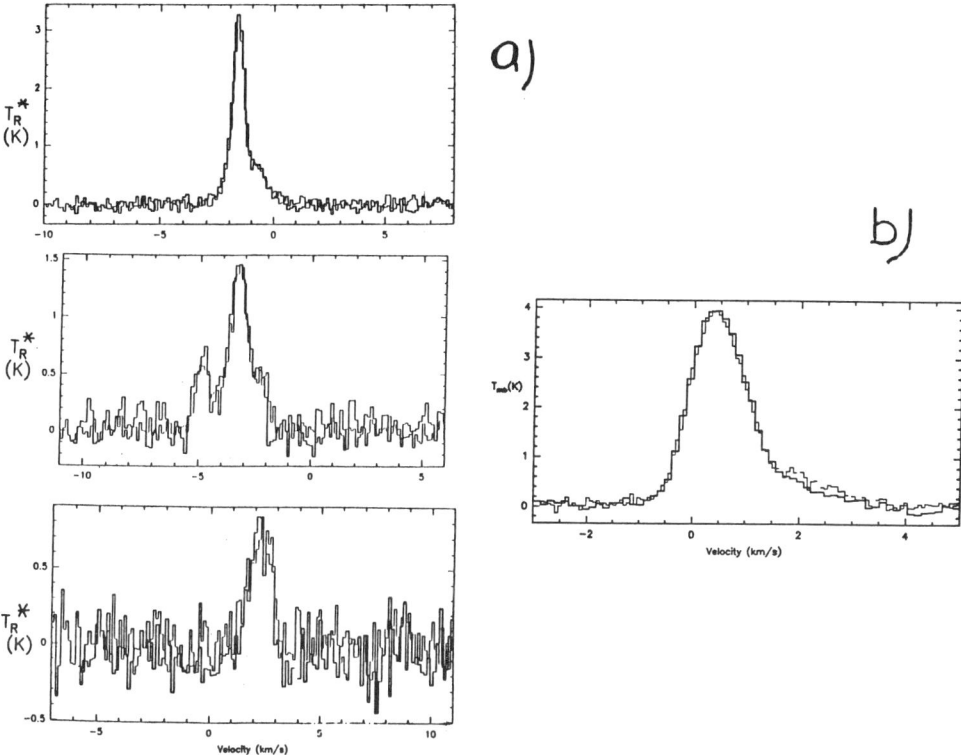

Fig. 2. (a) ^{13}CO(J=1-0) (thin line) and (J=2-1) (thick line) spectra of the integrated emission of three different fields in quiescent clouds. Their sizes are respectively 0.07×0.07 pc, 0.2×0.05 pc and 0.01×0.01 pc. The temperature scale is that of the ^{13}CO(J=1-0) line. The 2-1/1-0 line intensity ratios are 0.5, 0.4 and 0.45 respectively. (b) Same with the ^{12}CO lines over a different field (size 0.2 pc $\times 0.1$ pc). The line 2-1/1-0 line intensity ratio is 0.75 (main beam temperature).

The major element is the spectacular similarity of the entire J=1-0 and 2-1 profiles in ^{13}CO (Fig. 2a) and ^{12}CO (Fig.2b). It means that the line temperature ratio is uniform over a large range of $N(CO)/\Delta v$ and/or line intensities. Another element is the uniformity of the ^{12}CO(J=3-2) to (J=2-1) line integrated intensity ratio over one order of magnitude of line intensity (Falgarone et al. 1991). These results basically suggest that photon trapping does not play a role in the line excitation. For ^{13}CO, several possibilities exist, namely subthermally excited transitions and warm gas, or thermal excitation in colder gas. But for ^{12}CO, which is optically thick, thermalization is clearly inferred. According to existing excitation models, the local density is therefore of the order of 10^4 cm^{-3} (or even larger) but the exact value is model dependent. The apparent disagreement between the gas density derived by this method and that derived from line absorption profiles in the UV and visible (see van Dishoeck, this volume) is uncomfortable and has to be understood.

It may be argued that the kinetic temperature is not well known. However, an indirect argument (again) favors low kinetic temperatures for the CO emitting regions. The maxima of CO antenna temperature are always low in quiescent regions. Histograms of peak antenna temperatures of several maps show clear maxima between 7K and 9K in the ^{12}CO(J=1-0) line and between 3K and 5K in the ^{12}CO(J=2-1) line (Clemens and Barvainis, 1988). These values suggest kinetic temperatures of the order of 10K or less, if CO emission is thermalized although higher T_k are expected in the least shielded regions.

These results favor a picture in which the CO emission arises in an ensemble of tiny, dense and cold structures, distributed on a fractal set immersed in a lower density and warmer component. The ultimate scale has to be small, $\sim 1000 AU$, because of the observed smoothness of the line profiles (Tauber, Goldsmith and Dickman, 1991; Falgarone et al. 1992).

It is even possible that such tiny dense structures exist in atomic gas. Very Long Baseline Interferometry (VLBI) measurements of HI absorption against several extragalactic sources (Diamond et al. 1989) reveal line opacity variations over angular scales $\sim 0.05"$ possibly originating in dense structures of atomic hydrogen $n_{HI} \sim 10^4 - 10^5 \, \text{cm}^{-3}$ smaller than $\sim 25 \, AU$. In the summary of their review on the HI emission, Dickey and Lockman (1991) note that "... aperture synthesis observations do show some variations in 21cm opacity on all angular scales".

4. Velocity structure

Quiescent gas is highly turbulent and the observed internal motions have several well-defined properties. The first moments of the velocity distribution scale as $\delta v \propto l^\alpha$ with $\alpha \sim 0.4$ (Falgarone et al. 1992), a scaling law similar to that found by Larson (1981) among molecular clouds of all kinds. This scaling is not governed only by gravity since it is found in a large sample of mostly unbound entities, of sizes ranging from 0.05 pc to 50 pc. It is very likely governed by MHD turbulence (Myers and Goodman, 1988a and b). The second characteristic of the velocity field, as provided by the lineprofiles, is the existence of high velocity wings which are not due to stellar outflows. It has been proposed that these wings are due to cloud evaporation (Magnani et al. 1988) to cloud collisions (Keto and lattanzio 1989) or Alfvén waves (Elmegreen 1990). But the scale invariance of the linewings and the fact that the gas which emits in the wings is as dense and cold as that which emits in the cores suggest that the CO line profiles trace the zones of intermittency of the turbulent velocity field (Falgarone and Phillips 1990). They are zones of ephemeral enhanced vorticity and shear which are also those in which the dissipation of the turbulent kinetic energy is concentrated. Intermittency is observed in laboratory flows (Anselmet et al. 1984, in atmospheric clouds (van Atta and Park 1971) and in numerical simulations (Vincent and Meneguzzi 1991). It is characterized by non-Gaussian distributions of the velocity derivatives and increments.

Important consequences can be drawn. The limit between macroscopic and microscopic processes tends to disappear in these regions if one considers that the dissipation scale is only a few times larger than the mean free path for a neutral neutral collision at densities between $10^2 \, \text{cm}^{-3}$ and $10^3 \, \text{cm}^{-3}$. Also the dissipation zones have been estimated to be very hot. The kinetic temperature depends on their volume filling factor. One finds $T_k \sim 100K - 10^3 K$ for $f_v \sim 10^{-3}$ for a low and high ionization degree respectively. This locally high kinetic temperatures do increase the rate of some chemical reactions, in particular radiative associations (Lignières and Falgarone 1992).

5. Concluding remarks

Quiescent clouds have a scale invariant structure with fractal edges of dimension $D_B \sim 1.4 - 1.5$ independent of the tracer used (HI, CO, or 100 μm) over decades of linear scales (< 0.1 pc to 50 pc). Their structure is very likely the result of MHD turbulence.

CO emitting gas in these clouds is cold, concentrated in tiny units $< 1000 AU$. The local density is high $\sim 10^4$ cm^{-3} or even larger, the actual value depending on the collisional excitation rates.

The gas is turbulent and exhibits the signatures of intermittency, in particular the non-Gaussian distribution of the velocity increments. Intermittency might be at the origin of local and transient zones of very high kinetic temperature.

A long list of open questions should follow but the page number limitation has already been overstepped. We anticipate that our understanding of the complex structure of quiescent gas might help elucidating the most flagrant mysteries of chemistry, like the formation of CH$^+$.

References

Anselmet, F., Gagne, Y., and Hopfinger, E. J. 1984, *J. Fluid Mech.*, **140**, 63.
Bazell, D. and Désert, F. X. 1988, *Ap. J.*, **333**, 353.
Boulanger, F., Falgarone, E., Puget, J.L., Helou, G. 1990, *Ap. J.*, **364**, 136.
Clemens, D.P. and Barvainis, R. 1988, *Ap. J. Suppl.*, **257**, 27.
Diamond P.J. et al. 1989, *Ap. J.*, **347**, 302.
Dickey J.M. and Lockman F.J. 1990, *Ann. Rev. Astr. Astrophys.*, **28**, 215.
Dickman, R.L., Horvath, M.A., and Margulis, M. 1991, *Ap. J.*, **365**, 586.
Elmegreen, B.G. 1990, *Ap. J. Letters*, **361**, L77.
Falgarone, E., and Pérault, M.: 1987, *Physical Processes in Interstellar Clouds*, eds. G.E. Morfill and M. Scholer.
Falgarone, E., and Pérault, M. 1988, *A.&A.*, **205**, L1.
Falgarone, E., and Phillips, T. G. 1990, *Ap. J.*, **359**, 344.
Falgarone, E., Phillips, T. G., and Walker C. 1991, *Ap. J.*, **378**, 186.
Falgarone, E., and Puget, J.L. 1986, *A&A*, **162**, 235.
Falgarone, E., Puget, J.-L., and Pérault, M. 1992, A&A in press.
Guhathakurta, P. and Tyson J.A. 1989, *Ap. J.*, **346**, 773.
Heithausen, A., and Thaddeus, P. 1990, *Ap. J. Letters*, **353**, L49.
Joncas, G., Boulanger, F. and Dewdney, P.E. 1992: *Ap. J.* in press.
Keto, E.R., and Lattanzio, J.C. 1989, *Ap. J.*, **346**, 184.

Larson, R.B. 1981, *Monthly Notices Roy. Astron. Soc.*, **194**, 809.
Lee, Y., Snell, R.L. and Dickman, R.L. 1990, *Ap. J.*, **355**, 536.
Lignières, F. and Falgarone E. 1992 in preparation.
Lovejoy, S. 1982, *Science*, **216**, 185.
Magnani L., Blitz L., and Wendel A. 1988, *Ap. J.(Letters)*, **331**, L127.
Myers, P.C., and Goodman A. 1988a, *Ap. J. Letters*, **326**, L27.
Myers, P.C., and Goodman A. 1988b, *Ap. J.*, **329**, 392.
Pérault, M., Falgarone, E., and Puget, J.L. 1985, *A.&A.*, **152**, 371.
Scalo, J.M. 1990 *Physical Processes in Fragmentation and Star Formation*, eds R. Capuzzo-Dolcetta et al., Kluwer Academic Publ.: Dordrecht.
Sreenivasan, K. R. and Méneveau, C. 1986, *J. Fluid Mech.*, **173**, 357.
Tauber J.A., Goldsmith P.F. and Dickman, R.L. 1991, *Ap. J.*, **375**, 635.
van Atta, C. W. and Park, J. 1971, *Statistical Models and Turbulence*, eds. M. Rosenblatt and C. van Atta: Springer.
Vincent, A. and Meneguzzi M. 1991, *J. Fluid Mech.*, **225**, 1.
Wakker, B.P., 1990, Ph.D. Dissertation, University of Leiden.

QUESTIONS AND ANSWERS

J.C.Pecker: You admit that turbulence (the Kolmogorov-spectrum, I presume) governs the fractal behavior of density distribution and "edge" struture. But in some cases, the Kolmogorov spectrum is known to be valid only in a certain scale interval of size (I am refering to Muller and Roudier study of the solar turbulent field). So my question: don't you think that, at unresolved scales, the fractal coefficient D might perhaps change its value (from about 2 to ...less, or to about 3!)? And, if so, could not models admit a possible range of variation of D between the resolved and the unresolved structures?

E.Falgarone: Yes, but there is not even agreement among theorists on the possible different links between the fractal dimension of turbulent interfaces (and dissipation zones) and the slope of the velocity-size power law correlation in laboratory incompressible turbulence. Turbulence in interstellar clouds is still much more complex because of the presence of magnetic field and we are far from being able to model the entire hierarchy.

A.Leger: Can you form CH^+ in your $T = 10^2 - 10^3\ K$ regions?

E.Falgarone: On energetical grounds only, CH^+ formation is therefore possible there. But detailed computations are needed which would include all the time and density dependences, very similar to what has been done in shocks.

Abundance of DCO⁺ in Nearby Molecular Clouds

Harold M. Butner[†]
Space Science Division, NASA Ames Research Center, California 94035.

ABSTRACT. DCO⁺ is one of the most common deuterated molecules in cold ($T_K \sim 10$ K) molecular cloud cores such as TMC-1. We report the results of a survey for DCO⁺ and H^{13}CO⁺ emission regions among a sample of low mass cloud cores. We compare the derived DCO⁺/HCO⁺ ratio (0.046±0.014) with current chemistry models for deuterium fractionation.

Deuterium Fractionation Chemistry

Watson (1977) pointed out that isotopic fractionation of molecules can be important for the conditions typically found in interstellar clouds. In particular, the abundance of DCO⁺ can be enhanced significantly over that expected from the cosmic [D/H] ratio. The reason for this fractionation is the ancestral reaction

$$H_3^+ + HD \longleftrightarrow H_2D^+ + H_2 + \Delta E$$

to the formation of DCO⁺. The energy difference, ΔE, causes the abundance of H_2D^+ to be temperature dependent. Estimates for ΔE lie between 180 K (Wootten *et al.* 1982) and 240 K (Herbst 1982), making the forward reaction dominant in the cold ($T_K \sim 10$ K) gas found in low mass dense cores. Since DCO⁺ is produced primarily by

$$H_2D^+ + CO \longleftrightarrow DCO^+ + H_2,$$

the DCO⁺ abundance is also temperature dependent. At low temperatures, the abundance of DCO⁺ is significantly enhanced over the cosmic abundance of [D/H] $\sim 1.7 \times 10^{-5}$ (York and Rogerson 1976).

Additional reactions suggested by Dalgarno and Lepp (1984) and revised H_3^+ recombination rates (Smith and Adams 1984) have complicated the simple picture developed by Herbst (1982). In addition, our knowledge about possible reaction networks that could occur has increased dramatically in the past ten years. There has also been increased interest in the possible time-dependent variations of molecular abundances (Millar *et al.* 1988). However, there have been few attempts to incorporate deuterium chemistry in the new reaction models.

For this reason, Millar *et al.* (1989) have calculated the expected relative abundance for a number of deuterated molecules over a range of physical conditions, including those found in TMC 1 (cold, $T_K \sim 10$ K, dense, n $\sim 10^4$-10^5 cm^{-3}). They used a reaction network, based on the models of Millar, Leung and Herbst (1987) which describes the chemistry of dense molecular clouds using a pseudo-time-dependent model. In cores with similar physical conditions to TMC 1, Millar *et al.* (1989) find that the DCO⁺/HCO⁺ ratio should be between 0.02 and 0.05.

Observations

Based on the Myers and Benson (1983) NH$_3$ observations of low mass dense cores, the NH$_3$

[†]NAS/NRC Resident Research Associate

cores identified in their survey should be an excellent laboratory to measure the amount of deuterium fractionation and thereby to constrain the chemical models better. The cores are all similar to TMC 1 in their density and temperature (n $\sim 10^4\text{-}10^5$ cm^{-3}, $T_K \sim$ 10-15 K based on NH$_3$ observations).

We observed the NH$_3$ cores using the DCO$^+$ and H^{13}CO$^+$ J=1→0 lines with the NRAO 12 meter and the DCO$^+$ J=2→1 line with the MWO 5 meter. We also observed the DCO$^+$ J=3→2 line with the NRAO 12 meter. H^{13}CO$^+$ was observed instead of the more abundant HCO$^+$ because the HCO$^+$ transitions are expected to be optically thick in most cases. Maps of 9 cores revealed that 2' was the average DCO$^+$ J=1→0 emission region, which is similar to the NH$_3$ core size seen by Myers and Benson. All line intensities were corrected to T_R, where a source size of 2' was assumed for η_c. We detected DCO$^+$ and H^{13}CO$^+$ in 24 out of 28 cores surveyed. We restrict our discussion to the 24 cores for which we have detections of both species.

The Deuterium Fractionation Ratio

To estimate the DCO$^+$/HCO$^+$ ratio, we used the column density derived from the DCO$^+$ and the H^{13}CO$^+$ J=1→0 observations with the assumption that $T_{ex} = T_K$. The H^{13}CO$^+$ abundance was converted to an HCO$^+$ abundance by assuming that the HCO$^+$/H^{13}CO$^+$ ratio is 75 (Wilson et al. 1981). Since the typical optical depth of the H^{13}CO$^+$ J=1→0 line is $\sim$ 0.2, the typical HCO$^+$ J=1→0 line will have optical depths in excess of 10, justifying our choice of H^{13}CO$^+$. There was no apparent difference in the DCO$^+$/HCO$^+$ ratio between cores with associated embedded infrared sources and cores without embedded sources. **The average [DCO$^+$]/[HCO$^+$] = 0.046±0.014 over all the cores is in excellent agreement with current models for cloud chemistry and deuterium fractionation in cold dense cores.**

The Physical Properties of the Cores: DCO$^+$ versus NH$_3$

From the DCO$^+$ J=1→0, J=2→1 and J=3→2 lines, it is possible to estimate the densities of the emission regions. Using an LVG model and assuming a gas kinetic temperature of 10 K, we derive an average density of 10^5 cm^{-3}. This is somewhat higher than the estimates derived form NH3, suggesting the DCO$^+$ is coming from a higher density region on average. The DCO$^+$ density estimates are in reasonable agreement with density estimates from other LVG studies using CS (Zhou et al. 1989) or C$_3$H$_2$ (Cox et al. 1989).

Another analomy compared to the NH$_3$ results of Myers and Benson is the observed DCO$^+$ linewidth. There are several potential explanations of the difference. However, the optical depth effects that have been proposed to explain large CS or C^{18}O linewidths (see Zhou et al. 1989 and references therein) do not appear to work for the DCO$^+$ lines. The mechanisms require either line scattering by an envelope, producing a large observed core size relative to the NH$_3$ core size, or large line optical depths. Neither is seen in the case of the DCO$^+$ J=1→0 line.

References

Cox, P., Walmsley, C. M., and Güsten, R. 1989, Astr.Ap., **209**, 382.
Dalgarno, A., and Lepp, S. 1984, Ap.J.(Letters), **287**, L47.
Herbst, E. 1982, Astr.Ap., **111**, 76.
Millar, T. J., Bennett, A., and Herbst, E. 1989, Ap.J., **340**, 906.
Millar, T. J., Defrees, D. J., McLean, A. D., and Herbst, E. 1988, Astr.Ap., **194**, 250.
Millar, T. J., Leung, C. M., and Herbst, E. 1987, Astr.Ap., **183**, 109.
Myers, P. C., and Benson, P. J. 1983, Ap.J., **266**, 309.
Smith, D., and Adams, N. G. 1984, Ap.J.(Letters), **284**, L13.
Watson, W. D. 1977, in CNO Isotopes in Astrophysics, ed. J. Audouze (Dordrecht: D. Reidel), p. 105
Wootten, A. 1987, in IAU Symp. No. 120: Astrochemistry, ed. M. S. Vardya and S. P. Tarafdar (Dordrecht: D. Reidel), p.311
Wootten, A., Loren, R. B., and Snell, R. L. 1982, Ap.J., **255**, 160.
York, D. G., and Rogerson, J. B. Jr. 1976, Ap.J., **203**, 378.
Zhou, S., Wu, Y., Evans, N. J. II, Fuller, G. A., and Myers, P. C. 1989, Ap.J., **346**, 168.

Molecular Abundance Variations Among and Within Cold, Dark Molecular Clouds

MASATOSHI OHISHI
Nobeyama Radio Observatory, Nobeyama, Mimamimaki, Minamisaku, Nagano 384-13, Japan

WILLIAM M. IRVINE
Five College Radio Astronomy Observatory, 619 Lederle GRC, University of Massachusetts, Amherst, MA 01003, U.S.A.

and

NORIO KAIFU
National Astronomical Observatory, 2-21-1, Osawa, Mitaka, Tokyo 181, Japan

September 15, 1991

Abstract. The latest table of molecular abundances in the cold, dark clouds TMC-1 and L134N is presented. Molecular abundance variations between TMC-1 and L134N, those within TMC-1 and L134N, and those among 49 dark cloud cores surveyed by Suzuki *et al.* (1991) are interpreted as an effect of chemical evolution.

Key words: Interstellar Molecules - Molecular Abundance - Chemical Evolution

1. Interstellar Molecules in Cold, Dark Clouds

The cold, dark molecular clouds are formation sites for low-mass stars. These clouds often contain several dense cores with $T_K \sim 10$ K, $n(H_2) \sim 10^4 - 10^5$ cm^{-3}, and the mass of one to a few M_o. Such physical conditions together with lack of embedded high-luminosity sources make the cold, dark molecular clouds ideal testing sites for models of gas-phase ion-molecule chemistry.

The recent development of large millimeter-wave telescopes like the 45-m telescope at Nobeyama and the 30-m dish of IRAM, and highly sensitive submillimeter-wave telescopes have resulted in detections of many new interstellar molecules. Some 90 interstellar molecules so far detected are summarized by Irvine, Ohishi and Kaifu (1991) (very recently CCO and SiN have been detected, and H_3O^+ has been confirmed.). They also list molecules detected in cold, dark clouds. As is well known, many radicals and molecular ions which have very short lifetimes under terrestrial conditions are often found in cold, dark clouds. Most of the molecules listed in Table II of Irvine, Ohishi and Kaifu were detected at the cyanopolyyne peak of TMC-1 (Taurus Molecular Cloud 1) that is located about 140 pc from the Sun. Characteristic molecules in TMC-1 are the carbon-chain molecules (C_nX ; X=H, N, O and S) and their derivatives. Several chemical models, e.g. Herbst & Leung (1989), show that these molecules' abundances peak in the "early time" ($\sim 3 \times 10^5$ years) and decrease rapidly as clouds reach the steady state.

Such molecules are not abundant in all cold, dark clouds which we can observe at present. One good example is provided by a comparison of molecular abundances between TMC-1 and L134N. Both clouds have very similar physical environments. But the molecular abundances have great differences, as are summarized in Table I. Carbon-chain molecules are much more abundant in TMC-1 (abundance ratios are greater than 2), while NH_3, SO, SO_2, CH_3OH, and NO are more abundant in L134N. Some molecules such as CO, CS, HCO^+ and H_2CO do not show any significant differences between the two clouds. The carbon-chain molecules and their derivatives are, as we have stated before, abundant in the "early time" of cloud evolution. NH_3, SO, SO_2, CH_3OH, and NO show a common characteristic in their formation chemistries : they include endothermic reactions or neutral-neutral ones. CO, CS, HCO^+ and H_2CO are widespread and usually very abundant in most or all molecular clouds.

There are a few published ideas to explain such abundance differences. One is that TMC-1 is carbon-rich while L134N is oxygen-rich. Another one is that TMC-1 is in the early stage of its chemical evolution and L134N is more evolved. Considerable effort to explain the abundance differences with these ideas has been made. But no one has found a clear answer for this basic question.

2. Abundance Variations within TMC-1 and L134N

One important approach for the above mentioned question would be to investigate abundance variation within a cloud. It is very natural for the chemistry to be strongly affected by the physical conditions (kinetic temperature, density, elemental abundance, radiation field, etc.) of the cloud.

Hirahara *et al.* (1991) have compared molecular distributions of several carbon-chain molecules (CS, C_2S, C_3S, HC_3N, HC_5N, and C_4H) with other ones (NH_3, HCS^+, SO and N_2H^+) in TMC-1, and found that carbon-chain molecules peak at the cyanopolyyne peak while NH_3, SO and N_2H^+ peak around the ammonia peak. The difference is prominent and, surprisingly, it is similar to that found in Table I between the two clouds TMC-1 and L134N ! They analyzed the data for C_2S and derived that the number density of molecular hydrgen at the cyanopolyyne peak is about 10^4 cm^{-3} and that for the ammonia peak is about 10^5 cm^{-3}. There is no clear evidence for current star-formation in the vicinity of TMC-1. These facts together with small core sizes ($\sim$ 0.02 pc) suggest that the cores may be younger than the time when the cloud would reach the steady state of its chemical evolution. Therefore Hirahara *et al.* favored the conclusion that the cyanopolyyne peak is chemically younger than the ammonia peak.

Swade (1989) made extensive mapping observations for L134N with $C^{18}O$, CS, C_3H_2, SO, $H^{13}CO^+$, NH_3, and so on. Although physical conditions do

TABLE I
Measured Molecular Abundances in Dark Clouds

Species	N(Species)/N(H_2) × 10^9		TMC-1/L134N	Note
	TMC-1	L134N		
CO	8000	8000	1	
C_2O	0.06			
C_3O	0.01	< 0.005	> 2	
C_2	5			a
OH	30	7.5	4	b
CH	2	1	2	b
C_2H	5-10	< 5	> 1	
C_3H	0.05			
C_4H	2	0.1	20	
C_5H	0.03			
C_6H	0.01			
CH_3CCH	0.6	< 0.12	> 5	
CH_3C_4H	0.02			
CN	3	< 0.3	> 10	
C_3N	0.1	< 0.02	> 5	
CH_3CN	0.1	< 0.1	> 1	
CH_3C_3N	0.05			
HCN	2	0.4	5	
HNC	2	0.6	3.3	
$HCNH^+$	0.19	< 0.31	> 0.6	
HC_3N	0.6	0.018	30	
HC_5N	0.3	0.01	30	
HC_7N	0.1	< 0.002	> 50	
HC_9N	0.03			
$HC_{11}N$	0.01			
CH_2CHCN	0.02	< 0.01	> 2	
CS	1	0.1	10	
HCS^+	0.06	0.006	10	
C_2S	0.8	0.06	13.3	
C_3S	0.1	< 0.02	> 5	
CH_2C_2	0.03			
CH_2C_3	0.08			
HNCO	0.02			
N_2H^+	0.05	0.05	1	
NH_3	2	20	0.1	
HCO^+	0.8	0.8	1	
H_2CO	2	2	1	
H_2CCO	0.1	< 0.07	> 1.4	
OCS	0.2	0.2	1	
SO	0.5	2	0.25	
SO_2	< 0.1	0.4	< 0.25	
CH_3CHO	0.06	0.06	1	
C_3H_2	1	0.2	5	
$c - C_3H$	0.06	0.03	2	
CH_3OH	0.2	0.3	0.66	
HC_2CHO	0.02			
CH_2CN	0.5	< 0.1	> 5	
H_2S	< 0.05	0.08	< 0.6	c
H_2CS	0.3	0.06	5	
NO	< 3.0	6.0	< 0.5	d
HCOOH	< 0.02	0.03	< 0.66	

Notes. Values assume column densities $N(H_2) = 1 \times 10^{22} cm^{-2}$ in both clouds. Positions are:

TMC - 1 : $\alpha(1950) = 4^h 38^m 38.6^s$, $\delta(1950) = 25°35'45''$;
L134N : $\alpha(1950) = 15^h 51^m 30.0^s$, $\delta(1950) = 2°43'31''$.

a : 20 arcmin from std. position. b : beam size ≫ that for heavier species. c : TMC-1 (detected at the ammonia peak (-4',+6')); L134N (3 × stronger at 1' west). d : Values refer to position in note a.

not seem to vary greatly, these molecules show very different distributions : for example, NH_3 shows two peak positions while $H^{13}CO^+$ has a single maximum between these two positions of NH_3, and furthermore SO shows several peaks away from the peaks of NH_3 and/or $H^{13}CO^+$. Thus it is clear that the molecular abundances listed in Table I are not entirely representative for both clouds. Swade concluded that the distributions reflect variation of the carbon to oxygen ratio within L134N. But his argument is based on the steady state chemistry. Many chemical model calculations show that it takes about 10^7 years to reach the steady state, which is usually longer than the clouds' lifetime. This means that the chemistry is not in the steady state, and therefore an argument based on the steady state chemistry may not be valid.

3. Variations among Cloud Cores and their Chemical Evolution

Another approach to study the abundance variation is to survey many dark cloud cores with some selected key molecules. Suzuki et al. (1991) have surveyed 49 dark cloud cores in the Taurus and in the Ophiuchus regions with C_2S, C_3S, NH_3 and related molecules. Although their purpose by comparing chemical network calculations and observational facts was to investigate the formation mechanism of the C_2S radical, they found that the abundance of the C_2S radical has a strong correlation with the chemical evolution of the cloud.

They also suggest that the abundance ratio of C_2S to that of NH_3 will be a good indicator of the chemical evolution as well as the physical evolution. Fig. 1 plots fractional abundances of C_2S of the cores surveyed by Suzuki et al. as a function of the column density ratio of C_2S to NH_3. These plots are classified into two categories : cores without IRAS point sources and those with IRAS. Because IRAS point sources are regarded as candidates for proto-stars, the cores with IRAS sources would be physically more evolved than those without IRAS sources.

As we can see from Fig. 1, the cores without IRAS sources tend to be distributed in the upper-left region of the diagram, while those with IRAS sources tend to stay in the oppsite side. When we overlay an "evolutionary track" of the chemical evolution of the core (Suzuki et al. 1991), the upper-left portion of the figure corresponds to a cloud age of $4 - 10 \times 10^5$ years and the lower-right part corresponds to $10 - 20 \times 10^5$ years (in the case of $n(H_2) = 10^4$ cm^{-3}, $T_K = 10$ K). The cyanopolyyne peak of TMC-1 is located in the upper-left region, the amminia peak is in the central area, and L134N locates at the furthest right side among the three cores. This clearly means that the cyanopolyyne peak of TMC-1 is the chemically youngest core, the ammonia peak of TMC-1 is chemically older than the cyanopolyyne peak, and L134N is the most evolved. Although the absolute

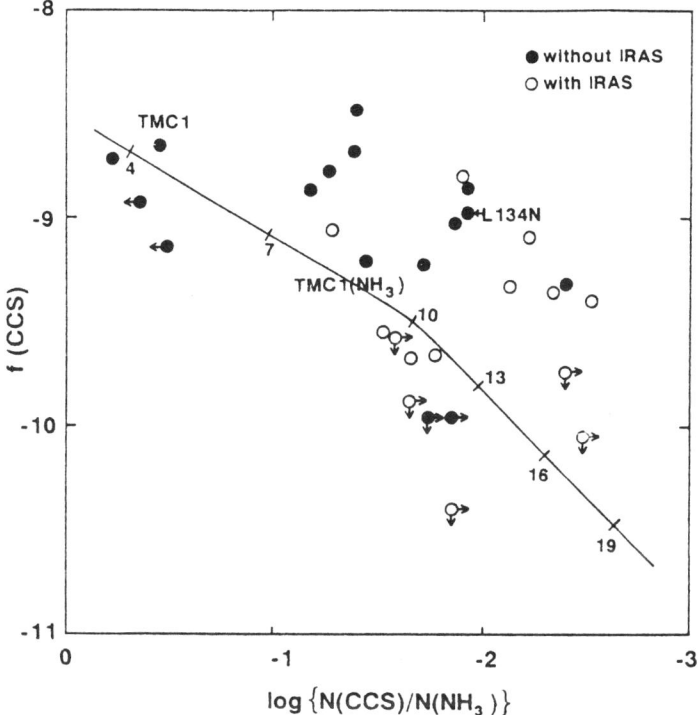

Fig. 1. Variation of the fractional abundance of C_2S to H_2 as a function of the abundance ratio of C_2S to NH_3. Filled circles represent dark cloud cores without IRAS point sources, and open circles are cores with IRAS sources. Arrows indicate limits. Solid line is the "evolutionary track" in the case of $n(H_2) = 10^4$ cm^{-3}, $T_K = 10$ K. Numbers along the line represent "chemical age" of the cloud in unit of 10^5 years.

values of the "chemical age" are not reliable, this diagram clearly shows a correlation of the chemical evolution and the physical evolution of the dark cloud cores. Finally we note that the simulation did not assume any elemental abundance variation, and we believe it will be possible to explain most of the abundance variations of molecular clouds by this idea of "chemical evolution".

4. Summary

Recent extensive millimeter-wave and centimeter-wave observations have revealed that the molecular abundances of cold, dark cloud cores vary not only from cloud to cloud but also from core to core inside a single cloud. Several ideas have been proposed to explain such variations. We propose that the variations can be reconciled considering the chemical evolution of the cloud core. More active research including the survey observations of dark cloud cores and reaction network calculations, are neccesary to understand the relation between the chemistry and the physics of dark cloud cores.

Acknowledgements

We thank S. Saito, K. Kawaguchi and S. Yamamoto for valuable discussions. Y. Hirahara has kindly provided us his maps of TMC-1 prior to publication. M. O. and W.M.I. is grateful for the financial support from the Japan-US Cooperative Science Program.

References

Herbst, E. & Leung, C. M. (1989) ApJS, **69**, 271.
Hirahara, Y. *et al.* (1991) submitted to ApJ.
Irvine, W. M., Ohishi, M. and Kaifu, N. (1991) Icarus, **91**, 2.
Suzuki, H. *et al.* (1991) submitted to ApJ.
Swade, D. A. (1989) ApJS, **71**, 219.

QUESTIONS AND ANSWERS

Qin Zeng: You did very good job to measure the column densities, but how to obtain the abundances from the column densities?

M. Ohishi: Fractional abundances of molecules are defined by $f(mol.) \equiv N(mol.)/N(H_2)$, where $N(mol.)$ and $N(H_2)$ represent the column densities of the molecule and of H_2, respectively.

RESULTS FROM A THREE POSITION SPECTRAL SCAN IN THE SGR B2 MOLECULAR CLOUD CORE

P. BERGMAN, Å. HJALMARSON, Onsala Space Observatory, Sweden
P. FRIBERG, Joint Astronomy Centre, Hilo, HI, USA
W.M. IRVINE, Five College Radio Astronomy Observatory, Amherst, MA, USA
T.J. MILLAR, UMIST, Manchester, England
M. OHISHI, Nobeyama Radio Observatory, Japan
S. SAITO, Department of Astrophysics, Nagoya University, Japan

We report on results from an ongoing spectral scan of four nearby positions in the Sgr B2 molecular cloud using SEST (Swedish-ESO Submillimeter Telescope). The antenna beam size is approximately 22" in the frequency range 226–245 GHz presently covered. This high angular resolution allows detailed studies of the physical and chemical conditions in the warm and compact cores discovered (Vogel et al. 1987, Goldsmith et al. 1987) inside the region previously surveyed in the 3 mm band with lower angular resolution (Cummins et al. 1986, Turner 1989, beam sizes of 1.5–2.9' and 1–2', respectively). The Sgr B2(OH) position used by these investigators is located about 30" south of our M position, and hence the cores M and N will contribute to the observed spectral line emissions to a larger or lesser extent.

The spectral differences between the positions we observe are very striking, see Figure 1. While the line density is high in the warm cores M and N the lines are rare in the 2'N position (Table 1). Some 30 molecules have been observed toward the M and N cores, while only 12 species have been detected toward the 2'N position chosen since HOCO$^+$ peaks here (Minh et al. 1988). Also the NW position, where Nobeyama observations indicate a large column of gas, exhibits few emission lines. Some 100 U-lines remain to be identified for the N core and some 20 toward M.

Similar to SO (Goldsmith et al. 1987) the dominant SO$_2$ emission emanates from core M, which may be an outflow region (Vogel et al. 1987). In contrast the emissions from CH$_3$OH, HCOOCH$_3$, (CH$_3$)$_2$O, as well as CH$_3$CN, C$_2$H$_3$CN and C$_2$H$_5$CN are strongly peaking in the warmer, and more massive core N.

Table 1. Adopted Sgr B2 source positions and the observed average line densities

Source	α(1950) h m s	δ(1950) ° ' "	No. of lines per GHz
M	17 44 10.4	−28 22 02	12
N	17 44 10.1	−28 21 16	30
2'N	17 44 10.4	−28 20 17	2
NW	17 44 06.6	−28 21 18	2

References
Cummins, S.E., Linke, R.A., Thaddeus, P. 1986, ApJS, 60, 819
Goldsmith, P.F., Snell, R.L., Hasegawa, T., Ukita, N. 1987, ApJ, 314, 525
Minh, Y.C., Irvine, W.M., Ziurys, L.M. 1988, ApJ, 334, 175
Turner, B.E. 1989, ApJS, 70, 539
Vogel, S.N., Genzel, R., Palmer, P. 1987, ApJ, 316, 243

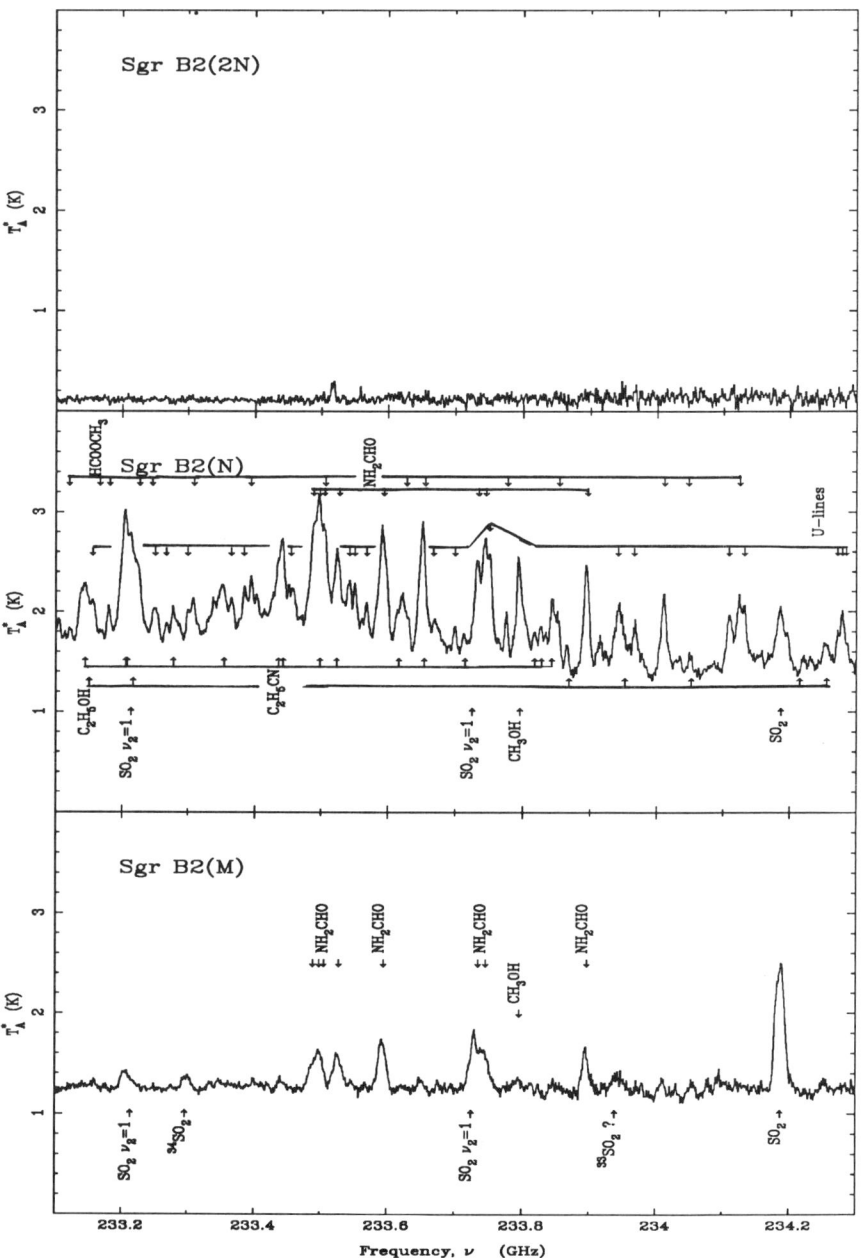

Figure 1. Sample spectra around 234 GHz toward three positions in the Sgr B2 molecular cloud

WHAT SPECIES REMAIN TO BE SEEN?

B.E. Turner
National Radio Astronomy Observatory
Charlottesville, Virginia, U.S.A.

ABSTRACT. We review what species remain to be seen for several types of astrochemistry: Thermochemical Equilibrium (TE) in circumstellar envelopes (CSEs); photo- and ion-molecule chemistry in CSEs; ion-molecule chemistry in cold interstellar clouds; grain chemistry (passive, catalytic, disruptive); and shock chemistry. In CSEs, a rich Si gas-phase chemistry is now recognized, and two predicted species (SiN, SiH_2) have been seen. Others are predicted. In the ISM, a global picture of refractory-element chemistry predicts that compounds of Mg, Na, Fe, and possibly Al occur with detectable gas-phase abundance. Predicted species require laboratory synthesis and spectroscopy. Reactions of hydrocarbon ions with neutral species dominate the formation of the families C_nH, HC_nN, H_2C_n, and C_nO in both interstellar (TMC-1) and circumstellar (IRC10216) cases, and readily explain the favored values of n in each case as well as predicting which higher-n species remain to be seen. Confirmation of H_3O^+ (interstellar) is discussed.

1. Circumstellar Molecules

1.1 TE CHEMISTRY

In their innermost regions, temperatures and densities of CSEs are high enough to insure TE, the products of which are assumed to flow outward in the expanding shell rapidly enough to minimize further reaction by the time they reach the outer envelope. Such simple models agree quite well with observations, even for refractory elements, which also condense into solids. In O-rich objects the most abundant molecular species involving second and third row elements are predicted (Tsuji 1973) to be SiO∗, SiS∗, SiH_4, FeO, FeS, $Fe(OH)_2$, MgOH∗, AlOH∗, AlF, AlCl, CaOH∗, NaCl, PS, KCl, TiO. Here, ∗ denotes a dominant species with detectable predicted abundance. In C-rich envelopes, existing predictions (Tsuji 1987; Turner et al 1990; Tsuji 1991) are SiS∗, SiO∗, Si_2C, SiH_4, NaCN∗, NaCl, HCP, TiO. The elements Mg, Na, and Fe are predicted to remain almost entirely atomic. Predicted condensates (Tarafdar 1987) are FeS, Fe_2SiO_4, SiO, SiS, $MgSiO_3$, Mg_2SiO_4, Al_2O_3, AlCl for O-rich envelopes, and Fe_3C, SiC, AlN, TiC, TiN for C-rich objects.

Observationally, SiO and SiS dominate the Si species, as predicted. SiH_4 is also seen in IRC10216, at surprisingly high abundance. The metallic compounds AlF, AlCl, NaCl, KCl are seen in IRC10216 at 10 times or so smaller abundances than predicted in O-rich envelopes. CP is observed, presumably as a photodissociation product of the predicted HCP. CaOH is not seen at levels roughly predicted in both O- and C-rich objects, while TiO is undetected at levels now somewhat below those predicted. Species which should provide important tests when measured frequencies become available are MgOH, AlOH, and possibly $Fe(OH)_2$ (O-rich envelopes), and NaCN (C-rich envelopes). As of now, TE models for Si, Al, K, Na appear quite satisfactory for IRC10216, when (uncertain) allowance is made for loss of these elements to condensation or subsequent adsorption of their gas-phase molecules onto grains. The predictions for Mg, Fe, Ca, Ti are not yet tested.

1.2. PHOTO- AND ION-MOLECULE CHEMISTRY OF THE OUTER ENVELOPE

A rich gaseous Si chemistry of the outer envelope of IRC10216 is indicated by the recent detection of SiN (Turner 1991b) and possible detection of SiH_2 (Turner and Steimle 1991). These detections substantiate the Si ion-molecule chemistry proposed by Glassgold et al (1986) and Herbst et al (1989) in which Si^+ reacts rapidly with NH_3 to produce SiN (and HNSi), and with C_2H_2 to produce SiC_2. If SiN and SiC_2 are destroyed at the same rate, and if NH_3/C_2H_2 is the same in the outer and inner (TE) envelopes, then we expect $SiN/SiC_2 \sim$ 1/8, and we observe $\sim$ 1/30, in reasonable agreement. HNSi (iminosilicon) is also expected, though the SiN/HNSi ratio is uncertain. The source of Si^+ is taken to be SiH_4, which can be shown from its IR observation in the inner envelope ($SiH_4/Si \sim 10^{-3}$) to have a column density in the outer envelope comparable with that of SiO + SiS (2×10^{15} cm^{-2}) if the same SiH_4/Si ratio persists. SiH_4 photodissociates more readily than SiO or SiS, first to SiH_2 (with 75% efficiency), then to Si^+ or Si with unknown efficiency. The possible detection of SiH_2 (2 transitions; $N \sim 10^{15}$ cm^{-2}) provides a suitable source of Si^+ to drive the outer envelope chemistry, and also means that the fraction of Si lost to the ISM in gas phase (SiS + SiO + SiC_2 + SiN + SiH_4) is $\sim 7\times10^{-3}$, so that the traditional picture of the gas/dust ratio of $\sim$ 150 representing condensation of *all* refractories needs no modification. No other Si compounds of significant abundance are expected.

The ions Mg^+, Na^+, and possibly Al^+ are expected with abundances comparable with that of Si^+ in the outer envelope, and their reactions with C_2H_2, if exothermic, should produce MgC_2, NaCCH (sodium acetylide), and :AlCCH. MgC_2 and NaCCH are known to be stable. No microwave spectra exist. Reaction of the ions with NH_3 to produce MgN (MgNH), NaN (NaNH) are possible also. Detection of such species would be important in establishing whether gas-phase chemistry is similar among the refractory-element ions.

Turning to first-row elements, the recent detections of H_2C_4 and HC_2N in IRC10216 have drawn attention to the patterns observed in the C_nH, HC_nN, and H_2C_n families. The dominance of even values of n in the C_nH sequence stems from the dominance of reactions of $C_2H_2^+$ and C_2H_2, both highly abundant. For example, C_4H is produced from $C_2H_2^+$ + $C_2H_2 \to C_4H_3^+$, and C_6H from $C_4H_3^+$ + C_2H_2. Species with n = 8,10,... are expected. Odd-n species occur by various, relatively inefficient branching routes, and $n \geq 7$ is doubtful, based on current observations. The HC_nN family favors odd n, because these species start with $C_2H_2^+$ + HCN $\to HC_3N$, and successive reactions of HC_nN ($n \geq 3$) with $C_2H_2^+$ lead to higher-n members. Even-n members require reaction of a hydrocarbon ion other than $C_2H_2^+$ with N, neither of which have significant abundance. For the cumulene family, we predict that even-n members of H_2C_n dominate ($C_2H_2 + C_2H_2^+ \to C_4H_3^+ \to C_2C_4$, and successive reactions with $C_2H_2^+$ add two C atoms each). Finally, the C_nO family is not expected in IRC10216, even though it forms from reactions of the form $C_nH_m^+$ + CO $\to C_{n+1}OH_m^+ + h\nu$, because the required radiative association reactions (Adams et al 1989) have negligible rates in the warm gas of the outer envelope.

2. Interstellar Ion-Molecule Chemistry in Cold Clouds

2.1. HYDROCARBON SPECIES AND THEIR DERIVATIVES

Predictions for these species are much more difficult in the interstellar context. We refer to a diagram of hydrocarbon reaction pathways given by Turner (1989), and based on mostly measured reactions existing in the literature. The preference for even-n in C_nH and odd-n in HC_nN and C_nO results from the details of this reaction scheme, not from a dominance of a certain reactant (e.g. C_2H_2 in the case of IRC10216). The observations clearly imply large amounts of $C_3H_3^+$ in the interstellar case, which means small amounts of $C_3H_2^+$ (competing pathway) and hence C_3H. There are several paths to C_4H from

$C_3H_3^+$, but after C_4H only one path leads to C_5H (not seen in TMC-1) while several lead to C_6H (detected in TMC-1). Predictions to higher n depend on as yet unknown pathways. Similarly, HC_3N forms from $C_3H_3^+ + N$, and several pathways via $C_2H_2^+$ or C_2H_2 lead to HC_5N etc. HC_2N (recently identified in IRC10216) is not seen in TMC-1, signalling a low abundance of $C_2H_3^+$. HC_4N cannot form appreciably from $C_3H_2^+ + HCN$ (low abundances), but is possibly expected via $C_4H_3^+ + N$ and will provide a useful diagnostic of the higher order hydrocarbon abundances. Likewise, the cumulene H_2C_3 is an isomer of the cyclic C_2H_2 and forms from $C_3H_3^+$, while H_2C_4 (not seen in TMC-1) forms via $C_4H_3^+$, which is evidently much less abundant than $C_3H_3^+$. It is likely that HC_nN and H_2C_n will have similar abundance patterns as a function of n, and also similar to those of $C_{n+1}H$, as each likely has the same hydrocarbon ion precursor. Finally, the C_nO family may form from the radiative association reactions of hydrocarbon ions with CO in the cold interstellar gas. The precursor ions are CH_3^+ and $C_2H_3^+$ for the C_2O (recently detected) and C_3O species. If C_5O (Turner 1990a) is confirmed, it will imply a significant abundance of $C_4H_2^+$ (or $C_4H_3^+$). A definite prediction either way is C_4O (high $C_3H_3^+$ abundance and high reaction rate with CO), while C_6O is also likely if C_5O is seen. Of course, the associated species H_mC_nO will also be present.

A sensitive search has been made for vinyl (CH_2CH), which should form simply from $C_2H_2^+$ via radiative association with H_2, while C_2H forms directly from $C_2H_2^+$ by electron recombination. Thus $CH_2CH/C_2H \sim 1$ is expected from the given reaction rates, while $CH_2CH/C_2H < 0.06$ is observed. Results such as this underscore the limited predictability of the overall reaction scheme currently in use. Each reaction rate and branching ratio is open to question a priori.

2.2 OTHER SPECIES

Two additional transitions of H_3O^+ have now been seen, the 3_2-2_2 at 364.9 GHz (Wootten et al 1991) and the 3_0-2_0 at 396 GHz (van Dishoeck et al 1991). The abundance ratio is $1500 < H_2O/H_3O^+ < 6000$ in Orion(KL) and SgrB2, as compared with model values ranging from 2000 (Millar 1990) to 10^5 (Viala 1986). Here, the H_2O abundance is derived from observations of $H_2^{18}O$. The models assume a branching ratio of $H_3O^+ + e \rightarrow H_2O$ of 35 %, an upper limit to the measured value. If significant quantities of H_2O arise from grain mantle evaporation, then smaller values of the branching ratio will be required.

Additional Si compounds are predicted in the ISM by the chemistry model of Herbst et al (1989). SiN is not among them, as it is quickly destroyed by atomic O, but the closed shell species HNSi may be expected, in view of the detection of SiN in IRC10216. A preliminary search reveals a line in W51N and SgrB2 corresponding to the J = 6-5 transition. If confirmed, then the Herbst et al model predicts abundances of several other Si species relative to HNSi: $HOSi^+$ (4), $HSiC_2$ (54), $SiCH_2$ (35), $Si_2C_2H_2$ (13), $SiCH_3$ (8), and SiC_3H (5), if the chemistry is early-times. If it is steady state, only $HOSi^+$ (0.4) and $HSiC_2$ (0.6) have abundances comparable with HNSi. The small dipole moments of many Si species make sensitive searches difficult. $HOSi^+/HNSi < 3.8$ has been established.

3. Interstellar Refractory-Element Chemistry

Ninety percent of all refractory-element grains condense in the innermost envelopes of evolved stars. The order of condensation is Al, Ti, Ca, Fe, (Mg,Si), P, (Na,K), S from highest to lower condensation temperature, so grains are layered. The remaining 10% of refractory elements are ejected into the ISM in atomic form, in fast winds from early-type stars, or supernovae.

Grains are ejected first into the warm diffuse ISM (T $\sim 10^4$ K, n ~ 0.3 cm^{-3}, $\tau_{shock} \ll \tau_{acc}$, $\tau_{ph} \sim \tau_{acc}$; here τ's are timescales for shock erosion of grains, accretion, and photoprocessing). Measured extinctions $\delta = \log(X/X_{solar})$ of refractory elements show values of $\sim$ -1.0 for Al, Fe, Ti, Ca, but smaller values (-0.1 to -0.6) for P, Mg, Na, Si, exactly as expected if shocks erode the outer grain layers down to the (Si,Mg) layers but leave the inner (Al,Ti,Ca,Fe) cores intact. Since $\tau_{ph} \sim \tau_{acc}$, simple hydrides forming on the surfaces quickly desorb and are photodissociated.

In the cold diffuse ISM (T $\sim$ 100 K, n $\sim$ 30 cm^{-3}, $\tau_{shock} \gg \tau_{acc}$, $\tau_{ph} \sim \tau_{acc}$, $\tau_{mix} \sim \tau_{acc}$), binding energies and photodesorption dominate the fate of accreted refractory elements. Binding energies are similar for the hydrides of O,C,N,P,S and possibly Mg, higher for Si, and much larger for Fe, Ti,Ca and likely Al. Accretion of refractory ice (and organic) mantles proceeds in this phase. Observed extinctions are all greater than in the warm medium, ranging from -0.7 for S, P, Mg, Na, to -3.6 for Ca. Si is at -1.6.

In cold dense clouds, which evolve from the cold diffuse ISM, T $\sim$ 10 K, n $\geq 10^3$ cm^{-3}, and $\tau_{chem} \sim \tau_{acc}$. Based on unsuccessful searches for PN and SiO, one has $(R/NH_3)_{cold} \ll (R/NH_3)_{warm}$, where R stands for refractory- element compounds, NH_3 for first-row compounds, which are readily seen in cold dense clouds. Eventually stars form in cold dense clouds, and evaporate the grain mantles. "Warm" refers to these regions. These two relations for cold dense clouds imply one of the following: i) all molecules desorb in some way (and gas-phase chemistry does not form SiO and PN at T = 10 K); ii) first-row species desorb while refractories do not; iii) no molecules desorb, and the presence of first-row species implies a longer accretion time, as a result of their chemistry. We favor iii).

A 3-layer picture of grains is usually envisioned (Leger et al 1985): a refractory, heavy-element core; a refractory ice mantle, mostly of H_2O ice, laid down by efficient binding of polar molecules during the cold diffuse ISM phase; and finally a volatile outer mantle (CO, N_2, O_2,...) accreted during the cold dense cloud phase. Known binding energies of many substances indicate that atoms in general, as well as O_2, N_2, CO, CH_4 will accrete efficiently only during the cold dense phase, while all saturated hydrides (SiH_4, PH_3, H_2S, NH_3, H_2O...) and other polar molecules will accrete earlier and form the refractory ice mantle. Thus the non-polar group represents a gas phase "reservoir" with a long accretion timescale which allows continuation of the gas-phase chemistry observed in cold dense clouds. The cold dense cloud model of Herbst and Leung (1989) at t = 3×10^5 yr shows the following "reservoir" amounts: N (99%), P(80%), O (57%), S (30%), Si (1%). All Si forms SiO while only 20% of P forms PN (or PO). This is why Si (SiO) is not observed, while the presence of many S-species implies that P compounds should be observable in cold clouds. We predict that PO (the principle P species formed by ion-molecule reactions) should be observable, and that the reaction PO + N $\rightarrow$ PN + O invoked to explain PN in warm star-forming regions does not proceed at T = 10K.

In warm star-forming regions, refractory mantles (mostly H_2O ice will start to evaporate at T $\sim$ 90 K, the outer volatile mantle earlier. All SiO and PN (PO) are accreted into the refractory ice mantle in the cooler phases. A central point is that the abundances of both SiO and PN, as observed in star-forming regions, are quantitatively explained in terms of the cosmic abundances (δ_c) of Si and P, the observed extinctions (δ_d) in the cold diffuse ISM, the conversion of 99% Si to SiO and 20% P to PN, *and the evaporation of 4% of the refractory ice mantle in the case of Ori(KL)*. The relative abundances of molecules as observed in cold dense clouds and warm star-forming regions can be explained quantitatively on this picture by placing all Si, P, S species in the refractory ice mantle, and all N species in the outer volatile mantle, as expected on this picture. Then the fractional abundance X of any refractory element in Ori(KL) is given by X = 0.04×10$^{-\delta_c - \delta_d}$. Values of $\delta_c + \delta_d$ are as follows: Mg (5.24), Si (6.13), Na (6.48), Fe (7.10), P (7.17), Al (7.72), Ca (9.38), Ti (9.88). Since Si and P species are observed, we predict that Mg species will be easily observable, and Na, Fe species also. Al is possible, while Ca, Ti are very unlikely.

Nothing is known of the gas-phase chemistry of Mg, Na, Fe, Al but in analogy with that of Si (Herbst et al 1989) we expect oxides to dominate although they may convert to nitrides. MgO, NaO, and possibly AlO are not detected at expected levels. Here we assume surface molecules are destroyed in the gas phase after evaporation, and that ion-molecule processes start with the resulting atoms. If there are few ions in the dense, hot gas near massive protostars, surface molecules may survive; then full hydroxides are likely.

The detection of Mg (and Na,Fe) compounds, a strong prediction of this picture, would not only provide the first information on the chemsitry of metals in the ISM (also poorly known terrestrially), but would provide insight into the lifecycles of grains, whose existence in the ISM is not clearly understood (Seab 1987).

4. Grain Chemistry

There is now much evidence of the importance of grain chemistry, of several types. Several molecular species are seen with abundances much greater than can be explained by gas-phase chemistry: i) complex species such as CH_3CHO, $(CH_3)_2O$, EtOH, CH_3OHCO, NH_2CHO, VyCn, EtCN in dense star-forming regions (Turner 1991a); ii) NH, recently identified in diffuse clouds (Meyer and Roth 1991); iii) fully saturated species such as H_2O, NH_3, HCN, CH_3CN in Orion(KL); iv) H_2CO in molecular cloud envelopes whose extinction is < 4 magn (Federman and Allen 1991). Other species such as D_2CO (Turner 1990b) are not producible at all via gas-phase chemistry, but have a natural explanation in terms of catalytic surface chemistry (Tielens 1983).

No detailed model of catalytic grain chemistry is currently possible, but a simple picture by Millar et al (1988) illustrates potentially useful predictions. For a molecular cloud with [H] > [O], (n < 5×10^4 cm^{-3}), mobile surface C atoms typically accrue H atoms and exist as H_3C-, which grow unidirectionally at low T. Thus H_3C-CH_3 and H_3C-O-CH_3 form readily. If the surface contains much CO, then H_3C-CO-CH_3 (acetone) is also likely. If, conversely, [H] < [O] (n > 5×10^4 cm^{-3}), then C-, HC-, H_2C- dominate the surface carbon, and these permit linear growth in two dimensions. It is also possible that on icy surfaces the mobilities $\mu(N) < \mu(O) < \mu(C)$ (Iguchi 1976). Then in decreasing order of likelihood the following backbones may form: C-C-C or C-C-O; C-C-N; O-C-N or N-C-O. None of these will saturate with H atoms. Thus C-C-C may have the form H_3C-CH=CH_2 (propylene) and C-C-O the form EtOH. C-C-N forms EtCN or VyCN. Observationally, EtCN < EtOH ~ $(CH_3)_2O$ in SgrB2 implies both H-rich and H-depleted regions occur within the observing beam. In the Orion hot core, EtCN >> EtOH, $(CH_3)_2O$ implies H-depletion. In the Orion compact ridge, $(CH_3)_2O$ >> EtOH, EtCN implies an H-rich region.

Of course, reactive species such as C_2H_2 should be plentiful on grains, and may hydrogenate to form H_3C-CH_3, or oxidize directly to form EtOH (Tielens, 1990) but *not* $(CH_3)_2O$. If $\mu(N) < \mu(O)$, we expect EtCN < EtOH, contrary to the Orion case. The presence of surface carbon radicals rather than CO, C_2H_2 seems necessary to provide a non-restrictive surface chemistry, as suggested by the observations.

Deuterated forms of complex molecules likely provide the best test of surface chemistry. D-species reflect $(D/H)_{gas}$, if the addition and abstraction reactions of H and D are similar on surfaces. $(D/H)_{gas}$ is largest for smallest [H]. Thus we predict D-EtCN, D-EtOH will be more abundance than D-$(CH_3)_2O$. In addition, for p = D/(D+H), we predict EtOD/EtOH = p/(1-p), CH_2DCH_2OH/EtOH = CH_2DCH_2CN/EtCN = 3p/(1-p), while CHDCHCN/VyCN = 2p/(1-p), CH_2CDCN/VyCN = p/(1-p). The CH_2DOH/$CH3OD$ ratio is also important in this context.

5. Shock Chemistry

Turner and Lubowich (1991) recently established directly the presence of hot molecular gas other than H_2 in a shocked region, namely H_2CO in IC443G. Subsequently, attempts to test both non-dissociative and dissociative shock models have involved observations in IC443 of several species (SiO, SO, SO_2, CN, HCN) whose abundances are predicted to be similar in both types of shock, and other species (HCO^+, H_2CO, C_3H_2) whose abundances are predicted to be very different. The first group is used to derive beam-filling factors, which are applied to the second group to deduce the nature of the shock. Both non-dissociative shocks (to explain H_2CO) and dissociative shocks (to explain HCO^+ and C_3H_2) are implied, similar to the requirements for H_2 (Burton et al 1990). One new species, SO^+, is predicted at detectable levels in the dissociative shock models of Neufeld and Dalgarno (1989). SO^+ forms from SO and H^+, and the low abundance of H^+ in non-shocked regions explains why $[SO^+]$ is expected in shocked regions at 100 to 1000 times its normal interstellar value. Hence it provides one of few unambiguous tests of shock chemistry, as distinct from SiO or H_2CO, which may well be evaporated from grain mantles. In fact, [SiO] $\sim 6\times10^{-9}$ in IC443G, some 10^3 to 10^4 times smaller than predicted by shock models.

References

Adams, N.G., Smith, D., Giles, D., and Herbst, E. 1989, A.&A. 220, 269
Burton, M.G., Hollenbach, D., Haas, M., and Erickson, E. 1990, ApJ. 355, 197
Federman, S.R., and Allen, M. 1991, ApJ. 375, 157
Glassgold, A.E., Lucas, R., and Omont, A. 1986, A.&A. 157, 35
Herbst, E., and Leung, C.M. 1989, ApJS 69, 271
Herbst, E., Millar. T.J., Wlodek, S., and Bohme, D.K. 1989, A.&A. 222, 205
Iguchi, T. 1975, Publ. Astron. Soc. Jpn. 27, 515
Leger, A., Jura, M., and Omont, A. 1985, A.&A. 144, 147
Meyers, D.M., and Roth, K.C. 1991, ApJL. (August 1)
Millar, T.J., Olofsson, H., Hjalmarson, A., and Brown, P. 1988, A.&A. 205, L5
Millar, T.J. 1990, in *From Ground-Based to Space-Borne Sub-mm Astronomy*, Ed. B. Kaldeich (Nordwijk: ESTEC), p. 233
Neufeld, D.A., and Dalgarno, A. 1989, ApJ. 340, 689
Seab, G.C. 1987, in *Interstellar Processes*, Eds. D.J. Hollenbach and H.A. Thronson (Dordrecht: Reidel), p.491
Tarafdar, S.P. 1987, in *Astrochemistry*, Eds. M.S. Vardya and S.P. Tarafdar, D. Reidel
Tielens, A.G.G.M. 1983, A.&A. 119, 137
Tielens, A.G.G.M. 1990, in *Chemistry and Spectroscopy of Interstellar Molecules*, Ed. N. Kaifu, Univ. of Tokyo Press
Tsuji, T. 1973, A.&A. 23, 411
Tsuji, T. 1987, in *Astrochemistry*, Eds. M.S. Vardya and S.P. Tarafdar, D. Reidel
Tsuji, T. 1991, private communication
Turner, B.E. 1989, Space Sci. Reviews 51, 235
Turner, B.E. 1990a, in *Chemistry and Spectroscopy of Interstellar Molecules*, Ed. N. Kaifu, Univ. of Tokyo Press
Turner, B.E. 1990b, ApJL 362, L29
Turner, B.E. 1991a, ApJS 76, 617
Turner, B.E. 1991b, ApJL submitted
Turner, B.E., Tsuji, T., Bally, J., Guelin, M., and Cernicharo, J. 1990, ApJ. 365, 569
Turner, B.E., and Lubowich, D.A. 1991, ApJ. (Oct. 20)
Turner, B.E., and Steimle, T.C. 1991, in preparation
van Dishoeck, E., and Phillip, T.C. 1991, this symposium
Viala, Y.P. 1986, A.&A. Suppl. 64, 391
Wootten, H.A., Mangum, J.G., Turner, B.E., Bogey, M., Boulanger, F., Combes, F., Encrenaz, P.J., and Gerin, M. 1991, ApJL in press

OBSERVATIONS OF HNO

L. E. Snyder[1], J. M. Hollis[2], L. M. Ziurys[3], and Y.-J. Kuan[1]

[1]Astronomy Dept., Univ. of Illinois; [2]Space Data & Computing Div., NASA/GSFC; [3]Chemistry Dept., Arizona State Univ.

ABSTRACT. Interferometric observations of the HNO 3.68 mm line in Sgr B2 show clumping and some distribution similarities with HNCO. A search for the HNO 1.23 mm line with the NRAO 12-m telescope was unsuccessful.

1. Introduction and Observations

HNO studies are important because some compounds with the N-O bond appear to be underabundant compared to ion-molecule predictions, and the exact location of N compounds in cloud clumps is not well known. The identification of interstellar HNO was confirmed recently by FCRAO and BIMA array observations (Hollis et al. 1991; Snyder et al. 1991). New BIMA array maps of the HNO 1_{01}-0_{00} line at 81,477.56 MHz (3.68 mm) around Sgr B2 are presented here (Figs. 1 and 2); these new maps used 20 baselines (5 configurations in B-array and 3 in C-array).
    In addition, the NRAO 12-m telescope equipped with a DSB SIS receiver was used to search Sgr B2(M) for the HNO 3_{03}-2_{02} line at 244,364.09 MHz (1.23 mm). These observations were conducted in March and April, 1991.

2. Results and Conclusions

At 3.68 mm, 5 HNO emission clumps and 1 absorption source (Sgr B2(M)) were detected with the BIMA array. We note that HNO observations with a 6-m BIMA single antenna (2' HPBW) show that the emission peaks at 69 km/s in Sgr B2(M), but 30" south in Sgr B2(OH) it peaks at 64 km/s. This suggests that a typical single-dish Sgr B2(M) HNO spectrum (not shown) is dominated by clumps with higher velocity than those near Sgr B2(OH). The main HNO features generally agree with BIMA maps of the HNCO 4_{04}-3_{03} line at 87,925.24 MHz near 68 km/s, but the HNCO is much stronger.
    The HNO 3_{03}-2_{02} line was not detected in Sgr B2(M), with an upper limit of $T_A^* < 0.05$ K (3σ). Because this limit is comparable to the HNO intensities at lower frequencies, it suggests weak excitation.

Support came from NASA PO S-46462-E, and the Laboratory for Astronomical Imaging with funds from the University of Illinois and NSF AST 90-24603.

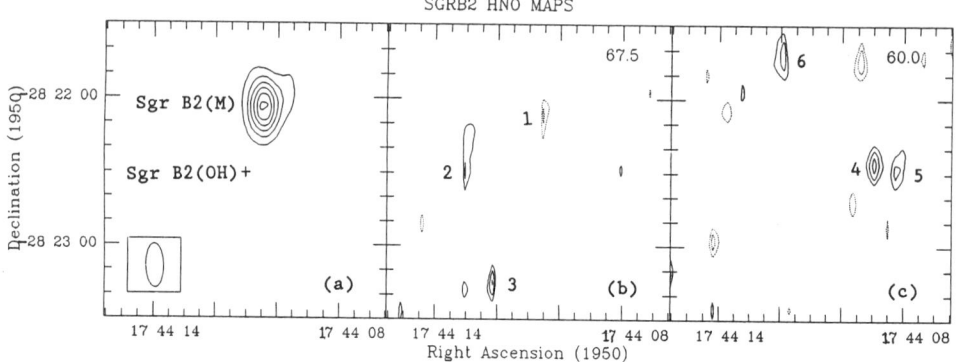

Figure 1. (a) Sgr B2(M) 3.68 mm continuum. Contour intervals: 1.0 Jy/bm. Peak = 6.11 Jy/bm. The inset shows the 17".95 x 7".41 beam. (b) HNO at 67.5 km/s (averaged over 10 km/s intervals). Contour intervals: -0.25, -0.30 (dots); and 0.25, 0.30, 0.35 Jy/bm (solid). Peak = 0.36 Jy/bm. (c) HNO at 60.0 km/s (averaged over 7 km/s intervals). Contour intervals: -0.3, -0.4 (dots); and 0.3, 0.4, 0.5 Jy/bm (solid). Peak = 0.56 Jy/bm.

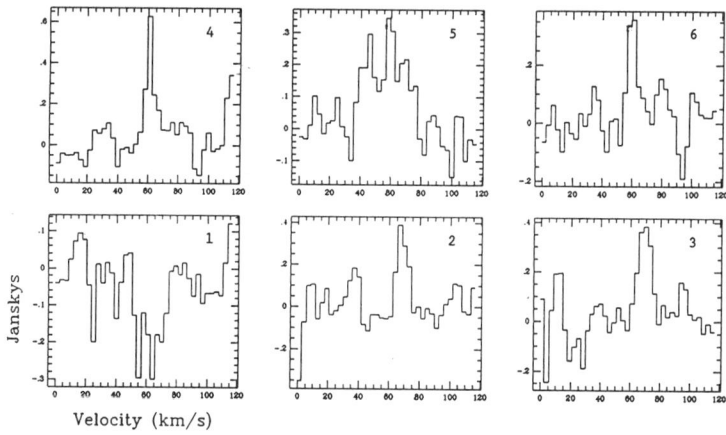

Figure 2. Spectra of the HNO 3.68 mm line constructed from data cubes corresponding to numbered clumps in Figure 1. <u>Abscissa</u>: LSR radial velocity (km/s). <u>Ordinate</u>: intensity (Jy/beam).

5. References

Hollis, J. M., Snyder, L. E., Ziurys, L. M., and McGonagle, D. (1991) 'Interstellar HNO: confirming the identification', in A. D. Haschick and P. T. P. Ho (eds), Skylines, ASP Conference Series 16, San Francisco, pp. 407-12.

Snyder, L.E., Kuan, Y.-J., and Pratap, P. (1991) 'Millimeter wavelength molecular maps of the clumped gas around Sgr B2(OH)', Ibid., pp. 191-6.

ON THE POSSIBLE DETECTION OF SOLID O_2 IN INTERSTELLAR GRAINS

P. EHRENFREUND [1,2], R. BREUKERS[1], L. d'HENDECOURT[2], J.M. GREENBERG[1]
[1] Huygens Laboratory Astrophysics, Leiden, The Netherlands
[2] Groupe de Physique des Solides, Univ. Paris 7, Paris,France

ABSTRACT. In various models of the chemistry of interstellar grains, solid O_2 is formed by accretion as well as by surface reactions. In dense molecular cloud models, at a later stage of evolution of an interstellar grain, solid O_2 becomes a major grain mantle constituent at the expense of water ice abundance. If molecular oxygen is embedded in a "dirty" ice" matrix, the forbidden fundamental vibration of O_2 at 1550 cm^{-1} may become observable.

1. Introduction

Oxygen, the cosmic most abundant element after H and He plays an important role in interstellar chemistry. O_2, a diatomic homonuclear molecule shows no transitions in the infrared (IR). Therefore no direct estimates of the abundance can be obtained. We have studied the fundamental band of solid molecular oxygen at 1551 cm^{-1} in various matrices and discuss the detectability of solid O_2 in interstellar grains and its photolysis product, O_3.

2. Results and Discussion

We could detect the weak vibrational transition of molecular oxygen at 1559 cm^{-1} in a CO_2 matrix at 10 K and confirm the isotopic shift, using isotopically labelled $^{18}O_2$ (1469 cm^{-1}). Using the well defined integrated absorbance A_m of the bending mode of CO_2 at 15.2 μm (Sandford et al., 1988) we could estimate this value with some accuracy for molecular oxygen: A_m (cm.mol^{-1}) = 3 x 10^{-18}. Fig. 1 shows the IR spectrum between 4000-500 cm^{-1} of a gas mixture containing $H_2O : CO : O_2 : CO_2$ (2 : 2 : 1 : 0.5). The cross section A_m (cm.mol^{-1}) is a factor 30 weaker in this complex mixture than in pure and highly diluted CO_2. The interaction of polar and polarizable molecules with molecular oxygen in a matrix may be responsible for the enhancement of the weak vibrational transition of O_2, a process which can also occur in interstellar grain mantles. Another way to estimate the abundance of O_2 on interstellar grains is from a study of its photolysis product O_3. Ozone is easily produced by the photodissociation of O_2 in the ISM and the v_3 strongest vibrational transition becomes clearly visible at 1042 cm^{-1} (9.6 μm). Though this band can be obscured by the strong absorption band of silicates at 10 μm the sharp feature could be detectable, as well as the overtone of O_3 at 2110 cm^{-1}.

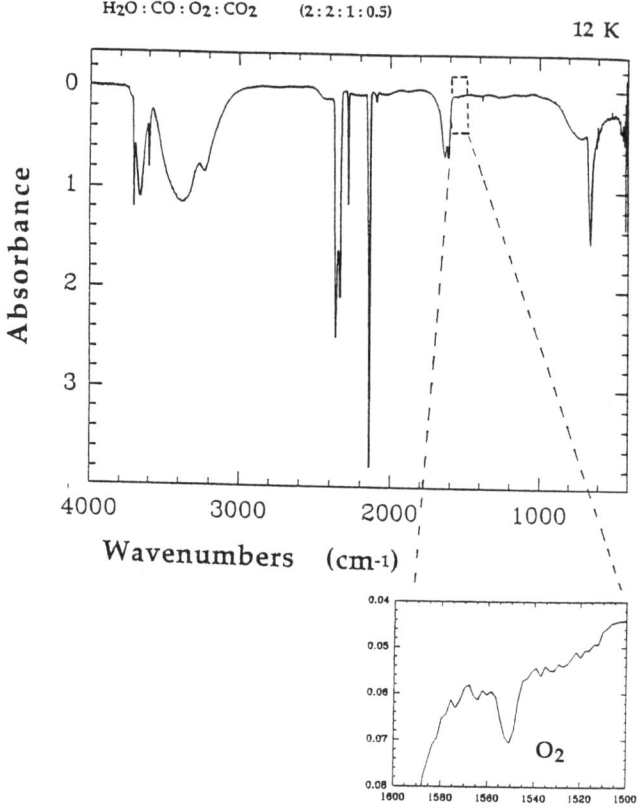

Figure 1. IR spectrum of a "dirty" ice mixture $H_2O : CO : O_2 : CO_2$ (2 : 2 : 1 : 0.5) at 10 K. The weak fundamental transition of molecular oxygen at 1551 cm^{-1} is shown in detail.

We want to point out that the results presented here indicate the possible detection of solid O_2 in interstellar space in the mid-IR. The calculated integrated absorbance is weak, but can be enhanced by interactions with molecules in the environment, disturbing the symmetry of molecular oxygen. In the theoretical models, O_2, becomes a major grain mantle constituent at later times in the evolution of an interstellar grain in dense molecular clouds (Breukers, 1991). The search for the fundamental transition of O_2 will probably be successful in astronomical targets like dense molecular clouds, with high extinction and where other grain mantle constituents have already been identified. Furthermore the photolysis product ozone can very likely be observed.

References
Breukers, R., thesis, Univ. Leiden (1991)
Grimm, R.J.A., d'Hendecourt, L. (1986), A & A, 167, 161
Herbst, E., Leung, C.M. (1989), Ap.J.Suppl. 69, 271
Sandford, S.A., Allamandola, L., Tielens, A.G.G.M., Valero, G.C. (1988), Ap. J.329, 498

INTERSTELLAR H_3O^+

T.G. Phillips[1], Ewine F. van Dishoeck[2] and Jocelyn B. Keene[1]

[1] Div. of Physics, Math. & Astronomy, Caltech 320-47, Pasadena, CA 91125, USA
[2] Leiden Observatory, P.O. Box 9513, 2300 RA Leiden, The Netherlands

1. Introduction

The H_3O^+ ion is a key species in the oxygen chemistry leading to H_2O, OH and O_2. Chemical models predict O_2 and H_2O to be the dominant oxygen-bearing molecules in interstellar clouds. However, neither of them can easily be observed in the bulk of the interstellar medium because of blockage from the Earth's atmosphere. Determination of the abundance and distribution of the precursor H_3O^+ ion might thus provide an important indirect measure of their abundances.

H_3O^+ has a complicated energy spectrum (see Bogey et al. 1985). It is isoelectronic with NH_3, but its inversion splitting is much larger, about 55 cm^{-1}, and comparable to the rotational splitting. As a result, the lowest transitions lie at submillimeter wavelengths, and only four lines at 307, 364, 388 and 396 GHz are accessible with ground-based telescopes.

Possible detections of the 307 GHz line of H_3O^+ have been presented by Wootten et al. (1986) and Hollis et al. (1986) in Orion OMC-1 and Sgr B2, but definite identification is difficult due to the huge complexity of lines in these sources. Recently, Wootten et al. have observed the 364 GHz line as well in these sources using the Caltech Submillimeter Observatory (CSO). Of the two remaining lines, only the 396 GHz line lies in a reasonably transparent region of the atmosphere. We have concentrated our searches on this line using the CSO. Encouraged by our success in detecting the 396 GHz line in a number of sources other than Orion and Sgr B2, we also searched for the 364 GHz and 307 GHz lines in the same regions, to secure the identifications.

2. Observations

The most impressive detection of the 396 GHz line is toward W3 IRS5, where the line is about $T_A^* \approx 0.4$ K. In contrast with Orion and Sgr B2, this spectrum shows virtually no other features over a 500 MHz bandwidth. The 364 GHz line is also seen toward W3 IRS5 at about one third of the strength of the 396 GHz line. Figure 1 shows a blow-up of the two lines. A marginal feature, probably better described as an upper limit with $T_A^* \lesssim 0.08$ K, is found at 307 GHz. By contrast, no detections were made toward W3 IRS4 and only a marginal 396 GHz feature is present toward W3 OH, even though the total hydrogen column densities are fairly similar in the three cases. Thus significant variations in the abundance and/or excitation of H_3O^+ appear to occur on small scales.

The 396, 364 and 307 GHz lines are all seen toward G34.3 +0.15, whereas the 396 and 307 GHz lines are only possibly detected toward W51. Toward Orion/KL, a strong feature with $T_A^* \approx 2$ K is present at 396 GHz, but toward Sgr B2, the 396 GHz line is only 0.2 K. No H_3O^+ lines were seen toward NGC 2024 FIR5, ρ Oph A and B2 and IRAS 16293 −2422, and toward a number of oxygen-rich stars.

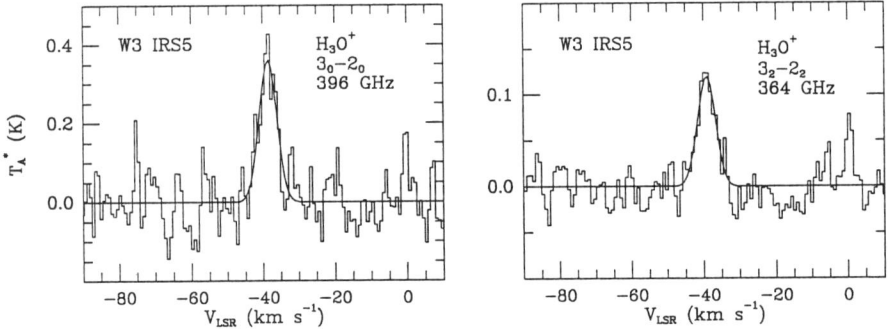

Fig. 1. The H_3O^+ 396 and 364 GHz lines toward W3 IRS5 obtained at the CSO

3. Excitation and Abundance

Excitation calculations have been performed for H_3O^+ under a variety of conditions using collisional cross sections based on a simple model potential (Offer, private communication). The 396 and 364 GHz lines are found to have comparable strengths (within a factor 2) under conditions typical for warm, dense molecular clouds such as W3 ($T \approx 50$–100 K, $n \approx 10^6 - 10^7$ cm^{-3}). In contrast, the 307 GHz line is very weak in such models, consistent with the observations. Only for very high densities, $n > 10^7$ cm^{-3}, does the 307 GHz line become comparable in strength to the 396 GHz line. The different H_3O^+ line ratios observed toward Sgr B2 are most easily explained in a low density model ($n \approx 10^4 - 10^5$ cm^{-3}), in which the 364 GHz line is preferentially excited by far-infrared radiation.

The derivation of H_3O^+ column densities and abundances is hampered by uncertainties in the collisional rates and in the H_2 column densities. We derive $N(H_3O^+) \approx 10^{14}$ cm^{-2} toward W3 IRS5, resulting in an abundance with respect to H_2 of about 10^{-9}. Similar abundances around $10^{-10} - 10^{-9}$ apply to the other sources.

4. Conclusions

We have definitely detected H_3O^+ toward W3 IRS5 and G34.3 +0.15, and have obtained plausible identifications toward W3 OH, W51, Orion/KL and Sgr B2. The inferred abundances are consistent with gas-phase chemistry models within an order of magnitude (e.g. Herbst and Leung 1989; Millar et al. 1991), although they favor the lower values. The abundance of the ion toward W3 IRS5 implies an O_2 abundance of about 10^{-6}–10^{-5}. Searches for lines of $^{16}O^{18}O$ in these same sources are currently being performed (Keene et al. in preparation), and should provide important constraints on the oxygen chemistry.

Bogey, M., DeMuynck, C., Denis, M., & Destombes, J.L. 1985, A&A, 148, L11.
Herbst, E., & Leung, C.M. 1989, ApJS 69, 271.
Hollis, J.M., Churchwell, E.B., Herbst, E., & De Lucia, F.C. 1986, Nature, 322, 524.
Millar, T.J., Rawlings, J.M.C., Bennett, A., Brown, P.D., & Charnley, S.B. 1991, A&A Suppl., 87, 585.
Wootten, A., Boulanger, F., Bogey, M., Combes, F., Encrenaz, P.J., Gerin, M., & Ziurys, L. 1986, A&A, 166, L15.
Wootten, A., Mangum, J.G., Turner, B., Bogey, M., Boulanger, F., Combes, F., Encrenaz, P.J., & Gerin, M. 1991, ApJ, 380, L79.

CARBON ISOTOPIC CHEMISTRY

W. D. LANGER
MS 169-506
Jet Propulsion Laboratory
California Institute of Technology
Pasadena, CA 91109 USA

ABSTRACT. Isotopic molecular abundances are used to interpret Galactic chemical evolution and the properties of interstellar clouds. The isotopic chemistry of carbon plays an important role in the interpretation of these measurements. This paper reviews the recent measurements of the carbon twelve to thirteen ratio across the Galaxy and the isotopic chemistry.

1. Introduction

For twenty years isotopic abundances of molecules in interstellar clouds (ISC) have been measured by radio and optical techniques. These observations are important to: (1) determine Galactic chemical evolution (nucleosynthesis, star formation rate, injection and mixing of gas in the interstellar medium); (2) interpret physical conditions in ISCs; (3) test models of interstellar chemistry; and, (4) study the chemical evolution of the solar system. Carbon isotopic species are especially important because carbon molecules are pervasive and important probes of ISCs, the chemistry is reasonably well understood, and twelve and thirteen carbon trace primary and secondary nucleosynthesis, respectively. Here I review carbon isotopic ratio measurements and discuss the basic elements of the carbon isotopic chemistry

2. Observations of Carbon Isotopic Ratios

Across the Galaxy carbon isotopic ratios have been measured in the radio, mainly with CO and H_2CO, and locally (< 1 kpc) also measured in the optical with CH^+ and CN. CO is pervasive and easily observed in emission, however the most abundant isotopes have large optical depths and good determinations of a ratio depend on detecting the very weak $^{13}C^{18}O$ and correcting for $^{12}C^{18}O$ opacity. Formaldehyde is observed in absorption towards strong continuum sources in its ^{12}C and ^{13}C isotopes. To use this species one must deal with the weakness of $H_2^{13}CO$, photon trapping corrections for $H_2^{12}CO$, and the uncertain effects of clumpiness on the trapping correction. In addition chemical processes can enhance isotopic ratios in these molecules. The optical absorption studies generally have poor velocity resolution, though some high velocity resolution measurements have been made (limited to very bright nearby sources). The determination of ratios is complicated by baseline subtraction and line fitting (cf. Vladilo and Centurion 1991). Isotopic enhancement is probably not important for CH^+ formation towards these sources, but may be a factor for CN and CO.

To address the issue of the carbon isotopic ratio across the Galaxy, Langer and Penzias (1990; hereafter LP90) recently presented a study of nine interstellar clouds observed in the rare and doubly rare isotopes of carbon monoxide, $^{12}C^{18}O$ and $^{13}C^{18}O$. The use of low abundance species minimizes radiative transfer effects, and observations towards dense shielded cores reduces complications from isotopic chemical and photodissociation enhancements. They found a systematic gradient in the $^{12}C/^{13}C$ ratio across the Galaxy from ≈ 30 in the inner part at 5 kpc to ≈ 70 at 12 kpc. This trend is similar to that derived from formaldehyde observations (Henkel, Wilson, and Beiging 1982), however the formaldehyde ratios are generally higher across the disk, and show more scatter than the CO measurements. Both the CO and formaldehyde indicate a Galactic Center value about 24. Figure 1 displays the isotope ratio from formaldehyde and carbon monoxide across the Galaxy as a function of distance, D, from the Galactic center) along with the corresponding linear fits ($^{12}C/^{13}C$ proportional to 5.6 D(kpc) for H_2CO and 5.9 D(kpc) for CO

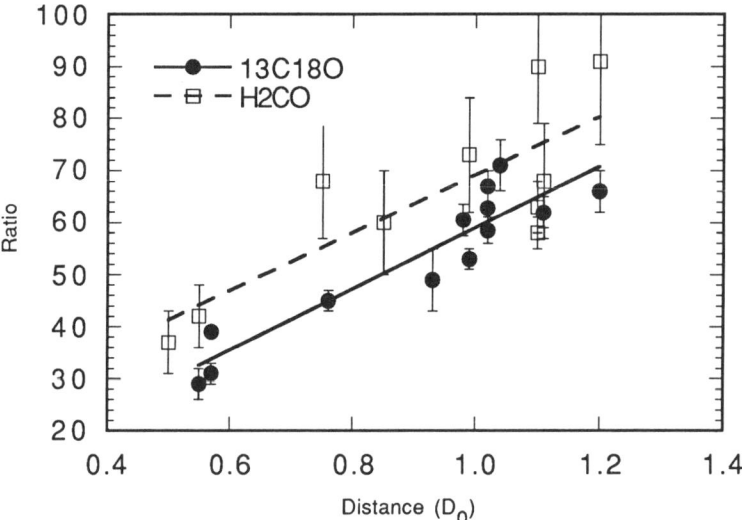

Fig. 1. Carbon 12 to 13 isotope ratio across the Galaxy using H_2CO (Henkel et al. 1982) and CO (LP90 and LP91) data, where distance is with respect to Solar distance D_0.

In the local solar neighborhood LP90 derived an average value $^{12}C/^{13}C = 57\pm3$ from the Galactic gradient, although individual sources ranged from 49 to 67 (excluding 79 in the Orion A K-L region because the strong nearby HII region produces photodissociation enhancement effects). Because there are a wide variety of techniques and species available for measuring the ratio in the solar neighborhood accurate determinations in this region are particularly important for fixing the local value, thereby establishing absolute values across the Galaxy.

To improve the value for the carbon isotope ratio in the local solar neighborhood Langer and Penzias (1991; hereafter LP91) extended their $^{13}C^{18}O$ and $^{12}C^{18}O$ study to three sources not known to be associated with HII regions or strong UV radiation fields: B5, B335, and L134N (also known as L183). In addition they observed another position in Orion far from the HII region near K-L. A sample spectrum from B5 is shown in Figure 2. In the four sources the ratio ranges from 58 to 71 with a weighted average of 61±2 (plotted in Figure 1), slightly larger than the value of 57 suggested in the earlier work on the Galactic gradient. The local value of 61±2 derived from CO is considerably smaller than the formaldehyde result of 80±7, as well as the terrestrial value of 89.

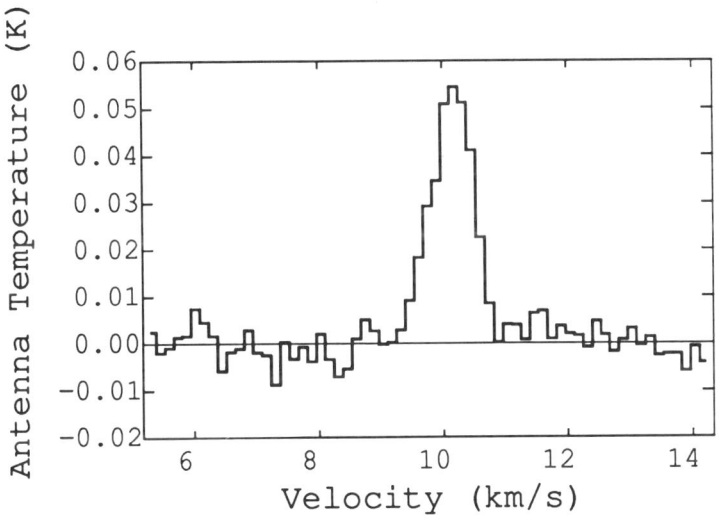

Fig. 2 Spectrum of $^{13}C^{18}O$ in the source B5.

The optical measurements of the isotope ratio have improved in recent years but also show considerable differences among various observations of CH+ and CN. Kaiser, Hawkins, and Wright (1991) summarize recent optical observations in the local solar neighborhood (see their Table 1). The isotope value derived from CH+ is not without controversy. Stahl et al. (1989), Crane et al. (1991), and Stahl and Wilson (1991) find a value about 67±2, in agreement with earlier (but less accurate) measurements. While the measurements of Hawkins et al. (1985) and Hawkins and Jura (1987) yield 42±7. CN absorption line observations of Crane and Hegyi (1988) yield 47±5. Furthermore, Hawkins and Jura (1987) derive a local value of 43±4 from averaging several CH+ sources within a kpc. Recently Vladilo and Centurion (1991) suggested that the data analysis approach of Hawkins and co-workers could overestimate the equivalent width for ^{13}CH+ by a factor of two, perhaps explaining the lower values. It also differs somewhat from the best values derived from the CH+ absorption measurements of the nearby cloud Zeta Oph, 71±3 (Stahl and Wilson 1991) to 67± 2 (Crane et al. 1991).

These radio and optical results raise the following important questions. 1) Why are there differences between the CO and H_2CO? (2) Why do CN and CH^+ isotope ratios differ? The answer may lie in the details of the isotopic chemistry.

3. Carbon Isotopic Chemistry

Isotopic reactions are important both in shielded regions where cosmic ray ionization drives the ion molecule chemistry of carbon and in translucent or PDR regions where UV radiation is important. In dense shielded regions cosmic rays produce H_3^+ ions which through their interactions with O and C initiate molecule production. Most of the carbon ends up as CO (assuming there is more gaseous oxygen available than carbon), with small amounts found in other molecules such as formaldehyde, CS, and CN. A wide variety of neutral-neutral, electron recombination, and ion-molecule reactions are involved in the chemistry once the initiating steps are taken (cf. Prasad and Huntress 1980; Langer et al. 1984; Langer and Graedel 1989). Helium ions destroy the stablest molecules and in the case of CO produce carbon ions. The balance between H_3^+ converting carbon and oxygen atoms into molecules and He^+ destroying the most stable molecules CO, O_2, and H_2O, for example, determines the relative abundance of atomic to molecular form. Isotopic exchange reactions can significantly alter the isotopic abundance of these molecules. For carbon the isotopic exchange reaction (Watson, Anicich, and Huntress 1976),

$$^{13}C^+ + {}^{12}CO \leftrightarrow {}^{13}CO + {}^{12}C^+ + \Delta E(=35K),$$

can enhance the ^{13}CO abundance, particularly at low temperatures, but only if this reaction is an important pathway for producing CO. In dense shielded regions almost all available carbon is in CO and the CO isotope ratio reflects the elemental isotopic ratio, $^{12}CO/^{13}CO \cong {}^{12}C_{true}/^{13}C_{true}$. This result does not necessarily hold true for the other carbon molecules. The destruction rate of $^{13}C^+$ is larger than that for $^{12}C^+$ and therefore less thirteen carbon (in the form of ions and neutrals) is available to form other molecules through the non-isotope exchange reactions (cf. detailed discussion in Langer et al. 1984). Therefore, $H_2{}^{12}CO/H_2{}^{13}CO$, $H^{13}CN/H^{13}CN$, etc. < $^{12}C_{true}/^{13}C_{true}$. Model calculations (Langer and Graedel 1989) show that formaldehyde can be enhanced in the twelve carbon species by up to a factor of two even when CO is not (or only slightly) enhanced with respect to thirteen carbon species. This aspect of the isotopic chemistry may explain the difference in the isotopic ratios derived from CO and formaldehyde in the Galactic surveys, but more observations are needed to confirm this explanation.

In UV dominated regions (cloud edges, PDRs) there is sufficient UV to keep some carbon ionized and therefore the isotopic exchange reaction can be very important for CO. In addition, CO is photodissociated by line absorption and self-shielded to different degrees depending on their relative abundances (cf. Bally and Langer 1981; van Dishoeck and Black 1988). Under such conditions the less abundant species may be isotopically underabundant relative to the more abundant species and the observed isotope ratios will not reflect the true elemental isotopic ratio. This regime of the interstellar chemistry has been extensively modeled by van Dishoeck and Black (1988). Their models confirm that combinations of isotopic exchange and self-shielding lead to varying degrees of enhancement in both ^{12}CO and ^{13}CO.

In the UV regions this variability of CO enhancement can also impact the isotopic abundances of the other carbon molecules. Any isotopic enhancement of CO (no matter which isotope) results in the reverse enhancement in the exchange ion C$^+$ and hence its direct products, such as CH and CH$_2$. In addition, for CN there is the possible exchange reaction,

$$^{13}C^+ + {}^{12}CN \leftrightarrow {}^{13}CN + {}^{12}C^+ + \Delta E(=34K),$$

which will be important when it competes with photodestruction of CN. A simple model of CN formation from reactions of N with CH, CH$_2$, C$_2$, and C$_2$H (whose sum is denoted by CM) leads to the following relationship for the carbon isotopic ratio of CN,

$$[^{12}CN/^{13}CN] \approx [^{12}CM/^{13}CM] \frac{[\Gamma(CN) + k_f n(^{12}C^+) + k_0 n(O)]}{[\Gamma(CN) + k_r n(^{12}C^+) + k_0 n(O)]}$$

where, Γ, is the photodestruction rate, k_f and k_r are the forward and reverse isotope exchange reaction rate coefficients, k_0 the neutral-neutral reaction rate coefficient with oxygen, and n(X) the density of species X. From this expression it can be seen that the CN isotope ratio can be enhanced either in twelve or thirteen C depending on conditions. The isotope exchange chemistry for CN may explain the variability in the CN observations and the differences between the CN and CH$^+$ isotope ratio in the same sources. CH$^+$ is not expected to be isotopically enhanced since it is thought to be formed under high temperature conditions where the difference in isotopic rates is unimportant.

4. References

Bally, J. and Langer, W. D. 1981, *Ap. J.*, **255**, 143.
Crane, P. and Hegyi, D. J. 1988, *Ap. J. (Letters)*, **326**, L35.
Crane, P., Hegyi, D. J., and Lambert, D. L. 1991, *Ap. J.*, **378**, 181.
van Dishoeck, E. and Black, J. 1988, *Ap. J.*, **334**, 711.
Hawkins, I., Jura, M., and Meyer, D. M. 1985, *Ap. J. (Letters).*, **294**, L131.
Hawkins, I. and Jura, M. 1987, *Ap. J.*, **317**, 926.
Henkel, C., Wilson, T. L., and Beiging, J. 1982, *Astr. Ap.*, **82**, 41.
Henkel, C., Walmsley, M., and Wilson, T. L. 1980, *Astr. Ap.*, **109**, 344.
Kaiser, M. E., Hawkins, I., and Wright, E. 1991, *Ap. J.*, in press.
Langer, W. D., Graedel, T. E., Frerking, M. A., and Armentrout, P. B., 1984, *Ap. J.*, **277**, 581.
Langer, W. D. and Graedel, T. E., 1989, *Ap. J. Suppl.*, **69**, 241.
Langer, W. D. and Penzias, A. A. 1990, *Ap. J.*, **357**, 477 (LP90).
Langer, W. D. and Penzias, A. A. 1991, submitted to *Ap. J.*, (LP91).
Prasad, S. S. and Huntress, W. T., Jr. 1980, Ap. J. Suppl., **43**, 103.
Stahl, O., Wilson, T. L., Henkel, C., and Appenzeller, I. 1989, *Astr. Ap.*, **221**, 321.
Stahl, O. and Wilson, T. L. 1991, *Astr. Ap. (Letters)*, in press.
Vladilo, G. and Centurion, M. 1991, *Astr. Ap.*, in press.

CHEMICAL MODELS OF ACTIVE REGIONS

S.B. CHARNLEY[†]
Space Science Division, NASA Ames Research Center, California 94035.

ABSTRACT. Theoretical models of the chemistry in regions of low-mass star formation are reviewed.

1. Introduction

I will interpret active regions as those regions in which young, low-mass stars are forming. Other regions where dynamical processes directly influence the chemistry are described in the contributions by Walmsley, Prasad and Ziurys. The physical conditions pertaining to regions of low-mass star formation have been reviewed recently by Cernicharo [5].

Theoretical models of molecule formation in dense interstellar clouds have tended to concentrate on the chemistry and to neglect the physical evolution of the gas (Millar [23] and references therein). 'Pseudo-time-dependent' [21], or static, models assume that the density and temperature of the gas remain constant throughout the age of a cloud and neglect, apart for H_2 formation, the gas-grain interaction. Under these assumptions a chemical steady-state is attained after about 10^7 years, however, detailed comparison with observed molecular abundances show that the abundances at times $\sim 10^5$ years appear to be in better agreement. Ice mantles (H_2O, CO, CH_3OH) containing significant fractions of the available heavy elements are widely observed in molecular clouds [29]. The characteristic time for complete accretion of heavy gas phase species on to dust grains is typically $\sim 10^6$ years, and so, in the absence of a continous desorption mechanism [12], one must have a means of returning solid material to the gas phase. Static models that do allow for the formation of molecular ices, and do not return mantle material, show that a complete freeze-out occurs well before a steady-state is reached, and also that the chemical evolution at earlier times is different in a non-trivial way [4].

Molecular clouds are observed to be the sites of star formation and exhibit a complex, clumpy, structure with material co-existing in several density phases. The effect of star formation is to disrupt the ambient medium via the associated outflows and shocks. It has been suggested that the periodic removal of mantle material by processes associated with star formation, on a time-scale comparable with the local star formation rate, can return mantle material to the gas phase and maintain a large C/CO ratio [30]. In the dark cloud Barnard 5 (B5) it has been shown that the outflows from embedded young stars are interacting with the surrounding clumpy medium from which they originally formed [16]. Estimates suggest that the gas density changes on a time-scale of about 6×10^5 years, shorter than the time-scales of accretion and chemical steady-state. The important conclusion is that the chemistry in dense interstellar clouds is of a nonequilibrium nature [16]. Chemical kinetic models of these regions should therefore attempt to include the gas-grain interaction, and also the density and temperature variations associated with collapse, expansion,

[†]*NAS/NRC Resident Research Associate*

and shocks [11]. In this review I briefly describe recent work on modelling the chemistry in star-forming regions, with particular reference to B5.

2. Theoretical Models

A theory of the physics of clumpy molecular clouds regulated by star formation was developed by Norman & Silk [26]. In the Norman-Silk model, gas and dust are continously cycled between dense ($n \sim 10^4 \mathrm{cm}^{-3}$) clumps and the interclump medium ($n \sim 10^3 \mathrm{cm}^{-3}$). The winds from young stars create bubbles and disrupt the surrounding medium. The chemistry in these wind-blown bubbles has been modelled [6] and included such processes as the ablation and entrainment of clump material by stellar winds, chemistry in the reverse shock at the bubble edge, and depletion on to grains as the cool shell is disrupted, fragments and reforms the next generation of clumps. The assumed presence of ionised wind material (H^+ and He^+) in the mass-loaded wind [18] can limit the conversion of heavy atoms into hydrides in the post-shock gas. The high pre-shock fractional ionisation, x_e ($\sim 10^{-2}$), means that the shock is J-type [13].

Two types of model have been developed specifically for the nonequilibrium chemistry in B5 [7]. In the first, based upon a scenario in which gas and dust is rapidly cycled between clump and interclump phases on a time-scale of 6×10^5 years [16], clumps form at the free-fall rate and the shocks which occur are C-type [13]. In a second model a much slower clump formation rate and longer lifetime was considered which led to a cycling time-scale of 6×10^6 years. Partial ionisation of stellar winds by strong bow shocks and entrainment of clump material into the winds was also considered. In both models the formation and removal (by shock waves) of molecular mantles was important. Calculations showed that the chemistry was sensitive to the adopted dynamics and that the models could produce repeatable abundances between cycles. In these studies the grain surface formation of saturated molecules had a significant effect on gas phase chemistry.

The chemistry of nitrogen-bearing molecules in star-forming clouds has been studied in the context of the slow cycling models described above [25]. The abundances of several molecules in cold clumps were found to be determined by the post-shock chemistry (e.g. HCN) and by formation on grains (e.g. NH_3).

A simple model has been developed to study the chemistry which may occur at wind-clump interfaces as hot, ionised, wind material mixes with the cold gas at the surfaces of clumps embedded in the flow. Calculations indicated that it may be possible to elucidate chemical diagnostics of the interface physics. For example, when both mixing and heating occur with high efficiency the CH molecule is predicted to become very abundant in the interface region [8].

The formation and removal of ice mantles is a fundamental feature of the cycling models and they predict spatial and temporal variation in the mantle size throughout the cycle associated with a particular dynamical model. Ice mantles have been detected in B5 towards the infrared source IRS1 [9]. The time-scale required to form the observed column density of ice has been used to constrain the clump formation time-scale and hence also the dynamics of the region. Fitting of the $3\mu m$ feature has been used to infer the composition and thermal history of the ice mantles during a star formation cycle [10].

To test the validity of various aspects of the theoretical work outlined above, multi-molecular studies of star-forming regions are required. For the case of B5 such a study would be particularly useful as the chemistry in this cloud has been the focus of a substantial amount of theoretical work [11]. Such studies have been performed for other star-forming regions. In B1, significant differences in the abundances of methanol, SiO and complex hydrocarbons, compared to other clouds, have been found [1]. Observations of ρ Oph [22] have illustrated the clumpy nature of this region and the stong spatial variation in the emission of several molecules. Recent maps of NGC 2071(N) also show dramatic spatial differences in molecular emission and it has been proposed that these arise from differences in molecular abundances due to the effects of local star formation [17].

It would be interesting to model the above sources in the future.

It has recently been suggested that the chemistry of TMC-1, long considered the typical cold, quiescent cloud, is the result of local dynamical interactions, due specifically to collapse of diffuse gas in Heiles Cloud 2 [31]. An alternative dynamical origin for TMC-1 has been proposed to account for its apparently peculiar chemistry : the cloud is a shock remnant [27]. Appreciable abundances of methane and acetylene can be produced in diffuse cloud shocks [24, 28] and provide the initial conditions for the synthesis of large hydrocarbons by ion-molecule reactions [19]. Models of TMC-1 in which diffuse gas is shocked, followed by collapse to the required density, can produce large abundances of cyanopolyynes [3].

3. Discussion

Barnard 5 has been observed in ^{12}CO, ^{13}CO, $C^{18}O$, CS, NH_3 and HCN [2, 15, 16], as well as in ice, and so one may begin to compare the theoretical predictions. The observed HCN abundance of 2.8×10^{-9} [15] in the core IRS1, taken with the observed ice column density in this line of sight [9], are particularly interesting as they place constraints on the published cycling models of nitrogen chemistry. These calculations have only been performed for the slow cycling models in which ablation and mixing are important. The HCN abundance in the cold clump phase reflects its value in the post-shock gas, as the clump ages gas phase HCN is lost by accretion on to grains. The minimum time, t_{min}, to form the observed ice mantles (roughly the minimum clump age) in each of the fast and slow cycling models is 7.6×10^4 and 7.6×10^5 years [9]. Relative to molecular hydrogen the *minimum* HCN abundance obtained in the clumps of the slow cycling models is 1.2×10^{-8} at 3×10^6 years [25]. Prior to one million years of the clump phase the HCN abundance is never less than 1.0×10^{-8} and is $\sim 10^{-7}$ at t_{min}. Studies of N-chemistry in the rapid cycling models of B5 have not yet been performed.

Two possibilities can be inferred from the above. First, the cycling time in B5 may be longer than 5×10^6 years: continued accretion in the clump phase would eventually reproduce a low HCN abundance similar to that observed. The second possibility is that $x_e < 10^{-3}$ in the pre-shock gas, this could be due to inefficient mixing or to significant recombination having occurred in the mass-loaded wind. The J-shocks of the slow cycling models with mixing have peak post-shock temperatures in the range 3000-4000K and produce copious amounts of HCN [25]. When the pre-shock x_e is lower the shocks will be C-type, have significantly lower maximum post-shock temperatures $\sim$1000K, and may be expected to produce less HCN. Observations of other molecules such as SO, SO_2 and H_2S which are predicted to exhibit strong time dependence in their clump abundances [25] would help to further constrain the models.

A fundamental feature of the cycling models is the periodic return of mantle material to the gas by shock waves. In all the published models this was assumed to occur instantaneously with unit efficiency by an unspecified physical process. Although some calculations have recently been done in which the mantle removal efficiency is less than 100% (Nejad & Williams; this volume), this aspect has not been treated consistently and is a critical weakness of the models. Shock speeds considered in the models are typically in the range 10-20 km s^{-1}. In the slow cycling models, where the shocks are J-type, the pre-shock H nucleon density, n_o, is 400 cm^{-3}. Calculations show that under these conditions no mantle removal is possible [20]. If x_e is 10^{-3} or less the shocks will be C-type, however, calculations show that mantle removal by grain-neutral sputtering will be inefficient for the model parameters considered [14]. Betatron acceleration, grain-grain collisions, or ion-grain collisions, could be important for mantle removal in the slow models but these processes have not been investigated quantitatively. Mantle removal in wind-clump interfaces has been suggested (e.g. [11]) but this remains speculative at present. It therefore appears that mantle removal in the slow cycling models is an inefficient process: these models should eventually deplete the heavy component of the gas phase in a similar manner as static ones [4].

In the C-type shocks of the rapid cycling models, the preshock ionisation and density are about 10^{-7} and 10^4cm^{-3} respectively. For the expected magnetic field strengths substantial mantle removal can occur via grain-neutral sputtering for shock speeds in excess of about 18 km s^{-1} [14]. The rapid cycling models appears to be the only ones in which mantles can be efficiently removed and more work is required to determine whether this process is viable in the slow models.

4. Conclusions

Theoretical models of star-forming regions must attempt to include a realistic treatment of the associated dynamical processes. Models of molecular clouds developed in this manner can plausibly provide natural answers to problems connected with more conventional, static ones e.g. mantle removal, C/CO ratio, existence of 'young' chemistries as opposed to steady-state ones. Several theoretical problems do exist with the current models. In particular, a more consistent treatment of shock structure and the physics of mantle removal are required. Multi-molecular studies of particular star-forming regions are needed, especially of species which may act as diagnostics of the dynamics.

References

1. Bachiller, R., Menten, K. & Rio-Alvarez, S. 1990, *Astron. Astrophys.*, **236**, 461.
2. Benson, P.J. & Myers, P.C. 1989, *Ap. J. Suppl.*, **71**, 89.
3. Brown, P.D. & Charnley, S.B. 1989, unpublished.
4. Brown, P.D. & Charnley, S.B. 1990, *M.N.R.A.S.*, **244**, 432.
5. Cernicharo, J. 1990, in *Physics of Star Formation and Early Stellar Evolution*, NATO Advanced Study Institute, Crete, 1990.
6. Charnley, S.B., Dyson, J.E., Hartquist, T.W. & Williams, D.A., 1988, *M.N.R.A.S.*, **231**, 269.
7. Charnley, S.B., Dyson, J.E., Hartquist, T.W. & Williams, D.A., 1988, *M.N.R.A.S.*, **235**, 1257.
8. Charnley, S.B., Dyson, J.E., Hartquist, T.W. & Williams, D.A., 1990, *M.N.R.A.S.*, **243**, 405.
9. Charnley, S.B., Whittet, D.C.B. & Williams, D.A. & 1990, *M.N.R.A.S.*, **245**, 161.
10. Charnley, S.B., Whittet, D.C.B. & Williams, D.A. 1990, in *Molecular Clouds*, Eds. R.A. James & T.J. Millar, Cambridge University Press, p321.
11. Charnley, S.B., & Williams, D.A. 1990, in *Molecular Astrophysics*, Ed. T.W. Hartquist, Cambridge University Press, p313.
12. d'Hendecourt, L.B., Allamandola, L.J. & Greenberg, J.M., 1985, *Astron. Astrophys.*, **152**, 130.
13. Draine, B.T. 1980, *Ap. J.*, **241**, 1021.
14. Draine, B.T., Roberge, W.G. & Dalgarno, A. 1983, *Ap. J.*, **264**, 485.
15. Fuller, G.A., et al. 1991, *Ap. J.*, **376**, 135.
16. Goldsmith, P.F., Langer, W.D. & Wilson, R.W. 1986, *Ap. J.*, **303**, L11.
17. Goldsmith, P.F., Margulis, M., Snell, R.L., & Fukui, Y. 1991, preprint.
18. Hartquist, T.W. & Dyson, J.E. 1988, *Astrophys. Space. Sci.*, **144**, 615.
19. Herbst, E. & Leung, C.M. 1989, *Ap. J. Suppl.*, **69**, 271.
20. Hollenbach, D.J. & McKee, C.F. 1979, *Ap. J. Suppl.*, **41**, 555.
21. Leung, C.M., Herbst, E. & Huebner, W.F. 1984, *Ap. J. Suppl.*, **56**, 231.
22. Loren, R.B. & Wooten, A. 1986, *Ap. J.*, **306**, 142.
23. Millar, T.J. 1990, in *Molecular Astrophysics*, Ed. T.W. Hartquist, Cambridge University Press, p115.
24. Mitchell, G.F. 1983, *M.N.R.A.S.*, **205**, 765.
25. Nejad, L.A.M., Williams, D.A. & Charnley, S.B. 1990, *M.N.R.A.S.*, **246**, 183.
26. Norman. C. & Silk, J. 1980, *Ap. J.*, **251**, 533.
27. Olano, C.A., Walmsley, C.M. & Wilson, T.L. 1988, *Astron. Astrophys.*, **196**, 194.
28. Pineau des Forêts, Flower, D., Hartquist, T.W. & Millar, T.J. 1987, *M.N.R.A.S.*, **227**, 993.
29. Tielens, A.G.G.M. & Allamandola, L.J. 1987, in *Physical Processes in Interstellar Clouds*, Eds. G.E. Morfill & M. Scholer., Kluwer, p.333.
30. Williams, D.A. & Hartquist, T.W. 1984, *M.N.R.A.S.*, **210**, 141.
31. Williams, D.A. & Hartquist, T.W. 1991, *M.N.R.A.S.*, **251**, 351.

QUESTIONS AND ANSWERS

J.M.C.Rawlings: I would just like to point out that cosmic-ray induced photodissociation reactions (not included in the cyclic models of Nejad et al.) can have an important effect on the gas-phase chemistry (and consequently the initial mantle composition) in dynamical models in which freeze-out onto grains is significant. Abundance of saturated molecules (eg. NH_3, H_2O) can be suppressed by over an order of magnitude. This is to be expected as the timescale for clump collapse, freeze-out and photodissociation are similar.

S.B.Charnley: It may be interesting to study cyclic models incorporating this process. The abundance of HCN in the slow models is 10 times that observed by Fuller et al., and is probably due to the reaction of CN with H_2 in the postshock gas.

EVOLUTIONARY MODELS OF INTERSTELLAR CHEMISTRY

SHEO S. PRASAD[1]
Lockheed Palo Alto Research Laboratory (O/91-20, B255)
3251 Hanover Street, Palo Alto, CA 94304, USA

ABSTRACT. Evolutionary chemical models are ultimately unavoidable for a full understanding of interstellar clouds. They include not only the chemical processes but also the dynamical processes by which the modeled object came to be the way it is. From an evolutionary perspective, dark cores may be ephemeral objects and dynamical equilibrium an exception rather than norm. Evolutionary models have numerous advantages over "classical" fixed condition equilibrium models. They have the potential to provide more elegant explanations for the observed inter-cloud and intra-cloud chemical differences. The problem of the depletion of gas phase molecules by condensation onto the grain may also be less serious in evolutionary models. Hence, these models should be actively developed.

1. EVOLUTIONARY MODELS ARE ESSENTIAL:

Impressive strides are being made in the observations of molecular clouds. *Do we have matching quality theoretical tools for data interpretation? Unfortunately not.* Currently fixed condition equilibrium models are the most widely used tool for elucidating observed molecular abundances. In these models the physical conditions of density, temperature and visual extinction are kept fixed. Table 1 compares these models with evolutionary models which consider, not only the chemistry, but also the mediating dynamical processes by which the objects came to be the way they are. In evolutionary models, therefore, the physical conditions are allowed to change with time in response to known quantifiable dynamical processes and/or external environmental conditions. Due to lack of space, consider just the non-equilibrium nature of the chemistry in even dark clouds away from active star forming region (Herbst and Leung 1986, 1989) and/or the survival of the molecules in the gas phase in those clouds despite their efficient absorption onto the cold grains from which efficient desorption is still uncertain. Their explanations in fixed condition "classical" models involves mixing currents (Chièze and Pineau des Forêts 1987, Chièze et al 1991). But this has not worked well. The latest mixing model finds it necessary to introduce elements of evolutionary modeling. This need was fulfilled in an improvised manner, *e.g.*, exponential evolution of density and power law growth of visual extinction with time (Chièze et al 1991). Given this experience with fixed-condition models, realistic evolutionary models are ultimately unavoidable. Fixed condition "classical" equilibrium models do not mean much, especially when arbitrary elements are introduced to generate agreement with observations.

---

[1]Adjunct affiliation with the Departments of Physics & Astronomy, University of Southern California, Los Angeles, CA 90089-1341, USA.

Table 1. Comparison of the "classical" and evolutionary models

Model Type	Merits	Demerits
"Classical" or Fixed-Condition equilibrium models [Density, temperature & visual extinctions, and their spatial variations in the cloud are held fixed]	● Simplicity ● Chemical production & loss processes easily understood ● First order feel for which molecule trace what regions of the clouds	● Non-equilibrium chemistry, inter-cloud & intra-cloud chemical variations, and depletion of molecules by the grains have *ad hoc*, rather artificial solution ● Cloud structure taken for granted. Lots of interesting (intermediate) phenomenon lost ● Diffuse, translucent and dark objects unrelated
Evolutionary Models [Cloud Structure changes with time in response to known quantifiable dynamical processes]	● All of the above minus simplicity ● Includes equilibrium models as a subset ● (i) Better understanding of the non-equilibrium chemistry and of the inter-cloud & intra-cloud variations, and (ii) the grain absorption problem less serious ● Diffuse, translucent & dense clouds need not be unrelated	● <u>None</u> of the above but ● Computationally demanding ● Analysis of results more laborious (although rewarding)

2. TYPES OF EVOLUTIONARY MODELS:

Evolutionary models can be driven in a number of ways, all depending upon the objects and their environment. Following two classes of evolutionary models are relatively more developed:

(i) In the region of active star formation (*e.g.*, M17 SW, Bernard 5) evolution of interstellar gas between dense and diffuse phases and the non-equilibrium nature of the chemistry can be sustained by the stellar winds. Williams and his collaborators (*e.g.*, Williams and Hartquist 1984, Charnley et al 1988a,b) have made evolutionary models of interstellar chemistry driven by stellar winds. Their models have four evolutionary phases: (1) collapse of dense cores and the freeze-out of molecules onto the grains, (2) ablation of cores by stellar winds, (3) arrest of wind flow by weak reverse shock, and (4) accumulation of dense post shock gas into the core. These have been reviewed by Charnley (1992).

(ii) The other class of evolutionary models, on which we have focussed, are for quiescent dark cloud cores mostly isolated from energetic processes associated with nascent stars. These objects are either far away from concentrations of young stars, or the concentration of stars near them is too small to be important. L134N is currently thought to be an example of such an quiescent object, as evidenced by its narrow line widths indicating the absence of shocks or supersonic velocities. High latitude clouds (or, the cirrus) are additional examples. Our aim has been to understand inter-cloud and intra-cloud variations of molecular compositions in this class of objects. We were also motivated to provide a frame work for a more refined understanding of the epochs (i) and (iv) of the evolutionary models dominated by stellar winds.

In addition to the above two categories, evolutionary models may also be based on the formation of denser clouds by cloud-cloud collisions (Henriksen and Turner 1984, Elmegreen 1987), or on the formation and decay of clumps of moderate densities by fluctuations in the uv radiation field (Chièze and Boisanger 1991).

3. MILESTONES IN SIMPLIFIED EVOLUTIONARY MODELS OF DARK CLOUD CORES:

This class of models have come a long way. In 70s, free fall or arbitrarily retarded free fall of isothermal interstellar gas was used to model the molecule rich dense cloud cores (Kiguchi et al 1974, Suzuki et al 1976). Gerola and Glassgold (1978) model was an exception to these rather crude modeling, because it involved both gravity and pressure gradient forces in the equation of motion which was solved simultaneously with the continuity and heat budget equations. Unfortunately, this model was never implemented beyond the single case of a quite massive (1000 solar mass) cloud which was further restricted to being already dense to start with (initial density = 10^3 cm^{-3}). These restrictions may have been the consequence of the then prevailing notion that low mass diffuse clouds generally do not collapse gravitationally.

By mid-80s, a major advance occurred when we showed that even low mass warm diffuse clouds could easily contract to form dense cores (Tarafdar et al 1985). Due to the extinction of the background interstellar uv radiation with depth, the model clouds were always warmer at the outer edges compared to the inner regions. There was, therefore, a pressure gradient force that assisted gravity in initiating the contraction of even low mass warm diffuse clouds. These models, however, generated one serious concern. Formation of dense cores in this way, it was argued by the critics, would lead to star formation rate in excess of the observed. We responded to this concern by including forces that may oppose gravity, and by including lower initial densities.

The forces that may oppose gravity and increase in strength with the density were mimicked by tangled frozen-in magnetic fields. They lead to outward directed magnetic pressure gradient force whose magnitude increases as the core density increases. The gravitational contraction is now no longer always monotonic. Depending on the initial magnetic field strength, gravitationally contracting clouds may follow a star forming track (monotonic contraction to a protostellar stage)or a non-star forming track along which they would revert back to the diffuse state after reaching high core densities. Frozen-in magnetic field is not the only mechanism to reverse a gravitational collapse. Lower initial densities can produce similar effects in non-magnetic clouds (Tarafdar et al 1989). Thus, the key points are: (i) contraction is reversible under a variety of conditions, and (ii) this reversal eliminates conflict with the observed star formation rate.

4. EVOLUTION MAY BE THE NORM:

It now appears that dark cloud cores may be transient, not long-lived, objects even in the absence of harsh stellar influences. They may be just one epoch in the incessant dynamical evolution. Contrary to the common belief, dynamical equilibrium may be exception rather than the norm. Indeed, the notion of all pervading dynamical equilibrium has serious problems. Cox (1990) and Turner et al (1991) have pointed out one difficulty with respect to diffuse clouds, viz., "How the heating mechanism manages to provide just the right amount of energy so that diffuse clouds have thermal pressures that can be easily confined"? Dark cores present similar problems. "While it is easy to determine a set of conditions that will hold a dark core in equilibrium, it is much more difficult to ensure that exactly those conditions will be seen by the core in the course of its formation" (Prasad et al 1991). Evolutionary models do not have these difficulties. In addition, they have the potential to provide a more natural explanation of the observed inter-cloud and intra-cloud chemical differences. Prasad et al (1991) give an example of this potential in the context of the cloud-to-cloud variation of the HC_3N abundance (see their Figure 17)

5. RESPONSE TO THE COMMON OBJECTIONS TO EVOLUTIONARY MODELS OF DARK CORES:

As with any new idea, objections may continue, for sometime, to be raised against evolutionary models of dark clouds. One objection is that the dark cores obey the virial equations. This objection is not serious, because the dark cores obey the equations only crudely. The second objection argues that no one has ever observed a contracting/expanding dark core. This may, however, be a detectibility issue (A. Dalgarno, private communication). No one can deny that stars are forming, and therefore at least some clouds are gravitationally contracting. Why, then, even those are not detected? The answer is that realistic dynamically evolving dark cores have very small systematic velocity in the region where the molecular tracers of high density reside. This is easily verified for NH_3 by inspecting figures 4 and 13 of Prasad et al (1991). Non-detection of contracting/expanding cores, therefore, can not stand as an argument against evolutionary models.

6. LIMITATIONS OF THE PRESENT EVOLUTIONARY MODELS OF DARK CORES:

The present evolutionary models have limitations, because we considered the cloud to be homogeneous and ignored the observed clumpiness. Even so, the models are applicable to select high latitude dense clouds, some dark clouds (e.g., L134N) which are not excessively clumpy, and possibly to individual clumps of a clumpy cloud depending on the clumps' mass, environment and origin.

7. SUMMARY:

Evolutionary models are essential for a proper understanding of interstellar clouds. In contrast to the more common fixed condition equilibrium models, the evolutionary models allow the conditions in the modeled objects to change with time in response the mechanisms by which the objects came to be way they are. These models imply that dark cloud cores may be dynamically evolving short-lived objects, even in the absence of stellar influences. There are probably no very solid reasons to doubt the that dynamically evolving objects may the norm and dynamical equilibrium the exception. Furthermore, the concept of ubiquitous equilibrium has some problems both in the dense and diffuse regions. Dynamically evolving models have the potential to provide a more natural explanation of the observed inter-cloud and intra-cloud chemical variations. These models should, therefore, be vigorously developed so as to reduce their limitations and increase their applicability.

ACKNOWLEDGEMENTS:

This research has been supported by a contract from the ISR Branch of NASA's Astrophysics program to Lockheed Missile and Space Company.

REFERENCES:

Charnley, S. B. 1992, *The Astrochemistry of Cosmic Phenomena* (IAU Symposium #150), ed. P. D. Singh (Dordrecht:Kluwer Academic), p.
Charnley, S. B., Dyson, J. E., Hartquist, T. W., & Williams, D. A. 1988a, *MNRAS*, 231, 267.
---------. 1988b, *MNRAS*, 235, 1257.
Chièze, J. P., & de Boisanger, C. 1991, in *Fragmentation of Molecular Clouds and Star Formation*, ed. E. Falgarone et al (Dordrecht:Kluwer Academic) p. 197.
Chièze, J. P., & Pineau des Forêts, G. 1987, *Astron. Astrophys.*, 183, 98.
Chièze, J. P., Pineau des Forêts, G., & Herbst, E. 1991, *Ap. J.*, 373, 1991
Cox, D. P. 1990, in *The Interstellar Medium in Galaxies*, ed. H. A. Thronson and J. M. Shull (Dordrecht:Kluwer Academic), p. 181
Elmegreen, B. G. in *Interstellar Processes* ed. D. J. Hollenbach and H. A. Thronson (Dordrecht:Kluwer Academic) p.259
Gerola, H. & Glassgold, A. E. 1978, *Ap. J. Suppl.*, 37, 1.
Henriksen, R. N., and Turner, B. E. 1984, *Ap J*, 287, 200.
Herbst, E. & Leung, C. M. 1986, *Ap. J.*, 310, 378.
----------- 1989, *Ap. J. Suppl.*, 69, 271.
Kiguchi, M., Suzuki, H., Sata, K., Miki, S., Tominatsu, A., & Nakagawa, Y. 1974, *Proc. Astron. Soc. (Japan)*, 26, 499.
Prasad, S. S., Heere, K. R., & Tarafdar, S. P. 1991, *Ap. J.*, 373, 123.
Suzuki, H., Miki, S., Sata, K., Kiguchi, M., & Nakagawa, Y. 1976, *Progr. Theoret. Phys. (Japan)*, 56, 1111.
Tarafdar, S. P., Ghosh, S. K., Heere, K. R., & Prasad, S. S. 1989, *Highlights Astr.*, 8, 345.
Tarafdar, S. P., Prasad, S. S., Huntress, W. T., Jr., Villere, K. R., & Black, D. C. 1985, *Ap. J.*, 289, 747.
Turner, B. E., Xu, L., & Rickard, L-J. 1991, "On the Nature of the Molecular Cores in the High Latitude Cirrus Clouds: II. Structure, Stability, and Physical Conditions from $C^{18}O$ Observations, Preprint submitted to *Ap. J.*
Williams, D. A., & Hartquist, T. W. 1984, *MNRAS*, 210, 141.

QUESTION AND ANSWER

B.E.Turner: While polytropic hydrostatic equilbrium models of index N=-3 can explain $C^{18}O$ observations (J=2-1,1-0) in many cirrus cloud cores (Turner et al.1991), they cannot explain $C^{18}O$ observations in cold dark clouds of higher density. These latter have non-equilbrium chemistry, as distinct from the equilbrium chemistry that describes cirrus cloud cores, hence require a mixing within the cloud material. The idea of Prasad et al. that magnetic fields can produce such "turning" made out of the core region, thus renewing the chemistry, also produces a flatter density distribution with radius than can hydrostatic equilbrium or most dynamical collapse models. Such a flatter distribution is precisely what is needed to explain the $C^{18}O$ observations of these cold dense clouds.

S.S.Prasad: Thank you, Barry, for your constructive comment. The possibility that your observational data can differentiate between hydrostatic model, on the one hand, and various dynamical models, on the other hand, is quite significant. This should provide the much needed stimulus to build better models with useful physics content.

The Formation of Deuterated Molecules in Dense Clouds

T J MILLAR

Department of Mathematics, UMIST, PO Box 88, Manchester, UK

November 13, 1991

Abstract. In this article, I list the deuterated species detected so far, together with an estimate of their abundance ratios relative to their hydrogen bearing parent for a variety of astronomical regions, discuss the basic chemical processes which fractionate deuterium together with some simple estimates of the degree of fractionation which can result from these processes. I then discuss in detail the deuteration of hydrocarbon molecules and show that the effects of chemical reactions which can recycle parent species can appreciably affect estimates of fractionation. Finally I shall discuss how structural effects may inhibit fractionation in the cyanopolyynes and related species.

1. Introduction

Molecules containing deuterium have been observed in interstellar clouds for many years. Except for the detection of HD in diffuse clouds (Rogerson et al. 1973), all D-bearing species are found in dense molecular clouds having a variety of temperatures ranging from $\sim$ 10 K in cold, dust clouds, to $\sim$ 30-70 K in giant molecular clouds, to $\sim$ 100-200 K in hot molecular cores. Table I lists the 14 deuterium-bearing molecules detected in interstellar clouds, together with a rough estimate, R, of its abundance relative to its hydrogen-bearing form, for a sample of sources. Note that this list excludes H_2D^+ for which Phillips et al. (1985) find one line in one source; this identification, although probably correct, needs confirmation. Since the last Astrochemistry meeting in Goa (Wootten, 1987), there have been detections of C_3HD (Bell et al., 1986; Gerin et al., 1987), C_4D (Turner, 1989) and D_2CO (Turner, 1990), the first doubly-deuterated molecule observed. In general the abundances of deuterated molecules relative to their hydrogenated forms are enhanced, or fractionated, over the HD/H_2 abundance ratio, HD being the reservoir of deuterium in dense clouds. There have also been some significant theoretical advances in this period. In particular, it has been realised that for the hydrocarbon species, it is difficult to decouple the fractionation reactions from the large set which govern the chemistry because several different chemical pathways can contribute to the formation of any particular hydrocarbon. Detailed time-dependent calculations are required and have been carried out by Brown and Rice (1986) and Millar, Bennett and Herbst (1989). Indeed even for a simple species such as DCO^+, the simple analyses performed by, for example, Watson (1976, 1977), Guélin et al. (1977) and Snyder et al. (1977), do not give the correct ratio at so-called 'early-time' which is thought to give the best agreement with observations (Millar et al.,

1989). In addition, it has been recognised that the fractionation observed in hot molecular cores does not result from a high temperature chemistry but must be a relic of a low temperature chemistry involving both gas-phase and grain-surface reactions (Brown and Millar, 1989a,b; Turner, 1990).

The basic reactions which initiate the fractionation of many species involve the abstraction of D atoms from HD:

$$H_3^+ + HD \longleftrightarrow H_2D^+ + H_2 - \Delta E_1 \qquad (1)$$
$$CH_3^+ + HD \longleftrightarrow CH_2D^+ + H_2 - \Delta E_2 \qquad (2)$$
$$C_2H_2^+ + HD \longleftrightarrow C_2HD^+ + H_2 - \Delta E_3 \qquad (3)$$

where ΔE_1, ΔE_2 and ΔE_3 are equivalent to about 200 K, 370 K and 550 K, respectively. One can estimate the enhancement factor, S, that is the factor by which the abundance ratio $R(XD^+)$ is enhanced over the deuterium abundance ratio in the reservoir, R(HD) - here R(XD) = [XD]/[XH], where the square brackets refer to fractional abundances. At high temperatures, $\widetilde{>}$ 25 K for H_2D^+, $\widetilde{>}$ 60 K for CH_2D^+ and C_2HD^+, the enhancement factors decrease exponentially with increasing temperature. The larger abundance of H_3^+ usually ensures that reaction (1) dominates fractionation at very low temperatures, the maximum value of S_1 being $\sim (1\text{-}2)\ 10^3$.

In addition to fractionation via HD, the dissociative recombination of DCO^+ leads to an enhancement in the atomic deuterium to atomic hydrogen ratio, and fractionation via reaction of D atoms with species such as H_3^+, HCO^+, N_2H^+ and OH may occur (Dalgarno and Lepp, 1984; Croswell and Dalgarno, 1985; Adams and Smith, 1985). The free energies associated with these reactions are much larger than those of reactions (1)-(3) and would imply large enhancement factors at high temperatures. This is not observed for N_2D^+, for example, indicating that D atoms are not the major reservoir of gas phase deuterium in dense clouds. Even so, appreciable fractionation of OD has been predicted (Croswell and Dalgarno, 1985; Millar et al., 1989; Brown and Millar, 1989a).

2. Deuteration of hydrocarbon molecules

The chemistry of the hydrocarbons have been discussed in detail by many workers. Turner (1989) pointed out that the chemistries of C_3H, $c\text{-}C_3H_2$, C_4H and HC_3N are inter-related, mainly through the $C_3H_3^+$ ion, and that the fractionation of these ions might be simply related to one another. Consider the formation of C_4H from C_3H_2:

$$C^+ + C_3H_2 \longrightarrow C_4H^+ + H \qquad (4)$$
$$C_4H^+ + H_2 \longrightarrow C_4H_2^+ + H \qquad (5)$$
$$C_4H_2^+ + e \longrightarrow C_4H + H \qquad (6)$$
$$\longrightarrow \text{products} \qquad (7)$$

where f is the fraction of the dissociative recombinations of $C_4H_2^+$ which give C_4H as a product. The C_4D molecule may be synthesised from C_3HD in similar fashion, but with an efficiency reduced by statistical considerations.

If one considers only these formation processes, as Turner (1989) does, then $R(C_4D) = \frac{1}{6} R(C_3HD)$. Similarly one can show $R(DC_3N) = \frac{1}{2} R(C_3HD)$ if HC_3N is a result of the $C_3H_3^+ - N$ reaction. This leads to the ratio $R(C_4D):R(DC_3N):R(C_3HD) = 1:3:6$ whereas the observed ratio in TMC-1 is 1:3.5:18.6 (Turner, 1989), although $R(C_3HD)$ is very large and uncertain (Bell et al., 1988, Turner, 1989). A more detailed approach is to recognise that proton transfer reactions, involving ions such as H_3^+, HCO^+ and H_3O^+, and charge transfer reactions can destroy deuterated and hydrogenated species at different rates. For example proton transfer reactions with C_4H lead to destruction of C_4H only on a fraction $(1-f)$ of collisions, while the same reactions with C_4D destroy this species on $(1-\frac{1}{2}f)$ of collisions. Hence

$$R(C_4D) = \frac{1}{6} \frac{(1-f)}{(1-\frac{1}{2}f)} R(C_3HD).$$

Likewise

$$R(DC_3N) = \frac{1}{2} \frac{(1-f)}{(1-\frac{1}{2}f)} R(C_3HD).$$

If $f = \frac{4}{5}$ in both reaction (6) and in the recombination of HDC_3N^+ (but see below) we find a ratio $R(C_4D):R(DC_3N):R(C_3HD) = 1:3:18$, close to that observed. We can also predict $R(C_3D)$ on the assumption that C_3H is formed in the recombination of $C_3H_3^+$ - note that the reaction $C_3H^+ + H_2 \longrightarrow C_3H_2^+ + H$ is endothermic by ~ 550 K - as

$$R(C_3D) = \frac{1}{2} \frac{(1-f)}{(1-\frac{1}{2}f)} R(C_3HD).$$

Such an analysis is uncertain due to several problems:- (i) we have ignored deuteration from H_2D^+ which is important at low temperatures - although not if $R(C_3HD)$ is indeed very large, and (ii) we have ignored the effects of the linear and cyclic forms of $C_3H_3^+$, C_3H and C_3H_2, which need to be further studied in the laboratory to determine the relevant rate coefficients and branching ratios. Finally the deuteration of the cyanopolyynes is sensitive to the structure of the protonated cyanopolyyne ion, a topic we now discuss in detail.

3. Deuteration of the cyanopolyynes and of HCN and HNC

Microwave spectrscopy of protonated cyanoacetylene has shown that the most stable state is linear, HC_3NH^+ (Lee and Amano, 1987). Thus deuteration of HC_3N by ions such as H_2D^+, N_2D^+ and DCO^+ may not lead to

DC_3N as has previously thought. That is the reaction $H_2D^+ + HC_3N$ leads to linear HC_3ND^+. If we now assume that in the dissociative recombination of HC_3ND^+ the D-atom cannot be rearranged sufficiently to bond to the carbon atom at the other end, then the deuterium in DC_3N, and by extension, DC_5N, DC_7N, etc, cannot arise from deuteron transfer, but in the formation process itself. It is also worth noting that if HC_3NH^+ and HC_3ND^+ are the correct structures, then reactions of H_3^+, HCO^+, H_3O^+, etc. will recycle HC_3N and DC_3N at the same rate, so that $R(DC_3N) = \frac{1}{2} R(C_3HD)$ independent of f, the fraction recycled.

The actual value expected depends on various assumptions concerning the formation of the formaldehyde-type ions $H_2C_3N^+$ and HDC_3N^+ from the reactions of $C_3H_3^+$ and $C_3H_2D^+$ with N atoms. In order to see the effects of including the various isotopomers of HC_3N and HC_3NH^+, D A Howe and I have adapted the model of Millar et al. (1989) using the so-called 'old' dissociative recombination branching ratios and calculated the pseudo-time-dependent behaviour of abundances and fractionation ratios for conditions typical of TMC-1 and of the Orion ridge clouds. For our purposes, we shall concentrate on the TMC-1 model, for which $R(DC_3N)$ falls from $1.9 \; 10^{-2}$ in the model of Millar et al. (1989) to $4.3 \; 10^{-3}$ at early time, compared with an observed value of $(1.5 \pm 0.5) \; 10^{-2}$ (Suzuki, 1987). At steady-state, the values of $R(DC_3N)$ are larger, but the absolute abundances of HC_3N and DC_3N are much less than those observed.

Like HC_3NH^+, the $HCNH^+$ ion is linear and H_2D^+ and related ions may also not be efficient in synthesising DCN from HCN and DNC from HNC. If the deuterium atom cannot move during dissociative recombination, then the possibility arises that HCN is related to DNC and HNC to DCN, by the reaction scheme:

$HCN + H_2D^+ \longrightarrow HCND^+ + e \longrightarrow$ DNC, not DCN
$HNC + H_2D^+ \longrightarrow HNCD^+ + e \longrightarrow$ DCN, not DNC.

Other products are possible, of course. It is possible that because $HCND^+$ and $HNCD^+$ are much smaller than the cyanopolyyne ions, rearrangement of the D atoms could occur.

Detailed calculations on this suggestion are needed but are complicated by the presence of H_2NC^+ formed in the reaction of C^+ with NH_3 (Allen, Goddard, and Schaefer 1980), and the lack of reliable branching ratios in the dissociative recombinations of $HCNH^+$ and H_2NC^+ with electrons.

References

ADAMS, N G and SMITH, D: 1985, *Ap. J.* **294**, L63
ALLEN, T, GODDARD, J D and SCHAEFER, H F: 1980, *J. Chem. Phys.* **73**, 3255
BELL, M B, AVERY, L W, MATTHEWS, H E, FELDMAN, P A, WATSON, J K G, MADDEN, S C and IRVINE, W M: 1988, *Ap. J.* **326**, 924
BELL, M B, FELDMAN, P A, MATTHEWS, H E and AVERY, L W: 1986, *Ap. J.* **311**, L89

BROWN, P D and MILLAR, T J: 1989a, *Mon. Not. R. astr. Soc.* **237**, 661
BROWN, P D and MILLAR, T J: 1989b, *Mon. Not. R. astr. Soc.* **240**, 25P
BROWN, R D and RICE, H N: 1986, *Mon. Not. R. astr. Soc.* **223**, 429
CROSWELL, K and DALGARNO, A: 1985, *Ap. J.* **289**, 618
DALGARNO, A and LEPP, S: 1984, *Ap. J.* **287**, L47
GERIN, M, WOOTTEN, H A, COMBES, F, BOULANGER, F, PETERS, W F III, KUIPER, T B H, ENCRENAZ, P J and BOGEY, M: 1987, *Astron. Astrophys.* **173**, L1
GUÉLIN, M, LANGER, W D, SNELL, R L and WOOTTEN, A: 1977, *Ap. J.* **217**, L165
LEE, S K and AMANO, T: 1987, *Ap. J.* **323**, L145
MILLAR, T J, BENNETT, A and HERBST, E: 1989, *Ap. J.* **340**, 906
PHILLIPS, T G, BLAKE, G A, KEENE, J, WOODS, R C and CHURCHWELL, E B: 1985, *Ap. J.* **294**, L45
ROGERSON, J B, YORK, D G, DRAKE, J F, JENKINS, E B, MORTON, D C and SPITZER, L: 1973, *Ap. J.* **181**, L110
SNYDER, L E, HOLLIS, J M, BUHL, D and WATSON, W D: 1977, *Ap. J.* **218**, L61
SUZUKI, H: 1987, in M S Vardya and S P Tarafdar, ed(s)., *Astrochemistry*, Dordrecht:Reidel, p.199
TURNER, B E: 1989, *Ap. J.* **347**, L39
TURNER, B E: 1990, *Ap. J.* **362**, L29
WATSON, W D: 1976, *Rev. Mod. Phys.* **48**, 513
WATSON, W D: 1977, in J Audouze, ed(s)., *CNO Processes in Astrophysics*, Dordrecht: Reidel, p.105
WOOTTEN, A: 1987, in M S Vardya and S P Tarafdar, ed(s)., *Astrochemistry*, Dordrecht:Reidel, p.311

TABLE I
Deuterium-bearing molecules detected in interstellar clouds. For each species we give an estimate, R, of its abundance relative to its hydrogen-bearing form, for a sample of sources. HD is detected in diffuse clouds with an abundance ratio $HD/H_2 \sim 10^{-6}$.

Molecule	TMC-1[a]	Orion[b]	Orion Hot Core[c]
HDO			>.002
DCO^+	.015	.002	
N_2D^+	<.045		
DCN	.023	.02	.005
DNC	.015	.01	
C_2D	.01	.045	
NH_2D	<.02		.003
HDCO	.015	.02	.14
c-C_3HD	.08		
C_4D	.004		
DC_3N	.015		
DC_5N	.013		
CH_3OD			.03
D_2CO			.02

[a]TMC-1 is a dark cloud with kinetic temperature ~ 10 K. [b]Orion is a giant molecular cloud with kinetic temperature ~ 50-70 K. [c]Orion Hot Core is close to a site of star formation and has kinetic temperature ~ 200 K.

MOLECULAR LINE STUDIES OF DENSE CORE MOTIONS

YUEFANG WU
CCAST(World Laboratory), P.O. Box 8730, Beijing 100080
Geophysics Department, Peking University, Beijing 100871

ABSTRACT. Molecular lines have revealed various supporting motions in dense cores. Line widths and emission region sizes of NH_3 and CS in the same kind of cores or of the same line in cores with or without sources are different and can not be explained with the line width- size relationship. Outflows in dense cores show rich characteristics which can account for the NH_3 emission difference between the two kinds of the cores; CS emission is consistent with the chemical effects in shocked regions. Rotation exists in both kinds of cores and may be related to the observed polarities and collimations of outflows.

1. Introduction

Millimeter lines and infrared studies indicate that small visually opaque regions are the locations of low mass star formation and are usually 0.1 -0.3 pc in size, 10^4-10^5 cm^{-3} in density and have T_k about 10K (Myers *et al.* 1983, Myers and Benson 1983, Zhou *et al.*, 1989). About 50% of these dense cores have associated IRAS sources of which more than half have no optical counterparts and may be the potential protostars(Beichman *et.al.* ,1986). All of the cores with sources have luminosities ≤ 100L_0. Cores with and without sources have different properties. In this paper we analyse the inner motions, outflows and rotations, and make comparisons between these two classes of cores.

2. Core Motions

Myers *et al.*(1983) have measured 90 dense cores with the molecular pair ^{13}CO and $C^{18}O$. They have compared the sizes, temperatures and densities of the observed cores with the conditions for the equilibrium and stability of a pressure-bounded isothermal sphere, and found that if these cores were supported by turbulent motions and take the typical core with Δv ($C^{18}O$) = 0.6 kms^{-1} or T_D = 230K. equilibrium appears possible and may be stable in this case. It is also consistent with the Larson's model (1981). If the line width partly reflects the supporting motion, many cores are also consistent with turbulent contraction.

With another tracer NH_3 and the same method for ^{13}CO and $C^{18}O$ data analysis, Myers and Benson(1983) investigated the stability of the cores and found that the equilibrium of these cores may be supported by 10K thermal motions plus either a subsonic microturbulence (0.14kms^{-1}) or an early collapse with an age of less than 10^5 yr.

Zhou *et a l*. (1989) observed CS [J=(2-1), (3-2) and (5-4)] line in 27 cores of which 13 have IRAS sources. Comparison between the result and NH_3 data shows that the line width differ by a factor of 2 or so. Both the CS lines of J=2-1 and J=3 - 2 are wider than NH_3 lines. This situation means that thermal motions can not dominate in the CS regions. The CS line broadening can not be explained with optical depth effect, and it can not be accounted by the emission region size, either.

On the other hand all the millimeter measurements, indicate that line widths in cores with IRAS sources are wider that those in cores without sources(Table 1).

Table 1 Line Widths and Intensities

Core Class	Line Width (kms^{-1})				T_k (K)	
	$C^{18}O^a$	NH_3^d	CS^b	CO^c	$C^{18}O^a$	NH_3^d
With Sources	0.66+0.04	0.42+0.05	1.05+0.06	11.1±0.2	11.0±0.6	10.8±0.4
Without Sources	0.47+0.03	0.27+0.02	0.58+0.04	10.4±0.9	10.4±0.6	10.7±0.5

Data from: a. Myers *et al*, 1983. b. Zhou *et al*, 1989. c. Wu *et al*, 1990. d. Myers and Benson, 1983

The possible responsible reasons would not be the thermal motion since as Table 1 shows, all CO, $C^{18}O$ and NH_3 observations indicate the temperatures are nearly the same. Using the relation of the intrinsic line width and the turbulent width: $\Delta v_t = [(\Delta v_i)^2 - 8(\ln 2)kT_K/m]^{1/2}$ and the observed data of these molecular species we got Δv_t and size for the two groups of cores, listed in columns 2 - 7 of Table 2. These parameters show that the law $\Delta v \propto R^{1/2}$ is broken for all of the molecular line measurements, particularly for NH_3 and CS. The correlation of the size-line width has been found from various line-line in the same clouds and the same line for cloud - cloud or region - region . The derived power-law exponents range from 0.3 -1.0(Myers, 1983). For all the cores of the two groups the relation between line width and size is out of this range. This phenomenon may concern with the role of the young stellar objects.

Table 2 Turbulent Widths and Sizes

Core Class	$C^{18}O$		NH_3		CS	
	Δv_t(kms^{-1})	R(pc)	Δv_t(kms^{-1})	R(pc)	Δv_t(kms^{-1})	R(pc)
With Sources	0.64+0.05	0.36+0.03	0.41+0.06	0.17+0.03	1.01+0.15	0.34+0.05
Withoout sources	0.45+0.03	0.31+0.05	0.22+0.02	0.11+0.02	0.55+0.04	0.22+0.07

3. CO Outflows

The outflows in dense cores are usually weaker than those in high mass star formation regions. The energies of these outflows range in 10^{41}-10^{45} erg, 2 orders lower than that of high mass sources. The mass loss rates are 10^{-6}-$10^{-9} M_\odot yr^{-1}$, also lower than the high mass ones. Nevertheless, these outflows may have some characteristics. a). They have rather high detected rates(Myers *et al*.1988, Wu *et al*. 1990). If we take this kind of outflows within the distance of 500pc, statistics show that the occurence rate is 5.5×10^{-5}/$\tau pc^{-2} yr^{-1}$, which may be very close to the star birth rate near the sun(Ostriker *et al*. 1974, Mezger and Smith, 1977), where τ is the lifetime of outflows and is about 7×10^4 year on the average(Wu 1990). Their dynamic evolution is rather long: 1×10^4-1.9×10^5 yr(Myers *et al*. 1988, Heyer *et al*. 1987). The relative long time scale may increase the chance for finding these outflows. b). Morphology: These outflows are more polar-like and less isotropic. All 4 isotropic sources(S140, M8E, NGC 7538 and MWC 1080) are found in the high mass star formation

regions. Their collimation may be better than that of the high ones(Mao et al. 1989, Wu et al. 1991).
c).Optical Phenomena:The most striking feature about low mass outflows is the associated optical phenomenon. HH-objects are thought to be tracers of high velocity gas and to associate with T - Tauri stars almost exclusively. Recently one of the faintest HH objects was found to be associated with the CO blue peak position in L1582B(Wu et al. 1990). So far over 22 optical jets were found and more than half of them have CO outflows. Stocke et al. (1988) have found that in L1551 the jet HH objects are bow shock interface between two winds coming from IRS5 region. The second wind has an inferred velocity of about 160 kms^{-1}. Molecular outflow results in the momentum flux of this not very high velocity but pervasive wind acting on the material in cores (Mundt, 1988, Stocke et al.1988). d). Roles for the cores: The outflows in dense cores can still put significant momentum and energy in the surrounding gas though they are rather weak. Myers et al. (1988) found that almost in all cases P_{flow} is larger than or equal to P_{core}, and in more than half sources which they analysed, $P_{flow}/P_{core} = 1 \sim 2$ For the line width difference of the two kinds of cores listed in Table 2, there should be no porblem for NH$_3$ emission regions since $P_{flow} \geq P_{core\ w.} > P_{core\ w.} - P_{core\ w.o.}$

For CS regions, theaverage size is a factdr of 1.5 larger than that of the NH$_3$ regions and the CS line width is greater by a factor of 2. Besides the density is also higher than that of NH$_3$ cores(Zhou et al. 1989). Therefore the momentum of CS regions will be larger than that of NH$_3$ cores by a factor of at least 5, while P_{flow} is only 1 - 2 times of P_{core} of the NH$_3$ cores.. Therefore the flow momentum may not be able to enhance the line width in the CS regions. Similar analysis shows that it can not account for the momentum increases of the CS emission regions to the NH$_3$ emission regions, either

For C^{18}O emission regions, we calculate the velocity dispersion σ are 0.33 and 0.27kms^{-1} respectively, taking T_K=10K, for 20 cores with and 23 cores without sources (Myers et al. 1983). The corresponding average core masses are 27.4±3.7 and 19.1±4.0 M_θ respectively. The average momentum difference is 3.9 $M_\theta$kms^{-1}. For 26 low mass outflows (Wu, 1990), the average momentum is 3.3 $M_\theta$kms^{-1}. Thus the outflows could couple momentum to the gas to increase the line widths of the C^{18}O emission in cores with sources almost completely. Here the problem about the CS kinetics remains, and it may also exist in the high mass star formation regions. It seems that the gas traced by CS obtains more energy from the stellar winds coming from the center sources. It may be owing to this part of the gas which is located in the inner region bears the brunt when the wind blows out. In L43, the two positions of the maximum CS line widths seem that at these positions the gas is plowing directly into the surrounding gas (Mathieu et al .1988). CS high velocity wings were detected in a number of high mass star fomation regions(Thronson and Lada, 1984, Hayashi et al. 1985). Another effect of the shock is that it heats the gas in post shock regions. According to Hartquist et al.(1980), the reaction S+H$_2$ → HS +H could occur at the high temperature of this kind of regions. And consequently, CS is formed in the presence of HS: C+HS → CS+H CS is removed by the reverse reaction of it and by the reactions: CS+OH→OCH+S and CS+O→CO+S Their calculations for the molecular fractional abundances show that the CS abundance increases from 4.7x10^{-12} at 3x10^{9}cm behind the shock to 4.7x10^{-8} at a distance of 10^{13}cm, where the CS abundance reaches a plateau of this value since there the abundance of OH and O decrease. The velocity of the shock that they considered is 8kms^{-1}. It may be met generally in star forming regions.

4. Rotation

Table 3 which is devided into 3 parts lists the rotation parameters. Ω and R_{colli} are quoted from different observations or different authors(Boss, 1987, Clark and Johnson, 1981, Wu et al. 1991).
All cores with sources and high mass sources listed in Table III have bipolar outflows. The direction

Table 3 Rotation and High Velocity Gas

Source Name	Ω (10^{-14} rad·s^{-1})	Polarity of HVG	R_{coll}	Direction of velocity gradient & flow project axis	Ref.
Low Mass Cores without Sources					
L183	5.3, 5.6	possible pedestal			1, 22
L134N	2.7, 2.9	Bluewing			1, 2
L1709A	~10	Redwing			2
Low Mass Cores with Sources					
L1455	11, 8.0	Bipolar	3.7	~⊥	3, 4
L1551	25, <6.7	Bipolar	5.2	Cross Angle >45°	5, 6, 4
HH1-2	15, <1.2	Bipolar	2.1		3, 7, 4
HH26IR	1.7	Bipolar	2.7	⊥	8, 4
HH24	<0.51	Bipolar	1.0		9, 4
L43B	6.5	Bipolar	2.5	Cross Angle<45°	17
L723	10, <6.7	Bipolar	4.1	⊥	21, 3, 4
L778	<2.2	Bipolar	2.3		3
B335	2.0, 7.0	Bipolar	2.8	Cross Angle~45°	11, 3, 4
High Mass Star Formation Regions					
GL490	<11	Bipolar	1.9		18, 4
GL437	<1.1	Bipolar	1.6		3, 4
Orion-KL	40	Bipolar	1.0	⊥	12, 4
NGC2071	5.6, 13	Bipolar	3.7	⊥	13, 14, 4
NGC2261	<0.83	Bipolar	2.0		19, 4
MonR2	2.8, 15	Bipolar	3.0	~⊥	3, 8, 4
G35.2N	23	Bipolar	2.6	⊥	15, 4
CRL2591	5.7	Bipolar	1.9	Cross Angle>45°	20, 4
Cep A	8.9, 23	Bipolar	?		16, 10, 4

1 Clark & Johnson, 1981. 2 Wu *et al*, 1990. 3 Heyer *et al*, 1986. 4 Boss, 1987. 5 Kaifu *et al*, 1984. 6 Batrla & Menten,1985. 7 Torrelles *et al*, 1985. 8 Torrelles *et al*, 1983. 9 Matthews & Little, 1983. 10 Torrelles *et al*, 1986a. 11 Menten *et al*, 1984. 12 Hasegawa *et al*, 1984. 13 Bally, 1982. 14 Lichten, 1982. 15 Little *et al*, 1985. 16 Gusten *et al*, 1984. 17 Mathieu *et al*, 1988. 18. Kawabe *et al*, 1984. 19 Canto *et al*, 1981. 20 Takano *et al*, 1986. 21 Torrelles *et al*, 1986b. 22. Frerking and Langer 1982.

of the gradients of 7 flows among 11 with known directions are perpendicular to the flow axes. It suggests that in most flow sources, the rotation axis is coincided with the flow lobe axis. Column 4 of Table 3 lists the collimation factors which are generally better for the flows with high Ω than those with low Ω. It is also coincident with the trend that the R_{coll} is higher for low mass sources than that of high mass ones. These are the important tests of the models suggested by Boss(1987).

Motions in dense cores are related to the forming and the activity of the stars, and the interaction of the stellar objects and the surrounding materials. Observations of the cores at high spatial and spectral resolution is important for latter determination of these processes in the foreseeable future.

The project was supported by NSFC. I thank IAU for partially supporting my presence at IAU Symposium No. 150.

References

Bally, J. 1982, Ap. J., **261**, 558.
Batrla, W., and Menten, K. M. 1985, Ap. J..(Letters), **298**, L19.
Beichman, C. A., Myers, P. C., Emerson, J. P., Harris, S., Mathieu, R., Benson, P. J., and Jennings, R. E. 1986, Ap. J., **307**, 337.
Boss, A.P. 1987, Ap. J., **316**, 721.
Canto, J., Rodriguez, L. F., Barral, J. F., and Carral, P. 1981, Ap. J., **244**, 102.
Clark, F. O., and Johnson, D. R. 1981, Ap. J., **247**, 104.
Frerking, M. A., and Langer, W. D. 1982, Ap. J., **256**, 523.
Gusten, R., Chini, R., and Neckel, T. 1984, Astr. Ap., **138**, 205.
Hartquist, T. W., Oppenheimer, M., and Dalgano, A. 1980, Ap. J., **236**, 182.
Hasegawa, T., Kaifu, N., Inatani, J., Morimoto, M., Chikada Y., Hirabayashi, H., Iwashita, H., Morita, K., Tojo, A., and Akabane, K. 1984, Ap. J., **283**, 117.
Hayashi, M., Omodaka, T., Hasegawa, T., and Suzuki, S. 1985, Ap. J., **288**, 170.
Heyer, M. H., Snell, R. L., Goldsmith, P. F., Strom, S. E., and Strom, K. M. 1986,Ap., J., **308**, 134.
Heyer, M. H., Snell, R. L., Goldsmith, P. F., and Myers, P. C. 1987, Ap. J., **321**, 370.
Kaifu, N., Suzuki, S., Hasegawa, T., Morimoto, M., Inatani, J., Nagane, K., Miyazawa, K., Chikada, Y., Kanzawa, T., and Akabane, K. 1984, Astr. Ap., **134**, 7.
Kawabe, R., Ogawa, H., Fukui, Y., Takano, T., Takaba, H., Fujimoto, Y., Sugitani, K., and Fujima, M. 1984, Ap. J. (Letters), **282**, L73.
Larson, R. B. 1981, M. N. R. A. S., **194**, 809.
Lichten, S. M. 1982, Ap. J., **253**, 593.
Little, L. T., Dent, W. R. F., Heaton, B., Davies, S. R., White, G. J. 1985, M. N. R. A. S.,.**217**, 227.
Mao, X., Wu, Y., Hao, J., and Hou, M.1989, Acta.Scientiarum Naturalium, U. Pekinensis, **25**,505.
Mathieu, R. D., Bensan, P. J., Fuller, G. A., Myers, P. C., and Schild, R. E. 1988, Ap. J. **330**, 385.
Matthews, N., and Little, L.T. 1983, M. N. R. A. S., **205**, 123.
Menten, K. M., and Walmsley, C. M. 1985, Astr. Ap., **146**, 369.
Mezger, P. G., and Smith, L. F. 1977, in IAU Symposium No. 75, P. 133.
Mundt, R.1988 in *Formation and Evolution of Low Mass Stars*, ed.A. K.Dupree, M.T. V.Lago,P257
Myers, P. C. 1983, Ap. J., **270**, 105.
Myers, P. C., and Benson, P. J.1983, Ap. J., **266**, 309.
Myers, P. C., Heyer, M., Snell, R. L., and Goldsmith, P. F. 1988, Ap. J., **324**, 907.
Myers, P. C., Linke, R. A., and Benson, P. J. 1983, Ap. J., **264**, 517.
Ostriker, J. P., Richstone, D. O., and Thuan, T. X. 1974, Ap. J.(Letters), **188**, L87
Stocke, J.T., Hartigan, P. M., Strom, S. E., Strom, K. M., Anderson, E. R.1988,Ap. J. Suppl.,**68**,279
Takano, T., Stutzki, J., Fukui, Y., and Winnewisser, G. 1986, Astr. Ap., **158**, 14.
Thronson, Jr., H. A., and Lada, C. J. 1984, Ap. J., **284**, 135.
Torrelles, J. M., Canto, J., Rodriguez, L. F., Ho, P.T.P., Moran, J. M.1985, Ap. J.(Letters) **294**,L117
Torrelles, J. M., Ho, P. T. P., Rodriguez, L. F., and Canto, J. 1986a, Ap. J., **305**, 721.
Torrelles, J. M., Ho, P. T. P., Moran, J. M., Rodriguez, L. F., and Canto, J. 1986b, Ap. J., **307**, 787.
Torrellas, J. M., Rodriguez, L. F., Canto, J., Carral, P., Marcaide, J., Moran, J. M., and Ho, P. T. P. 1983, Ap. J., **274**, 214
Wu, Y. 1990, *Progress in astronomy*, **8**, 291
Wu, Y., Zhou, S., and Evans, N. J. II 1990, in preparation.
Wu, Y., Huang, M., He, J.1991, in *The Stellar Populations of Galaxies*, ed. by B. Barbuy, in press.
Zhou, S., Wu, Y., Evans, N. J. II, Fuller, G. A., and Myers, P. C. 1989, Ap. J., **346**, 168.

QUESTIONS AND ANSWERS

A.Leger: If you had no IR data to detect stars, would you really be able to decide whether a cloud is forming star or not from radio data?

Y.Wu: We can consider the line width density which may be rather large or high for the star formation regions; the most strong evidences of the existence of a stellar source from radio observations are bipolar outflow maser line and continuum emissions; we can also see if there is any molecular species overabundant to obtain the representations of the activity of the forming star. Systematic large molecular line velocity shift may also be the evidence of a stellar source under formation.

MOLECULAR ABUNDANCES IN THE SGR A MOLECULAR CLOUD

Y. C. MINH[1], W. M. IRVINE[2], and P. FRIBERG[3,4]
1 Inst. of Space Sci. and Astronomy, Daejon 305-348, Korea
2 FCRAO, Univ. of Massachusetts, Amherst, MA 01003, USA
3 Onsala Space Observatory, S-439 00 Onsala, Sweden
4 Joint Astronomy Centre, Hilo, HI 96720, USA

ABSTRACT. We have obtained column densities for HCO^+, HCO, HCS^+, C_3H_2, HC_5N, SiO, OCS, HCOOH, CH_3CH_2OH, and CH_3CCH toward Sgr A. The fractional abundance of SiO relative to molecular hydrogen in Sgr A is comparable to that for the Orion plateau, $\sim 10^{-7} - 10^{-8}$, which may be a typical value for hot clouds. The abundances of HCO, CH_3CH_2OH and CH_3CCH all appear to be enhanced relative to other molecular clouds suc as Sgr B2.

The molecular abundances and chemistry of the Sgr A molecular cloud have not been as well characterized as those of Sgr B2, but the strong activities and shocks of the Galactic center could affect the clouds in Sgr A more efficiently because of their greater proximity. This may result in a unique chemistry of the Sgr A clouds, such as has been suggested from observations of HCO_2^+ (Minh et al. 1991a; Paper 1).

Data were obtained in 1988 June with the Swedish-ESO 15 m telescope in Chile. Telescope parameters and observing method were included in Paper 1. Observed molecules, transitions, and rest frequencies are listed in Table 1. We have obtained data for the clouds observed in NH_3 (Gusten et al. 1981) and in HCO_2^+ (Paper 1). Column densities were determined assuming optically thin emission and an apparent background radiation temperature of 10 K (cf. Paper 1 and Minh et al. 1991b).

In Figure 1 we plot the fractional abundances relative to H_2 for th molecules observed toward M-0.13-0.08, and also those for TMC-1 and Sgr B2, and for Orion(KL) from the tabulations of Irvine et al. (1987), and Blake et al. (1987), respectively, for comparison. The fractional abundance of SiO at Sgr A is derived to be $\sim 10^{-7} - 10^{-8}$ relative to molecular hydrogen which is similar to that of the Orion plateau. The high SiO abundance could be explained by high-temperature or shock chemistry (Ziurys et al. 1989). It is also possible, however, that an enhanced abundance of elemental Si comes from the disruption of silicat grains by shocks in the Galactic center region, which can lead naturall to an enhanced SiO abundance.

TABLE 1. Observed molecules.

Molecule (Trans.)	Rest Frequency (GHz)
$H^{13}CO^+$ (1-0)	86.75429
$HC^{18}O^+$ (1-0)	85.16226
HCO (1_{01}-0_{00} [a])	86.67082
HCS^+ (2-1)	85.34790
C_3H_2 (2_{12}-1_{01})	85.33889
HC_5N (32-31)	85.20135
SiO (2-1)	86.84700
^{29}SiO (2-1)	85.75913
OCS (7-6)	85.13911
HCOOH (4_{14}-3_{13})	86.54613
CH_3CH_2OH (6_{06}-5_{15})	85.26547
CH_3CCH (J=5-4 K=0)	85.45730
(J=6-5 K=0)	102.54798
CH_3OH (3_1-4_0 A^+)	107.01385

[a] For the (3/2-1/2 2-1) trans.

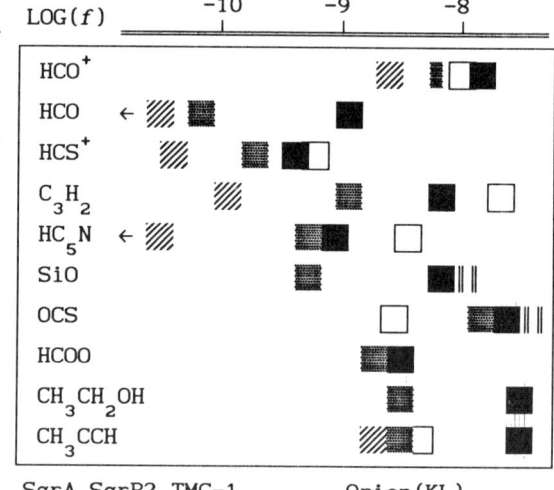

Figure. 1 Abundances relative to molecular hydrogen on a logarithmic scale for Sgr A, Sgr B2, TMC-1 and the Orion extended ridge and the plateau. Data for Sgr B2 and TMC-1 from Irvine et al. (1987), and for Orion(KL) from Blake et al. (1987).

The fractional abundances of several molecules observed here, in particular HCO, CH_3CH_2OH, and CH_3CCH, appear to be enhanced relative to values for other sources (Figure 1). It is interesting that the production of CH_3CH_2OH and CH_3CCH probably involves relatively hydrogenated species such as C_2H_4 or CH_4 (Millar et al. 1991; Millar & Freeman 1984); this might suggest the influence of grain chemistry or high temperature reactions.

We conclude that a rich chemistry exists in Sgr A, which could partly be a result of the energetic processes of the Galactic center region, such as shocks, UV radiation, and also the possible interaction of the neutral and the ionized gas around the nucleus.

Blake G.A., Sutton E.C., Masson C.R., Phillips T.G., 1987, ApJ 315, 621
Gusten R., Walmsley C.M. Pauls T., 1981, A&A 103, 197
Irvine W.M., Goldsmith P.F., Hjalmarson A., 1987, *Interstellar Processes* eds. D.J. Hollenbach, H.A. Thronson, Jr., D.Reidel, p. 561
Millar T.J., Freeman A., 1984, MNRAS 207, 405
Millar T.J., Herbst E., Charnley B., 1991, ApJ 369, 147
Minh Y.C., Brewer M.K., Irvine W.M., Friberg P., Johansson L.E.B., 1991a, A&A 244, 470 ; Paper 1
Minh Y.C., Irvine W.M., Friberg P., 1991b, submitted to A&A
Ziurys L.M., Friberg P., Irvine W.M., 1989, ApJ 343, 201

Molecular Line Observations towards W58 and GL490

D.G. Roh[1], H.R. Kim[1], Y.C. Minh[1], B.R. Auh[1] and B.C. Koo[2]
[1] Radio Astronomy Section
Institute of Space Science and Astronomy, Korea
[2] Harvard-Smithsonian Center for Astrophysics, U.S.A.

ABSTRACT. We observed 3 mm transitions of several interstellar molecules toward star-forming regions in W58 and GL490 using the 14 m Daeduk Radio Telescope (Korea). We derive molecular abundances for the "broad" components observed with a high spectral resolution, which could represent the abundances of outflowing materials.

Several molecular transitions in Table 1 have been observed toward the strong galactic continuum source W58 having four compact HII regions A(K3-50), B, C(ON3), and D(NGC6857), and the infrared source GL490 showing a high velocity CO outflows. They are intensively studied star-forming regions in our Galaxy (cf. van Gorkom et al. 1981; Mundy & Adelman 1988; and references therein). All observations were carried out with the 14m telescope of the Daeduk Radio Astronomy Observatory which locates about 150 km south from Seoul, Korea during March to May 1991. A cryogenic Schottky diode mixer was employed with the 1024 channel autocorrelator (a total bandwidth of 20 MHz) and the 256 channel filter bank (a resolution of 250 kHz/channel). Observing method and telescope parameters are included in Roh et al. (1991).

CO and ^{13}CO line intensity maps of W58 show that the observed compact HII regions are embedded in a CO cloud elongated to the NE-SW direction, and locate at the maximum molecular density where the HI cloud and the HII region complex may interact (Israel 1980). The HCN line intensity toward GL490 peaks about 1 arcmin south from those of HCO$^+$ and ^{13}CO which may result from the different excitation or chemistry for these species.

Figure 1 shows sample spectra obtained with the autocorrelator and the dotted lines are gaussian fit results for the "broad" components ($\Delta V = 10\text{-}20$ km s^{-1}) which may represent outflowing materials in these regions. Beam averaged column densities in Table 2 are derived assuming optically thin emission and rotational temperatures (T_{rot}) of 10-15 K (5-10 K) and 30-60 K (10-30 K) for the CO J=1-0 line (for other molecular transitions in Table 1) of "broad" and "quiescent" components, respectively. Cosmic background radiation was corrected only for the case of $T_{rot} = 5$ K (see discussions in Roh et al. 1991). The molecular emissions in outflowing materials may arise from many small optically thick clumps in the gas. Our preliminary results indicate that the molecular abundances in outflowing gases are comparable to those for quiescent components.

TABLE 1. Observed molecules

Molecules (Tran.) (J-J')	Frequency (GHz)
CO (1-0)	115.271
^{13}CO (1-0)	110.201
C^{18}O (1-0)	109.782
CS (2-1)	97.981
C^{34}S (2-1)	96.413
HCO$^+$ (1-0)	89.189
H^{13}CO$^+$ (1-0)	86.754
HCN (1-0, F=2-1)	88.632
SiO (2-1, v=0)	86.847

TABLE 2. Observed column densities (cm^{-2}).

Source Mol.		A(K3-50)	W58 B	C(ON3)	D(NGC6857)	GL490
CO	Q	8.9±2.5(18)	1.1±0.3(19)	1.6±0.5(19)	1.3±0.8(19)	2.0±0.6(19)
	B	1.8±0.2(18)	9.6±1.2(17)	≥1.7±0.2(17)	7.6±0.8(17)	2.2±0.2(18)
CS	Q	7.1±2.1(13)	1.3±0.4(14)	2.5±0.7(14)	8.3±2.4(13)	9.0±2.6(13)
	B	1.5±0.3(14)	1.6±0.4(14)	...	...	...
HCO$^+$	Q	2.0±0.7(13)	2.7±1.0(13)	2.0±0.7(13)	2.5±0.9(13)	2.0±0.7(13)
	B	1.1±0.1(13)	3.1±0.4(13)	...	...	7.1±0.8(12)
HCN	Q	1.8±0.6(14)	7.9±2.8(13)	7.7±2.7(13)	4.0±1.5(13)	7.3±2.6(13)
SiO	Q	≤ 3.3(13)	≤ 2.5(13)	≤ 3.8(13)	≤ 3.8(13)	...

Notes: Q: "Quiescent" component; B: "Broad" component. Upper limts are 3 σ. Numbers in parentheses are power of 10.

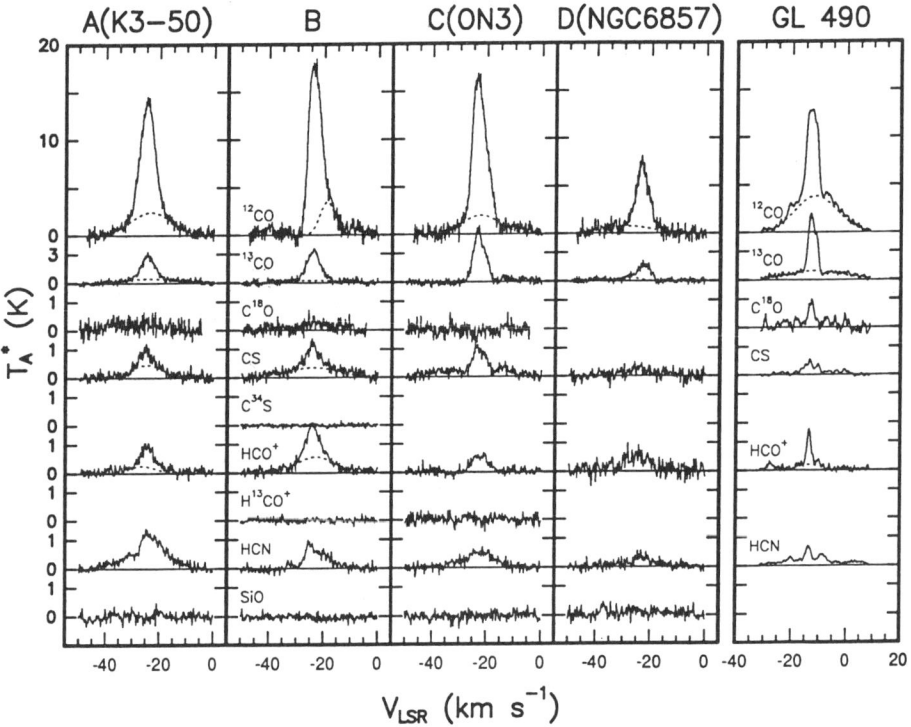

Figure 1. Sample spectra obtained toward W58(A, B, C and D) and GL490

REFERENCE :
van Gorkom,J.H. 1981, A. &Ap., **94**, 259.
Israel,F.P. 1980, Ap.J., **236**, 465.
Mundy,L.G. and Adelman,G.A. 1988, Ap.J., **329**, 907.
Roh,D.G., Kim,H.R., Minh,Y.C., Auh,B.R. and Koo,B.C. 1991, in preparation.

Nitrogen Sulfide (NS) In Star Forming Regions

Douglas McGonagle   William Irvine
Young Minh *

Five College Radio Astronomy Observatory
University of Massachusetts at Amherst, USA

Gas phase models of ion molecule chemistry have been rather successful in matching the observed abundances of small interstellar molecules containing carbon, hydrogen, and oxygen. However, the situation is somewhat less clear for nitrogen–containing species, partly because the important initiating reaction $N^+ + H_2$ is slightly endothermic; and for sulfur–containing molecules, where it remains uncertain whether it is necessary to invoke surface reactions on grains to match the observed abundances. As a relatively simple species, the abundance of nitrogen sulfide should provide a good test of the models of the coupled chemistry of nitrogen and sulfur. Until very recently only two molecules containing both these elements were known in the interstellar medium, NS and HNCS, and both have been observed only in Sgr B2. We have therefore undertaken a survey for interstellar NS in Galactic molecular clouds using the FCRAO 14–meter telescope. The $^2\Pi_{1/2}$, $J = 5/2 \to 3/2$, transition has in fact been detected in many regions of massive star formation (see table).

The three dominant hyperfine components of the lower frequency Λ–doublet component were detected in almost all cases. The ratios of the component intensities differ from the predicted intrinsic relative intensity ratios, thus suggesting that in several sources the emission is not optically thin, which in turn implies that the NS source size is small compared to the beam size (45"). The beam average column densities for NS derived from these observations are $N \sim 10^{13} - 10^{14}$ cm^{-2}. Using a canonical molecular hydrogen column density of 10^{23} cm^{-2} we estimate the fractional abundance to be $f \sim 10^{-10}$ relative to H_2, e.g. in Orion–KL. These abundances are 2 – 3 orders of magnitude larger than the predictions of the gas phase models for an Orion type cloud by Miller and Herbst (A&A, 1990, **231**, 466).

---

*Now at DRAO, South Korea.

Nitrogen Sulfide in Star Forming Regions

Source	α(1950) hh mm ss.s	δ(1950) dd mm ss.s	V_{lsr} (km s^{-1})	Ta^* (K)	I (K km s^{-1})	N_{NS} (cm^{-2})
DR21(OH)	20 37 15.0	42 12 8.0	-2.9	0.08	1.06 ± 0.08	3×10^{13}
G34.3+0.2	18 50 46.0	1 11 10	+57.8	0.17	2.04 ± 0.16	6×10^{13}
NGC7538(E)	23 11 53.0	61 10 57.9	-58.4	0.05	0.34 ± 0.08	1×10^{13}
NGC7538(N)	23 11 36.6	61 11 48.0	-66.7	0.05	0.61 ± 0.13	2×10^{13}
NGC7538(S)	23 11 36.6	61 10 48.0	-55.8	0.09	0.92 ± 0.08	3×10^{13}
ORION(S)	5 32 44.8	-5 25 60.0	+6.4	0.07	0.74 ± 0.21	2×10^{13}
ORION(KL)	5 32 47.0	-5 24 23.0	+7.6	0.14	1.96 ± 0.23	6×10^{13}
S140	22 17 41.2	63 3 41.0	-8.0	0.05	0.38 ± 0.08	1×10^{13}
SGR(B2M)	17 44 10.6	-28 22 5.0	+60.6	0.17	4.19 ± 0.52	1×10^{14}
W3(OH)	2 23 17.3	61 38 57.9	-46.7	0.10	0.82 ± 0.10	2×10^{13}
W51(MS)	19 21 26.4	14 24 42.0	+55.6	0.19	2.64 ± 0.26	8×10^{13}
W51(N)	19 21 22.0	14 25 20.0	+59.8	0.09	1.66 ± 0.26	5×10^{13}

NOTES:

1) The integrated intensity, I, is over the three major hyperfine components of the $^2\Pi_{1/2}$, $J = 5/2 \rightarrow 3/2$ transition. Not corrected for beam efficiency ($\eta_b = 0.55$).

2) The total column density, N_{NS}, is corrected for beam efficiency and the unobserved portion of the lambda doubling and hyperfine structure. For all sources, $T_{rot} = 30\ K$ is assumed.

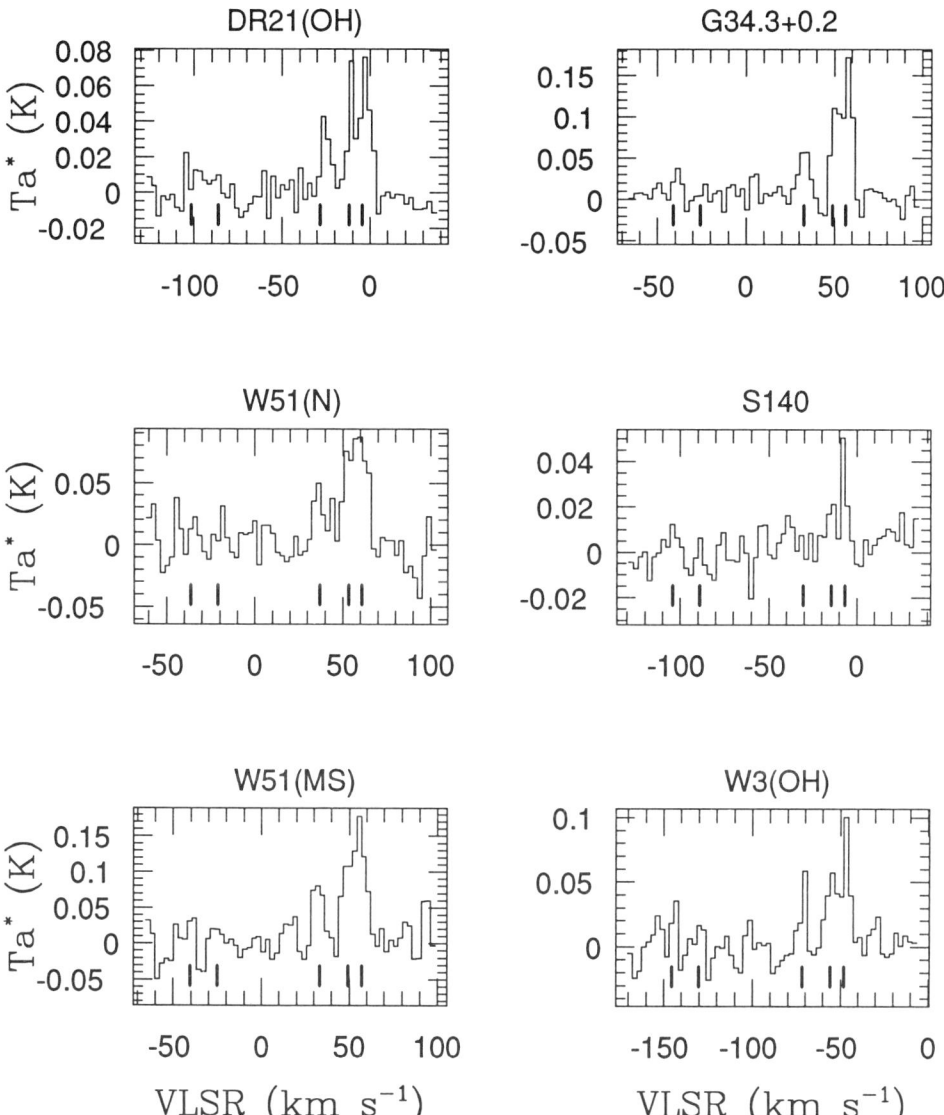

Figure 1: Spectra of the $^2\Pi_{1/2}$, $J = 5/2 \to 3/2$ transition of NS towards the sources listed in table 1. The spectral resolution is 1 MHz, and the spectral band includes the transitions three major hyperfine components.

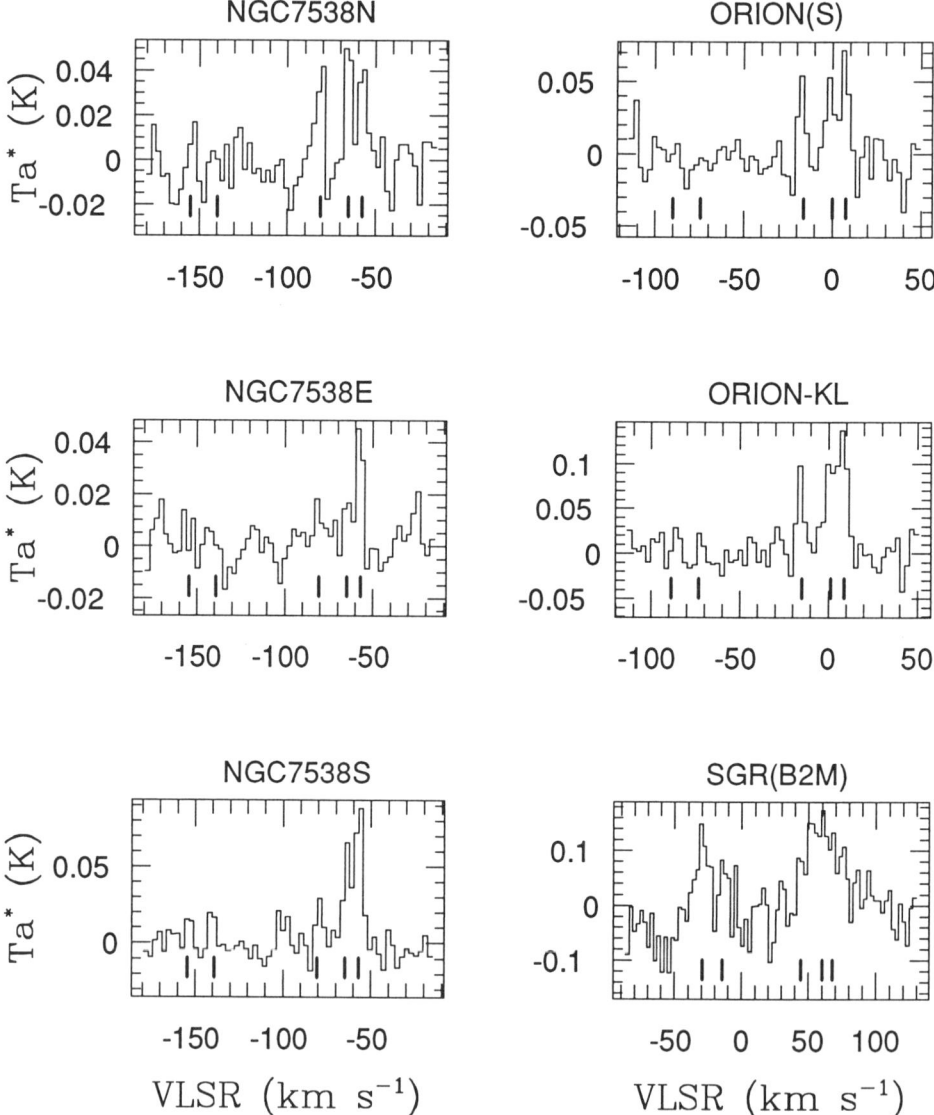

Figure 1: (cont) Spectra of the $^2\Pi_{1/2}$, $J = 5/2 \to 3/2$ transition of NS towards the sources listed in table 1. The spectral resolution is 1 MHz, and the spectral band includes the three major hyperfine components.

RADIATION TRANSFER AND PHOTOCHEMICAL EFFECTS IN INHOMOGENEOUS DENSE CLOUDS

S. Aiello[1], C. Cecchi Pestellini[1,2], S. Tiné[1]
1. Dip. di Astronomia e Scienza dello Spazio, Università di Firenze, Italy
2. Harvard-Smithsonian Institute of Astrophysics, Cambridge, Massachusetts, USA

ABSTRACT: UV radiation leads to photodissociation and photoionization of molecular species in diffuse clouds and in the extended envelopes of dark regions. In the inner part of heavily-obscured objects, the residual external field may be negligible; alternative mechanisms, such as cosmic-ray-induced H_2 emission, appear to be efficient in photodestruction processes. We evaluate the relative importance of these radiation sources, and give some estimates of their effects on the mean lifetime of interstellar molecules, taking into account the cloud structure that significantly affects the penetration depth of UV radiation.

1. INTRODUCTION

Evaluation of the radiation field inside interstellar clouds is a crucial step in theoretical investigations of the physical and chemical evolution of interstellar matter. Indeed, the radiation field affects the ionization of interstellar gas and its temperature and the temperature of dust particles, which in turn affects the formation rates of some molecules and the growth of mantles. Finally, it affects the lifetime of the molecules against photodissociation and photoionization.
    In diffuse clouds and in the outer parts of dense clouds, photodissociation by interstellar UV photons is the dominant destruction mechanism of the neutral molecules, while in deeper regions inside the clouds, UV radiation is generally assumed to play a negligible role because of screening by dust particles. However, how efficiently and how deeply the UV photons penetrate inside the clouds depends upon the extinction and scattering properties of the dust.
    In this work we investigate the level of ambient UV fields as they result from the transfer of interstellar radiation and from the production of cosmic-ray-generated photons. Preliminary results regarding the photodissociation rates of some molecules of astrophysical interest are given.

2. SOURCES AND TRANSFER OF UV PHOTONS

We assume that the cloud is exposed to an external isotropic UV radiation field with intensity (Van Dishoeck, 1985):

$$N_0(\lambda) = 3.24 \times 10^{15}(\lambda^{-3} - 1.58 \times 10^3 \lambda^{-4} + 6.26 \times 10^{-5} \lambda^{-5}) \text{ photons cm}^{-2}\text{s}^{-1}\text{Å}^{-1}$$

Furthermore, we consider the photons generated inside the cloud by the impact excitation of hydrogen molecules by cosmic rays (Prasad and Tarafdar, 1983), taking into account the contribution of both primary protons and secondary electrons. The primary proton spectrum is the one measured by Webber and Yushak (1983), extrapolated down to 1 Mev. The resulting ionization rate is:

$$\zeta = 4 \times 10^{-17} \text{s}^{-1}$$

The total photon flux (photons cm^{-2}s^{-1}) in a point at distance r from the center of the cloud, except near the boundaries, is given by the superimposition of the residual interstellar field, $T(\lambda)N_0(\lambda)$, and the cosmic-ray induced field:

$$N(\lambda_{v'v''}) = (1-\omega)^{-1}[1-T(\omega,g)]C_{0v'}A_{v'v''}(\Sigma A_{v'v''})^{-1}(N(H_2)/A_{vis})/E(\lambda)$$

Where ω and g are the albedo and asymmetry factor of dust particles, T is the transmissivity of the medium at λ, $A_{vv'}$ are the transition probabilities (Allison & Dalgarno, 1970), C_0 are the total collisional coefficients, $E(\lambda) = A(\lambda)/A(\lambda_{vis})$ is the extinction scale. The gas to visual extinction ratio is derived from Dickman (1978).

We carried out calculations for spherical clouds and for both homogeneous and inhomogeneous density distributions. More specifically, we considered clouds with a density distribution dropping off as r^{-n} from a constant density core. Because many dark clouds show similar density distribution (Snell, 1981; Arquilla & Goldsmith, 1985), it appears that inhomogeneous models can provide a more realistic picture of the radiation penetration into clouds than the homogeneous ones can. In both cases, the edge-to-center visual optical depth is 10. In the inhomogeneous model, the core radius is one tenth of the total cloud radius. For the homogeneous model, the transmissivity has been computed by using the Spherical Harmonic Method (SHM) as developed by Flannery et al. (1980). For the inhomogeneous model, we used the extension of SHM worked out by Tiné et al. (1991) in order to handle the variable-density model.

The extinction law has been studied with regard to a large number of stars located in different galactic regions and astrophysical environments. The Mean Extinction Curve (MIEC, Savage & Mathis, 1979), usually adopted for correcting the observations for reddening and in computing the radiation transfer inside interstellar clouds, can actually be considered as representative of the extinction properties of dust in the diffuse medium and in old associations (Aiello et al., 1988). However, in dense clouds and in regions of recent star formation, dust grains are likely to have different properties: in particular, their size distributions appear to be biased towards large radii, which results in a more or less marked flattening of the extinction curve in the far UV spectral region. Therefore, the use of MIEC for computing the radiation field inside the clouds could be completely inappropriate and lead to large errors (Aiello et al., 1984; Mathis, 1990).

Cardelli et al. (1989) found some analytical expressions which relate the extinction scale to the observed values of R, the ratio of the total-to-selective extinction. This latter is considered to be one of the indicators of local properties of dust particles. Indeed, there is quite a strong correlation between the level of extinction and the values of R: the extinction curves towards dense clouds and regions of recent star formation are in general associated with values of R which are larger than the ones associated with MIEC.

In the present computations, therefore, we shall consider two borderline cases: R = 3.2, a value characteristic of the diffuse medium, R = 4.2, as representative of dense regions.

Albedo and g are poorly known in the UV. The results of different studies are conflicting, in particular with regard to g. For some guidelines we can turn to the existing dust models. The values of ω and g computed for the MNR model (Draine, 1986) for the far UV are in the ranges 0.4 - 0.5 and 0.5 - 0.6, respectively. Close values are suggested by some observational studies (Lillie & Witt, 1976; De Boer, 1986). We adopt ω = 0.4 and g = 0.6 as plausible choices for a low R regime.

Because of the total lack of observational data, the UV scattering characteristic of dust particles inside dense clouds can only be a matter of speculation. From a size distribution biased towards large radii, high values of albedo and g are expected. In Case 2 we tentatively adopt ω = 0.5 and g = 0.7 for the high R limit.

3. RESULTS

The results of our transfer computations have been applied to a determination of the photodissociation rates of various interstellar molecules. The cross-sections for photodissociation were derived from Lee (1984) and from a compilation kindly provided by E. Herbst. Figures 1-4 show the rates, as a function of the radial distance from the cloud center, for 4 molecules of particular interest from the point of view of prebiotic evolution: H_2O, HCN, NH_3, CH_4. Figure 5 shows the contribution of external radiation and internal photons to the photodissociation rate of CH_2O_2.

The effects of cloud structure, as well as of the internal photons, are evident. The latter dominate in the innermost part of the clouds in all the models considered, and are capable of slowing down the decay of photodissociation rates. However, it is evident that the internal field dominates in the deeper part of the cloud in all the models considered; but, the "region of dominance" narrows with a steepening in the density gradient and with an increase in R.

FIGURES

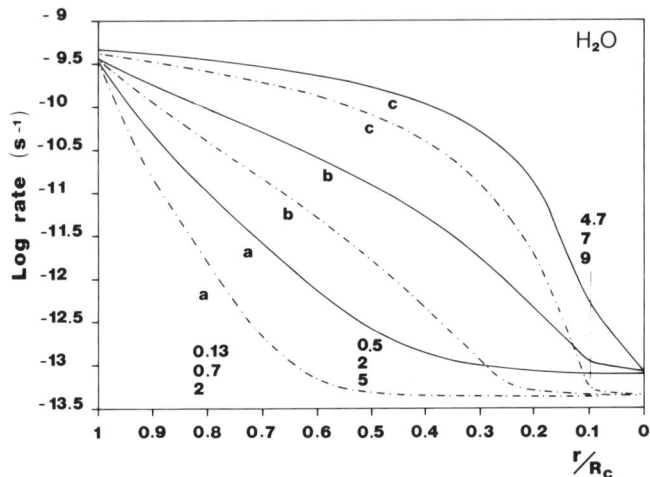

Figure 1. Photodissociation rates for H_2O as a function of the distance from the cloud center: R_c = cloud radius. Dotted line - R = 3.2. Continuous line: R = 4.2. a-cloud with uniform density; b-n = 1; c-n = 2. The visual optical depths corresponding to r/R_c = 0.8, 0.5, 0.1 (core edge) are also indicated.

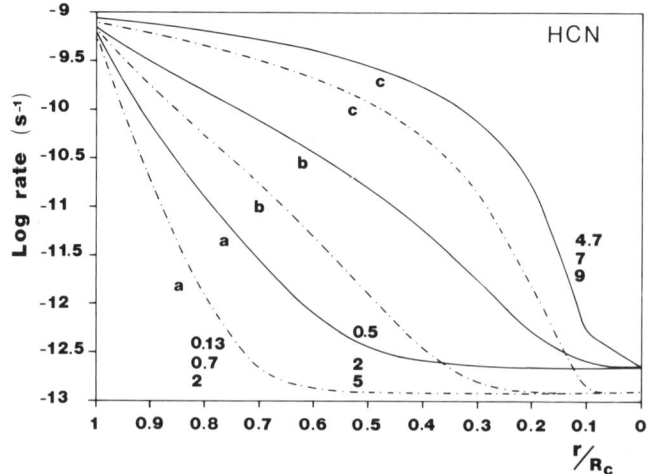

Figure 2. The same as Figure 1 for HCN

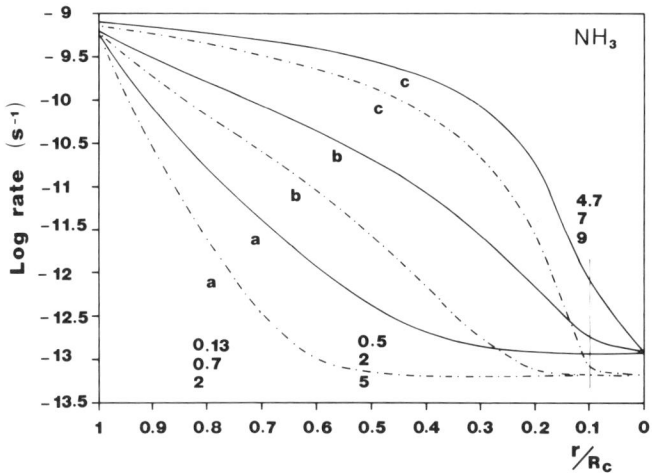

Figure 3. The same as Figure 1 for NH_3

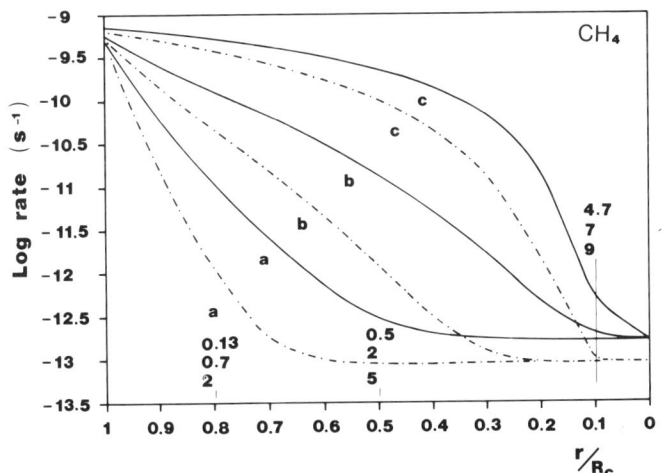

Figure 4. The same as Figure 1 for CH_4

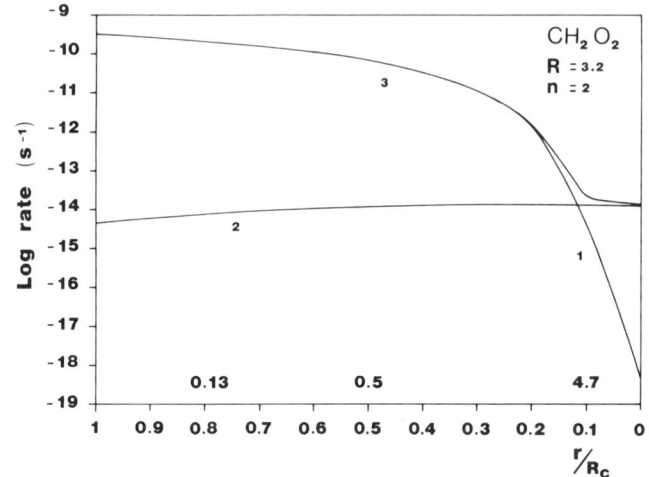

Figure 5. The contribution to the photodissociation rates by external radiation and internal photons for the CH_2O_2 molecule. 1) external radiation; 2) cosmic-ray-induced photons; 3) total rates.

4. REFERENCES

Aiello S., Rosolia A., Barsella B., Ferrini F. and Iorio D., 1984, Nuovo Cimento, 7, 840.
Aiello S., Barsella B., Chlewicki G., Greenberg J.M., Patriarchi P. and Perinotto M., 1988, Astr.Ap., Suppl. Series, 73, 195.
Allison A.C. and Dalgarno A., 1970, Atomic Data, 1, 289.
Arquilla R. and GOldsmith P.F., 1985, Ap.J., 297, 436.
Cardelli J.A., Clayton G.C., Mathis J.S., 1989, Ap.I., 345, 245.
de Boer K.S., 1986, In: New Insights in Astronomy, ESA SP 263, 551.
Dickman R.L., 1978, Ap.J.Suppl., 37, 407.
Draine B.T., 1985, Private Communication
Flannery B.P., Roberge W.G. and Rybicki G.P., 1980, Ap.J., 236, 598.
Lee, L. C., 1984, Ap.J., 282, 172.
Lillie C.F. and Witt A.N., 1976, Ap.J., 208, 64.
Mathis I.S., 1990, Ann.Rev.Astr.Ap., 28, 37.
Prasad S.S. and Tarafdar S.P., 1983, Ap.J., 267, 603.
Snell R.L., 1981, Ap.J. Suppl., 45, 121.
Tiné S., Aiello S., Belleni A. and Cecchi-Pestellini C., IQSRT, in press.
van Dishoeck E.F., 1987, in IAU Symposium No. 120, Astrochemistry, ed. M.S. Vardya and S.P. Tarafdar (Dordrecht: Reidel, pag. 51).
Webber W.R. and Yushak W.M., 1983, Ap.J., 275, 391.

OBSERVATIONS OF SHOCKED REGIONS

L. M. Ziurys
Department of Chemistry
Arizona State University
Tempe, AZ 85287-1604

ABSTRACT. Recent observations of molecular clouds perturbed by interstellar shocks are reviewed. Effects of the shock waves on the chemical composition of these regions are discussed.

Introduction

Interstellar shocks occur when the bulk fluid velocity of a gas exceeds the local sound speed. Shocks heat, compress, and accelerate the gas, changing its physical and chemical characteristics substantially. Since the local sound speed in a molecular cloud is usually ≈ 0.3 km/s, shocks should be common phenomena.

There are various types of shocks, including fast, dissociate "J-type" ones. These satisfy the so-called "jump" conditions, and generally destroy molecules. Lower velocity C-type shocks are thought to involve magnetic precursors (e.g., McKee, Chernhoff, Hollenbach 1983). In the latter type, molecules aren't dissociated, and thus shocked regions studied by molecular-line radio astronomy are thought to be of this kind.

The passage of a shock wave through cold, unperturbed gas is thought to affect its chemical composition. Not only can dust grains be destroyed, releasing "frozen-out" species, but the rise in temperature and pressure can also significantly change ambient gas phase abundances. At low temperatures typical of cold, quiescent molecular clouds, i.e., T~10 K, endothermic reactions, and ones with any significant activation barrier, cannot occur on a reasonable time scale. A rise in temperature, anywhere from 100 to thousands of degrees as a result of a shock wave, can open these new chemical channels and critically alter expected "ion-molecule"-type abundances.

Many "shock chemistry" models have been made to examine the chemistry behind a shock wave (e.g., Hartquist, Oppenheimer and Dalgarno 1980; Mitchell 1984; Neufeld and Dalgarno 1989). Although the details of individual models differ, there are some common conclusions: CO is thought to be unaffected chemically by a shock, while abundances of certain oxygen-containing molecules such as OH, H_2O, and SiO, and certain sulfur species (H_2S, SO, and SO_2) are enhanced due to the occurrence of reactions with energy barriers.

Observationally, shocked regions are discovered by the presence of unusually broad, often asymmetric line profiles found in molecular spectra. Line profiles from quiescent gas are generally quite narrow. A classic example of a molecular line which is thought to show the presence of a shock wave is the J=1→0 line of CO, observed toward the B cloud near the supernova remnant IC443 (DeNoyer and Frerking 1981). A narrow "spike"-type feature is found near $V_{LSR} \approx +5$ km/s, and is thought to represent the cold, unperturbed "pre-shock" gas. Nearby in velocity is a broad, asymmetric line that extends from V_{LSR}=-5 to -60 km/s, thought to trace material that has been hit by a shock wave from the supernova explosion. This gas presumably has been heated and accelerated, hence the abrupt change in LSR velocity, as well as the presence of emission at velocities far larger than that of the quiescent gas.

Besides broad line profiles, there are other possible tracers of shock waves. These include the presence of vibrationally excited H_2 emission, as well as the appearance of highly-excited lines of molecules such as CO, HCN, and OH, particularly in the far IR. However, there may be other mechanisms that could be exciting these tracers, such as imbedded infrared sources.

Because shock waves are readily generated when stars form, as well as when they destruct, observations of shocked regions should be in abundance. However, only a few such objects have been looked at in great detail. These include the Orion-KL high mass star-forming region, a few cloud complexes which have been subjected to a supernovae blast (IC443, W28, W44), and a few loss-mass star-forming objects, which include L1551 and Barnard 1.

Shocked Gas in a Region of High Mass Star-Formation: Orion-KL

The shocked gas in Orion is associated with various outflows present in the region. A young ~25 $M_\odot$ star, presumably IRc2, has formed in the quiescent cloud OMC-1. In the course of its evolution, it has undergone mass loss, initiating several outflows that shock and heat the surrounding gas (e.g., Plambeck et al. 1985). Several kinematic components and chemical components exist in this region, introducing a rather complicated morphology. There is the ambient, quiescent material of OMC-1, which is called the "spike" or "ridge." Then there are the outflow regions, which are centered near IRc2. There is thought to be high velocity, "extended" bipolar outflow, and a lower velocity one, sometimes called the "doughnut" or "plateau." These outflows probably represent shocked gas since they are rather hot (T_k~100K), and dense, the high velocity plateau exhibiting $n(H_2) \approx 10^5$ cm^{-3}, and the low velocity one having $n(H_2) \geq 10^6$ cm^{-3}.

Another region of interest in Orion-KL is the "hot core," a dense clump or clumps of material a few arcseconds in distance from IRc2, with T_k~200 K and $N(H_2)$ ~10^7 cm^{-3} (e.g. Plambeck et al.1985). This region of hot dense gas may or may not actually be heated as a result of shock waves. In the initial stages of the mass loss from IRc2, the hot core may have been subjected to a shock wave blast, which elevated its temperature and density (Masson et al. 1985). This is supported by the presence of a so-called cavity around IRc2 (Wynn-Williams et al. 1984). Alternately, the hot core may be gas that is primarily heated by IRc2, or a highly imbedded, young star (Plambeck et al. 1985).

There have been numerous studies of the shocked regions in Orion, including several spectral line surveys carried out at mm-wavelengths (e.g., Johansson et al. 1984; Sutton et al. 1985; Jewell et al. 1989; Ziurys and McGonagle 1991). Such surveys are useful because the hot core and plateau regions , i.e. the shocked material in Orion, are usually distinguishable from the quiescent gas in a single line profile. Thus, it is possible to assign a species as arising from shocked or quiescent gas, or both.

The molecules that appear to be quite abundant in the "plateau" outflows are SO, SO_2, H_2S, SiO, and SiS. (e.g. Johansson et al. 1984; Blake et al. 1987; Minh et al. 1990). Sulfur and silicon molecules do appear to be preferentially produced in shocked gas, at least in Orion. SO, SO_2, and H_2S are abundant in the hot core, as well, but have much reduced concentrations in the ridge (e.g., Minh et al. 1990; Ziurys and McGonagle 1991). In contrast, silicon species do not usually appear in the quiescent gas (Ziurys, Friberg, Irvine 1989).

There are many species present in the hot core, including NH_3, HCN, etc., that are also abundant in the ridge as well. Ethyl cyanide, however, appears to be only in the hot core (e.g. Johansson et al. 1984; Sutton et al. 1985; Ziurys and McGonagle1991). It, therefore, may be exclusively produced by shocks. Blake et al. (1987) attribute the behavior of EtCN to grain processing, which hydrogenates double and triple carbon bonds. Yet, HC_3N, which contains carbon-carbon triple bond, is prominent in the hot core as well.

The dominant presence of EtCN in the hot core is curious. In contrast, other large organic species such as dimethyl ether (($CH_3)_2O$) and methyl acetylene ($H_3C-C\equiv CH$), are exclusively observed in the ridge (e.g., Ziurys, Wilson, and Waterlout 1991). The chemistry of large organic compounds in interstellar gas is thus very selective, and has yet to be satisfactorily explained.

Shocked Gas Near Supernova Remnants

Supernova remnants which demonstrate a clear-cut interaction of molecular cloud and blast wave are relatively rare (e.g., Wootten, 1978). One case where there does appear to be a definite interaction is the molecular cloud complex associated with the young SNR IC443. There may also be some interaction in SNR W28, and perhaps in W44, but the evidence is less convincing.

Shocked gas was initially detected in IC443 through the observation of molecular line profiles with extremely large velocity dispersions in several nearby clouds (e.g., DeNoyer 1979; DeNoyer and Frerking 1981). These spectra were measured at several positions near the SNR, identifying small, perturbed regions of gas referred to as clouds or "clumps" A, B, and C. Broad line profiles were observed in CO, HCO^+, HCN, H_2CO, OH, and CS.

More recent studies by Dickman et al. (1991) reveal an even more interesting structure. Large scale mapping of the $J=1\rightarrow 0$ transition of HCO^+ shows the presence of at least six more highly perturbed clumps. Together with clouds A, B, and C, they form a roughly elliptical ring about 9 pc across on the face of the optical SNR. This is illustrated in Figure 1. The bulk of the red-shifted emission is observed on the northern section of the ellipse and the blue-shifted emission on the southern part, which suggest the clumps lie along the edge of a tilting, expanding ring.

Of the nine clumps associated with IC443, only two have been examined in a variety of molecules. These are the most perturbed regions, clumps B and G. Ziurys, Snell and Dickman (1989) studied the previously-detected species (HCN, HCO^+, CO, and CS), and observed several new molecules towards these regions, including SiO, SO, N_2H^+, HNC, CN, and NH_3. They also detected the $J=3\rightarrow 2$ transition of HCO^+ in both clumps. The multi-transition study of HCO^+ showed that the densities in both clumps was at least 3×10^5 cm^{-3}. Analysis of the (1,1) and (2,2) inversion transitions of NH_3 showed that the gas kinetic temperature was $T_k \geq 40K$. These densities and temperatures strongly indicate shock heating and compression.

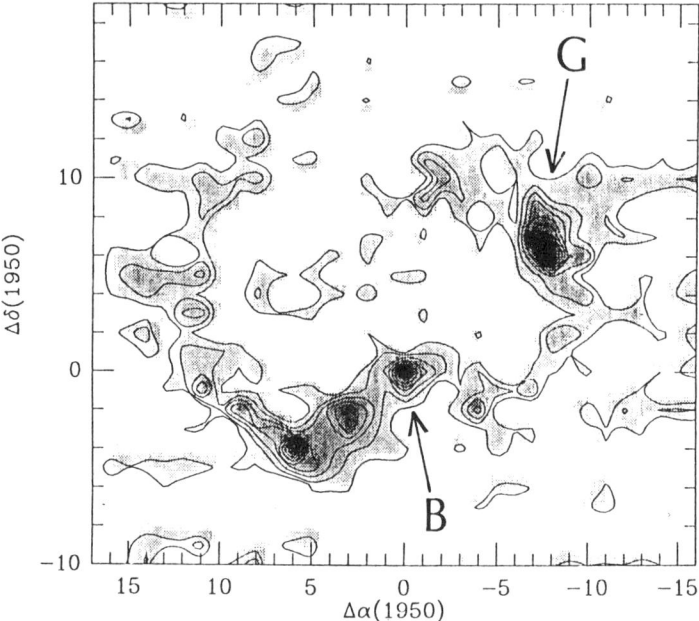

Figure 1. Contours of integrated J=1→0 HCO$^+$ emission, showing a "ring" of perturbed clumps in SNR IC443. The (0,0) position is $\alpha = 06^h14^m15^s$; $\delta = 22°26'50"$ (1950.0). Positions of clumps B and G are indicated (from Dickman et al. 1991).

The chemical abundances in clumps B and G are thus likely to be affected by shock waves. A summary of chemical abundances measured toward these two regions is given in Table 1. For comparison, abundances established toward TMC-1 are also shown. Because of the large dipole moments of the species studied, relative to that of CO (1-4D vs. 0.1D), all abundances are referenced to HCN ($\mu = 2.98$D). As the table shows, the only molecule whose abundance is significantly enhanced in the shocked gas of IC443 is SiO, whose concentration is 100 times higher than the upper limit found for TMC-1. In contrast, the SO abundances in IC443 are comparable to those of TMC-1, as well as for most of the other species studied. There appears to be a slight enhancement in the HCO$^+$ abundance, as well. HNC, however, is an order of magnitude less abundant in IC443 vs. TMC-1. This is not unexpected, as HNC is a metastable isomer. Elevated temperatures allow it to convert to its more stable form, HCN.

Broadened line profiles in molecules have also been observed towards the supernova remnants W28 and W44 (Wootten 1981; DeNoyer 1983). Towards W44, however, the moderately broadened profiles ($\Delta V_{1/2} \approx 10$ km/s) can be explained by the presence of multiple velocity components along the line of sight, rather than resulting from shock acceleration. Towards W28, the line profiles in species such as HCO$^+$ exhibit $\Delta V_{1/2} \approx 20$ km/s, which perhaps are due to the passage of a shock. Again, CO shows at three individual velocity components present over the same broad velocity interval as the HCO$^+$ emission. The evidence for the presence of shocked material is therefore uncertain in this object.

Table I Abundances in shocked regions relative to TMC-1

Molecule		X/HCN[a] IC443	X/HCN[a] Orion Plateau	X/HCN[b] L1551E	X/HCN[b] L1551W	X/HCN[a] TMC-1
HCO^+	B	1.2-1.5	-	1.1	0.65	0.4
	G	0.80-1.3	-	-	-	-
N_2H^+	B	0.035	-	-	-	0.025
	G	0.013	-	-	-	-
CS	B	0.8	0.08	3.4	7.7	0.5
	G	0.7	-	-	-	-
SO	B	0.82	1.87	<0.45	7.7	0.25
	G	0.55	-	-	-	-
SiO	B	0.038	0.1	<0.07	<0.22	<0.0005
	G	0.063	-	-	-	-
HCN	B	1	1	1	1	1
	G	1	-	-	-	-
HNC	B	0.14	0.005	-	-	1
	G	0.11	-	-	-	-
CN	B	1.7	-	-	-	1.5
	G	0.89	-	-	-	-
NH_3	B	1.3	0.5	2.6	4.6	1
	G	0.4	-	-	-	-
CO	B	20,000	430	-	-	4,000
	G	7,500	-	-	-	-

a) From Ziurys et al. (1989); Ziurys (1990)
b) From Plambeck and Snell (1991)

Shocked Gas in Regions of Low Mass Star Formation

There are several objects where low-mass star-formation is occurring, and an associated outflow has affected nearby ambient gas. Two such sources where the chemical abundances has been investigated in detail are L1551 (Plambeck and Snell 1991) and Barnard 1 (Bachiller, Menten, and del Río-Alvarez 1990).

The outflow associated with L1551 was initially detected in broad CO line profiles (e.g. Snell, Loren and Plambeck 1980). Subsequent studies have shown that this outflow is bipolar, and is centered on the infrared source IRS-5, presumably a young star in its mass loss stage. More recent observations suggest that a complete shell of accelerated molecular material surrounds IRS-5 (Schloerb and Snell 1985). The bipolar lobes of the outflow are regions where the shell has impacted on surrounding quiescent clumps. Hence, the two lobes associated with the L1551 outflow are likely to be regions that have been shocked.

Recently, Plambeck and Snell (1991) have measured the abundances of various molecules in the L1551 bipolar lobes, L1551 E and L155W. As Table I illustrates, several of the species of interest in shocked gas were not detected. These include SiO, SO (in L1551E), and SO_2. Of the molecules detected, HC_3N and NH_3 have concentrations comparable to TMC-1. The abundance of HCO^+ may be slightly enhanced, however, as was found for IC443. CS, on the other hand, is about an order of magnitude more abundant in both lobes of L1551 vs. TMC-1. A similar enhancement is found for SO, but only in the western lobe. This contrasts the abundances measured in IC443, where those of both CS and SO were found to be quite similar to those in TMC-1. Unfortunately, upper limits placed on the SiO abundance in L1551 are not significant.

Barnard 1 is another dark cloud which shows evidence of an outflow associated with a young low mass star. Recent studies of this object by Bachiller, Menten, and del Río-Alvarez (1990) show several clumps in CS that contain low-luminosity young stellar objects detected by IRAS. The clumps exhibit $T_k \geq 12K$. Toward one such object, IRAS03301+3057, high velocity wings with a dispersion up to 40 km/s have been observed on $J=2 \rightarrow 1$ spectra of CO. Similar, less pronounced wings were also detected on SO and SiO spectra (see Figure 2). The wings are thought to arise from a high velocity flow originating with mass loss from IRAS03301+3057. The gas tracing the outflow is therefore likely to be shocked.

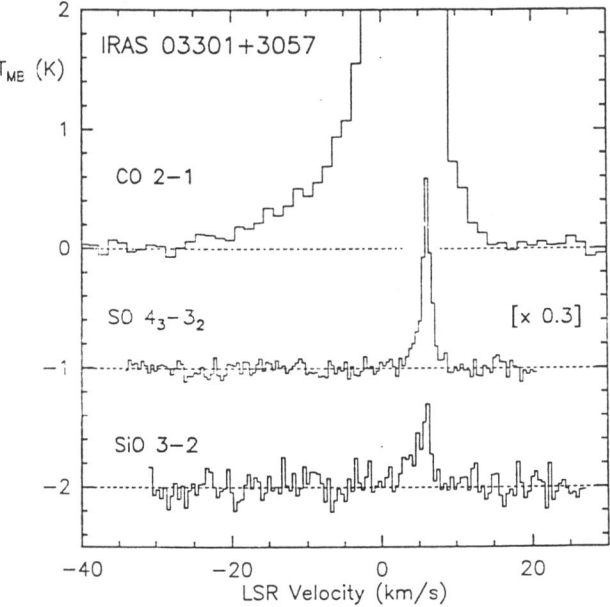

Figure 2. CO, SO, and SiO spectra observed towards B1 (from Bachiller et al. 1990).

Bachiller et al. measured molecule abundances towards the outflow. Unfortunately, they did not observe HCN, so their abundances are referenced to H_2. Curiously, their CS and SO concentrations are similar to TMC-1. SiO, on the other hand, is significantly enriched. The abundance of HCO^+ is slightly higher in B1 than TMC-1, but those of HC_3N and HC_5N are significantly lower.

Conclusions

The chemical affects of shocks in the regions examined here do vary. There appears to be a trend towards enhanced abundances of certain sulfur-bearing species such as SO, SO_2, H_2S, and, in some cases, CS. However, as was found for IC443, this may not always be the case. HCO^+ tends to show a small abundance enhancement in most of the shocked sources examined here, but additional studies need to be performed before this is a certainty. Curiously, certain large organic species show evidence of shock enhancement, in particular ethyl cyanide, while others prefer more quiescent gas such as dimethyl ether and methyl acetylene. This is difficult to explain. One species, however, seems to always be present in hot, shocked gas: SiO. Why SiO is abundant in shocked regions may be due to high temperature reactions (Langer and Glassgold 1990), and/or grain destruction. Clearly, additional studies need to be carried out before the chemical effects of a shock wave can be satisfactorily evaluated.

References

Bachiller, R., Menten, K. M., and del Río-Alvarez, S. (1990), Astr. Ap., 236, 461.
Blake, G. A., Sutton, E. C., Masson, C. R., and Phillips, T. G. (1987), Ap. J. 315, 621.
DeNoyer, L. K. (1979a), Ap. J. (Letters), 232, L165.
DeNoyer, L. K. (1983), Ap. J., 264, 141.
DeNoyer, L. K. and Frerking, M. A. (1981), Ap. J. (Letters), 246, L37.
Dickman, R. L., Snell, R. L., Ziurys, L. M., and Huang, Y.-L. (1991), Ap. J., submitted.
Jewell, P. R., Hollis, J. M., Lovas, F. J., and Synder, L. E. (1989), Ap. J. S., 70, 833.
Johansson, L. E. B. et al. (1984), Astr. Ap., 130, 227.
Langer, W. D. and Glassgold, A. E. (1990), Ap. J., 352, 123.
Masson, C. R., et al. (1985), Ap. J. (Letters), 295, L47.
Minh, Y. C., Ziurys, L. M., Irvine, W. M., and McGonagle, D. (1990), Ap. J., 360, 136.
McKee, C. F., Chernhoff, D., and Hollenbach, D. (1984), Galactic and Extragalactic Infrared Spectroscopy, ed. M. Kessler and J. Phillips (Dordrecht: Reidel) p. 103.
Mitchell, G. F. (1984), Ap. J., 287, 665.
Neufeld, D. A. and Dalgarno, A. (1989), Ap. J. 340, 869.
Plambeck, R. L., Vogel, S. N., Wright, M. C. H., Bieging, J. H., and Welch, W. J. (1985), Symposium on MM and sub MM Astronomy, ed. J. Gomez-Gonzales (URSI), p. 235.
Plambeck, R. L. and Snell, R. L. (1991), Ap. J., submitted.
Schloerb, F. P. and Snell, R. L. (1980), Ap. J., 295, 490.
Snell, R. L., Loren, R. B., and Plambeck, R. L. (1980), Ap. J. (Letters), 239, L17.
Sutton, E. C., Blake, G. A. Masson, C. R., and Phillips, T. G. (1985), Ap. J. S., 58, 341.
Wootten, H. A. (1978), Ph.D. Thesis, University of Texas at Austin.
Wootten, H. A. (1981), Ap. J. 245, 105.
Wynn-Williams, C. G., Genzel, R., Becklin, E. E., and Downes, D. (1984), Ap. J. 281, 172.
Ziurys, L. M., Snell, R. L., and Dickman, R. L. (1989), Ap. J., 341, 857.
Ziurys, L. M., Friberg, P., and Irvine, W. M. (1989), Ap. J., 343, 201.
Ziurys, L. M. (1990), Evolution of the Interstellar Medium, ed. L. Blitz (San Francisco: A.S.P.), p. 229.
Ziurys, L. M., and McGonagle, D. (1991), Ap. J. S., submitted.
Ziurys, L. M., Wilson, T. L., Waterlout, J. (1991), in preparation.

QUESTIONS AND ANSWERS

B.E.Turner: Si species, although observed to be enhanced in shocked regions, may simply represent the evaporation of refractory ice mantles from grains, which occours at $\sim 90K$. We find SiO abundances in IC 443G to be 100 to 1000 times lower than predicted by either non-disssociative or dissociative models. Further, line widths of SiO in most sources are not unusual, but would be expected to be wider if involved in the shocks. There is no direct evidence that the Si-species themselves are "hot" in the sense of large rotational temperatures, as would be expected if shock produced. One only needs a small fraction of refractory ice mantles to be evaporated, even in Orion(KL), to produce the observed SiO abundances (cf. Turner 1991, Ap.J. in press). In grain also, only a small fraction of total Si is in SiO, contrary to shock models.

L.M.Ziurys: Ziurys and Friberg back in 1987 (Ap.J.(Letters) 314,L49) noted that only a small fraction of silicon was needed to account for all gas phase silicon molecules in Orion. It could well be that enhancement of Si compounds in shocked regions may result from grain evaporation. However, SiO does exhibit high velocity wings quite often (i.e., see Dounes et al. 1982,Ap.J.(Lett),252,L29). In fact, it was the presence of an asymmetric wing in SiO that indicated the existence of the outflow at Orion-S (Ziurys and Friberg 1987). Asymmetric wings are also seen in SiO spectra towards BI (Bachillller et al.(1990),Astr.Ap. 236, 461). Also there are various indications that SiO in outflows is hot. Towards Orion, the *brightness* temperature in SiO is near 230 K! This is clealy hot! Also, LVG modeling of the $J = 2 \rightarrow 1 / J = 5 \rightarrow 4$ SiO transitions towards the Orion-S outflow indicates a kinetic temperature near 100 K (Ziurys et al. Ap.J.(Lett),356,L25,1990).

Y.Wu: Some molecular species are indeed enhanced in shock region, it can include CS; but CS is also overabundant in dense cores without source, where there is no shock caused by stellar wind. Have you any idea about this?

L.M.Ziurys: It is not clear yet whether CS is actually enhanced in shocked gas. It could appear to be overabundant in dense cores due to some low temperature reaction that hasn't been properly included in chemical models. About 50% of reaction rates used in theoretical models have never been measured experimentally at *room temperature*, let alone at $T \sim 10K$.

M.Guelin: There is another category of sources where we have a clear evidence of shock chemistry: the envelopes of some evolved stars (e.g. protoplanetary nebulae). There is a high velocity $v_{exp} > 200\,km/s$ outflow in CRL 618 where half a dozen of molecules have been detected (CO, HCN, HC_3N, HCO^+, CN).

L.M.Zuirys: CRL 618 would be another interesting source to examine chemical abundances in.

S.P.Tarafdar: Dr.Ohishi showed yesterday that HC_3N and NH_3 regions are different whereas you have put both species originating in the hot core of Orion-KL. Is it due to difference in resolution or HC_3N and NH_3 in Orion-KL do originate from the same region?

L.M.Zuirys: Both HC_3N and NH_3 are present in the Orion hot core, plateau, and ridge components, examining single dish spectra. However, their detailed distributions do seem to differ, if one looks at VLA NH_3 maps vs Hat Creek HC_3N maps. Assigning species to various Orion components is somewhat of a generalization.

H_2O MASER PUMPING BY SHOCK WAVES

Jacques R.D. Lepine
Instituto Astronomico e Geofisico, Universidade de São Paulo
CP 9638, 01065 São Paulo, Brazil

Astrid Heske
Sterrewacht Leiden, Postbus 9513, NL-2300, The Netherlands

ABSTRACT. We discuss a simple H_2O maser pumping mechanism in which the population inversion of the masing levels takes place during the quick cooling of the gas behind a shock wave. The population of the rotational energy levels in the initial hot state and final cool state of the molecular gas, and the decay paths between levels are analysed to calculate the average number of 22 GHz photons emitted per H_2O molecule in the cooling process.

1. The model

1.1. Shock Waves

We describe briefly the effect of a shock wave propagating in a dense (10^{10} cm^{-3}) molecular gas in a circumstellar or interstellar environment, with velocity 20-60 kms^{-1}. A sudden rise in temperature occurs at the shock front; the maximum temperature reached depends on the square of the velocity discontinuity, being about 8000 K for 20 kms^{-1}. Dissociation of molecules, excitation and even ionization of atoms occur, producing recombination lines emission.

The chemical reactions in the post-shock gas are almost independent of density, the chemical equilibrium between molecules being almost only a function of the temperature (Draine,1984). Due to endothermic reactions envolving O and OH, H_2O becomes the most abundant molecule after H_2, reaching a relative abundance of about 3 10^{-4}. Because of its large electric dipole moment, H_2O radiation is also the dominant cooling agent (see Neufeld and Melnik ,1987,for the cooling efficiency of H_2O). According to Draine et al.(1983), about one half of the power is radiated by rovibrationnal H_2 emission, one half is radiated by H_2O, and less than 2% is radiated by CO, OH and OI.

Considering a reference frame moving with the shock front, we are concerned with what happens in a thin layer of the post-shock cooling region, in which the molecular gas enters at a temperature of about 1000 K and leaves at about 200 K, this being the temperature interval at which most of the 22 GHz H_2O masing effect takes place.

1.2. H_2O de-excitation

The rotational energy levels of the ground state of ortho -H_2O are shown in figure 1 (this is a well known figure). Let us compute the fraction of H_2O molecules which pass through the 6_{16} state when the gas cools from 1000 K to 200 K. The fraction of H_2O molecules with $J \geq 6$ is:

$$f(J \geq 6) = Z\ (J \geq 6)\ /\ Z$$

where

$$Z = \Sigma \omega_i\ e^{-E_i/kt}$$

is the partition function, $\omega = 2J+1$, and $Z(J \geq 6)$ is the partition function obtained with the summation performed only on the levels with $J \geq 6$. The fractions $f(J \geq 6)$ and $f(J \geq 7)$ are shown as a function of temperature in figure 2.

Since some levels of the J=6 and J=7 ladders can decay to J=5 levels without passing through 6_{16}, and conversely some J=5 levels can decay to 6_{16}, the fraction of molecules decaying through 6_{16} is given by:

$$f(6_{16}) = \Sigma \omega_i\ b_i\ e^{-E_i/kT}/Z$$

where b_i are branching ratios representing the probability of a given level to use a decay path passing through 6_{16}, taking into account the transition probabilities involved in each possible decay path.

We have computed the probability of the various decay paths for each level, in several cases: radiation-dominated decay, collision-dominated decay, and the decay paths when the infrared transitions are optically thick. In all cases a fraction of the order of 50% of the ortho-H_2O molecules, or 30% of all H_2O molecules, pass through the 6_{16} state while they cool from 1000 K to 200 K.

Once the 6_{16} level is reached, we need to know the fraction of molecules which take the 6_{16} -5_{23} decay path, emitting a 22 GHz photon. The spontaneous transition rate strongly favors the 6_{16}-5_{05} transition (the Einstein A coefficient is $10^{-9}s^{-1}$ for 6_{16} -5_{23} and $10^{-1}s^{-1}$ for 6_{16} -5_{05}). But:
 –The radiative 6_{16}-5_{05} decay is inhibited by the optical thickness of this line.
 –Collisional 6_{16}-5_{05} transitions in fact tend to transfer population from the lower to the upper level, in a short-lived pumping mechanism caused by the hot-H_2-cold H_2O situation. During the cooling process the population of the"backbone" levels is controlled by collisions with H_2.
 –Due to stimulated emission, the effective 6_{16} -5_{23} transition rate is much larger than the spontaneous emission rate. In the presence of population inversion, the maser emission grows exponentially until saturation is reached.

We make use of the following basic principle: in a hot H_2-cold H_2O situation, all the available collisional transitions are used in the transfer of energy from H_2 to H_2O. The collisional transition rates depend on the population of the levels and on the electric dipole matrix elements, in a way similar to the radiative transition rates, so that in first approximation all the radiative transitions of H_2O are enhanced by a same factor, by the supply of energy from H_2.

The very simple result of our calculations can be summarized as follows: on the average *one H_2O molecule of the post-shock region emits about 600 photons in the 6_{16}-5_{23} transition*. In order to compute the output power of a maser, one has the to calculate the number of H_2O molecules reached by the shock. Even if they are dissociated by the shock, the molecules are formed again in the post-shock chemical reactions in about the same quantity.

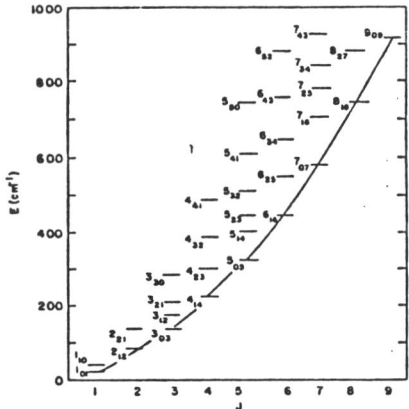
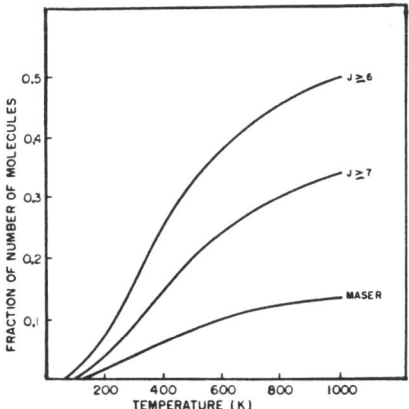

Figure 1: Rotational energy levels of ortho-H_2O (for v=0).

Figure 2: The fraction of H_2O molecules which are in excited levels with $J\geq 6$ and with $J\geq 7$, as a funtion of excitation temperature. The "maser" line indicates the fraction of molecules which pass through the 6_16 level, when the temperature decreases to less than 200 K.

2. Comparison with observations

The sites in which the H_2O maser are found fall in three main categories: 1)envelopes of Miras and M supergiants 2) the vicinity of young stellar objects in star-forming regions 3) the circum-nuclear region of some galaxies. All the three categories of H_2O sources are well known to be permeated by shock waves. We have compared the observed photon emission rate in well studied objects of each category, and we find very good agreement with our model. This discussion will be presented elsewhere.

References

Draine, B.T., Roberge, W.G., Dalgarno, A., 1983, Ap.J. **264**, 485

Neufeld,D.A., Melnik,G.J, 1987, Ap.J. **322**,266

The C:O ratio in dark clouds with cyclic star formation

L A M NEJAD

Department of Mathematics and Physics, Manchester Polytechnic, All Saints, Manchester, UK

and

D A WILLIAMS

Department of Mathematics, UMIST, PO Box 88, Manchester M60 1QD

August 28, 1991

Observations (Goldsmith et al., 1986) indicate that the gas in dark clouds with embedded low-mass stars experiences a cycle, driven by stellar winds, between low and high density phases. The cycle time is sufficiently short that the chemistry never attains steady state. Icy mantles accumulating on grains during the collapse and dense core phases are thought to be removed in each cycle by low velocity shocks that terminate the low density phase. Models of this type give detailed point-by-point descriptions of both gas and solid phases in molecular clouds (Charnley et al., 1988; Nejad et al., 1990).

In this paper we consider the consequences that follow if mantles are not entirely removed in the shock phase of each cycle. The dark cloud model used includes collapse, ablation and post-shock phases together with an extensive gas phase and surface chemistry of 84 species reacting in about 1200 reactions, with freeze-out of molecules on dust surfaces. We explore computationally several cases, distinguishing especially between those in which chemical differentiation occurs in the desorption process, and those in which all desorption is not selective. Results are presented here for two cases:

A In the shock, 90% of all mantle molecules are retained, and 10% are returned to the gas (non-selective desorption).

B In the shock, 90% of all H_2O mantle molecules are retained, and 10% are returned to the gas, while 100% of all other mantle molecules are returned to the gas (selective desorption).

We show that in the non-selective desorption case (A), the C:O ratio tends to reduce, whereas in the selective desorption case (B) the C:O ratio rises as the number of cycles increases, and the gas may eventually become carbon-rich. The molecular cloud chemistry is then markedly different in the two cases. The results are affected by the cosmic ray ionization rate (see Table I).

The major results of this study are:

1. The C:O ratio in cyclic chemistries is not a given quantity but is determined by the mantle desorption process, the cosmic ray ionization rate, and the cycle number.

2. The C:O ratio approaches a limit as cycle number increases.
3. Observations of the major species (CO, O, H_2O, C) will enable the C:O ratio to be determined directly, and will lay constraints on the nature of the desorption process.
4. The cycling process may create a carbon rich environment if a selective desorption process operates. If so, the formation of polyynes may be achieved more easily.
5. The behaviour of some species in clump and interclump gas is quite distinct in the two model cases A and B.

TABLE I

The C:O ratio in cyclic models of molecular clouds with (A) non-selective and (B) selective desorption of ice mantles in shocks, for several values of the cosmic ray ionization rate, ζ.

CASE A

Cycle	$\zeta(s^{-1})$ 5×10^{-18}	10^{-17}	10^{-16}
1	0.52	0.55	0.40
2	0.47	0.44	0.29
3	0.57	0.42	0.24
4	0.51	0.42	0.23

CASE B

Cycle	$\zeta(s^{-1})$ 5×10^{-18}	10^{-17}	10^{-16}
1	1.42	1.20	0.66
2	1.77	1.36	0.71
3	1.66	1.30	0.75
4	1.56	1.37	0.79
5	1.49	1.40	0.84
6	1.45		0.87
7	1.43		0.92
8			0.93
9			0.95

References

GOLDSMITH, P F, LANGER, W D, and WILSON, R W: 1986, *Astrophys. J.* **303**, L11
CHARNLEY, S B, DYSON, J E, HARTQUIST, T W and WILLIAMS, D A: 1988, *Mon. Not. R. astr. Soc.* **235**, 1257
NEJAD, L A M, WILLIAMS, D A and CHARNLEY, S B: 1990, *Mon. Not. R. astr. Soc.* **246**, 183

Hot cores and cold grains

C.M. WALMSLEY and P. SCHILKE

Max-Planck-Institut für Radioastronomie, Auf dem Hügel 69, D-5300 Bonn 1, F.R. Germany

September 4, 1991

Abstract. A review is given of the evidence for depletion of heavy elements in dense molecular cloud clumps onto dust grain surfaces. Particular attention is given to the recent controversy over the nature of the dense clumps in NGC 2024. We also discuss recent ammonia observations of "hot core" regions and evidence that the NH_3 abundance is greatly enhanced.

Key words: dust mantles - molecules - abundances -depletions - hot cores

1. Introduction

In dense molecular cloud regions, a considerable fraction of the heavy element content of the material is in the form of ices frozen on dust grain surfaces. This is clear on the basis of the studies of solid state absorption features seen typically at near infrared wavelengths. Recent reviews of dust mantle composition and properties are by Tielens (1989) and by Whittet and Duley (1991). There is little doubt that deposition of heavy elements onto grain surfaces is of importance for the chemistry in high density clumps or cores within molecular clouds.

On the other hand, there is also little doubt that close to newly formed massive stars, dust temperatures can be sufficiently high that the greater part of such ice mantles are evaporated and their content is returned to the interstellar medium. The classic example is the "hot core" in Orion. Discussions of this region are given by Genzel and Stutzki (1989) and by Wilson and Walmsley (1989). It seems very plausible that what one is seeing in this region is essentially freshly evaporated grain mantle material. Models by Brown et al. (1988) and by Brown and Millar (1989) show that the observational data are basically consistent with this idea although there are a lot of open questions. A consequence is that one learns something about grain mantle composition by observing the Orion and other "hot cores".

In this short review, we give a brief summary of recent developments in this field. It could perhaps be subtitled "getting on and off grains". One would clearly like to find examples of regions where a large fraction of the heavy elements have been deposited onto dust grains and section 2 discusses some recent attempts to identify such regions. The other side of the coin is as mentioned above when one studies hot cores. Section 3 considers what is known about the properties of these regions with particular emphasis on the rather little that is known about sources other than Orion. Finally, section 4 gives a few tentative conclusions and suggestions for the future.

2. Evidence for freeze out in NGC 2024

Evidently, if one believes that the gas seen in "hot core" regions is the product of evaporation of dust grain mantles, it follows that there are other regions where a large fraction of the heavy element content in the gas has condensed out onto dust grain surfaces. Identifying such regions has proven to be difficult. This is illustrated by several recent studies of the NGC 2024 molecular cloud where claims and counter-claims have been made concerning the fraction of molecules (and by inference heavy elements) in the dense gas surrounding the HII region NGC 2024 (Orion B) (Mezger et al. 1988, Schulz et al. 1991, Gaume et al. 1991, Mezger et al. 1991, Mauersberger et al. 1991). No completely clear picture has emerged from all these studies but a few points have been established.

As a preliminary, we discuss what one expects to observe in regions where a large fraction of heavy elements have condensed out onto dust grain surfaces. Here, we mean that a large fraction is condensed out relative to the "normal molecular cloud composition". Equivalently, the depletion factor δ of heavy elements (essentially C,N,O) relative to solar abundances is larger than (say) 10-100. In molecular clouds, it is useful to use CO as a standard since CO is the sole *observed* (apart perhaps from SiO which is a special case) interstellar molecule to contain the greater part of the available gas phase supply of one of it's constituent elements (in this case carbon). Thus, the estimated [CO/H$_2$] abundance ratio, which is typically 10^{-4} in normal low density molecular clouds might reasonably be taken as a measure of normalcy and evidence for a distinctly lower ratio (say below 10^{-5}) as evidence that molecules have condensed out. Other molecules clearly cannot be expected to have abundances proportional to the degree of C,N,O depletion. In fact, a recent study by Pineau des Forêts et al. (1992) shows that at a density of 10^4 cm^{-3}, NH$_3$ and HCN have abundances of around 10^{-8} relative to hydrogen independent of δ as long as δ is less than 1000. Such model studies may not represent reality in detail but they do demonstrate the dangers of using the abundances of trace species as proportional to the gas phase abundance of the heavy elements which they contain.

But let us suppose that there are high density clumps where the depletion factor is much higher even than 1000 and one can say that the heavy elements are frozen out. The most obvious observational criterion for freeze-out is a high ratio of the intensity of dust emission at, say, a wavelength of 1.3 mm to the intensity of some isotopic substitution of CO at a similiar wavelength. With this in mind, Mauersberger et al. (1991) have compared C^{18}O(2-1) maps of NGC 2024 with the earlier 1.3 mm dust continuum maps of Mezger et al. (1988). One sees in both cases a ridge of material running roughly north-south with some "condensations" apparently superposed. The compact features are however much more clearly present in the dust emission map which additionally shows less extended material. Such differences may partly be due to the different dependence of dust and $C^{18}O$ radiation on temperature. However, Mauersberger et al. conclude that there is evidence in their data for CO depletion in the high density condensations seen in the dust map (where they estimate that the temperature is approximately 20 K, the density 10^8 cm^{-3}, and the column density of order 10^{25} cm^{-2}). Incidentally, the mass (of order 10 solar masses) and density of these clumps are similar to those estimated for the Orion hot core. Thus, it is feasible that such condensations could eventually evolve into regions with parameters similar to those of the hot core if a massive star were to form in their interiors.

The claim that there is evidence for molecular depletion is contested by Schulz et al. (1991) who present maps of CS(5-4) and CS(7-6) and compare them with the dust data. Perhaps more important than the differences between molecular line and dust emission maps are their similiarities. In the dense ridge gas (where n(H$_2$) is of order 10^6 cm^{-3}), the distribution of molecular line emission and dust emission is very similiar. Differences make themselves apparent only around the embedded clumps. The CS and CO abundances in the general ridge gas appear to be normal and hence, one clearly has no great degree of depletion under these conditions. This itself is at first sight surprising given the expected freeze out timescale (roughly 10^4 years for a sticking coefficient of 0.1, see Walmsley 1991). The dynamical timescale for the region should be roughly $l/\Delta v$ where l is size and Δv is linewidth and thus it is also of order 10^4 years. It is difficult to imagine that the NGC2024 ridge is younger than this and hence we conclude that there is reason to believe that the freeze-out time is being somewhat underestimated. The most obvious explanation of this is that, due to grain coagulation, the mean grain cross-section per hydrogen atom is less than in diffuse interstellar material. A relatively modest reduction in this parameter seems capable of explaining the present observational situation.

In the denser condensations mentioned earlier, the situation is much less clear. However, one important result is that VLA ammonia observations with 3″ resolution detect clumps which in several cases are coincident with the dust emission clumps (Mauersberger et al. 1991). The apparent ammonia abundance is in the range $10^{-10} - 10^{-9}$ depending on whether one uses the virial mass or the dust emission data to estimate the H$_2$ column density. This is lower than "normal" but not drastically so and certainly the higher end of the range is within the uncertainties of current gas phase models. It is also significant that the temperatures inferred from the VLA NH$_3$ observations are similar to those (20 K) derived from the dust observations. This is much less than Schulz et al. obtain on the basis of their CS data (35 K). One explanation might be that these clumps are being heated from the outside and that consequently there are sharp temperature gradients from outside to inside. In this scenario, the VLA ammonia refers to an interior region with higher density but lower temperature than the gas seen in the single dish CS data. Obviously, it will be useful to obtain interferometric observations of species such as CS and $C^{18}O$ in order to verify this picture.

One should realise however that the observed ammonia abundance in the high density clumps (roughly 10^{-9} relative to H_2) is compatible with the possibility that essentially all heavy elements have condensed out onto dust grain surfaces but that Lyman photons produced in these clumps as a secondary product of cosmic ray ionisation maintain a small fraction of molecules in the gas phase (see Hartquist and Williams 1990, Prasad and Tarafdar 1983 for discussions of this). In this situation, molecules such as NH_3, which are relatively enhanced in abundance in dust grain mantles, may be as abundant as CO. Following Hartquist and Williams, one finds that the number density $n(NH_3)$ of ammonia molecules in such a steady state situation is given by :

$$n(NH_3) = 0.075 \left(\frac{\zeta_H}{10^{-17}}\right) S^{-1} \left(\frac{2\,10^{-21}}{\overline{\sigma_g}}\right) f_{ph}\, f(NH_3) \text{ cm}^{-3} \qquad (1)$$

Here, ζ_H is the hydrogen cosmic ray ionization rate and S is the ammonia sticking coefficient. f_{ph} is the quantum efficiency for release of mantle molecules per incident UV photon and $f(NH_3)$ is the fraction of mantle molecules which are ammonia. Finally, $\overline{\sigma_g}$ is the mean grain cross section per hydrogen which in diffuse interstellar gas is estimated to be $2\,10^{-21}$ cm^2. The various efficiency factors in this equation are highly uncertain. However, one can certainly concoct plausible parameter combinations which are compatible with the observed VLA ammonia number density towards NGC2024 FIR3 which is roughly 0.01 cm^{-3} according to Mauersberger et al. (1991). Reasonable parameters for instance might be : $f(NH_3) = 0.03, f_{ph} = 0.1, S = 0.1$, $\overline{\sigma_g} = 2\,10^{-22}$.

The best test of this hypothesis is to observe other molecules. Relative abundances should be similiar to those observed in the solid state (see e.g. Tielens) or in hot cores and quite different from normal molecular cloud material. In any case, one concludes paradoxically that gas phase molecules may be observable even when *everything* has frozen out. One needs to observe with an interferometer however in order to filter out normal gas!

3. Hot core sources in the galaxy

The Orion hot core is certainly not unique in the galaxy. Since one of its defining characteristics is a high ammonia column density in a region roughly 10^{17} cm in extent, one can use results from various recent ammonia surveys in order to isolate objects of similiar type. In table 1, we summarize data for an

TABLE I
Parameters of galactic hot core sources

Source	Temperature	cDiameter	$N(NH_3)$	$\log(N(NH_3)/N(H_2))$	
Ori-KL	160	0.025	$1.3\ 10^{18}$	-6.2	HWWH
SgrB2-N	200	0.2	$1.0\ 10^{20}$		VGP
W31C	100	0.026		-6.3	MHWW
W51d	250	0.06	$2.0\ 10^{19}$	-5.5	MHWW
N7538	220	0.015	$2.5\ 10^{18}$		MHWW

updated list of such sources. Data for table 1 has been taken from Hermsen et al. (1988,HWWH), Mauersberger et al. (1986,MHWW), and Vogel et al. (1987, VGP). The abundances given in table 1 should be treated with some scepticism. The main uncertainty is usually the molecular hydrogen column density in the "cores" and in several cases given in table 1, this has been estimated by assuming the virial theorem to hold in the clump and using the observed line width and size. This procedure is doubtful but not necessarily worse than the alternative of using measurements of the dust emission at millimeter and sub-millimeter wavelengths. In this latter case, the problem is the highly uncertain dust emissivity at these wavelengths. One notes for example that Sievers et al. (1991) deduce a column density of $8\ 10^{24}$ cm^{-2} of H_2 towards W51d from their 870 μm measurements which compares with $6\ 10^{24}$ from the virial estimate. However, the latter result refers to a region roughly 0.06 pc. in diameter as compared to 0.25 pc. in the case of the dust emission measurements and hence there is effectively an order of magnitude difference between the two estimates.

Somewhat less uncertain but still to be treated with caution are the direct column density and temperature estimates based upon the molecular data. It is interesting in this context to compare the recent studies of NGC7538 by Keto(1991) and Schilke et al. (1991) with the earlier work by Mauersberger et al. (1986,MHWW). All three studies use partially independent data sets to derive the column density and rotation temperature of the hot ammonia clump seen towards NGC 7538-IRS1. The derived temperatures vary between 125 K (Keto, 1991), 150 K (Schilke et al.), and 220 K (MHWW) whereas the NH_3 column densities are consistent to within a factor 3 (Schilke et al. $10^{18.7}$, Keto $10^{18.2}$, MHWW $10^{18.4}$). These differences are probably due to density and temperature gradients within the source which have the consequence that results depend upon the particular set of transitions used for the analysis.

We have recently (Cesaroni and Walmsley 1992) made a small survey of galactic HII regions in the (4,4) and (5,5) transitions of ammonia with the aim of expanding the list in table 1. This was based upon an earlier survey by Churchwell et al. (1990) where a large sample of galactic HII regions were examined in $NH_3(1,1)$ and $(2,2)$. Not too surprisingly, we find that there are generally small hot ammonia clumps associated with the compact HII regions. A surprise however was the observation of hyperfine satellites in the (4,4) and (5,5) lines towards G10.47+0.03 and G31.41+0.31. Simply interpreted, this implies very high line optical depths ($\tau(4,4) \sim 50-100$) and correspondingly high ammonia column densities (greater than $10^{19} cm^{-2}$). The situation is thus rather similiar to that found by MHWW towards W51d. The size of the emitting region must be of the order of 1-2″ or 10^{17} cm and thus we presume that the essential characteristics are similiar to those of the Orion hot core. The ammonia abundance must be extremely high and our best guess is around 10^{-5}. One should stress that, as usual, this is very uncertain and the main uncertainty is the hydrogen column density. Another interesting point is that in a separate study, torsionally excited methanol was found in these two sources (Cesaroni et al.1991). In view of the probability that methanol is abundant on grain surfaces (see e.g. Grim et al. 1991), this seems to substantiate the general hypothesis that the abundance distribution in "hot cores" reflects that found in grain mantles and that the molecules observed in hot cores have recently been evaporated from mantles.

An obvious extension of this single dish work is to observe NH_3 with an interferometer. Five of the $NH_3(4,4)$ sources found by Cesaroni and Walmsley (1992) have recently been observed with the VLA (Kurtz and Churchwell, priv.comm.). Results from this study are very preliminary but already show that the sources of $NH_3(4,4)$ emission are in general very compact and, in some cases at least, are emitted from a small clump very close to the compact HII region. The true ammonia brightness temperatures are clearly quite high (at least 40 K towards G10.47+0.03) which is important because it restricts the possibility that the ammonia transitions are in some way being cooled much below the kinetic temperature and that the high optical depths are partially a consequence of this "super-cooling".

4. Conclusions

The data to date support the contention that one requires high densities (above 10^6 cm^{-3}) before drastic depletion sets in. This does not prevent considerable depletion and indeed chemistry on grain surfaces at lower densities. At higher densities, one needs observations with arc second angular resolution in order to avoid confusion. The new millimeter interferometers which are becoming available should be a considerable help in this respect. This is true whether one is examining "hot cores" or clumps such as found in NGC 2024.

It is worth pointing out also that if one wants to distinguish effects of gas phase and dust chemistry, one needs information about the abundances of many species. Even in the Orion hot core, the present state of our knowledge is very vague due to confusion between the well known "zoo" of velocity components in and around Orion-KL. For example, there is still no reliable estimate for the HCO^+ abundance in the hot core. To resolve this and analogous problems, it would be useful to make a spectral line sweep with a sensitive millimeter interferometer. Thus, fortunately, there are problems left for the future.

References

Brown P.D., Charnley S.B., Millar T.J. 1988 *Monthly Notices Roy. Astron. Soc.* , **231**,409.
Brown P.D., Millar T.J. 1989 *Monthly Notices Roy. Astron. Soc.* , **237**,661.
Cesaroni R., Walmsley C.M., Churchwell E. 1991 *Astron. Astrophys.* (in press).
Cesaroni R., Walmsley C.M. 1992 *Astron. Astrophys.* (in press).
Churchwell E., Walmsley C.M., Cesaroni R. 1990 *Astron. Astrophys. Suppl.* ,**83**,119.
Gaume R., Johnston K.J., Wilson T.L. 1991 *Astrophys. J.* (in press).
Genzel R., Stutzki J. 1989 *Ann. Rev. Astron. Astrophys.* **27**,41.
Grim R.J.A., Baas F., Geballe T.R., Greenberg J.M., Schutte W. 1991 *Astron. Astrophys.* **243**,473
Hartquist T.W., Williams D.A. 1990 *Monthly Notices Roy. Astron. Soc.* , **247**,343.
Hermsen W., Wilson T.L., Walmsley C.M., Henkel C. 1988 *Astron. Astrophys.* **201**,285.
Mauersberger R., Henkel C., Wilson T.L., Walmsley C.M. 1986 *Astron. Astrophys.* **162**,199.
Mauersberger R., Wilson T.L., Mezger P.G., Gaume R., Johnston K.J. 1991 *Astron. Astrophys.* (in press).
Mezger P.G., Chini R., Kreysa E., Wink J.E., Salter C.J. 1988 *Astron. Astrophys.* **191**,44.
Mezger P.G., Sievers A.W., Haslam C.G.T., Kreysa E., Lemke R., Mauersberger R., Wilson T.L. 1991 *Astron. Astrophys.* (in press).
Pineau des Forêts G., Flower D.R., Millar T.J. 1992 *Monthly Notices Roy. Astron. Soc.* (submitted)
Prasad S.S., Tarafdar S.P. 1983 *Astrophys. J.* , **267**,603.
Schilke P., Walmsley C.M., Mauersberger R. 1991 *Astron. Astrophys.* **247**,487.
Schulz A., Güsten R., Zylka R., Serabyn E. 1991 *Astron. Astrophys.* **246**,570.
Sievers A.W., Mezger P.G., Gordon M.A., Kreysa E., Haslam C.G.T., Lemke R. 1991 *Astron. Astrophys.* (in press)
Tielens A. 1989 p239 in *Interstellar Dust*, Proceedings of IAU Symposium 135, (ed. Allamandola L.J., Tielens A.G.G.M. ; publ. Kluwer).
Vogel S.N., Genzel R., Palmer P. 1987 *Astrophys. J.* , **316**,243.
Walmsley C.M. 1991 in *Fragmentation of Molecular Clouds and Star Formation*, IAU Symposium 147, (ed. E.Falgarone et al.)
Whittet D.C.B., Duley W.W. 1991 *The Astron. and Astrophys. Review* , **2**,167.
Wilson T.L., Walmsley C.M. 1989 *The Astron. and Astrophys. Review* , **1**,141.

QUESTIONS AND ANSWERS

A.Dalgarno: To oversimplify, if there is no depletion, as you concluded, how can there be a grain chemistry?

C.M.Walmsley: When I said "no depletion", I meant "no drastic depletion" (i.e. loss of a major fraction of CNO) in the NGC 2024 ridge. There certainly is some depletion as evidenced by solid state features seen towards IR sources.

M.Ohishi: We have observed vibrationally excited CH_3OH up to $v_t = 2$ ($E \sim 700\,K$) in many active star forming regions (M.Ohishi et al. in prep.). Our results show T_{rot} in one vibrational state and T_{vib} among vibrationally state are very similar, and we think vibrationally excited CH_3OH is mainly excited by collision. But you mentioned that vib. excited CH_3OH traces FIR. I would like to know the reason.

C.M. Walmsley: Probably both collisions and radiation play a role. The collisional rates between torsional states in CH_3OH is quite unknown as far as I can tell and the relative importance of collisions and radiation depends on this.

A.Leger: What is the core life time so that we can compare it to the accretion time on grains?

C.M.Walmsley: The cores have a "dynamical lifetime" defined by size and linewidth which is $\sim 3 \times 10^4$ yrs. This is probably a lower limit.

N.Evans: There certainly is a way to get some of the molecules off the grains. There are several hundred young stars in the NGC 2024 region and these will heat grain mantles in at least the outer parts of the clumps.

C.M.Walmsley: The extinction in the ridge is $\sim$100 mag and the radiation from the stars has to penetrate that.

J.M.Greenberg: T_d is uncoupled from T_g for $n_H < 10^\alpha$ $\alpha > 10$. I see no problem in having $T_d < T_g$ at the gas densities you are talking about. A simple way of looking at this (which I will confirm in detail later) is that at 16 K the dust is radiating energy at a rate > 0.1 eV (energy density) x 3_c^- x 10^{10} (velocity) = 3×10^9 and if gas is at 30 K it can impart energy to the dust at a rate $\sim$0.003 eV (energy per H) x n_H x $v_H \simeq 3 \times 10^{-3} n_H 10^4$ cm/s $\geq$ radiation rate if $n_H > 10^8$ cm^{-3}.

C.M.Walmsley: I think the dust will cool the gas rather than the gas heating the dust. The question is what is heating the gas.

THE ENVIRONMENT OF HIGH–MASS YOUNG STELLAR OBJECTS

Jean-Pierre Maillard
Institut d'Astrophysique de Paris
98b Boulevard Arago
Paris 75014
France

George F. Mitchell
Department of Astronomy
Saint Mary's University
Halifax, N.S., B3H 3C3
Canada

ABSTRACT. Using high-resolution infrared CO spectroscopy new results have been obtained on the environment of high-mass young stellar objects (YSOs). In particular, a new class of neutral, warm and dense outflows has been discovered. The infrared outflows appear a general phenomenon of the activity of high-mass YSOs. With well-defined and often multiple velocities they seem to correspond to episodic and violent mass-loss events from the central source. In addition, a shell of warm, quiescent gas, is formed near the massive star. All these dynamical elements influence the chemistry inside the giant molecular clouds. Beside CO in solid and in gas phase, detection in the infrared has been attempted of simple molecules like CH_4, C_2H_2, H_2O and ion H_3^+ toward few of these sources.

1. Introduction

Two types of site for star formation are distinguished, the Giant Molecular Clouds (archetype Orion) and the Dark Cloud Complexes (archetype Taurus-Auriga) (Wilking 1989). The latter type forms exclusively low and intermediate mass stars ($M_\star < 2M_\odot$). With high-mass stars ($M_\star \geq 5M_\odot$) we are dealing with giant clouds in the core of which those stars are grouped in clusters of few units (e.g. S140) or are sometimes single (e.g. M8E-IR). Even if the probability of finding a star with $M_\star = 5M_\odot$ is 400 times smaller than that of a typical star according to current estimates of the Initial Mass Function (Scalo 1990), these massive stars, less numerous, will have a determining impact on the evolution of the giant clouds. They are also the most brightest infrared sources embedded in molecular clouds. From this point of view they present a special interest for probing the gas and dust content of clouds on their line of sight.

This paper will focus on results obtained by infrared high-resolution spectroscopy, in particular from the CO band at 5 μm. By detecting lines in absorption, the infrared range does not give access to a chemistry as rich as in the millimetric range. The method presents, however, some specific advantages. The line of sight is observed at high angular resolution (≤ 1"), which avoids the dilution of small scale structures, important to study the close environment of the sources. Very precise information on the kinematics and the column density of the absorbing gas are deduced. Finally, symmetrical molecules not detectable in millimeter such as CH_4, are only observable that way. The review extends the pioneering work on BN to several other high-mass sources, showing various degrees of evolution, to build a picture of the YSOs environment.

2. Observations

The CO observations reviewed here were taken using the Fourier Transform Spectrometer at the Cassegrain focus of the Canada-France-Hawaii Telescope on Mauna Kea, mainly between 1987 and 1988. Details of the FTS can be found in Maillard and Michel (1982). The main part of the vibration-rotation band of ^{12}CO and ^{13}CO ($\Delta v=1$) was covered, with a spectral resolution of 0.05 cm^{-1}, corresponding at 4.7 μm to a velocity resolution of 8 km s^{-1}.

3. Presentation of the sources

Sources of Brα and Brγ in emission toward molecular clouds (Simon et al. 1983), were the first direct evidence of embedded, hot luminous young stars. The comparison of the intensities of the lines was in addition a way to estimate the amount of dust in the line of sight. The bipolar outflows in millimetric CO (Bally and Lada 1983) centred on YSOs, proved to be another characteristic phenomenon of the first stages of stellar evolution.

The most-studied source toward which the two features are detected is BN. With more CO mappings made, some objects have been found to be source of large scale outflows although the HI recombination lines are not detectable (e.g. GL 2591), and some deeply embedded sources, with infrared luminosity typical of a massive star exhibiting neither a bipolar outflow nor Brα (e.g. W33 A). These observational facts suggest different ages of the central source. W33 A could be considered in the most primitive state, while BN is more evolved, having already created a developed HII region. GL 2591 is in an intermediate situation. At the latest stage, the massive star disrupts its parental cloud and becomes visible. That is LkHα 101. Table I presents the main parameters of 11 massive sources observed in infrared CO, grouped according to this sequence. Two papers (Mitchell et al. 1990, and Mitchell, Maillard and Hasegawa, 1991) present in more details the results of this study.

4. Gas and dust properties

Analysis of the numerous CO line profiles provided by the high-resolution spectra, preferably ^{13}CO to remain in the optically thin approximation, makes it possible to

Table I: PARAMETERS OF THE YOUNG HIGH-MASS SOURCES

Source Name	distance kpc	L_{IR} $10^5 \times L_\odot$	Bα $10^{-16} Wm^{-2}$	Bγ	$S_{4.7\mu m}$ Jy	bipolar outflow
GL 2136	1.7	1.8	no	no	120	no
W33 A	3.7	0.3	no	no	10	no
S140 IRS1	0.9	0.24	no	no	42	yes
GL 2591	1.5	0.16	no	no	240	yes
W3 IRS5	2.3	4.2	no	no	40	yes
NGC 7538 IRS9	3.5	0.3	no	no	28	yes
BN	0.5	1.2	194	22	122	yes
GL 490	0.9	0.014	39	9	35	yes
M8E–IR	1.5	0.25	16	no	66	yes
NGC 7538 IRS1	3.5	1.5	50	no	40	yes
LkHα101	0.8	0.14	1200	420	170	no

Table II: RESULTS IN INFRARED CO

Source Name	solid CO	quiescent gas ^{12}CO	^{13}CO	Infrared outflows	Nb of comp.	PCyg profile
GL 2136	×	×	×	no		
W33 A	×	×	×	no		
S140 IRS1		×	×	yes	1	yes
GL 2591		×	×	yes	> 4	no
W3 IRS5	×	×	×	yes	3	yes
NGC 7538 IRS9	×	×	×	yes	2	yes
BN		×	×	yes	3	yes
GL 490		×		yes	1	no
M8E–IR		×		yes	≥ 5	no
NGC 7538 IRS1	×	×	×	yes	1	yes
LkHα101		×		no		

determine the temperature of the medium in which the lines are formed. Combined with the radial velocity and the density, one can determine the properties of the various absorbing regions in the line of sight. Three regions can be discriminated: the ambient molecular cloud, a quiescent high-density shell, the outflowing gas. The main results are gathered in Table II.

4.1 Molecular cloud

The column density in ^{12}CO (assuming a solar ^{12}C/^{13}C ratio) is between 1 to 2× 10^{19} cm^{-2}, the density of the gas on the average 10^5 cm^{-3}. The temperatures deduced from the infrared CO are in the range 25 to 65 K, consistent with other temperature determinations. The same spectral range gives access also to the absorption band of solid CO at 2140 cm^{-1}. Using solid CO abundances of Sandford et al. (1988) a solid-to-gas phase ratio can be estimated to be ≤ 10%, i.e., *most of the CO is in gas phase*. These results give typical conditions of giant molecular clouds.

4.2 Quiescent high-density shell

On the same lines of sight, the high-J lines are indicative of the presence of a warm quiescent gas, with a temperature different for each source, ranging from 120 to 1000 K, likely heated by gas-grain collisions. The density is higher than in the ambient cloud ($\geq 10^7$ cm^{-3}), and the radial velocity can be different by a few km s^{-1} toward the blue. By analogy with the spherical VLA images of ultra-compact HII regions (Churchwell 1990) that most of these sources are forming, the quiescent gas can be thought to be on a shell where the gas blown away by the central star is piling up. Future high-resolution infrared imaging should be able to confirm this view.

4.3 Neutral outflowing gas

That is a general phenomenon we observed for all the sources with developed bipolar outflows. However, this outflowing gas is distinct from the millimetric outflows. The velocity of the gas is high, from 15 to 160 km s^{-1}, and the kinetic temperature can reach up to 500 K. Several velocity components are often present, which suggests episodic mass-loss events. Remain to elucidate how they relate to the formation of the large scale sub-millimetric bipolar flows, but their concurrent existence suggests a possible link between the two types of outflow.

5. Search for other molecules in the infrared

The existence of a high-density and warm gas shell in the line of sight of the massive YSOs should be favorable to the possible detection in the infrared of neutral molecules in absorption, in gas phase, which locked onto dust are released by warm grains. Search for molecules like NH$_3$, OCS, HCN, C$_2$H$_2$ in Orion, is reviewed by Evans (1991). Complements for H$_2$O, CH$_4$ and H$_3^+$ are given here.

5.1 H$_2$O

The detection of four weak lines of the ν_3 band at 2.7 μm is reported by Knacke and Larson (1991) toward BN from the KAO. A ratio [H$_2$O gas]/[ice] ≤ 0.05 implies that *most of H$_2$O is frozen on grains* in OMC.

5.2 CH$_4$

The first positive detection of interstellar gaseous and solid methane has been claimed by Lacy et al. (1991), who have searched for this basic molecule at 7.8 μm toward several of the sources we observed in CO. Another possible region is 3.1 μm. The observations, because of the telluric methane, are difficult and the most convincing detection is obtained only toward NGC 7538 IRS9. The tentative conclusion is that, *methane is predominantly in solid phase*, with an abundance of solid CH_4 comparable to that of solid CO.

5.3 H_3^+

The H_3^+ ion is fundamental for the ion-molecule chemistry. Its abundance is also a probe of the cosmic ray ionization in molecular clouds. Several observations have been conducted toward BN, NGC 2024 IRS2, NGC 2264 and GL 2591 (Geballe and Oka 1989, Black et al. 1990) in the ν_2 band at $4\mu m$ where 6 lines might be usable. No positive detection in any molecular cloud has been reported yet. From new observations toward GL 2591, assuming the canonical value $N[^{12}CO]/N[H_2] = 10^{-4}$, taking the cold gas temperature of 38 K and the ^{12}CO column density of 1.1×10^{19} cm^{-2} (Mitchell et al. 1989), we derived an improved upper limit of

$$N[H_3^+]/N[H_2] < 2.7 \times 10^{-9}$$

To estimate how this limit is already a test of ion-molecule chemistry, a calculation similar to Herbst and Leung's prediction (1986) made for TMC1 should be conducted for the conditions of a giant molecular cloud.

6. Conclusion

By high-resolution absorption spectroscopy of CO, a large number of new results have been collected, which can be extended to more sources to better generalize the conclusions drawn from about ten sources. For the other important interstellar neutral molecules like H_2O, CH_4, C_2H_2, NH_3, and ion H_3^+, the detection is more difficult, and this research is just beginning. High-mass YSOs are the only infrared sources which give a chance of detecting them. More sensitive high-resolution spectroscopy of these molecules, will provide new insights on the chemical composition of the giant molecular clouds in which massive stars form.

References

Bally, J. and Lada, C.J., 1983, *Ap. J.*, **265**, 824.
Black, J.H, van Dishoeck, E., Willner, S.P., and Woods, R.C., 1990, *Ap. J.*, **358**, 459.
Churchwell, E., 1990, *Astron. Astrophys. Rev.*, **2**, 79.
Evans, N.J., 1991, this symposium.
Geballe, T.R., and Oka, T., 1989, *Ap. J.*, **342**, 855.
Knacke, R.F., and Larson, H.P., 1991, *Ap. J.*, **367**, 162.
Herbst, E., and Leung, C.M., 1986, *Ap. J.*, **310**, 378.
Lacy, J.H., Carr, J.S., Evans, N.J., Baas, F., Achtermann, J.M., and Arens, J.F., 1991, *Ap. J.*, **376**, 556.
Maillard, J.P., and Michel, G., 1982, in *Instrumentation for Astronomy with Large Optical telescopes*, IAU Col. No. 67, C. M. Humphries ed, Reidel, p. 213.
Mitchell, G.F., Curry, C., Maillard, J.P., and Allen, M. 1989, *Ap. J.*, **341**, 1020.
Mitchell, G.F., Maillard, J.P., Allen, M., Beer, R., and Belcourt, K., 1990, *Ap. J.*, **363**, 554.
Mitchell, G.F., Maillard, J.P., and Hasegawa,T.I., 1991, *Ap. J.*, **371**, 342.
Sandford, S.A., Allamandolla, L.J., Tielens, A.G.G., and Valero, G.J, 1988, *Ap. J.*, **329**, 498.
Scalo, J.M., 1990, in *Windows on Galaxies*, A. Renzini, G. Fabbiano and J. Gallagher eds, Kluwer-Dordrecht, p. 125.
Simon, M., Righini-Cohen, G., Felli, M., and Fischer, J., 1981, *Ap. J.*, **245**, 552.
Wilking, B.A., 1989, *Pub. A. S. P.*, **101**, 229.

QUESTIONS AND ANSWERS

A.Leger: How do you understand $T_{rot} = 10^3\ K$ in Jupiter?

J.P.Maillard: That is a completely different problem, not relevant to interestellar medium. I have presented this emission spectrum of H_3^+ on Jupiter just to show what the H_3^+ spectrum looks like and the lines we are looking for in absorption in molecular clouds. The mode of excitation of H_3^+ in Jupiter is totally different. It is an auroral phenomenon.

S.Sandford: (1) What is the central peak position and full-width-at-half-maximum of the solid CO feature in NGC 7538 IRS 9? (2) Are the orientations of the bipolar outflows understood?

J.P.Maillard: (1) This information is in the data. We could look at, but we have been more interested on the abundance of solid CO to compare to gas phase CO than do the position and the shape of the band.

(2) There is a systematic study we could conduct, as how the line of sight intersect the large scale flows for instance. We have new maps in $^{12}CO(2 \to 1)$ I presented four of these sources which can help. However, the bipolar flows from these massive sources are generally very open, then there is a large probability that the line of sight is within a flow. We have no source so far with a flow like the prototype are for low-mass objects L1551-IRS 5 which is just oriented perpendicular to the line of sight. With such a case we could check if there is an influence on the absorption CO profile.

INFRARED MOLECULAR SPECTROSCOPY OF ORION

NEAL J. EVANS II
Astronomy Department,
The University of Texas, Austin Texas 78712, USA
E-mail NJE@ASTRO.AS.UTEXAS.EDU

Abstract. Recent results on infrared spectroscopy of the cluster of sources in the Orion molecular cloud are discussed. These results imply very high abundances of CO, C_2H_2, HCN, and OCS, suggesting that grain surface chemistry has been important.

Keywords : infrared spectroscopy, interstellar molecules, molecular clouds

1. Introduction

This review will consider new results on the infrared cluster in the Orion molecular cloud. This region of massive star formation has a rich chemistry (see Genzel & Stutzki 1989 for a review). There seem to be several different regions with distinctive kinematics and chemical abundances. Of particular interest to this meeting are the hot core, where evidence suggests that molecules have recently evaporated from grain mantles (Blake et al. 1987), and the plateau feature, thought to be an outflow from IRc2, which has enhanced emission from sulfur compounds. Because of space limitations in this article, I will restrict myself to infrared absorption spectroscopy of interstellar molecules, leaving discussion of emission from atoms, ions, or H_2 to other reviewers.

Infrared spectroscopy of rovibrational lines of interstellar molecules allows one to determine the populations of many rotational levels with observations at very similar wavelengths. Since infrared lines can be observed in absorption against embedded infrared sources, one can obtain an effective resolution equal to the size of the infrared source, although one is limited to observations along the lines of sight to such sources. Comparison of infrared absorption lines and millimeter emission lines can provide valuable information on geometry and kinematics, since only the gas in front of the source is seen in absorption in the infrared. Determination of column densities from absorption studies does not depend on intensity calibration, since the column density is determined from the equivalent width, a quantity unaffected by calibration errors. Finally, one can detect symmetric molecules such as C_2H_2 (Lacy et al. 1989a) and CH_4 (Lacy et al. 1991), which have no permanent dipole moment, rendering them undetectable by rotational spectroscopy.

On the negative side of the ledger, most of the vibrational bands lie at λ between 3 and 15 μm, where atmospheric attenuation allows observations in only a few windows. The best of these windows, around 10 μm, is also compromised by interstellar attenuation by the silicate feature. The first of these problems could be solved by high-resolution spectrometers ($\lambda/\Delta\lambda \gtrsim 1 \times 10^4$) on large space telescopes, but none of the currently planned space missions have the right characteristics to exploit this opportunity.

2. The Observations

2.1. THE SOURCES

The dense molecular cloud behind the Orion Nebula contains a complex group of infrared sources. Extinction has a major effect on the appearance of the region even at fairly long infrared wavelengths (see the maps in Wynn-Williams et al. 1984). The most prominent source at $\lambda \lesssim 20\mu$m is the Becklin-Neugebauer Object (BN), while IRc2, suspected to be the most luminous embedded source, is most apparent in the maps at 7.8 and 12.5 μm, local minima in the interstellar extinction (Draine and Lee 1983). Recent images of this region at 12.4 μm (Gezari et al. 1991) show IRc2, IRc7, and IRc4 to be strong and distinct sources, with IRc4 clearly extended and complex. Emission at all $\lambda \lesssim 30$ μm shows a distinct drop to the south of IRc2 and east of IRc4, a fact explained by extremely high extinction associated with the hot core component, as mapped by NH_3 (J,K)=(3,2) emission (see Genzel and Stutzki 1989).

2.2. THE BECKLIN-NEUGEBAUER OBJECT

The pioneering work in infrared absorption spectroscopy of molecules in star-forming regions was the study of CO toward BN (Scoville et al. 1983). In addition to band-head emission in the overtone bands near 2.3 μm, they observed discrete absorption features in the fundamental band near 4.6 μm at velocities of -18, -3, and $+9$ km s^{-1}. The last velocity component was naturally attributed to the ambient cloud (or the ridge component), while the blue-shifted features were assumed to be material in an outflow. Both the -18 and $+9$ km s^{-1} features were fitted by a single temperature of 150 K. More recent work on CO absorption toward other sources is reviewed by Maillard in this volume.

2.3. IRC2, IRC7, AND IRC4

Because of the heavy extinction to these sources, absorption spectroscopy has been difficult until high-resolution spectrometers were developed for longer wavelengths (Lacy et al. 1989b). Using a cryogenic echelle with a velocity resolution of about 20 km s^{-1}, Lacy et al. (1989a) were able to detect C_2H_2 toward IRc2, using the ν_5 vibrational band near 13.5 μm. More recently, CH_4 has also been detected, using the ν_4 band at about 7.6 μm, toward IRc2, as well as toward several other sources (Lacy et al. 1991). These were the first detections of these molecules in interstellar clouds.

A map of C_2H_2 R(5) equivalent width (Evans, Lacy, & Carr 1991) shows a strong peak on IRc2, with more extended absorption encompassing IRc7 and IRc4, but increasing to the southeast, in the direction of the hot core, suggesting that both the hot core and the plateau feature (centered on IRc2) may have enhanced

C_2H_2 abundances. In addition to the heavily blended Q branch, a total of 5 well-separated, unresolved R branch lines have been seen toward IRc2, with an average velocity of -3 km s^{-1}, suggesting that they arise in the approaching edge of the low velocity outflow centered on IRc2. Analysis of these lines indicates that the lower J lines are quite saturated and thus narrow (inferred FWHM linewidth of 2.7 km s^{-1}). The saturation is confirmed by the detection of several R branch lines and the Q branch of $^{13}CCH_2$. After correcting for this saturation, Evans et al. (1991) derive a temperature of about 150 K for the gas in the C_2H_2 absorbing region.

A second band of C_2H_2, the combination band $\nu_4 + \nu_5$, was also detected during the study of CH_4 at 7.6 μm. Analysis of five lines in this weaker band toward IRc2 again indicates some saturation. The correction which gives the best fit to a single temperature results in $T = 140$ K and about three times higher column density of C_2H_2 than was derived from the ν_5 lines. Compared to the lines in the ν_5 band these lines are less blueshifted (average velocity of +7 km s^{-1}) and wider (FWHM $\Delta v = 5.8$ km s^{-1}). These differences may be understood in the context of an accelerating flow centered on IRc2. Because the extinction is considerably less at 7.6 μm than at 13.5 μm, absorption can be seen from closer to the source at 7.6 μm. This picture would explain the larger column density, while the velocity differences could be explained if the flow is accelerating in the region probed at 7.6 μm, but not in the region probed at 13.5 μm.

In the C_2H_2 spectra toward IRc2, there are also lines of the ν_2 band of HCN. In all, five lines of HCN have been detected from levels up to $J = 17$. The average velocity of these lines is 0.8 km s^{-1}. The populations of the levels up to $J = 14$ are consistent with a thermal distribution, while the population of $J = 17$ is clearly sub-thermal. Indirect evidence, based on the strength of a blend of the C_2H_2 R(3) line with the HCN R(8) line, indicates that the HCN lines are also narrow and somewhat saturated. Consequently, the equivalent widths were corrected for saturation, assuming the same intrinsic width as the C_2H_2 lines. The resulting temperature, based on J up to 14, is about 150 K, consistent with the temperature determined from C_2H_2. Models of excitation, using a large-velocity-gradient code, indicate that densities of 3×10^6 cm^{-3} to 1×10^7 cm^{-3} can account for the thermalization of levels up to $J = 14$ while also explaining the sub-thermal population of the $J = 17$ level.

Bands of OCS (ν_1), CO, ^{13}CO, and probably $C^{18}O$ were detected toward IRc2 in spectra centered around 4.9 μm. Overlapping lines complicated the analysis, but with 13 lines detected, OCS is clearly present. Levels up to $J = 32$ in OCS appear to be thermalized at $T \sim 140$ K. Based on analysis of two ^{13}CO lines and an assumed isotope ratio of 50, the CO column density is 1.5×10^{20} cm^{-2}, but this number is very uncertain, partly because the ^{13}CO lines have P-Cygni profiles.

3. Abundances and Implications for Astrochemistry

Using estimates of the extinction to BN or IRc2 and a conversion from A_V to N_{H_2}, one can estimate abundances of CO and (for IRc2 only) the other species.

For CO, the resulting abundance is quite high ($\geq 5 \times 10^{-4}$), implying that all the carbon must be in gaseous CO toward both BN and IRc2 (though the estimate for $N(CO)$ in the latter source is quite uncertain). This conclusion may imply that N_{H_2} has been underestimated or that C>O in this region. Additional support for a carbon-rich chemistry comes from the high abundance of HCN (5×10^{-7} to 5×10^{-6}) toward IRc2, which can be matched by models with C>O (e.g., Langer and Graedel 1989). The C_2H_2 abundance (3×10^{-7} to 6×10^{-6}) is best understood in terms of models where molecules have been frozen on grain mantles, preserving early time abundances to be released when newly-formed stars heat the grains (e.g., Brown, Charnley, & Millar 1988). Such a model has already been suggested for the hot core (Blake et al. 1987) and may also be relevant for the plateau. Evidence for a more active role for the grains is provided by the observations of a high abundance of solid CH_4, which suggests CH_4 formation in the grain mantles (Lacy et al. 1991).

Portions of this research were supported by grants from the National Science Foundation and the Texas Advanced Research Program.

References

Blake, G. A., Sutton, E. C., Masson, C. R., and Phillips, T. G. 1987, *Ap. J.*, **315**, 621.
Brown, P. D., Charnley, S. B., and Millar, T. J. 1988, *MNRAS*, **231**, 409.
Draine, B. T., and Lee, H. M. 1984, *Ap. J.*, **285**, 89.
Evans, N. J., II, Lacy, J. H., and Carr, J. S. 1991, *Ap. J.*, in press (Dec. 20).
Genzel, R., and Stutzki, J. 1989, *ARA&A*, **27**, 11.
Gezari, D. 1991, *Ap. J.*, in prep..
Lacy, J. H., Evans, N. J., II, Achtermann, J. M., Bruce, D. E., Arens, J. F., and Carr, J. S. 1989a, *Ap. J. (Letters)*, **342**, L43.
Lacy, J. H., Achtermann, J. M., Bruce, D. E., Lester, D. F., Arens, J. F., Peck, M. C., and Gaalema, S. D. 1989b, *Pub. Astr. Soc. Pac.*, **101**, 1166.
Lacy, J. H., Carr, J. S., Evans, N. J., II, Baas, F., Achtermann, J. M., and Arens, J. F. 1991, *Ap. J.*, **376**, 556.
Langer, W. D., and Graedel, T. E. 1989, *Ap. J. Suppl.*, **69**, 241.
Scoville, N. Z., Hall, D. N. B., Kleinmann, S. G., and Ridgway, S. T. 1983, *Ap. J.*, **275**, 201.
Wynn-Williams, C. G., Genzel, R., Becklin, E. E., and Downes, D. 1984, *Ap. J.*, **281**, 172.

QUESTIONS AND ANSWERS

J.P.Maillard: (1) The unique temperature you estimate seems to me not in contradiction with what I reported toward luminous YSOs. In opposite, it confirms the existence of a quiescent warm gas enriched in neutral molecules released from the warm grains. (2) However, the column densities you determine may be criticized by the large difference between the width of line profiles (< 2 km/s) required for fitting the data and your resolution of 20-25 km/s. In addition, different widths are required depending on the molecule. How can that be physically possible?

N.Evans: (1) Yes, I agree. Essentially we see only the warm, enhanced abundance region (hot core and/or plateau) and not the ambient cloud (ridge) in Orion. (2) Again, I agree that it is important to improve our resolution and John Lacy is proposing to do this. For the moment, we have to rely on indirect arguments for narrow lines, but I think they are convincing. I would admit that there are uncertainties whenever lines are unresolved, but the column density estimates are still probably more certain that most of those obtained from emission lines, which have problems with unknown beam filling.

CHEMICAL GRADIENTS IN THE ORION MOLECULAR CLOUD

H. UNGERECHTS, E. A. BERGIN, J. CARPENTER, P. F. GOLDSMITH,
W. M. IRVINE, A. LOVELL, D. MCGONAGLE, F. P. SCHLOERB, AND R. L. SNELL
Five College Radio Astronomy Observatory
University of Massachusetts, Amherst, MA 01003, USA

ABSTRACT. We are mapping 29 rotational transitions of 21 chemical and isotopic molecular species in the central Orion molecular ridge with Nyquist sampling using the new 15-element focal plane array receiver QUARRY on the FCRAO 14 m telescope. Our goal is to obtain complete, unbiased data sets for a study of the interrelated physics and chemistry in GMC cores.

It has become commonplace in astronomy to use the strength of molecular rotational transitions measured at millimeter wavelengths to deduce physical parameters of interstellar clouds, such as temperature, density, and structure. In almost all cases the assumption is made that the chemical composition of the source being studied is uniform. On the other hand, models of the chemistry of molecular clouds predict that the composition will depend on a variety of variables including evolutionary history and radiation flux. Clearly, if chemical gradients are present, they may seriously compromise efforts to deduce physical parameters from the strength of molecular emission lines.

We are investigating this problem by mapping the central Orion molecular cloud in a variety of molecular transitions. Apart from the standard tracers of molecular gas at low and high densities, CO and its isotopic variants and CS, we chose to map important ions and radicals, the isomeric pair HCN and HNC, CH_3OH, and C_3H_2. Physical conditions in the clouds can be estimated from the CO lines (temperature, total column density), the HC_3N line ratio (density), and the CH_3CCH line ratio (temperature). The maps have 360 pixels on a 0.41' grid and cover an area of 4.5 by 12 arc minutes. They include the Orion-KL region in the center, the northern region where many ions and radicals peak, and the molecular ridge near the ionization front in the South.

These maps confirm qualitatively that there are significant differences among the distribution of various species. Most molecules, like CO and CS, show the Orion ridge extending north-south with a maximum toward Orion-KL as well as the crescent-shaped ridge at the ionization front in the South. By comparison, N_2H^+ and the radicals, CN and C_2H, are rather weak toward Ori-KL and stronger to the Northeast. The intensity of HCN is strongly enhanced near KL but its isomer, HNC, does not have such a dominant peak. The maps of SO and CH_3OH show two major maxima, the stronger toward KL and the weaker about 1.5' south; at much lower intensity, the emission from both species is extended along the ridge.

We have started similar surveys in the Cepheus A and M17SW molecular clouds, and we plan to extend this work to a larger sample of molecular clouds spanning a wide range of physical and chemical conditions.

Contour Maps

Line intensity integrated over velocity for selected lines. Position offsets are relative to $\alpha_{50} = 5^h 32^m 46.8^s$, $\delta_{50} = -5°24'28''$ (Orion-KL). The lowest contour and the separation between contours in units of $K\,km\,s^{-1}$ are 80 and 40 for CO $J = 1 \to 0$; both are 4 for ^{13}CO, 0.5 for $C^{18}O$, 4 for CS $J = 2 \to 1$, 0.5 for $C^{34}S$, 0.2 for CH_3CCH $J = 6 \to 5$ $K = 1$, 4 for HCO^+ $J = 1 \to 0$, 0.2 for $H^{13}CO^+$, 1 for N_2H^+ $J = 1 \to 0$ $F_1 = 2 \to 1$, 1.5 for CN $N = 1 \to 0$ $J = 3/2 \to 1/2$ $F = 5/2 \to 3/2$ and $F = 3/2 \to 1/2$, 0.5 for C_2H $N = 1 \to 0$ $J = 1/2 \to 1/2$ $F = 1 \to 1$, 10 for HCN $J = 1 \to 0$, 1 for $H^{13}CN$, 2 for HNC $J = 1 \to 0$, 1 for HC_3N $J = 10 \to 9$, 1.5 for SO $J_K = 3_2 \to 2_1$, and 0.5 for CH_3OH $J_K = 2_0 \to 1_0 A^+$. For HCN and $H^{13}CN$ the line intensity was integrated over all hyperfine components.

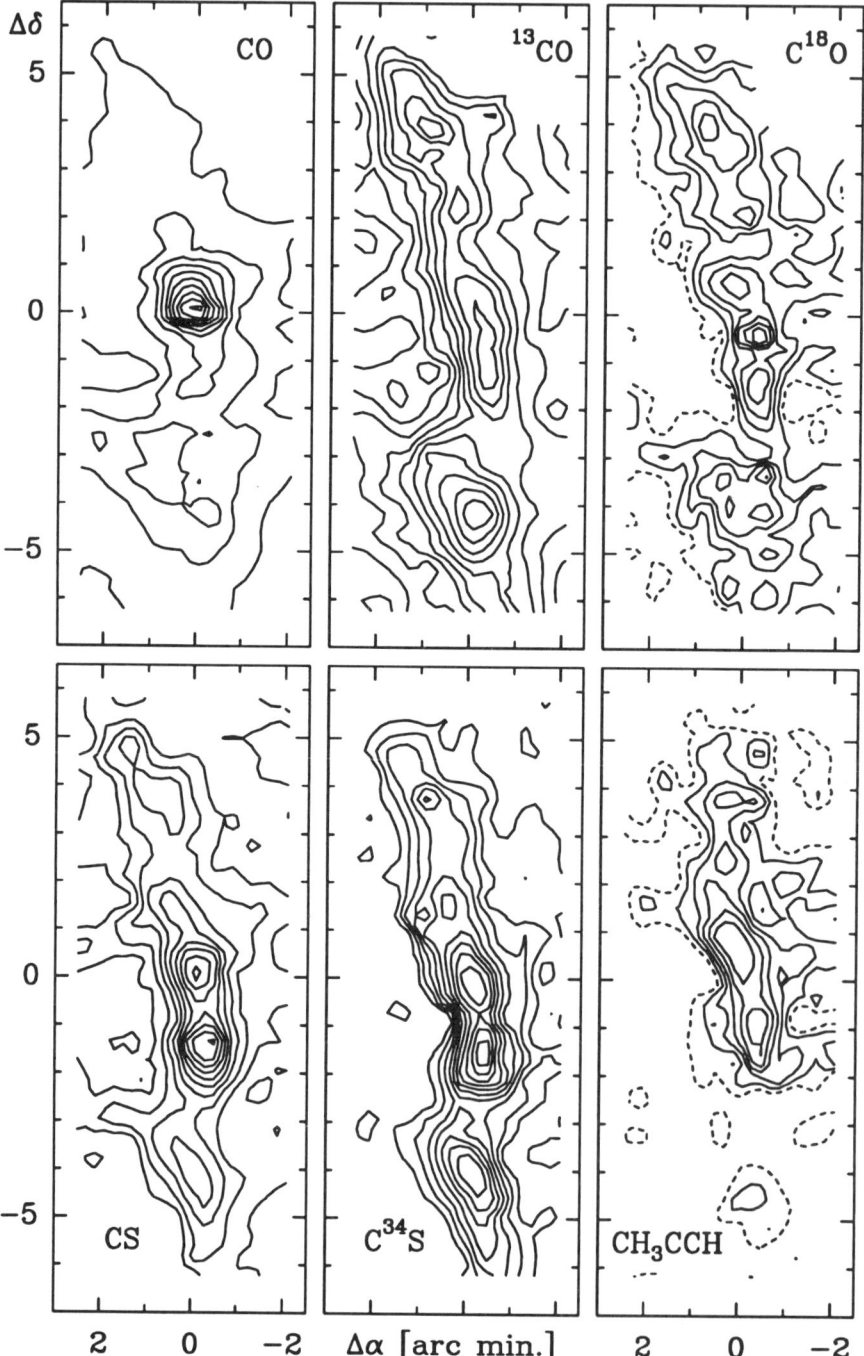

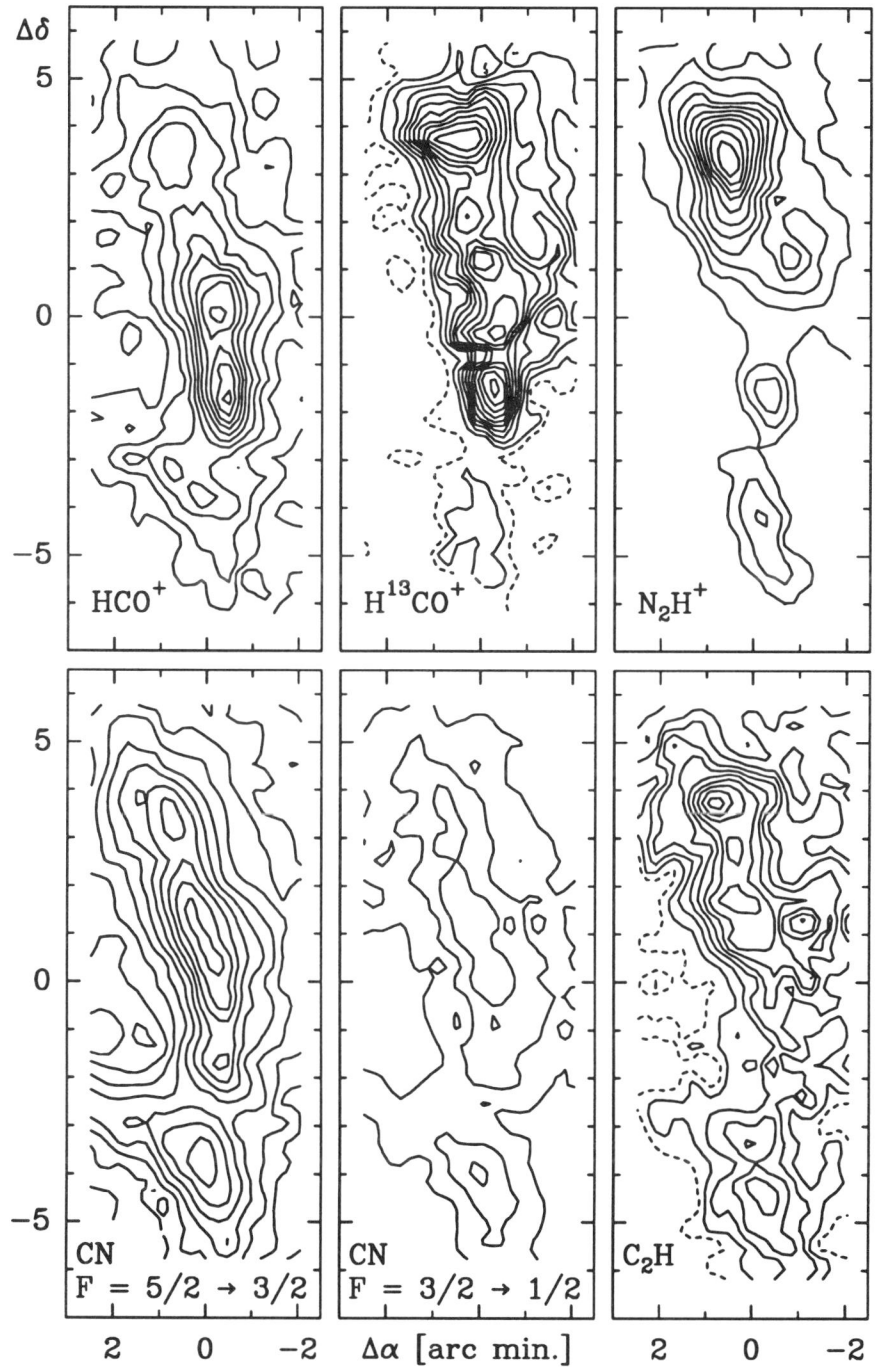

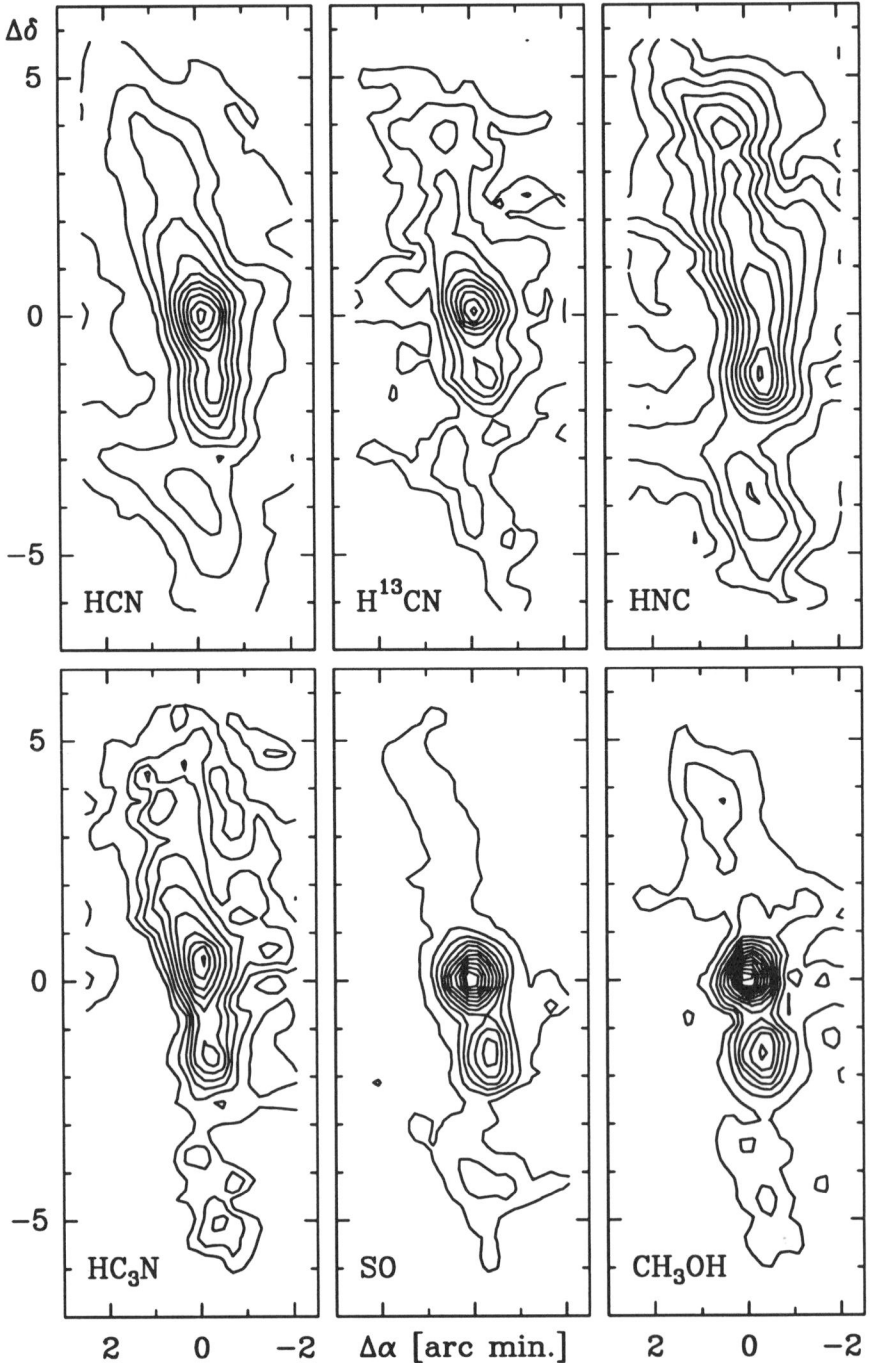

THE POLARIZED WATER MASER IN ORION

Z. Abraham[1] and J.W.S. Vilas Boas[2]
[1] IAG/USP, CP 9638, CEP 04301, São Paulo, SP, Brazil
[2] CRAAE/PTR, CP 8174, 05588, São Paulo, SP, Brazil

ABSTRACT. We present the results of 12 years of observations of the 8 km s^{-1} water maser in Orion. Four large flares in the flux density were observed, as well as variability in the degree and angle of polarization. The profile of the line is also variable, indicating the superposition of several sources.

OBSERVATIONS AND RESULTS

The observations obtained with the Itapetinga radiotelescope are presented in Fig. 1. The data up to 1987 were already discussed by Abraham et al. 1981, 1983 and by Vilas Boas and Abraham 1988. Since 1987 the intensity remained low but the line continues strongly polarized. Three small outbursts were observed in 1988, 1990 and 1991 as can be seen in Fig. 1. The shape of the line was not very different during the three outbursts, but the variation of the degree of polarization and polarization angle across the line changed drastically at the time of the outbursts. In April 1988 the degree of polarization decreased linearly within the line, from 53% at 7 km s^{-1} to 35% at 7.7 km s^{-1}. The polarization angle also changed linearly between these velocities, from -24° to +5°. Similar results were obtained in observations in 1983, 1986 and 1987 (Vilas Boas and Abraham 1988, Matveenko et al. 1988, Garay et al. 1989). In July 1990, when the second outburst occured, the degree of polarization was 65% at the center of the line and 55% at the half power width, the polarization angle remained constant across the line, at -5°. In April 1991, at the time of the third outburst, the degree of polarization decreased across the line between 60% at 7.2 km s^{-1} and 38% at 8 km s^{-1}. The polarization angle remained constant at -3°.

Several interpretations for the behaviour of the source were presented as new measurements became available (Abraham et al. 1986, Matveenko et al. 1988). Our results show that there is no correlation between the velocity of the components and the polarization angle, as claimed by Matveenko et al. (1986), since this relation changes at each outburst.

Aknowledgments: This work was partially supported by the brazilian agency CNPq.

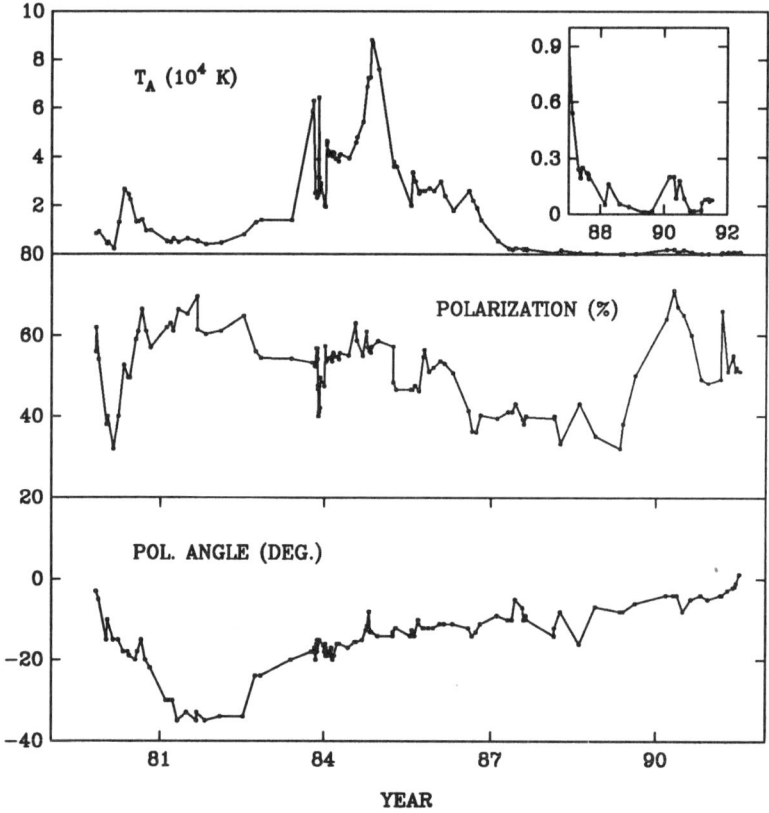

Figure 1. Antenna temperature, degree of polarization and polarization angle as a function of time for the 8 km/s water maser feature in Orion.

REFERENCES

Abraham, Z., Cohen, N.L., Opher, R., Raffaelli, J.C., Zisk, S.H. (1981), Astron. Astrophys. Letters **100**, L10
Abraham, Z., Vilas Boas, J.W.S., Del Ciampo, L.F. (1986) Astron. Astrophys. **167**, 311
Garay G., Moran, J.M., Haschick, A.D. (1989), Astrophys. J. **338**, 244
Matveenko, L.I., Graham, D.A., Diamond, P.J. (1988), Soviet Astron. Lett. **14**, 468
Vilas Boas, J.W.S., Abraham, Z. (1988), Astron. Astrophys. **204**, 239

OH MASERS IN OUTFLOW REGIONS

M. D. GRAY, D. FIELD AND R. C. DOEL
School of Chemistry
University of Bristol
Cantock's Close
Bristol BS8 1TS
U.K.

ABSTRACT. We combine a sophisticated model of maser propagation with a simple model of an accelerating molecular outflow and show that the observation of different OH maser frequencies is consistent with emission from different parts of the outflow.

1. Introduction

A combination of data on W3(OH) from Gaume & Mutel (1987) at 1720 MHz, Fouquet & Reid (1982) at 1665 and 1720 MHz and Baudry et. al. (1988) at 4765 Mhz show spatial coincidence of maser spots at these three frequencies. Demagnetized velocities for the strongest emission are: -42.0 kms^{-1} (1665 MHz); -43.2 kms^{-1} (4765 MHz) and -43.6 kms^{-1} (1720 MHz).

We attempt to model these observations as emission from different parts of an accelerating outflow from a young stellar object (YSO) by propagating model masers for short distances under different physical conditions which we calculate to be present along the flow.

2. The Model

2.1. MASER RADIATION TRANSPORT

Propagation of the partially coherent maser radiation is treated according to the semi-classical theory of Field and Gray (1988). This theory includes the processes of saturation and competitive gain, where different masers compete for the same molecular level populations.

2.2. PUMPING

The masers are assumed to be pumped by a combination of kinetic collisions, background continuum radiation from dust and by FIR line radiation, subject to line overlap, which is treated according to the theory of Doel, Gray and Field (1990). This theory parameterises overlap in terms of a constant velocity gradient and an 'overlap length', essentially a distance over which the velocity gradient is maintained. Transport of the pumping radiation is treated under the Sobolev (LVG) approximation (e.g. Castor 1970).

2.3. OUTFLOW

We model the outflow very simply, assuming it to be a plane-parallel flow, with a constant, accelerative velocity gradient in the direction of the flow. Model parameters such as the molecular hydrogen number density and the kinetic temperature are close to consistent with a Joule - Thomson expansion of the gas. The proportion of OH is assumed constant throughout and equality of the gas and local dust temperatures is maintained.

In a constant velocity gradient model, velocity shift along the outflow is proportional to distance along the outflow; the velocity shift in turn controls the amount of overlap. Increasing distance dilutes the radiation from an external dust component which is associated with the YSO.

3. Results

3.1. CONDITIONS

Table 1 shows the two sets of model parameters corresponding to positions A and B in the flow (see Figure 1) and a third set for a case with no FIR line overlap (see Figure 2).

Table 1. Physical Conditions			
Parameter	Region A	Region B	Figure 2
H_2 number density	6.0×10^6 cm^{-3}	3.0×10^6 cm^{-3}	6.0×10^6 cm^{-3}
External Dust Temperature	200K	200K	100K
External Dust Dilution	0.25	0.025	0.25
Local Dust Temperature	150K	75K	50K
Kinetic Temperature	150K	75K	50 K
OH Proportion	10ppm	10ppm	10ppm
Velocity Shift	0.8 kms^{-1}	2.5 kms^{-1}	0.0 kms^{-1}

The conditions A represent an early point in the flow, relatively close to the YSO, while

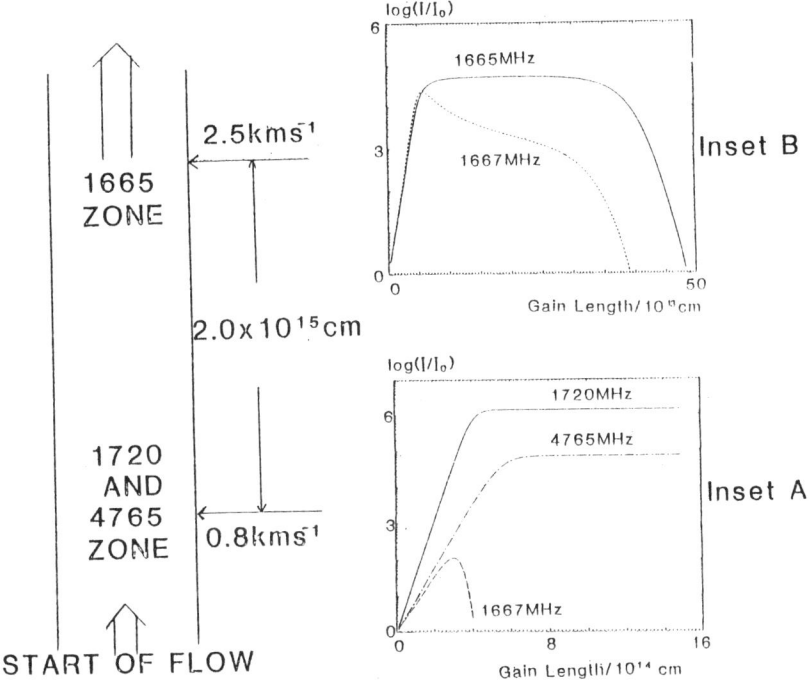

Figure 1: A schematic view of the flow, with maser growth curves appropriate to regions A and B inset

the conditions B correspond to a more distant region, with considerably more overlap.

3.2. ACTIVE MASERS

We see that at point A we get strong maser emission in the lines 1720 MHz and 4765 MHz while at point B, we see only 1665 MHz strongly (insets A and B in Figure 1). Note that strong 1667 MHz may readily form in the absence of line overlap (Figure 2) but it falls swiftly back into absorption in both sets of conditions A and B.

4. Discussion and Conclusion

The observed spatial correlation in the plane of the sky between masers at 1665 MHz, 1720 MHz and 4765 MHz is clearly reproduced by our model. We have also shown that the

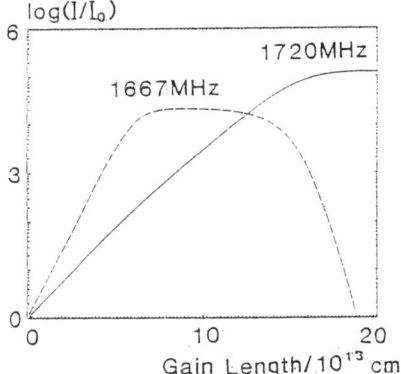

Figure 2: A maser growth curve showing strong emission at 1667MHz. Conditions are given in column 4 of Table 1.

1665 MHz emission appears from a physically different region at a different velocity from the other two lines again as observed. A general conclusion may be drawn that strong 1665 MHz spots are indicative of outflows whereas 1667 MHz spots are indicative of regions in which well defined velocity fields are absent.

5. References

Baudry, A., Diamond, P., Booth, R., Graham, D. and Walmsley, C. M.
1988. Astron. & Astrophys., **201**, 105.

Castor, J. I.
1970. Mon. Not. Roy. Astron. Soc., **149**, 111.

Doel, R. C., Gray M. D. and Field, D.
1990. Mon. Not. Roy. Astron. Soc., **244**, 504.

Field, D. and Gray M. D.
1988. Mon. Not. Roy. Astron. Soc., **234**, 353.

Fouquet, J. E. and Reid, M. J.
1982. Astronomical Journal, **87**, 691.

Gaume, R. A. and Mutel, R. L.
1987. Ap. J. Supp. **65**, 193.

ABUNDANCES OF REFRACTORY ELEMENTS IN THE ORION NEBULA

R.H. Rubin, E.F. Erickson, M.R. Haas, S.W.J. Colgan, and J.P. Simpson
NASA Ames Research Center, M.S. 245 − 6, Moffett Field, CA 94035 USA.
and R.J. Dufour
Dept. Space Physics &Astronomy, Rice University, Houston, TX 77251 USA.

ABSTRACT. We assess the gas-phase abundances of Si, C, and Fe from our recent measurements of Si^{++}, C^{++}, and Fe^{++} in the Orion Nebula by expanding on our earlier "blister" models. The Fe^{++} 22.9 μm line measured with the KAO yields Fe/H $\sim 3 \times 10^{-6}$ – considerably larger than in the diffuse ISM, where relative to solar, Fe/H is down by ~ 100. However, in Orion, Fe/H is still lower than solar by a factor ~ 10. The C and Si abundances are derived from new IUE high dispersion spectra of the C^{++} 1907, 1909 Å and Si^{++} 1883, 1892 Å lines. Gas-phase Si/C = 0.016 in the Orion ionized volume and is particularly insensitive to uncertainties in extinction and temperature structure. The solar value is 0.098. Gas-phase C/H = 3×10^{-4} and Si/H = 4.8×10^{-6}. Compared to solar, Si is depleted by 0.135 in the ionized region, while C is essentially undepleted. This suggests that most Si and Fe resides in dust grains even in the ionized volume.

1. Model for Orion Nebula

The model includes a detailed ionization and thermal equilibrium calculation for the ionized gas with an axisymmetric (2-dimensional) geometry. Details and results for the Orion Nebula blister model are in Rubin *et al.* (1991a = RSHE, 1991b). Here we emphasize the new work on fitting recent International Ultraviolet Explorer Satellite (IUE) and Kuiper Airborne Observatory (KAO) data to determine elemental abundances for C, Si, and Fe.

2. Abundances of Refractory Elements

2.1. Si/C RATIO

Silicon and carbon are major constituents of interstellar grains. Their gas-phase abundance ratio may be determined more reliably in nebulae than the ratio of either C or Si relative to H. This is the case presently for the ionized volume of Orion. In our model of the Orion Nebula, the dominant ionization states for C and Si are C^{++} and Si^{++} with fractional ionizations (RSHE) of 0.59 and 0.79. The important measurable UV lines from these species – Si III] 1883,92 and C III] 1907,09 Å – arise from energy levels comparably above ground at ~ 6.57 and 6.50 eV. Therefore the Si/C abundance ratio derived from observations of these lines is extremely insensitive to errors in the electron temperature, T_e, distribution. Also, because the critical densities for these lines are well above the highest density in the model, their volume emissivities have essentially the same dependence on density. Hence, the ratio is insensitive to errors in the density structure. Additionally, because of the proximity of the wavelengths, differential extinction corrections will play a negligible role in the determination of Si/C. In Figure 1, we show the best fit of Si/C = 0.016 to our new IUE high-dispersion data. These new observations were made at positions to avoid the bar to the SE (which the model does not address) and greatly expand what had been available. We note that there are virtually parallel arguments for using Si II] 2335-50 Å and C II] 2324-29 Å to derive Si/C. However, according to our model, the fractional ionizations for $<Si^+>$ and $<C^+>$ are 0.14 and 0.41. None of these Si II lines are measurable in the IUE spectra.

2.2. C/H RATIO

Important lines for determining C/H are C III] 1907,09 Å and the ratio of their sum to Hβ. We also used the ratio of the C II] 2326.1 Å ($^4P_{5/2} \rightarrow {}^2P_{3/2}$) line to H$\beta$. This line is the strongest of the 5 components that comprise the multiplet. At high dispersion, some of these are resolvable with IUE. Extinction is applied using C(Hβ) derived from Hα and Hβ imagery coincident with the actual IUE field observed and the Orion reddening function f(λ). This is from Walter (1991) and is similar in the optical to Torres-Peimbert *et al.* (1980) and in the UV with Bohlin & Savage (1981). Nevertheless, the differential extinction correction over such a large wavelength difference as well as the sensitivity to uncertainty in the T_e distribution renders the C/H abundance ratio more uncertain than Si/C.

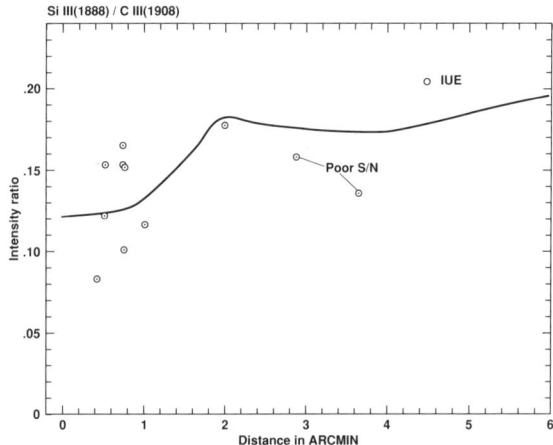

Figure 1. Model predictions for Si III (1883 + 1892 Å/ C III (1907 + 1909 Å) intensity ratio vs. projected distance from the center (θ^1 Ori C). We use the best-fitting model, as described in RSHE with the best fit to our new IUE observations determined for Si/C = 0.016. The data at distances larger than 2.5' have poor S/N and are not used for fitting.

Our C/H is determined predominantly on the basis of the C III data from the dominant state of C. There is a problem with the model in that the typical off-axis electron density, N_e, values are too high in the singly ionized region (RSHE). Furthermore, the agreement of the N_e-sensitive ratio C III] (1907/1909) with the data is excellent. Tentatively, our best fit obtains for C/H = 3×10^{-4}. This value of C/H in the model overestimates C II (2326.1)/Hβ. This could be explained in a number of ways: **a)** The model is overestimating T_e in the C II region (RSHE); **b)** There could be a larger average extinction for the 2326 Å line, which is concentrated farther from the observer than the average for the Balmer lines; **c)** There could be a lower fractional ionization for C II; **d)** There could be an actual decrease in the gas-phase C abundance due to the fact that the C II region is both closer to the presumed source of grains in OMC 1 and further from the exciting stars. Because a lower C/H ~ 1.7×10^{-4} is inferred, this might indicate that the difference is due to less destruction of grains in the C$^+$ zone compared with the C^{++} zone.

2.3. Si/H RATIO

Based on the Si III] 1883,92 Å lines and using the results from Si/C and C/H in the above sections, we find Si/H = 4.8×10^{-6}. We repeat that this ratio is not as reliable as Si/C. This value may be compared with Si/H inferred by Haas *et al.* (1986) from observations of [Si II] 34.8 μm near the Trapezium. When they attribute all of the 34.8 μm emission to the photodissociation region (PDR), they find Si/H = 2.6×10^{-6}. Using RSHE, and adjusting that calculation for Si/H = 4.8×10^{-6} in the ionized volume, this model predicts 0.085 of the 34.8 μm emission arises in the H II region. Applying this correction to the PDR abundance, we find Si/H = 2.4×10^{-6}. Taken at face value, the conclusion is that about a factor of 2 enhancement in gas-phase Si has occurred in the ionized volume. Because the PDR, as the interface with OMC 1, would be the source of fresh grain (and gas) material, this implies some of the dust is destroyed in the harsher environment of the H II region. Based on a solar Si/H of 3.55×10^{-5} (Anders & Grevesse 1989), the depletions in the PDR and H II region are 0.068 and 0.135. Hence, if the total Si abundance were solar in Orion, most of the Si is locked in grains even in the ionized volume.

2.4. Fe/H RATIO

By far the dominant species of iron *observable* in H II regions is Fe III. In the present Orion model, the fractional ionization of Fe^{++} is 0.41 (Rubin *et al.* 1991b). The [Fe III] 22.9μm line is the first Fe line detected in the far-infrared from an H II region (Erickson *et al.* 1989). The statistical equilibrium computation for the populations of the lowest 17 energy levels uses the collision strengths of Berrington *et al.* (1991). Based on recent KAO observations of [Fe III] 22.9 μm centered on θ^1 Ori C and the Orion model, we derive Fe/H. The tentative flux observed with the KAO using the cooled grating spectrometer with full width half maximum beam of 22″, is $\sim 1.5\times10^{-18}$ W/cm^2. The best fit to this line occurs with Fe/H $\sim 3\times10^{-6}$. We note that the 22.9 μm line is predicted to be the brightest Fe III line in Orion and that the derived Fe/H is very insensitive to T_e-structure uncertainty or extinction. This ratio may be compared with the solar value. According to Anders & Grevesse (1989), Fe/H = 4.68×10^{-5}, while Holweger *et al.* (1990) find a lower solar ratio of Fe/H = 3.02×10^{-5}. Thus in Orion, the depletion of iron is in the range 0.064 – 0.099.

In Orion, gas-phase iron is considerably more abundant than in the diffuse ISM. Van Steenberg & Shull (1988), using Fe II lines in IUE spectra along the line of sight to 12 stars in Ori OB1, find an average depletion for Fe of 0.0087. However, in Orion, Fe/H is still lower than solar by a factor ~ 10 which may indicate most Fe resides in dust grains even in the ionized volume.

References

Anders, E., & Grevesse, N. 1989, *Geochim. Cosmochim. Acta*, **53**, 197.
Berrington, K.A. et al. 1991, *J. Phys. B*, submitted.
Bohlin, R.C., & Savage, B.D. 1981, *ApJ*, **249**, 109.
Erickson, E.F., Haas, M.R., Simpson, J.P., Rubin, R.H., & Colgan, S.W.J. 1989, *BAAS*, **21**, 1156.
Haas, M.R., Hollenbach, D., & Erickson, E.F. 1986, *ApJL*, **301**, L57.
Holweger, H., Heise, C., & Kock, M. 1990, *Astron. Astrophys.*, **232**, 510.
Rubin, R.H., Simpson, J. P., Haas, M. R., & Erickson, E. F. 1991a, *ApJ*, **374**, 564 (RSHE).
Rubin, R.H., Simpson, J. P., Haas, M. R., & Erickson, E. F. 1991b, *PASP*, **103**, 834.
Torres-Peimbert, S., Peimbert, M., & Daltabuit, E. 1980, *ApJ*, **238**, 133.
Van Steenberg, M.E., & Shull, J.M. 1988, *ApJ*, **330**, 942.
Walter, D.K. 1991, *PASP*, **103**, 830.

PANEL DISCUSSION: THE CO/H_2 ABUNDANCE RATIO

E.F. van Dishoeck[1], A.E. Glassgold[2], M. Guélin[3], D.T. Jaffe[4], D.A. Neufeld[5],
A.G.G.M. Tielens[6] and C.M. Walmsley[7]

[1] *Leiden Observatory, P.O. Box 9513, 2300 RA Leiden, The Netherlands*
[2] *Dept. of Physics, New York Univ., New York 10003, USA*
[3] *IRAM, Univ. de Grenoble, 38406 St. Martin-d'Hères, France*
[4] *Dept. of Astronomy, Univ. of Texas, Austin, TX 78712, USA*
[5] *Dept. of Astronomy, Johns Hopkins Univ., Baltimore, MD 21218, USA*
[6] *NASA-Ames Res. Center 245-3, Moffett Field, CA 94035, USA*
[7] *MPI für Radioastronomie, Auf dem Hügel 69, D-5300 Bonn 1, Germany*

ABSTRACT. The observational and theoretical information on the CO/H_2 abundance in a variety of astrophysical regions including diffuse clouds, dense star-forming regions, shocked gas and circumstellar envelopes is discussed and reviewed.

1. Introduction

The determination of the total amount of molecular hydrogen from measurements of CO millimeter emission lines in galactic and extragalactic objects is one of the more controversial topics in astronomy. Most of the debate has centered on the so-called CO/H_2 conversion factor, which relates the measured integrated CO $J=1-0$ antenna temperature to the total column density of H_2, and many reviews and papers have appeared on this topic (see e.g. Dickman et al. 1986; van Dishoeck & Black 1987; Israel 1988; Bloemen 1989; Elmegreen 1989; Maloney 1990; Solomon & Barrett 1991; Henkel et al. 1991; Combes 1991; Young & Scoville 1991). The discussion has largely been of an empirical nature, however, with little attention to the variation of the actual CO/H_2 *abundance* in the various astrophysical environments. The implicit assumption is often made that the CO/H_2 abundance is more or less constant and equal to the "canonical" value of 10^{-4}. However, it is known (e.g. from the *Copernicus* satellite data) that this assumption is not always true and that there are regions where hydrogen is mainly molecular but CO has a very minor proportion of the solar carbon abundance.

Since astrochemists are in a better position to explore the chemical questions rather than the empirical relations, we focussed the panel discussion on the actual CO/H_2 abundance, with discussion of the empirical CO/H_2 conversion factor only where relevant. The panel members were asked to make a brief statement about the current observational evidence for the CO/H_2 abundance in a variety of regions and to address questions such as: What are the theoretical expectations? What are the prospects for future observational tests of the models? If CO is not a good tracer of H_2 in some regions, which other species could be useful? This paper is an edited (rather than verbatim) version of the comments made by the panel members and of some of the questions raised by the audience.

2. Direct and indirect measurements

2.1. Direct measurements

Direct observations of both CO and H_2 are limited to only a few specific interstellar regions.

TABLE 1. Direct measurements of CO/H_2 in interstellar clouds

Cloud	$N(CO)$ (cm^{-2})	$N(H_2)$ (cm^{-2})	CO/H_2	$CO/[C]^a$	Method	Ref.
π Sco	1(12)	2(19)	5(-8)	4(-6)	UV abs	1
ζ Oph	2(15)	4(20)	5(-6)	4(-3)	UV abs	2
NGC 2024 IRS2	8(18)	<1(23)	>8(-5)	>0.10	IR abs	3
NGC 2264 IRS	5(18)	<1(23)	>5(-5)	>0.06	IR abs	4
Orion/KL shock	4(17)	3(21)	1.2(-4)	0.16	FIR em	5

^{a} The reference abundance of carbon in all forms (gas and solid; atomic and molecular) has been taken to be the solar value of $[C]/[H]=4\times 10^{-4}$ (Grevesse et al. 1991). About 60±20% of the carbon is thought to be in solid form.

References: 1. Jenkins et al. 1989; 2. Morton 1975; 3. Black & Willner 1984; 4. Black et al. 1990; 5. Watson et al. 1985.

In diffuse clouds, both molecules can be seen by their absorption lines at far–ultraviolet wavelengths superposed on the spectra of bright background stars. Since the lines are saturated, high spectral resolution and high S/N such as provided by the *Copernicus* satellite are necessary to derive reliable column densities. The smallest CO column density detected with *Copernicus* is about 10^{12} cm^{-2} toward the star π Scorpii, whereas the largest column of about 2×10^{15} cm^{-2} was found toward ζ Ophiuchi. The observational data for these two stars are summarized in Table 1, which is an updated version of the table presented by van Dishoeck & Black (1987). The measured CO abundance, $N(CO)/N(H_2)$, varies by nearly two orders of magnitude for the various diffuse clouds. In terms of the total available carbon in the cloud, however, only a small fraction appears to be in the form of the CO. Even in the ζ Oph cloud, $CO/[C] \lesssim 0.01$. In the near future, high resolution ultraviolet observations of CO in thicker translucent clouds (A_V >1 mag) should be possible with the Goddard *High Resolution Spectrometer* on board the *Hubble Space Telescope* (HST). At the time of writing it was not yet clear, however, whether any direct observations of H_2 in such clouds would also be possible with HST.

Direct observations of CO and H_2 can also be made in thick molecular clouds through absorption lines at near–infrared wavelengths against embedded infrared sources. Measurements of CO lines in the $v=1\leftarrow 0$ and $2\leftarrow 0$ vibrational bands at 4.6 and 2.3 μm have been reported for a number of sources, but only for the cases of NGC 2024 IRS2 and NGC 2264 IRS have simultaneous limits been obtained on lines of H_2 in the $v=1\leftarrow 0$ band at 2.1–2.2 μm. The results are summarized in Table 1 and indicate CO/H_2 abundances consistent with the canonical value of $CO/H_2 \approx 10^{-4}$ usually assumed for dense clouds. With improved cryogenic echelle infrared spectrometers, the actual detection of these H_2 absorption lines seems possible in the near future. Extension of this method to a larger number of clouds will also be very important, since the regions probed so far are warm star–forming clouds, where molecules may have been released from the grains (see also §5).

Finally, both CO and H_2 can be observed by their high-J rotational and vibrational transitions in emission in very warm, disturbed gas such as the Orion/KL shocked region. The derived CO/H_2 abundance is 1.2×10^{-4}, consistent with the limits found above. This last determination is less "clean", however, since the high-J CO lines refer to gas with $T \approx 750$ K, whereas the H_2 $v=1\rightarrow 0$ lines arise in gas with $T \approx 2000$ K, so that it is not obvious whether the same volume of gas is sampled. Observations of the pure rotational lines of H_2 may provide a better comparison. In dense clouds, gaseous CO accounts for a substantial fraction of the total available carbon, CO/[C] $\gtrsim 0.1$.

2.2. Indirect measurements

In most cold interstellar clouds, H_2 is not directly observable so that tracer molecules like CO need to be used to determine molecular masses. Although the millimeter lines of CO are easily detectable, they are difficult to interpret in terms of column density because the lines are almost always optically thick. Various methods have been put forward to determine empirically a relation between the integrated ^{12}CO 1→0 line intensity I_{CO} and the H_2 column density $N(H_2)$. For completeness, we summarize them here briefly (see above references for more details and possible pitfalls):

- I_{CO} vs A_V: determine A_V from star counts; convert A_V to $N(H_2)$ using the gas to extinction ratio derived for diffuse clouds, assuming that this same relation also holds for denser clouds. Since the clouds may contain a non-negligible fraction of atomic hydrogen, an independent measurement of $N(H)$ is needed.

- $N(^{13}CO)$ vs A_V: determine ^{13}CO column densities from observations of optically thin lines assuming LTE. Determine $N(H_2)$ from A_V as above. $N(^{12}CO)$ can be found by adopting a $^{12}CO/^{13}CO$ abundance ratio. This conversion factor has been calibrated for clouds in which most gas-phase carbon is in CO.

- *Virial theorem:* assume that the cloud is in virial equilibrium, and use the observed line width and size to determine its mass. This leads not only to the proportionality factor, but also to the important result that $X = N(H_2)/I_{CO} \propto n_H^{1/2}/T_{ex}$, where T_{ex} is the excitation temperature of CO. A major problem in applying the virial theorem is the complicated morphology and small-scale structure of most clouds.

- *Gamma ray method:* compare the galactic distribution of diffuse gamma radiation, resulting from interactions of cosmic rays with hydrogen nuclei, with maps of H I and ^{12}CO $J=1\to 0$ emission to derive X. This method is biased toward GMCs in the inner Galaxy.

- *IRAS 100 μm:* compare the distribution of IRAS 100 μm emission, originating from dust grains in the clouds, with maps of H I and CO $J=1\to 0$ emission to determine X. This method assumes not only that the dust emissivity properties are known with sufficient accuracy, but also that they are similar in diffuse atomic and molecular gas.

A common characteristic of these methods (except the second one) is that the CO/H_2 abundance does not enter explicitly in the analysis, and that they refer to global scales of order one square degree or larger. Nevertheless, it is remarkable that these completely different and independent methods yield a similar conversion factor, $X = N(H_2)/I_{CO} \approx 2.5 \times 10^{20}$ cm^{-2} K^{-1} km^{-1} s within a factor of two. The largest deviations from this value are found for the Galactic Center (Blitz et al. 1985) and for some (but not all!) high-latitude clouds (de Vries et al. 1987), where X appears an order of magnitude smaller. For clouds in the outer Galaxy (Digel et al. 1990) and in the Magellanic Clouds (Cohen et al. 1988), the conversion factor may be significantly larger (see §4).

3. Diffuse and translucent clouds and PDRs

Table 1 illustrates the fact that carbon is still mostly in atomic form (as C^+) in diffuse clouds with $A_V \lesssim 1$ mag. For the thicker translucent clouds with $A_V \approx$1–5 mag, carbon is gradually transformed into CO, and the CO abundance increases by two orders of magnitude in this regime (van Dishoeck & Black 1988). As Figure 1 of van Dishoeck (this volume) shows, the exact location of the transition depends on the physical parameters of the cloud, such as temperature, density, column density and incident radiation field. Any conversion factor which assumes that the CO/H_2 abundance is constant at about 10^{-4} is therefore likely to fail for translucent clouds. This applies in particular to the ^{13}CO method, which has been demonstrated to lead to an underestimate of $N(H_2)$ by an order of magnitude in some cases (Gredel et al. 1991). Observational tests of this transition zone should be possible through absorption line studies with the *HST*.

Since the CO abundance varies so drastically for translucent clouds, it may be better to turn to other molecules whose abundances are nearly constant. A good candidate is CH, for which the column densities have been shown to scale linearly with those of H_2 up to A_V of a few (Danks et al. 1984; Mattila 1986).

A similar transition of carbon from atomic to molecular form occurs for denser, warmer photon–dominated regions (PDRs) (Tielens & Hollenbach 1985). The main difference with the translucent clouds is that the transition occurs deeper into the cloud, especially if the gas has a "clumpy" structure. Since such PDRs are ubiquitous and may contribute significantly to the total emission from galaxies, it is important to calibrate the CO/H_2 ratio for these regions separately. Unfortunately, such a measurement will prove extremely difficult, because of the varying abundances with depth into the PDR. In particular, hydrogen will become molecular around $A_V \approx 2$ mag, but the C^+–C–CO transition does not occur until $A_V \approx 4$ mag. These spatially distinct zones within the PDR will have vastly different temperatures (≈ 500 K versus 50 K) and this will complicate the derivation of the CO/H_2 ratio considerably. Probably the ideal source to examine is a nearby cloud–edge with a strong source of external UV radiation, where these zones within the PDR can be spatially resolved. Besides this, at least two other limitations confront the observations:

(1) *Direct observations of warm H_2.* The large energy–level spacing and lack of permanent dipole moment make H_2 simultaneously difficult to excite and difficult to observe. Observations of the ground–state pure rotational quadrupole lines provide the best method for determining the H_2 content of the PDR's. At current instrumental sensitivities, H_2 column densities of $< 10^{21}$ cm^{-2} can be detected in the (0–0) 3→1 and 4→2 transitions at 17 and 12 μm, respectively, if the gas temperature is ≥ 300 K (Parmar et al. 1991). With SIRTF, similar column densities will be detectable in gas at temperatures as low as 100 K.

(2) *Unambiguous detection of CO emission from the PDR.* In order to select CO emission from the hot PDR region where the H_2 emission arises and to discriminate against emission from cooler material deeper inside the PDR, one must observe an appropriately high CO transition. If we assume that the warm PDR region is reasonably dense and has a total column density of 10^{21} cm^{-2}, the J=9→8 transition at about 1 THz can be detected with current instrumentation as long as the CO abundance is $>10^{-6}$. Unfortunately, the predicted CO abundance in this warm region is at least an order of magnitude lower than this limit (Burton et al. 1990).

4. Global CO emission

How important are the regions in which CO makes up only a small proportion of the solar carbon abundance globally and what would be their effect on the overall conversion factor? Conversely, how can the conversion factor be nearly constant, when the CO abundance is varying considerably? Several explanations have been advanced to explain (and mostly comfort) the existence of the empirical proportionality factor. The first assumes that clouds are statistically identical and small enough not to shadow each other along the line of sight. In this case, the CO luminosity merely counts the number of clouds intercepted by the beam, and is proportional to the molecular mass. A more sophisticated explanation is

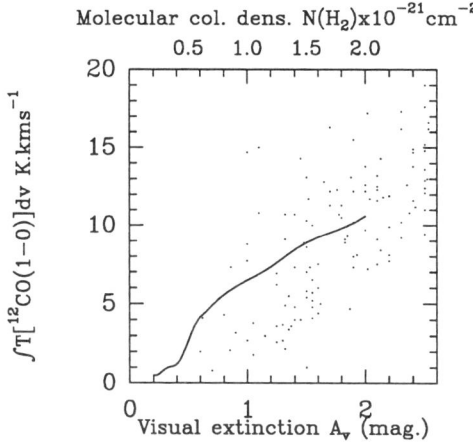

Figure 1. solid line: *The ^{12}CO (1–0) integrated intensity across a spherical cloud is plotted as a function of the H_2 column density along the line of sight (according to a Monte Carlo calculation by Guélin & Cernicharo 1988). The cloud, inspired by the dark cloud Heiles Cloud 2 in Taurus, has density/temperature ranging from 100 cm^{-3}/20 K at the periphery to 1000 cm^{-3}/10 K at the center and a radius of 1.5 pc. The CO fractional abundance is taken from calculations by van Dishoeck & Black (1988) for an attenuated IS radiation field; it varies from $5\ 10^{-6}$ to $4\ 10^{-4}$.* **dots:** *the intensities actually observed around Heiles Cloud 2 (from Guélin & Cernicharo 1987).*

the one advanced by Dickman et al. (1986), which assumes that the clouds (mostly dense GMCs) are in virial equilibrium and that CO is thermalized (see §2.2). A third explanation proposes that the molecular gas is dominated by relatively low density "halos" of the kind discussed above, where CO is subthermally excited (Guélin & Cernicharo 1987). Monte Carlo radiative transfer calculations for "realistic" halo + small core models, inspired by the clouds in the Taurus region, show that the CO intensity increases almost linearly with $N(H_2)$ along most lines of sight (see Figure 1); the derived proportionality factor is close to the observed one (Guélin & Cernicharo 1988). Basically, the bulk of the gas lies in regions with $n(H_2)=100-300$ cm^{-2}, half-way between the regime of thermalization where T_{ex}=constant, and the "low excitation" regime, where the Einstein coefficient divided by the optical depth is much larger than the collisional rate ($A/\tau >> C$) so that the optically thick line intensities are still proportional to the molecular column density.

Observations show that at large scales, the molecular mass lies mostly in cloud halos which are presumably far from virial equilibrium, thus supporting the last model (see Guélin & Garcia-Burillo 1991): the ^{12}CO and ^{13}CO J=2–1 and 1–0 line intensities observed in the Taurus region as well as in nearby galaxies, such as M 51 and NGC 891, imply the existence of a massive low density component (in the Taurus region, this component is directly observed through star counts, as e.g. for the cloud of Fig. 1). Recently, Wright et al. (1991) (see Wright, this conference) have reported intensities of the ^{12}CO J=1–0 through 5–4 lines averaged over the whole Galaxy, as observed by *COBE*. These intensities, despite their low signal–to–noise ratio and the lack of ^{13}CO data, also imply the presence of two molecular components, the most massive of which has $T_{ex} \approx 5$ K and is thus either extremely cold, or, in better agreement with the higher resolution studies, subthermally excited.

The halo model has important consequences for the constancy of the CO to H_2 conversion factor. Since halos are likely to be photon-dominated or "translucent" regions of the type discussed in §3, the relative abundances of H_2, ^{12}CO and ^{13}CO may vary considerably. In the model of Guélin & Garcia-Burillo, the $I(CO)$ will vary accordingly. Even if $N(H_2)/I(CO)$ could stay constant when averaged over large regions with similar global properties or entire galaxies, one may expect large changes when comparing intrinsically different regions such as arm and interarm regions or regions hosting active nuclei. A related question that needs further exploration is the extent to which the different regions show different conversion factors when CO $J=2-1$, $3-2$, ... are considered rather than the 1–0 line (see also §3).

The CO/H_2 abundance and the conversion factor are also not well-determined in the outer Galaxy, where the average density of the clouds is probably lower (Digel et al. 1990). Moreover, the presence of a $[^{12}C]/[^{13}C]$ gradient across the Galaxy may affect some of the results (Langer & Penzias 1990). In regions of lower metallicity such as the Magellanic Clouds, significantly higher conversion factors have been claimed (e.g. Cohen et al. 1988; Maloney & Black 1988; Israel & de Graauw 1991; Johansson 1991), although for the dwarf irregular galaxy IC 10 X appears close to "normal" (Wilson & Reid 1991). Because the transition of carbon from atomic to molecular form depends on the CO self-shielding, it will occur deeper into the cloud in metal-poor regions, because it takes longer to build up the necessary column density. Thus, metal-poor clouds have relatively less CO compared with clouds in our own Galaxy, *provided* they are not larger than the average clouds in our Galaxy. How do we measure the amount of H_2 in these cases, where all other molecules also have low abundances? One possibility would be to use the fine-structure emission of C II as a measure, since most of the carbon is probably in that form. Indeed, a large-scale correlation between C II and CO emission has already been established both observationally and theoretically for warm PDR gas, including that in luminous galactic nuclei (Stacey et al. 1991). The problem is that not only the "H_2" clouds, but also the H II regions and H I clouds are sampled in this way, and it is not clear how to distinguish the molecular part from the rest of the interstellar "mess", especially in regions of low C II surface brightness (see also Madden et al. this volume; Wright et al. 1991). Another possibility would be to use the dust IR emission, but this suffers from the same problem. Optical and UV absorption lines might be a third possibility, if strong enough background stars can be found.

5. Solid CO

The direct measurements of the CO abundance in dense clouds indicate that at least 10% of the carbon is in the form of gas-phase CO. The best estimate of the amount of carbon in solid form (such as graphite, PAH's, organic refractory) is 60±20% (van Dishoeck et al. 1991). These numbers do not exclude the possibility that the gas-phase CO abundance may vary by factors of 2–3 from cloud to cloud, depending on the nature of the dust (Greenberg 1991). How much of this CO can be depleted onto grains in the densest regions? It is well known that the time-scales for gas-phase molecules to collide with grains and stick on them is very short in dense clouds with $n_H \gtrsim 10^5$ cm^{-3}. What happens to the CO when it is depleted onto grains? Will it just sit there and eventually be released back into the gas phase, or will it be transformed into other molecules? How could this affect the gas-phase CO/H_2 abundance?

Evidence for molecules adsorbed on the surfaces of grains comes from infrared observations toward protostellar sources, which show in addition to the sharp gas-phase CO lines a prominent broad absorption feature near 2140 cm^{-1} (4.6 μm) generally attributed to solid CO (Lacy et al. 1984; Geballe 1986; Whittet et al. 1988, 1991). Analysis of the relative strengths of the features shows that the solid CO abundance varies from about 10^{-5} to less than about 10^{-7} with respect to total hydrogen. In all cases, this is significantly less than the canonical value for the gas–phase CO abundance of 10^{-4}. However, the shape of the feature may provide important information on the history and evolution of the CO.

Figure 2 shows that the solid CO band consists of a narrow ($\sim$ 5 cm^{-1}) feature centered at about 2140 cm^{-1} and a broader ($\sim$ 10 cm^{-1}) wing at about 2136 cm^{-1} (Tielens et al. 1991). Both the peak position and width of the narrower and generally stronger component vary from source to source. Laboratory studies suggest that the narrow CO band occurs in mixtures dominated by non–polar molecules (i.e., CO itself, CO_2, O_2, N_2, CH_4), while the broader component is due to polar mixtures such as H_2O (Sandford et al. 1989). Figure 2 compares the observed spectra at moderate resolution ($\lambda/\Delta\lambda$ =1200) towards NGC 7538 IRS 9 and AFGL 2136 with laboratory spectra of solid CO and a mixture of CO/H_2O=1/20. Note that AFGL 2136 is the only known spectrum in which the narrow 2140 cm^{-1} component is absent. Its broad feature at 2136 cm^{-1} is well fit by the H_2O–rich mixture. The spectrum of NGC 7538 IRS 9, on the other hand, is dominated by a narrow component, which is well fit by pure solid CO. In most other sources, the data are better fit by mixtures dominated by N_2 or CO_2, whereas in a few cases O_2 dominated mixtures provide the best agreement.

These results unambiguously demonstrate that there are (at least) two independent grain mantle components along many lines of sight. One component is dominated by a non–polar mixture and carries the narrow solid CO feature. The other is H_2O–rich and is probably the carrier of the broader solid CO component, as well as the 3.08 and 6.0 μm H_2O-ice bands. The solid CO/H_2O ratio is always much less than one ($\lesssim 0.1$). This dichotomy in grain mantle compositions may reflect chemical and/or physical variations during the accretion process. In view of the low abundance of CO in grain mantles and its high abundance in the gas phase, it is likely that most of the accreted CO has reacted with atomic H or O to form molecules such as H_2CO, CH_3OH, and CO_2 (Tielens & Allamandola 1987). In this model, the H–rich conditions required to form H_2O-rich ices also transform CO into CH_3OH. Non–polar grain mantles would then reflect accretion conditions with little atomic H available (i.e., high density; Tielens & Hagen 1982).

Alternatively, models based upon the difference in volatility between CO and H_2O can be developed. In such models, accretion takes place in a dark molecular cloud. A newly formed star in such an environment will heat up the surrounding dust, leading to partial evaporation. Close to the star only the cores remain. Farther out, where the temperature drops below 100 K, strongly H–bonding molecules (e.g., H_2O, CH_3OH, NH_3) can survive in a grain mantle, but more volatile molecules (i.e., CO) will evaporate. When the temperature drops below 50 K, a molecule such as CO_2 will also remain frozen out. Finally, at large distances from the star (i.e., in the dark cloud) the temperature is less than 15 K and highly volatile molecules such as CO (and O_2, N_2, CH_4) remain in the grain mantle. Thus, the birth of a new star will lead to a temperature stratification which in turn will lead to grain mantle separation.

Each of these classes of models has pros and cons. Heating of grain mantles followed by outgassing has undoubtedly played a role around luminous sources such as the Orion BN/KL region. Both the shape of the 3.08 μm ice band (Smith et al. 1989) and the observed gas phase abundances in the hot core and compact ridge (Walmsley 1989) attest to that. However, this model would predict that grain mantles in dark clouds far from embedded objects would be dominated by non–polar molecules such as CO. Yet, background stars behind the Taurus dark cloud show also separate H_2O–rich and non–polar grain mantle components (Tielens et al. 1991). In this case heating has played no role. It is likely that both types of mechanisms are of importance in interstellar clouds. Chemistry may dominate the observed variations in dark clouds, while volatility is important around (some) protostars. Further observations of solid CO as well as other grain mantle constituents will be very important to settle these issues.

In summary, the observations clearly demonstrate that CO does freeze out onto grains

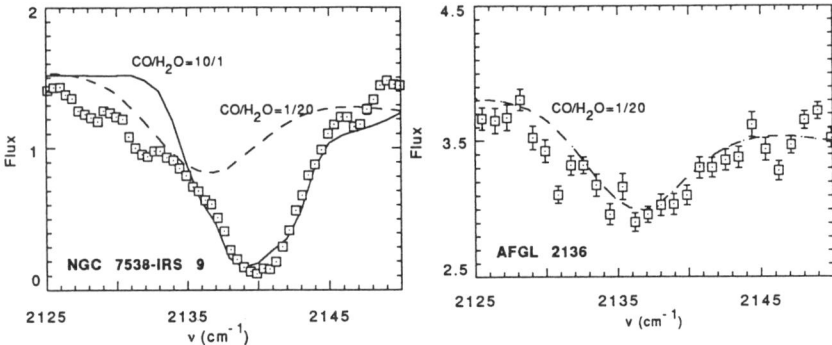

Figure 2. *The solid CO band toward the infrared source NGC 7538–IRS 9 is well fit in peak position and width by a mixture dominated by CO (solid line). Mixtures dominated by H_2O (dashed line) peak at lower frequencies and are broader than observed. In contrast, the spectrum of AFGL 2136 reveals only the broad component associated with H_2O-rich grain mantles (dashed line). See Tielens et al. (1991) for details. The error bars are smaller than the symbol, except where shown. Note also the weak gaseous CO feature.*

in dense interstellar clouds, but with different chemical results depending on the physical surroundings. The total amount of carbon depletion is not well known, however, especially in the coldest and densest protostellar condensations where the densities may be as high as $10^8 - 10^{10}$ cm^{-3} and where CO depletion could be significant. Comparison of submillimeter continuum observations with $C^{18}O$ maps of regions such as NGC 2024 (Mezger et al. 1988) and NGC 1333 (Sandell et al. 1991) may provide further insight into this question.

6. Dissociative shocks and protostellar outflows

Shock waves propagating through the interstellar medium at speeds faster than about 50 km s^{-1} are dissociative; they result in the temporary destruction of any molecules which enter them. The CO/H_2 abundance ratio behind dissociative shocks is presently not constrained by observation, primarily because such shocks are usually accompanied by slower, non-dissociative shocks which are far more luminous sources of molecular line emission and which therefore mask much of the molecular line spectrum which would be specific to fast shocks. The *theoretical* predictions for CO/H_2 in the dissociative case are outlined below. For non–dissociative shocks, the CO/H_2 abundance presumably does not deviate much from the "canonical" value of 10^{-4}, as found for Orion/KL (see Table 1).

Upon passing through a fast shock front, interstellar gas is rapidly heated to temperatures greater than $\sim 10^5$ K, resulting in the complete and rapid dissociation of molecules and the partial ionization of atoms via collisional processes (Hollenbach & McKee 1980). All information about the chemical state of the preshock gas is therefore lost. The shocked gas eventually cools, primarily via emission in optical and ultraviolet lines of atoms and atomic ions, and starts to recombine. Once the temperature falls below a few thousand degrees, molecule reformation ensues, starting with H_2 formation, which occurs in the gas phase via the H$^-$ intermediary and – if sufficiently cool grains survive – in reactions catalysed on grain surfaces. A further reaction of H_2 with atomic oxygen leads rapidly to OH formation, and reactions of OH with C or C^+ result in the production of CO. Detailed modeling has shown (Hollenbach & McKee 1989; Neufeld & Dalgarno 1989) that provided grain catalysis of H_2 formation is moderately efficient, hydrogen is almost entirely incorporated into H_2 and carbon into CO by the time that the shocked gas has cooled to a temperature of a few hundred degrees. The time required is $\sim 10^7 \text{yr}/n_0$, where n_0 cm^{-3} is the particle density in the preshock medium. The CO/H_2 abundance ratio will therefore simply reflect the gas–phase elemental abundance of carbon, which may have been somewhat modified as the result of carbonaceous grain destruction in the region behind the shock front.

If grain catalysis is ineffective – either because grains are absent or because they have been heated within the shocked gas to temperatures at which any adsorbed atoms evaporate before they have the chance to react – molecule reformation behind the shock is incomplete. Detailed studies have revealed (Neufeld & Dalgarno 1989) that in this circumstance the final H_2 abundance is only $\sim 10^{-3}$, while the fraction of gas–phase carbon in CO may be as large as ~ 0.1. In the absence of grain–catalysed H_2 formation, therefore, dissociative shocks may leave the interstellar medium in a rather exotic chemical state, in which the gas is largely atomic but possesses a substantial CO abundance, with a CO/H_2 abundance ratio of ~ 100 [C]$_{\text{gas}}$/[H].

The theoretical results for fast shocks are similar to those obtained for molecule formation in the ejecta of supernovae such as SN 1987A (Lepp et al. 1990) and in grain–free protostellar outflows (Glassgold et al. 1991). In the latter case, the deficiency of H_2 can be traced to the short time–scales in these winds. The CO/H_2 ratio is therefore sensitive to the mass-loss rate: most of the carbon will be transformed into CO for $\dot{M} \geq 3 \times 10^{-6}$ $M_\odot$ yr^{-1}, whereas most of the hydrogen will be molecular only for $\dot{M} \geq 10^{-4}$ $M_\odot$ yr^{-1}. Moreover, the CO quickly gets so cold that its millimeter transitions are unobservable until the wind interacts with the surrounding molecular cloud far from the protostar. Care should therefore be taken in inferring H_2 column densities and envelope mass–loss rates from radio observations of extremely high velocity (EHV) H I and CO flows near protostellar objects.

7. Circumstellar envelopes

The CO/H_2 ratio in the deep interior of circumstellar envelopes of evolved stars is largely determined by the temperature of the star (Glassgold & Huggins 1982). For $T_\star >$ 2500–3000 K, the hydrogen will be atomic; otherwise it is expected to be molecular. Carbon will be mostly in the form of CO for almost all evolved stars, simply because of the large binding energy of this molecule. Of course, if the star has an H II region, such as is found in the chromosphere of α Ori and in proto–planetary nebulae, the high temperatures and levels of dissociating radiation will alter this situation – and differently for CO and H_2. Thus, the CO/H_2 abundance ratio in the inner circumstellar regions of evolved stars may vary considerably from star to star.

In the outer part, the relative amounts of H and H_2 and of C^+ and CO, and the location of the transition zones, are determined by the effective line–shielding in the photodissociation of H_2 and CO (Glassgold & Huggins 1982; Mamon et al. 1988). Other effects such as ion–molecule reactions, fractionation processes and grain chemistry may also play a role in the outer part. If $T_\star < 3000$ K and if there are no internal sources of UV radiation, the CO distribution will be less extended than the H_2 distribution, simply because of the larger abundance of hydrogen. Without the above–mentioned restrictions on the nature of the star, other situations become possible.

Although CO emission from circumstellar envelopes is readily observed, measurements of H I and/or H_2 are very rare. For α Ori ($T_\star \approx 3500$ K), 21 cm radiation from the H I envelope has been detected (Bowers & Knapp 1987), and one can infer from other observations that carbon is more fully associated into molecules than is hydrogen, i.e.,

$CO/H_2 \gg 2\,[C]_{gas}/[H]$. For the best studied and chemically important AGB star, IRC +10216 ($T_\star \approx 2300$ K), interesting upper limits on the 21 cm emission have been obtained by Zuckerman et al. (1980) and by Knapp & Bowers (1983). These limits correspond to a mass of $\approx 10^{31}$ gr, close to the amount injected into the envelope from the upper atmosphere of the star (Glassgold & Huggins 1982). Keady & Ridgway (1991) have recently searched for the H_2 ro–vibrational lines at 2 μm and find that $\dot{M}(H_2) < 4 \times 10^{-5}\,M_\odot\,yr^{-1}$. Because the CO abundance and the mass–loss rate can both be determined from a detailed analysis of the spatial variation of the CO millimeter line emission – $CO/H_2 = 6 \times 10^{-4}$ and $\dot{M} = 3 \times 10^{-5}\,M_\odot\,yr^{-1}$ –, an upper limit to the H I/H_2 ratio for IRC +10216 is H I/$H_2 < 5 \times 10^{-3}$ (Huggins et al. 1988). For almost all other red giant winds, observers generally determine the mass–loss rate by assuming that all carbon is in CO and hydrogen in H_2, i.e. $CO/H_2 = 8 \times 10^{-4}$ for C stars and 3×10^{-4} for Miras (Knapp & Morris 1986).

8. Concluding remark

Although observations indicate a remarkable constancy in the conversion factor of integrated CO line intensity to H_2 column density on a global scale, various theoretical arguments and observational data suggest that the actual CO/H_2 abundance may vary substantially locally in certain astrophysical environments. Indeed, there are regions where CO/H_2 is significantly lower than the canonical value of 10^{-4}, such as diffuse clouds, PDRs, and possibly very dense protostellar condensations. On the other hand, several examples have been given where CO/H_2 may be substantially higher than 10^{-4}, including grain–free dissociative shocks, supernova ejecta, protostellar winds and some circumstellar envelopes. Care is required with the traditional analysis in these cases.

References

Black, J.H., & Willner, S.P. 1984, ApJ, 279, 673.
Black, J.H., van Dishoeck, E.F., Willner, S.P., & Woods, R.C. 1990, ApJ, 358, 459.
Blitz, L., Bloemen, J.B.G.M., Hermsen, W., & Bania, T.M. 1985, A&A, 143, 267.
Bloemen, J.B.G.M. 1989, ARAA, 27, 469.
Bowers, P.F., & Knapp, G.R. 1987, ApJ, 315, 305.
Burton, M.G., Hollenbach, D.J., & Tielens, A.G.G.M. 1990, ApJ, 365, 620.
Cohen, R.S. et al. 1988, ApJ, 331, L95.
Combes, F. 1991, ARAA, 29, 195.
Danks, A.C., Federman, S.R., & Lambert, D.L. 1984, A&A, 130, 62.
de Vries, H.W., Heithausen, A. & Thaddeus, P. 1987, ApJ, 319, 723.
Dickman, R.L., Snell, R.L. & Schloerb, P. 1986, ApJ, 309, 326.
Digel, S., Bally, J. & Thaddeus, P. 1990, ApJ, 357, L29.
Elmegreen, B.G. 1989, ApJ, 338, 178.
Geballe, T.R., 1986, A&A, 162, 248.
Glassgold, A.E. & Huggins, P.J. 1982, MNRAS, 203, 400.
Glassgold, A.E., Mamon, G.A., & Huggins, P.J. 1991, ApJ, 373, 254.
Gredel, R., van Dishoeck, E.F., de Vries, C.P., and Black, J.H. 1991, A&A, in press.
Greenberg, M. 1991, in *Cosmic Rays, Supernovae and the Interstellar Medium*, eds. M.M. Shapiro et al. (Kluwer, Dordrecht).
Grevesse, N. et al. 1991, A&A, 242, 488.
Guélin, M. & Cernicharo, J. 1987, in *Molecular clouds in the Milky Way and external galaxies*, eds. R.L. Dickman et al. (Springer).
Guélin, M. & Cernicharo, J. 1988, in *The physics and chemistry of interstellar molecular clouds*, eds. G. Winnewisser & J.T. Armstrong (Springer).
Guélin, M. & Garcia-Burillo, S. 1991, in preparation.

Henkel, C., Baan, W.A., & Mauersberger, R. 1991, Astr. Ap. Rev., 3, 47.
Hollenbach, D. & McKee, C. 1980, ApJ, 241, L47.
Hollenbach, D., & McKee, C. 1989, ApJ, 342, 306.
Huggins, P.J., Olofsson, H. & Johansson, L.E.B. 1988, ApJ, 332, 1009.
Israel, F.P. 1988, in Millimetre and Submillimetre Astronomy, eds. R.D. Wolstencroft & W.B. Burton (Kluwer, Dordrecht).
Israel, F.P. & de Graauw, T. 1991, in The Magellanic Clouds, eds. R. Haynes & D. Milne (Kluwer, Dordrecht).
Jenkins, E.B., Lees, J.F., van Dishoeck, E.F., & Wilcots, E.M. 1989, ApJ, 343, 785.
Johansson, L.E.B. 1991, in Dynamics of Galaxies and Their Molecular Cloud Distributions, IAU Symposium 146, eds. F. Combes and F. Casoli (Kluwer, Dordrecht).
Keady, J.J. & Ridgway, S.T. 1991, ApJ, in press.
Knapp, G.R. & Bowers, P.F. 1983, ApJ, 266, 701.
Knapp, G.R. & Morris, M. 1986, ApJ, 292, 640.
Lacy, J.H. et al. 1984, ApJ, 276, 533.
Langer, W.D. & Penzias, A.A. 1990, ApJ, 357, 477.
Lepp, S., Dalgarno, A., & McCray, R. 1990, ApJ, 385, 262.
Maloney, P. 1990, in Interstellar Medium in galaxies, eds. H.A. Thronson & J.M. Shull (Kluwer, Dordrecht).
Maloney, P. & Black, J.H. 1988, ApJ, 325, 389.
Mamon, G., Glassgold, A.E., & Huggins, P.J. 1988, ApJ, 328, 797.
Mattila, K. 1986, A&A, 160, 157.
Mezger, P.G., Chini, R., Kreysa, E., Wink, J.E., and Salter, C.J. 1988, A&A, 191, 44.
Morton, D.C. 1975, ApJ, 197, 85.
Neufeld, D.A., & Dalgarno, A. 1989, ApJ, 340, 869.
Parmar, P.S., Lacy, J.H. & Achtermann, J.M. 1991, ApJ, 372, L25.
Sandell, G., Aspin, C., Duncan, W.D., Russell, A.P.G., & Robson, E.I. 1991, ApJ, 376, L17.
Sandford, S.A., Allamandola, L.J., Tielens, A.G.G.M., & Valero, G.J., 1988, ApJ, 329, 498.
Smith, R.G., Sellgren, K., & Tokunaga, A.T., 1989, ApJ, 344, 413.
Solomon, P.M. & Barrett, J.W. 1991, in Dynamics of Galaxies and Their Molecular Cloud Distributions, IAU Symposium 146, eds. F. Combes and F. Casoli (Kluwer, Dordrecht).
Stacey, G.J. et al. 1991, ApJ, 373, 423.
Tielens, A.G.G.M. & Allamandola, L.J., 1987, in Interstellar Processes, eds. D. Hollenbach and H.A. Thronson (Reidel, Dordrecht), p. 397.
Tielens, A.G.G.M. & Hagen, W., 1982, A&A, 114, 245.
Tielens, A.G.G.M. & Hollenbach, D. 1985, ApJ, 291, 722.
Tielens, A.G.G.M., Tokunaga, A.T., Geballe, T.R., & Baas, F., 1991, ApJ, Nov. 1 issue.
van Dishoeck, E.F. & Black, J.H. 1987, in Physical Processes in Interstellar Clouds, eds. G. Morfill and M.S. Scholer (Reidel, Dordrecht).
van Dishoeck, E.F. & Black, J.H. 1988, ApJ, 334, 771.
van Dishoeck, E.F., Blake, G.A., Draine, B.T., & Lunine, J.I. 1991, to appear in Protostars and Planets III, eds. J.I. Lunine and M.S. Matthews (University of Arizona, Tucson).
Walmsley, C.M. 1989, in IAU Symposium 135 Interstellar Dust, eds. L.J. Allamandola and A.G.G.M. Tielens (Kluwer, Dordrecht).
Watson, D.M., Genzel, R., Townes, C.H., & Storey, J.W.V. 1985, ApJ, 298, 316.
Whittet, D.C.B., et al. 1988, MNRAS, 233, 321.
Whittet, D.C.B. & Duley, W.W. 1991, Astr. Ap. Rev., 2, 167.
Wilson, C.D. & Reid, I.N. 1991, ApJ, 366, L11.
Wright, E.L. et al. 1991, ApJ, 381, 200.
Young, J.S. & Scoville, N.Z. 1991, ARAA, 29, 581.
Zuckerman, B., Terzian, Y., & Silverglate, P. 1980, ApJ, 241, 1014.

OBSERVATIONS OF ATOMIC GAS IN PHOTODISSOCIATION REGIONS

GLENN J. WHITE
Department of Physics
Queen Mary & Westfield College
University of London
Mile End Road
London E1 4NS

RACHAEL PADMAN
Cavendish Laboratory
University of Cambridge
Madingley Road
Cambridge CB3 OHE

ABSTRACT. At the interface between an HII region and a molecular cloud, lies a neutral gas layer which is subject to both an intense radiation field, and to shocks arising from the expansion of the ionisation front of the HII region. The gas in these regions is highly excited, hot, and may be fairly dense. We present the first high resolution images of atomic carbon towards a sample of ionisation front sources. This study has relevance to our understanding of shock induced star formation, the formation and destruction of molecular species under extreme conditions, shock processes in the ISM, and the energy balance in molecular clouds.

1. Atomic Carbon

Carbon is the fourth most abundant element in the universe, and plays a major role in the chemistry of interstellar molecular clouds. Atomic carbon (CI) is formed when CO molecules are photo-dissociated in the interstellar ultraviolet (UV) field [CO $+h\nu \rightarrow$ C + O]. It has an ionisation potential which is similar to the dissociation energy of CO, and thus CI in turn is rapidly photo-ionised. At the outside edges of molecular clouds, carbon should exist mainly in the ionised form (CII), in a narrow transition zone near their edges, be in atomic form (as CII recombines with electrons), and inside them be bound into CO molecules (as molecules form). At submillimetre wavelengths, CI has two fine structure transitions, which are efficiently excited by the conditions encountered in interstellar molecular clouds. We present the first high angular resolution maps of molecular clouds in the CI $^3P_1$ - $^3P_0$ line, towards the high mass star formation region Orion A, the externally illuminated cloud S140, the edge-on ionisation front M17, and the Galactic Centre.

The observations were obtained at the James Clerk Maxwell Telescope in Hawaii. A new dual-polarisation InSb detector (Padman et al *in preparation*) was used at a frequency of 492.1607 GHz ($\lambda \sim$ 609 μm). The high altitude site, new receiver and excellent efficiency of the telescope allow observations which have hitherto been impossible. A new mapping technique was used: the local oscillator frequency was held constant, and the telescope raster-scanned in azimuth, whilst simultaneously beam-switching in the scan direction using the nutating sub-reflector. Several such dual-beam maps of each source, with different parallactic angles and chopper throws to give good coverage in 2-D Fourier space, were processed jointly using a maximum entropy algorithm (ref 1) to give the final RA-Dec maps. Single point position-switched spectra were also taken at a number of positions in each source: these were used to verify that the emission was indeed zero near the edges of the maps, and to check the maps.

From observations of Mars we find that the beam size is 9-arcseconds and the main beam efficiency is 0.39+/-.04: this efficiency is consistent with the effective surface error of ~34-microns (rms) deduced from 450-micron beam mapping of Mars, Jupiter and the moon, by R.Hills and J.Richer (*private communication*). We deduce from these beam maps that the error beam at 492 GHz should contain about 40 percent of the total power, and have a maximum amplitude ~2 percent that of the main beam, with a characteristic radius of ~50 arcseconds.

The MEM reconstruction process attempts to deconvolve the dual beam response function from the observed maps. Direct comparison of the dual-beam maps with the position-switched spectra shows that the former underestimate the total flux by up to 30 percent. This appears to be due to self-chopping of the extended error beam: the error beam was neglected in the MEM restoration because of memory constraints in the computer. The image of the northern part of Orion was re-processed to take the full error beam into account, using a much more computationally expensive algorithm: we find that the discrepancies with the single-point fluxes are reduced, but that otherwise the image is not qualitatively different from that presented in Figure 1a.

The Orion molecular cloud contains many indicators of on-going star-formation[2], including the ~ 50 $M_\odot$ pre-main sequence object, IRc2, which is surrounded by a dense rotating ring of gas. This is being eroded by the combined effects of radiation and an energetic molecular outflow which originates close to IRc2. IRc2 lies ~ 1 arc minute north-west of the well-known Trapezium star cluster, responsible for ionisation of surrounding gas to create the prominent optical nebula M42. This cluster also illuminates a dense molecular ridge (the 'Bright Bar') ~ 3 arc minutes south-east of IRc2, where the UV radiation intensity is ~ 10^5 times the average interstellar UV field, G_0 (G_0 = 1.7 x 10^{-3} erg s^{-1} cm^{-2}). The CI map obtained towards Orion is shown in Figure 1a.

In the north of the map, the CI appears as part of a clumpy broken cavity. The two most intense regions lie ~ 30 arc seconds (0.06 parsecs) north-east and south-west of IRc2, at the edges of the 'quiescent ridge', which contains several other star forming complexes. The CI intensity is weaker along the direction of the molecular outflow (south-east - north-west). While there is *general* agreement between the CI distribution and other tracers such as ^{13}CO or $C^{18}O$, any formal correlation is poor. CI traces the *edges* of the quiescent molecular ridge instead of following the distribution of other molecular species more closely. South-east of IRc2, the CI forms a narrow bar, containing several clumps embedded in diffuse emission, displaced back by ~ 30 arc seconds from the edge of the Orion HII region [3,4]. The northernmost clumps in the bar lie close to the positions of ionised knots, which resemble HH-objects [5].

Towards IRc2, the main beam brightness temperature, T_{MB} = 16.6 K, and the integrated emission is 79 K km s^{-1}. For an excitation temperature of 100K, N(CI) = 1.1 x 10^{18} cm^{-2}, and τ(CI) ~ 0.2 (see ref. 15 for details of CI abundance derivation). Assuming N(CO) = 1.6 x 10^{19} cm^{-2}(ref. 6), N(CI) / N(CO) = 0.07. At the strongest point on the Bright Bar, N(CI) / N(CO) = 1.22. The CI spectrum (Figure 2a) towards IRc2 is narrow (ΔV = 4.5 km s^{-1}), showing no evidence of out-flowing molecular gas. Subtracting emission from the ambient cloud, the optically thin upper limit for the outflow gas N(CI) / N(CO) ~ 0.005 (assuming T_{ex} = 100K). Thus CI appears less abundant in the outflow gas than at nearby cloud positions, as suggested from previous large beam studies [7].

S140 is an active region containing a cluster of forming B stars, which appear as compact infrared sources in its core. A nearby (~2 parsecs) B0 star photo-dissociates the outer layers of the S140 cloud (UV field ~ 150 G_0), exciting a $H\alpha$ emission nebula [8]. The CI map is shown in Figure 1b. This shows a

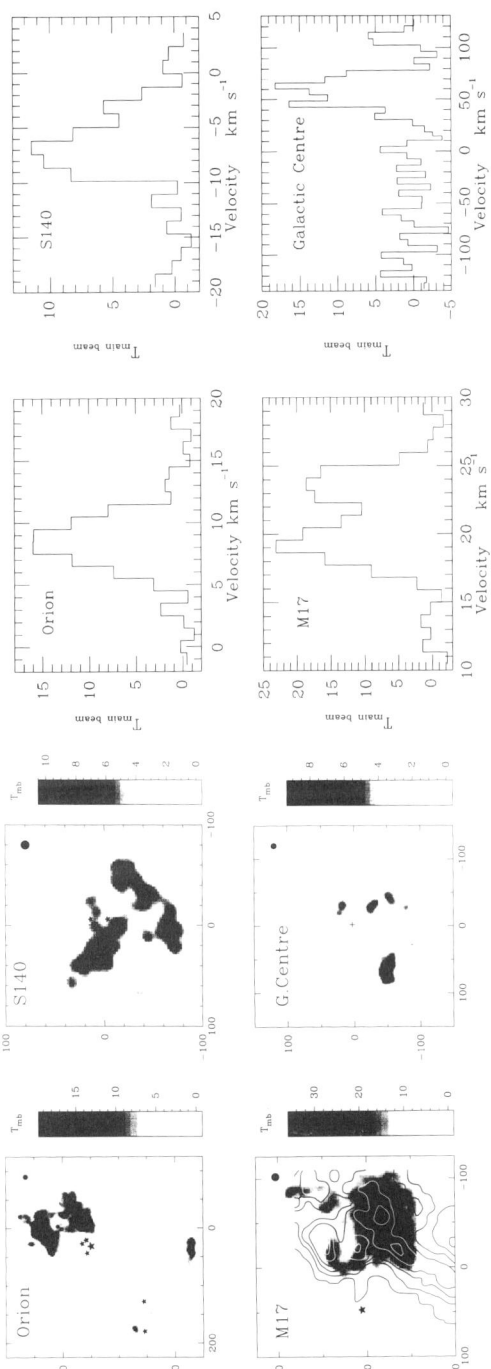

Figure 1

Figure 2

Figure 1 CI distribution towards the sources, obtained with a 1.3 km s^{-1} channel (half power width). The positions of infrared and some visible objects are indicated by stars. The centre positions and velocities of the maps are ; Orion:- R A (1950) -05^0 32^m 46.9^s Dec (1950) -05^0 24' 26" (+9 km s^{-1}) [the square and dot right of centre are IRc2 and BN respectively, the central four stars are the Trapezium Cluster, and the two at bottom are Θ^2 Orionis A and B], S140:- R A (1950) 22^h 17^m 42.0^s Dec +63^0 3' 45" (-8 km s^{-1}) [the stars show the positions of embedded B stars], M17:- R A (1950) 18^h 17^m 31.2^s Dec -16^0 13' 30" (+20 km s^{-1}) [contours of C^{18}O temperature for the same velocity interval as the CI data are overlaid; the star is the position of a bright SAO star], Galactic Centre:- R A (1950) 17^h 42^m 29.5^s Dec -28^0 59' 20" (+50 km s^{-1}) [the Centre is marked with a cross]. The beam size is shown as a filled circle in the top left corner. The grey scale is in units of Tmb.

Figure 2 Position switched CI spectra towards the sources, at the positions; Orion:- R A (1950) 05^h 32^m 47.0^s -Dec (1950) -05^0 24' 23", S140:- R A (1950) 22^h 17^m 42.0^s Dec 63^0 3' 45", M17:- R A (1950) 18^h 17^m 29.9^s Dec -16^0 13' 00", Galactic Centre:- R A (1950) 17^h 42^m 27.2^s Dec -28^0 59' 53".

clumpy bar-like distribution at the front edge of the molecular cloud, adjacent to the ionisation front. There is good correlation between this CI bar, running from offsets (+20,-70) to (-60,-10), and a ridge of hot CO[8]. The star-forming core and density peak lie ~ 50" north-east of the CI bar. As shown in Figure 1b, the infrared sources are adjacent to a clumpy CI ridge deeper inside the cloud. The tendency for CI to avoid the regions of highest column density resembles behaviour seen in Orion. The spectrum towards this core is shown in Figure 2b. N(CI) / N(CO) = 0.19 at this position, and N(CI) / N(CO) = 2.8 at 30 arcseconds west and 45 arc seconds south, on the ionisation front.

The M17 complex is the archetypal edge-on ionisation front, lying next to an O-B star cluster, which heats and provides UV illumination (~ 2 x 10^5 G_o) of the cloud edge. In Figure 1c, contours of the J=2-1 $C^{18}O$ line for a velocity interval which matches the CI data (Stutzki - *private communication*) are overlaid on the CI distribution. The CI appears clumpy and fragmented, dropping off sharply (< 1 beamwidth) at the eastern edge, next to the ionisation front. The CI peaks show significant position offsets from the $C^{18}O$ peaks; CI emission is *not* enhanced in the regions of highest column density, or on the ionisation front of M17 (see also ref. 9). No significant velocity variation is seen across the edge of the cloud. A spectrum is shown in Figure 2c.

The inner 10 parsecs of our Galaxy contains a rotating neutral molecular ring, surrounding a 2 parsec radius ionised cavity. The inner-edge of this ring is excited by a strong UV field (~ 7 x 10^4 G_o), and hot, dense gas has been reported [10]. The CI distribution is shown in Figure 1d, and a spectrum in Figure 2d. The CI is concentrated into several clumpy groups lying ~ 2 - 4 parsecs from the Centre. As in the other sources, the CI peaks lie close, but not co-incident, to those of other molecular tracers [11]. The ring has a rotation velocity of ~ 100 km s^{-1} , so the CI map velocity was chosen to match a dominant velocity seen in the optically thin $C^{18}O$ line.

The observations reported here represent a major advance on earlier observations of CI [12,13] which showed that N(CI) / N(CO) varied from >10 in clouds having low extinction (A_v), to ~ 0.1 in cores where A_v ~ 100 magnitudes. This contrasted with chemical models[14,15] where carbon at molecular cloud edges is almost all in the form of CII, with a layer of CI between A_v ~ 3-8, and the transition between C and CO at A_v ~ 5 [16,17], the CII/CI/CO interface existing in a narrow range of hydrogen column densities between 4 x 10^{21} and 10^{22} cm^{-2}. However CI would rapidly (< 10^6 years) convert back to CO, and multi-component models [18,19], in which dense clumps are interspersed with a more tenuous inter-clump medium, were developed. The CI abundances produced are relatively insensitive to the external UV field, the CI forming narrow shells around clump cores. In S140, we may be seeing evidence of both processes - direct photo-dissociation of the cloud edge giving the bright CI rim, and then deeper into the cloud penetration of UV photons through the interclump medium.The present observations show that;

2. Conclusions

a) Molecular clouds adjacent to ionization fronts often show strong CI emission from a thin surface layer. In S140 and the Orion bright bar photodissociation of CO leads to peak values of N(CI)/N(CO) greater than 1, even though this CI layer is only marginally resolved in our 9-arcsecond beam. Our observations also confirm previous conclusions based on lower-resolution data that CI is widely distributed (with less intensity) throughout molecular clouds, remaining abundant (N(CI)/N(CO) typically ~ 0.1) up to several parsecs away from the UV-illuminated edges. CI is not detected in the IRc2 outflow, where we find that N(CI)/N(CO) < .005, appearing to indicate that shock dissociation of CO in the vicinity of outflows is *not* a major source of CI.

b) In non-uniform molecular clouds such as M17, the CI peaks on the edges of the clumps revealed by column density tracers such as $C^{18}O$, and apparently avoids the densest regions, presumably because the UV radiation is too highly attenuated. Since the CI emission is distributed throughout the molecular cloud core, this is *direct* evidence for low filling-factor, high density-contrast clumping, as previously deduced from molecular line and lower resolution CI observations.

Acknowledgements

We acknowledge the work of many people at Cambridge University, Queen Mary College, London and the JACH, Hawaii who contibuted towards the instrumentation used in this work. The JCMT is operated by the Royal Observatory Edinburgh, on behalf of the SERC, the Netherlands Organisation for Pure Research, and the National Research Council of Canada.

References

1. Richer, J. *MNRAS in press* (1991)
2. Genzel, R and Stutzki, J, *Annual Rev. of Astronomy & Astrophysics*, **27**, 41 (1989)
3. Martin, A.H.M. and Gull, S.F., *M.Not.R.Astr.Soc.* **175**, 235 (1976).
4. Yusef-Zadeh, F., 1990, *Astrophys. J.* **361**, L19.
5. Taylor, K. and Münch, G. *Astr. Astrophys.* **70**,359 (1978).
6. Wilson, T.L., Serabyn, E., Henkel, C. and Walmsley, C.M. *Astr. Astrophys.* **158**, L1 (1986).
7. Hayashi, M., Omodaka, T., Hasegawa, T. and Suzuki, S. *Astrophys. J.* **288**, 170 (1985).
8. Beichman, C., Phillips, T.G., Wootten, H.A & Frerking, M.A. *Regions of Recent Star Formation, edited by R.Roger and P Dewdney,* D. Reidel Press (1982)
9. Stutzki, J. & Gusten, R. *Astrophys.J.* **356**, 513 (1990).
10. Harris, A.I., Jaffe, D.T., Silber, M. and Genzel, R. *Astrophys. J. Lett.* **294**, L93 (1985).
11. Mezger, P.G., Wink, J.E. and Zylka, R. *Astr.Astrophys.* **228**, 95 (1990).
12. Phillips, T.G., Huggins, P.J., Kuiper, T.B. and Miller, R.E. *Astrophys.J.* **238**, L103 (1980).
13. Phillips, T.G. & Huggins, P.J. *Astrophys. J.* **251**, 533 (1981).
14. Tielens, A. H. M. and Hollenbach, D. *Astrophys.J.* **291**,747, (1985b).
15. van Dishoeck, E.F. and Black, J.H. *Astrophys.J.* **334**, 771 (1988).
16. Tielens, A. H. M. and Hollenbach, D. *Astrophys.J.* **291**,722, (1985a).
17. Keene, J., Blake, G., Phillips, T.G., Huggins, P. & Beichman, C. *Astrophys.J.* **299**, 967 (1986).
18. Monteiro, T.S. *Astr.Astrophys.* **241**, L5 (1991).
19. Stutzki, J., Stacey, G.J., Genzel, R., Harris, A., Jaffe, D.T. & Lugten, J.B. *Astrphys. J.* **332**, 379 (1988)

QUESTIONS AND ANSWERS

D.A.Williams: Can you say what evidence there is of dynamics in these interfaces?

G.White: Sensitive spectra taken in strips running perpendicular to the interface zones are able to measure the expansion velocity of the HII region into the surrounding neutral gas, and hence estimate the incident shock speed. M 17 is a good example where the ionized gas impacting the M17 molecular cloud at $\sim 11\,km/s$, forcing gas to stream outwards along the line of sight. Close to interfaces, the material is observed to be highly fragmented, both spatially and in velocity space. One would hope to be able to examine whether turbulent mixing, systematic streaming of clump/interclump gas and clump/clump collisions are occuring. However we are really at the limit of what single dish telescopes can achieve in angular resolution, and the interferometers are for better suited to trying to examine these processes. One recent experiment which has set out to search for a dynamical signature of ambipolar diffusion - where ions and neutrals can achieve a relative drift velocity of a few km/s has provided a tentative suggestion that some small effect was detected - however the authors again remark that it is only when good higher resolution maps (specifically 2" resolution at M 17) that this can be unambiguously resolved.

E.vanDishoeck: There appears to be a lack of CI in M 17 close to the ionization front. Could you comment on this?

G.White: Rachel Padman and I were extremely surprised when we were at the JCMT making these observations not to see limb brightening across M 17. We spent a lot os time obtaining sensitive CI spectra across the interface region to really confirm this, and there is clearly no intensity enhancement, no any significant velocity shift across the interface. These results are not off the telescope, and we need to model this further. The two differences between M 17 and S 140 (where a strong CI enhancement is seen) is that the external UV illumination is 2 orders of magnitude greater, and that the neutral density and clumpy structure is much greater. Maybe the transition in M 17 is so abrupt that any CI transition zone is severely beam diluted by our 9 arc second beam.

A.E.Glassgold: In addition to the beautiful observation, I was pleased to hear your critical comments about the difficulty of measuring abundances in these regions. I wonder if you could comment on your thermometry and particularly how the temperature is determined at relatively low temperatures $\sim$100 K.

G.White: We have been fairly successful in using multi-transition CO data to thermometrically probe these interface zones. However the molecular ionised zone interfaces are usually spatially unresolved with 10 arc second beams, so we shall don't have the ability to probe the interesting regions at the neutral/HII boundaries where the gas temperature decouples from the dust. The molecule CH_3C_2H has an easy to observe k-split series across several of the ionisation fronts we have measured and can provide an alternative estimate of the rotational temperature to go with the CO estimates (the more commonly used temperature estimator CH_3CN which has similar k-splitting has been unsuccessful searched for across several interface regions). However these more complex molecular species are less robust than CO, and may give a misleading comparison under some circumstances. On your comments about the CI observations, I must thank my collaborators R.Padman and M.Griffin and others workers at Queen Mary College, Cambridge University and the Joint Astronomy Centre, Hilo, without whom these data would not have been obtained.

R.Rubin: With spatial resolutions achievable how (better than $\sim$ 9″), it may be necessary to consider models of the PDR that are not plane parallel when comparing with observations. This may be important for example for properly looking for species stratification. The boundary of HII region/PDR interface will vary with location along the line of sight. The boundary will be curved toward the exciting star(s).

G.White: All of the high angular resolution studies of interface regions have shown that reality lies far away from idealised plane parallel homogeneous models which have been developed so far. It will also be important to include in future modelling a) the possibility that not all the UV illumination comes from a single direction, b) the probability that dynamical effects such as clump rotation or clump obscuration as they move about on time scales less than that required to allow the chemistry to relax after UV stimulation (as proposed by Monteiro A&A 241,L5,1991), and c) that much of the chemistry will occur on the outer layers of dense clumps, may all modify both the chemistry, and also the spectroscopic signature of the region to a telescope. There are currently several groups working with interferometric data, PDR interfaces which may already have detected evidence for chemical stratification. It is clear that "archetypical" sources such as M17 may not be the best places to test more sophisticated chemical models, the S140 ionisation front looks far more suitable as the source of choice due to its relative simplicity compared with M17 or the Orion Bright Bar regions.

R.Opher: Shocks are associated with first order Fermi acceleration of particles, with an appreciable energy of the shock going into these particles. I thus doubt the possibility of treating the clouds near the shock by a single temperature without taking into account the details of the ionization caused by the accelerated particles in the clouds.

G.White: Yes, this is undoubtedly a complex and important issue. Temperature stratification is of course already well known across a region such as M17 (Gatley et al.,Ap.J. 233,575,1979), but little is known observationally about the detailed ionisation profiles. Innovative observational diagnostics of the ionisation state and the small scale magnetic fields would be very helpful in resolving this issue.

OBSERVATIONS OF CII, CI AND CO IN INTERFACE REGIONS

JÜRGEN STUTZKI
*I. Physikalisches Institut der Universität zu Köln, Zülpicher Staße 77,
D-5000 Köln 41, Germany*

September 3, 1991

1. Introduction

UV-radiation longward of the Lyman edge (912 Å) can escape from HII-regions. It photodissociates carbon monoxide ($\lambda < 1118$ Å) and photoionizes neutral atomic carbon ($\lambda < 1101$ Å). The resulting CII/CI/CO transition zone on the edges of molecular clouds (photodissociation region or photodominated region: PDR) has been studied in great detail theoretically (Tielens & Hollenbach, 1985; van Dishoek & Black, 1988; Sternberg & Dalgarno, 1989) and these investigations have recently been extended to cover a wide range of densities and UV-intensities (Burton, Hollenbach & Tielens, 1990; Hollenbach, Takahashi & Tielens, 1991; see also A. Sternberg, this volume).

Observations of the three species CII, CI and CO by means of optical and UV absorption studies are limited to diffuse clouds (E. van Dishoek, this volume). The study of PDR's on dense molecular clouds thus has to rely on the fine structure lines of [CII] (158 μm) and [CI] (371 μm and 609 μm) and the higher rotational transitions of CO in the far-IR and the submm. The energy range above ground state from a few 10 K up to about 150 K, as well as the critical density of the transitions, ranging between a few $\times\ 10^3$ cm^{-3} for the fine structure transitions, up to a few $\times\ 10^6$ cm^{-3} for the higher rotational levels of CO, make the transitions easy to excite in PDR's, and they in fact contribute significantly to the gas cooling.

Interpretation of the observed line intensities relies on the availability of reliable collision rates. The references for these are summarized in Table 1. In addition to the presentation here the observations have been summarized in two recent reviews (Genzel, Harris & Stutzki, 1989; Keene 1990) to which the reader is referred for further information. With the limited amount of space available, this review tries to outline the basic observational results and their astrophysical implication.

2. Instrumentation

The submm and far-IR transitions have only become observable with the rapid technical advance in instrumentation over the last ten years. In ad-

C^+ - H	Launay & Roueff, 1977. J.Phys.B 10, 879
C^+ - H_2	Flower & Launay, 1977. J.Phys.B 10, 3673
C^+ - e	Haye & Nußbaumer, 1984, A&A 134, 193
C^0 - H	Launay & Roueff, 1977, A&A 56, 289
C^0 - H_2	Staemmler & Flower, 1991, J.Phys.B 24, 2343 [1]
C^0 - He	Schröder et al. 1991, J.Phys.B 24, 2487
C^0 - e	Johnson et al. 1987, J.Phys.B 20, 2553
C^0 - p	Roueff & Le Bourlot, 1990, A&A 236, 515
CO - He	Green & Chapman, 1978, Ap.J.Suppl. 37, 169
CO - He	McKee, et al. 1982, Ap.J. 259, 647
CO - $H_2\ {}^{para}_{ortho}$	Flower & Launay, 1985, MNRAS 214, 271
CO - H_2 para	Schinke et al. 1985, Ap.J. 299, 839 [2]

TABLE I

Collision Rate References
1) suggested preferential J=2→0 excitation is actually unimportant, in contrast to [OI] (Monteiro & Flower, 1987, MNRAS 228, 101). 2) good agreement with McKee et al. (1982) once detailed balance factors are included properly (in contrast to the conclusion by the authors)

dition to the sophisticated instruments necessary, all these lines require extremely good atmospheric conditions. For ground based observations these are presently only available from the observatories on Mauna Kea, Hawaii. Far-IR observations have to be carried out from the Kuiper Airborne Observatory and from balloon platforms (Okuda 1991). The angular resolution is thus limited typically to the range of a few arcminutes down to 30 arcsecs. With the availability of the new large submm ground based telescopes, the angular resolution has recently been pushed down to the 10 arcsecs regime in the sub-mm spectral lines, thus making possible the first detection of extragalactic CO J=6→5 emission (Harris et al. 1991) and allowing to spatially resolve the clumpy structure of the [CI] emission (G. White, this volume).

The following discussion is still based on a limited number of observations towards typically the brighter sources with favourable geometry (e.g. M17 SW, S140 and Orion A). This is especially true for the submm lines. The situation is better in case of the [CII] observations, where e.g. the high sensitivity multi pixel Fabry-Perot instrument built by the MPE group now allows rapid mapping with adequate spectral resolution in many sources, including external galaxies (e.g. S. Madden et al., this volume).

3. Observational Results: Clumpy, UV-penetrated Clouds

With a few exceptions discussed below, both the spatial distribution and the observed line intensities are in rough agreement with the theoretical expectations from PDR models, i. e. bright emission in [CII], [CI] and warm CO originates from HII-region/molecular cloud interfaces. However, the simple theoretical picture of a semi infinite, plane parallel slab with a stratified layer of CII, CI and CO is not adequate. The observations provide strong support for a picture where the UV-radiation can penetrate substantially deeper into a molecular cloud due to the clumpy or filamentary structure of the molecular material with a high clump/interclump density contrast. While being attenuated and scattered by the dust in the clumps, the UV radiation can create PDR surfaces on clumps rather deep into the molecular cloud, and thus affect a much larger fraction of the cloud material. Evidence for this conclusion comes from basically three facts:

The observed intensity ratios imply rather high densities of the emitting material. In case of the [CII] 158 μm line this is inferred from the intensity ratio relative to the two [OI] lines at 63 μm and 145 μm (Watson 1985), that constrain the density and temperature of the emitting gas to $\sim 10^4$ cm^{-3} and a few 100 K. Optical depth corrections for the [OI] 63 μm line can change these values some. The observed brightness ratio of the two [CI] finestructure transitions $T_B(370~\mu m)/T_B(609~\mu m)$ of about unity implies densities above $10^{3.5}$ cm^{-3} at temperatures of $\sim$ 200 K and $> 10^4$ cm^{-3} at the minimum temperature of $\sim$ 60 K (e.g. Zmuidzinas et al. 1988). Similar conclusions are drawn from the comparison of far-IR and submm CO line intensity ratios (Harris et al. 1987), where, dependent on the column density of the warm CO, a minimum pressure n$\times$T of about $\leq 10^7$ K cm^{-3} has to be present. In all three cases comparison of the line intensities expected under these conditions and the observed intensities implies area filling factors of about 20 - 30 %, and hence small scale structure within the telescope beam.

A second, strong argument comes from the fact that none of the observations so far have succeeded in spatially resolving the CII/CI/CO transition region. The new, high angular resolution [CI] observation with the JCMT (G. White, this volume) show for the first time the [CI] emission peaking at the edge of the molecular clump as traced in $C^{18}O$ 2$\rightarrow$1. At lower angular resolution the emission of CII, CI and warm CO coincides both spatially and in the line shapes (Keene et al. 1985, Genzel et al. 1988, Boreiko et al. 1990, Stutzki et al. 1991, Stacey et al. 1991c). This is naturally explained by the assumption of many spatially unresolved PDR's on individual clumps in the beam, where the observed linewidth is due to the interclump velocity dispersion, whereas the clump intrinsic linewidth is much narrower. An additional component in the [CII] emission from M17 SW overlaps with the clumpy radio-continuum emission at the edge of the HII region.

The third evidence comes from the large spatial extent of the observed [CII] emission, out to an A_v of typically 40-100, estimated from the average density and linear scale. Clearly, no UV photons can travel that far through a homogenous medium. Only clumpiness, with sufficient density contrast so that the photons can travel far distances through the interclump medium, can explain the large observed extent. Modelling of the [CII] emission distribution from a clumpy medium along a cut through the M17 SW interface (Stutzki et al. 1988), and more recently for the two dimensional mapping results in W3 and NGC 1977 (Howe et al. 1991) can successfully reproduce the observed emission. A recent large scale [CII] map of the Orion B region (Jaffe et al. 1991) supports this picture by showing that the [CII] emission closely follows the ridge of high density clumps south of NGC 2023 seen in the survey by Lada et al. (1991).

4. CII/CO correlation: PDR's everywhere?

Also on larger scales does the [CII] 158 μm emission follow very closely the CO 1→0 emission tracing the molecular cloud distribution. Latitude cuts across the Galaxy (Shibai et al. 1991) show this correlation in great detail, confirming earlier results by Stacey et al. 1985. Apparently, PDR's due to the average interstellar radiation field significantly influence also the average molecular cloud, not only the material in the immediate neighbourhood of bright HII regions. The observed correlation between the [CII] and CO 1→0 line flux can indeed be reproduced by theoretical modelling (Wolfire et al. 1989). A similar conclusion is suggested by low-J ^{12}CO and ^{13}CO observations. Castets et al. 1990 reported ^{12}CO and ^{13}CO 1→0 and 2→1 line ratios throughout large parts of the extended Orion A cloud that indicate external heating. Similar line ratios are observed in other clouds (Weikard, et al. 1991, Herbertz, priv. comm). Recently, Gierens, Stutzki and Winnewisser (1991) showed that in a simple model of a spherically symmetric clump with an $n \propto r^{-3/2}$ density structure, and the radial temperature and chemical distribution given by a simple PDR model approximation, can reproduce the observed line ratios. This model includes as a necessary ingredient the ^{13}CO enrichment in the deeper PDR layers due to the $^{13}C^+ +^{12}CO \leftrightarrow\ ^{12}C^+ +^{13}CO + 35K$ exchange reaction.

At least for luminous galaxies, the observed [CII] emission also seems to be dominated by PDR's (Stacey et al. 1991a). In fact, the PDR modelling results can be used to infer from the observed line flux the typical density and UV intensity in the emitting gas (Wolfire et al. 1989, Stacey et al. 1991a).

5. Discussion

Despite of the remarkable success of the UV penetrated, clumpy cloud model in explaining the CII, CI and CO observations, two major problems should be noted: 1) The recent ^{13}CO 6→5 detection (Graf et al. 1990) and subsequent observations in many sources (Graf et al. 1991) indicate the presence of very large column densities of medium warm ($\geq$ 100 K molecular material, much more than can be explained by the UV-heating within PDR models. It may thus well be that our models still miss an important heating mechanism in molecular cloud cores. 2) At the moment it is unclear what mechanism might stabilize the clump/interclump gas, considering the large density contrast required for sufficient UV-penetration.

The basic observational results outlined above may be, at least to some extent, affected by optical depth effects or self absorption. The reported tentative detection of a rather strong [^{13}CII] fine structure line from Orion A (Boreiko et al. 1988), implying a [CII] 158 μm optical depth much too large to be consistent with PDR models, was not confirmed by more recent observations (Stacey et al. 1991b). The new observations detect [^{13}CII] emission on a much lower level, consistent with τ of around unity. Also, observed self reversed profiles as reported for [CII] by Boreiko et al. (1990) and tentatively reported for [CI] 609 μm by Keene (1990), may be the result of chopping onto an extended emission component, rather than originating from absorption in a low excitation foreground gas component. On the other hand, an extended low density and low temperature component, if present, could easily acquire an optical depth of close to unity in the fine structure transitions, and up to now most observations would easily miss its rather weak emission.

We should keep in mind that the present observations are limited to a few selected galactic regions and luminous external galaxies. Observations of lower luminosity regions and larger-scale surveys, which will become feasible with some of the planned space borne observatories and new ground based facilities, may significantly alter the picture outlined here. In fact, the recent spectral line results obtained by COBE (B. Wright, this volume) may be the first challenge towards this direction.

References

Boreiko, R.T., Betz, A.L., and Zmuidzinas, J., 1988, *Astrophys. J. Letters* **325**, L47.
Boreiko, R.T., Betz, A.L., and Zmuidzinas, J., 1990, *Astrophys. J.* **353**, 181.
Burton, M.G., Hollenbach, D.J., and Tielens, A.G.G.M., 1990, *Astrophys. J.* **365**, 620.
Castets, A., Duvert, G., Dutrey, A., Bally, J., Langer, W.D., and Wilson, R.W., 1990, *Astron. Astrophys.* **234**, 469.
Genzel, R., Harris, A.I., Jaffe, D.T., and Stutzki, J., 1988, *Astrophys. J.* **332**, 1043.
Genzel, R., Harris, A.I., and Stutzki, J., 1989, *Proc. of 22nd Eslab Symposium on Infrared Spectroscopy in Astronomy*, ESA SP-290, ed. B.H. Kaldeich, 115.
Gierens, K., Stutzki, J., and Winnewisser, G., 1991, *Astron. Astrophys.*, submitted.
Graf, U.U., Genzel, R., Harris, A.I., Hills, R.E., Russell, A.P.G., and Stutzki, J., 1990, *Astrophys. J. Letters* **358**, L49.
Graf et al. 1991, in preparation.
Harris, A.I., Stutzki, J., Genzel, R., Lugten, J., Stacey, G., and Jaffe, D.T., 1987, *Astrophys. J. Letters* **322**, L49.
Harris, A.I., et al., 1991, *Astrophys. J.* submitted.

Hollenbach, D.J., Takahashi, T., and Tielens, A.G.G.M., 1991, *Astrophys. J.*, in press.
Howe, J.E., Jaffe, D.T., Genzel, R., and Stacey, G.J., 1991, *Astrophys. J.* **373**, 158.
Jaffe, D.T. et al. 1991, in preparation.
Keene, J., Blake, G.A., Philips, T.G., Huggins, P.J., and Beichmann, C.A., 1985, *Astrophys. J.* **299**, 967.
Keene, J., 1990, in *"Carbon in the Galaxy: Studies from Earth and Space"*, NASA-CP 3061, eds. J.C. Tarter, S. Chang, and D.J. DeFrees.
Lada, E.A., Bally, J., and Stark, A.A., 1991, *Astrophys. J.* **368**, 432.
Okuda, H., 1991, *Infrared Phys.* **32**, 365.
Shibai, H. et al. 1991, *Astrophys. J.* **374**, 522.
Stacey, G., Viscuso, P., Fuller, C., and Kurtz, N., 1985, *Astrophys. J.* **289**, 803.
Stacey, G., Geis, N., Genzel, R., Lugten, J.B., Poglitsch, A., Sternberg, A., and Townes, C.H., 1991a, *Astrophys. J.* **373**, 423.
Stacey, G. et al., 1991b, *Astrophys. J.* submitted.
Stacey, G. et al., 1991c, preprint.
Sternberg, A. and Dalgarno, A., 1989, *Astrophys. J.* **338**, 197.
Stutzki, J., Stacey, G.J., Genzel, R., Harris, A.I., Jaffe, D.T., and Lugten, J.B., 1988, *Astrophys. J.* **332**, 279.
Stutzki, J., Genzel, R., Graf, U.U., Harris, A.I., Sternberg, A., and Güsten, R., 1991, in *"Fragmentation of Molecular Clouds and Star Formation"*, IAU, ed. E. Falgarone, p. 135.
Tielens, A.G.G.M. and Hollenbach, D.J., 1985, *Astrophys. J.* **291**, 722.
van Dishoek, E.F. and Black, J.H., 1988, *Astrophys. J.* **334**, 771.
Watson, D.M., 1985, *Physica Scripta*, Vol. **T11**, 33.
Weikard et al. 1991, in preparation.
Wolfire, M.G., Hollenbach, D.J and Tielens, A.G.G.M., 1989, *Astrophys. J.* **344**, 770.
Zmuidzinas, J., Betz, A.L, Boreiko, R.T., and Goldhaber, D.M., 1988, *Astrophys. J.* **335**, 774.

QUESTIONS AND ANSWERS

M.Guelin: Wouldn't carbon recombination lines provide same additional valuable information? They can be observed with a fair/good angular resolution and their intensities scales like n_e^2.

J.Stutzki: The first observations of what we now call PDR's where actually the carbon recombination lines are in Orion, NGC 2024 and a few other sources. At the low radiofrequencies where the recombination lines are bright, the angular resolution is, however, rather poor. In addition, due to the high S/N required to sort out the C recombination line on the shoulder of the hydrogen line, makes the observations rather difficult also.

V.Burdyusha: 1)After off shock waves it takes place very good conditions for thermal instability and a medium must fragmentate. Your results confirm this or not? 2)Why do you think that clouds rotate?

J.Stutzki: The picture of a clumpy, UV penetrated cloud implies that the clumpiness exists already before the UV radiation from one or several newly formed stars is turned on. The increased pressure due to the UV radiation, and possibly also the ionisation shock propagating into the clump cloud could then indeed trigger the collapse of at least some of the clumps and thus induce the formation of the next generation of stars.

L.F.Rodrigues: You showed the remarkable agreement between CO and C^+ in the galactic center. Can you say how much of the carbon is in CO and in C^+ for this region?

J.Stutzki: The galactic latitude cut I showed from the work by Shibai et al. is actually at a galactic longitude of 31°. The remarkable agreement between the [CII] and CO emission is, however, very typical and supports the picture of most of the [CII] emission coming from molecular cloud surfaces. The relative amount of [CII] and CO is thus determined by the amount of molecular material behind the interface, whereas the column density of CII is limited to an A_V of a few. I do not recall the precise numbers for the galactic center region, but I think it is not very different from other luminous regions, and is probably of the order of about 10%.

CH$^+$ in Shocks, Cloud-Intercloud Interfaces, and Dense Photodissociation Regions

W W DULEY

Physics Department, University of Waterloo, Waterloo, Ontario, Canada N2L 3G1

T W HARTQUIST

Max Planck Institute for Physics and Astrophysics, Institute for Extraterrestrial Physics, 8046 Garching, Germany

A STERNBERG

Department of Astronomy, Tel Aviv University, Tel Aviv, Israel

and

R WAGENBLAST and D A WILLIAMS

Department of Mathematics, UMIST, PO Box 88, Manchester M60 1QD, UK

September 5, 1991

1. Introduction

Substantial CH$^+$ abundances are found in at least three types of environment (Lambert and Danks, 1985): atomic regions with little H$_2$, N(CH$^+$) $\sim 10^{12}$ cm^{-2}; diffuse clouds such as that toward ζ Oph, N(CH$^+$) $\sim 10^{13}$ cm^{-2}; reddened ($2 \lesssim A_v \lesssim 4$) lines of sight to bright stars N(CH$^+$) $\sim 10^{14}$ cm^{-2}. We explore the view that several different mechanisms operate.

2. CH$^+$ in atomic gas

Chemistries based on H$_2$ are inoperative. We propose that CH$^+$ is the major erosion product of amorphous carbon grains (Jones, Duley and Williams, 1990) in shocked atomic gas. The outer layers of such grains are primarily alkane chains. Laboratory studies of such carbons show that small alkanes are released thermally and in sputtering. Alkanes in the interstellar medium are subjected to a variety of reactions, and ions tend to appear in the processing, and the final stage before dissociation to atoms is often CH$^+$.

The column density of CH$^+$ may be calculated and may be shown to have values comparable with those observed on lines of sight where H$_2$ is low in abundance.

3. CH$^+$ in warm interfaces of diffuse clouds

In diffuse clouds such as that towards ζ Oph H$_2$ is abundant. Shock models (Pineau des Forêts et al., 1986) have been developed in which CH$^+$ is formed in the warm post-shock gas via the endothermic reaction $C^+(H_2,H)CH^+$. It is unclear whether MHD shock models meet the observational constraints ($H_2(J)$, OH, and velocity shifts: Lambert et al., 1990). We explore the possibility that CH^+ is formed in a warm interface between the diffuse cloud and the ambient hot gas. We have examined a large number of such models. Models in least conflict with observations are warm (2000-4000 K) and of low intensity (radiation parameter $\chi \sim 3$). A model that meets all constraints is $\chi = 3$, T = 4000, nT = 10000 with k_1 twice the canonical value. Warm interfaces $\sim 10\%$ of cloud diameter are required to produce the observed CH^+. Turbulent boundary layers generally have this extent, and their temperatures are elevated (Hartquist and Dyson, 1988).

4. CH^+ in highly reddened regions

$N(CH^+) \sim 10^{14}$ cm^{-2} could be achieved in a single exceptionally large interface, or by contributions from several interfaces. Alternatively, the background stars are very luminous; if the cloud is close to the star a photodissociation region (PDR) will develop (Sternberg and Dalgarno, 1991) and large values of $N(CH^+)$ arise at high number densities and with intense radiation fields. We find that $N(CH^+)$ achieves values $\sim 10^{14}$ cm^{-2} only when conditions are similar to those associated with OH masers (Hartquist and Sternberg, 1989).

5. Conclusions

When H_2 is absent, CH^+ must arise from grains. Models of CH^+ production satisfying other observational constraints are viable, and consistent with the expected warm boundary layers expected around diffuse clouds. Very high CH^+ column densities in highly reddened regions may arise in boundary layers. Alternatively, PDRs can be locations where CH^+ is abundant.

References

HARTQUIST, T W and DYSON, J E: 1988, *Astrophys. Spac. Sci.* **144**, 615
HARTQUIST, T W and STERNBERG, A: 1991, *Mon. Not. R. astr. Soc.* in press,
JONES, A P, DULEY, W W and WILLIAMS, D A: 1990, *Qart. J. R. astr. Soc.* **31**, 567
LAMBERT, D L and DANKS, A C: 1986, *Astrophys. J.* **303**, 4
LAMBERT, D L, SHEFER, Y and CRANE, P: 1990, *Astrophys. J. Letters* **359**, 19
PINEAU DES FORÊTS, G, FLOWER, D R, HARTQUIST, T W and DALGARNO, A: 1986, *Mon. Not. R. astr. Soc.* **220**, 801
STERNBERG, A and DALGARNO, A: 1991, *Astrophys. J.*, in press

WARM MOLECULAR CLOUDS

D.T. JAFFE

Astronomy Department,
The University of Texas, Austin, TX 78712, USA
E-mail DTJ@ASTRO.AS.UTEXAS.EDU

Abstract. Warm molecular gas is important in a large range of astronomical contexts. We discuss here determinations of the temperature and mass of warm material in protostellar disks and cores, photon dominated regions, and molecular material shocked by protostellar outflows. We then compare these results to heating and cooling models. The models of dense cores and photon dominated regions are not adequate to explain the large amounts of warm material observed. This conclusion raises the possibility that there may be other heating mechanisms at work in these regions which theorists have not yet included in their models.

1. Introduction

Warm gas (at 100 K $\lesssim$ T $\lesssim$ 4000 K) exists in molecular clouds on a wide range of size scales with a variety of excitation mechanisms. The gas appears in disks and dense cores surrounding young stars or their accompanying ultracompact HII regions. Photon–dominated regions (PDR's) and recently shocked clouds also contain warm molecular material. On a global scale, warm gas may be the primary molecular component in the inner regions of starburst and other types of active galaxies. Emission from the warm gas usually reflects the presence of strong radiative or mechanical heating sources and traces the effect of these sources. The warm gas is important observationally since, when present, it often dominates the appearance of a region in many lines and the continuum, even if it accounts for only a fraction of the mass. The high temperatures in this gas also permit gas–phase and gas–grain chemistry to differ significantly from the chemistry in cool quiescent clouds. Other reviews in this volume discuss the chemistry and physics of the warm regions. I will summarize here the current observational results on the temperatures and amounts of warm material in the different kinds of regions. We can then compare the results to heating and cooling models to see if these models are adequate to explain the observed properties of the regions.

The simplest method for determining the kinetic temperature of a cloud employs measurements of the brightness temperature, T_{MB} of an optically thick line. For warm clouds, measured in appropriate transitions, the observed T_{MB} is equal to the kinetic temperature of the optically thick gas times the area filling factor of the cloud. A more widely useful method makes use of optically thin emission in several lines arising from different states of the same molecule. A model taking into account both the collisional and radiative excitation of the states and the effects of radiative coupling can be used to derive the relative populations in the measured states and also the kinetic temperature in the cloud. Abundant linear rotors, especially CO, are a frequent choice for temperature determinations using relative state populations. One can overcome problems with high opacity by using a rare isotopic variant of CO. In cases where there is a background continuum source, one can determine the populations of many levels at once using the infrared ro-vibrational transitions. In some sources where density dependent non-LTE population effects are important, it may be possible to use ro-vibrational absorption lines of molecules without permanent dipole moments like H_2 and C_2H_5 to determine the temperature.

Symmetric top molecules like methyl cyanide offer access in the millimeter band to a large number of states at a range of energies above ground, all at similar frequencies. Transitions arising from the same rotational state in different K ladders offer a relatively density–independent means of estimating kinetic temperatures. Molecules like CH_3CN suffer, however, from being difficult to excite and not very abundant.

2. Observations of Disks and Cores

The regions surrounding protostars and young stellar objects often include warm molecular components. Closest to the stellar objects are extremely warm (T$\sim$3500 K), dense (n$\sim 10^{10}$ cm^{-3}), zones producing emission in the 2 μm vibrational overtone bands of CO. In low to moderate luminosity regions, $\sim$25% of the sources searched show such emission (Carr 1989). More luminous sources also have detectable CO overtone emission (Scoville et al. 1979, Geballe and Persson 1987), but higher dust opacities may conceal the emission regions in many of these sources. The total masses of the emission regions are small (a few 10^{-8} to a few 10^{-6} $M_\odot$). A number of heating mechanisms including accretion luminosity and shocks due to infalling matter seem possible as explanations of the observed emission (Scoville et al. 1983, Carr 1989).

Multi-transition millimeter and submillimeter line studies of cores around newly formed luminous stars imply that these regions contain substantial amounts of dense, warm gas. Observations of CS toward regions containing ultracompact HII regions and/or H_2O masers show that there is typically 30-300 $M_\odot$ of material at densities of $10^{5.5}$ to $10^{6.5}$ cm^{-3}(Cesaroni et al. 1991, Plume 1991). NH_3 and CH_3CN observations yield temperatures for the dense material of 120-300 K (Mauersberger et al. 1986, Güsten and Fiebig 1988, Churchwell, Walmsley and Wood 1991, Mangum 1991).

What heats the warm gas in dense cores? The linewidths in the regions are similar to those in other quiescent clouds ($\Delta V \sim$5 km s^{-1}), making shock excitation unlikely. Are gas-grain collisions a plausible heating mechanism in these cores? Radiative transfer models of dense cores with a central luminosity source and a surrounding cavity show a hot layer at the inside where primary radiation strikes the cloud. Once outside of this layer, the dust temperatures are, at best, slightly below the unattenuated equilibrium temperature of the grains. The models predict dust temperatures of only $\sim$50 K at the typical $\sim 10^{17}$ size of the dense cores discussed here (Egan, Leung, and Spagna 1988, Butner et al. 1990). As a consequence, the source of the high gas temperatures in these cores remains unclear.

3. Photon Dominated Regions

The edges of clouds illuminated by far-ultraviolet radiation contain a layer of warm molecular gas. Models of the PDR's explain the temperature as the result of a balance of heating by electrons photoejected from grains and collisionally de-excitation of photo-excited high-v H_2 with cooling through O^o and C^+ fine structure lines, CO rotational lines, and gas-dust collisions (Tielens and Hollenbach 1985, Black and van Dishoeck 1987, Sternberg and Dalgarno 1988). At high densities, the models predict gas temperatures of several hundred K into the cloud to a column depth of $\sim 10^{21}$ cm^{-2}. Model temperature profiles agree well with the observations of the v=(0-0) rotational transitions of H_2 in the Orion Bar PDR (Parmar, Lacy, and Achtermann 1991). In addition, the models succeed in accounting for observations of strong emission in ro-vibrational H_2 lines arising in states with v=1 in sources that also show the emission in the higher v lines which is characteristic of fluorescence. Sternberg and Dalgarno (1989) conclude that the high temperature region necessary to produce the v=1→0 radiation arises from heating by collisional de-excitation of the higher vibrational states of H_2.

The theoretical models are somewhat less successful at explaining the high-J CO emission from PDR's. Observations of ^{12}CO imply column densities of > a few 10^{21} to 10^{22} cm^{-2} in H_2 of gas at 100–300 K, substantially higher than the observed dust temperatures. Models with high density clumps come close to satisfactorily accounting for this emission (Burton et al. 1990). However, more recent observations of ^{13}CO 6→5 implies that there are even larger column densities (a few 10^{22} cm^{-2}) of H_2 at temperatures $\gtrsim 100$ K in some regions where T_{dust} is not high enough for dust-gas heating to be effective (Graf et al. 1990).

4. Shock Heated Gas

In the cores of molecular clouds, high velocity protostellar winds are the ultimate source of the most prominent shock-heated material. The material cooling behind the shock in the best-studied region, Orion/KL, ranges in temperature from several thousand K down to only ~ 2 times the temperature of the ambient cloud (i.e., to 150–200 K). In the Orion region, models variously explain the line intensities of the numerous near-IR ro-vibrational transitions of H_2 as arising in gas affected by a C-shock, where the magnetic field drags along the ions and moderates the usual shock dicontinuites (Draine, Roberge, and Dalgarno 1983) and as arising in a pure J-shock (Brand et al. 1988). Each of these models has its own problems and advantages (see Hollenbach, Chernoff, and McKee 1988). The recent suggestion that one can combine both C and J shocks if the shocked material lies along a bow-wave on the leading edge of a supersonic gas clump offers some promise as a means of accounting for the observed emission in H_2 (Smith, Brand, and Moorhouse 1990).

Observationally, one of the big problems in studying the shocked protostellar regions has been that the best diagnostic lines- the near-IR H_2 lines and the far-IR and submm lines of CO- have not given us information about the same region within the shock. The H_2 lines almost all arise in gas at temperatures in excess of 1500 K while the temperatures derived from the CO observations are significantly lower. It has never been very clear how to de-couple the possible effects of excitation, chemistry, and extinction when comparing H_2 and CO observations. Observations of lower excitation H_2 pure rotational lines (Parmar et al. 1991) indicate that the differences between earlier H_2 and CO measurements were dominated by excitation. The 0-0 J=4-2 and 3-1 line observations of Peak 1 in Orion give a H_2 column density of 3×10^{21} and a temperature of 500 K (Parmar et al. 1991). A temperature of 750 K and a very similar column density provide the best fit to the far-IR CO lines (Watson et al. 1985), while the submm CO data require somewhat lower temperatures and higher column densities. The cooling post-shock gas in both C and J shock models with reasonable parameters can have sufficient column density at the appropriate temperature to explain the observations.

Emission from warm, post-shocked gas is more widespread and prominent than previoiusly thought. In very luminous star formation regions, high velocity gas dominates the emission in the CO J=7→6 line (Jaffe et al. 1989). Even in the well- studied Orion region, recent CO J=2→1 mapping implies that there is high velocity molecular emission present on scales of ~1 pc, covering an area as much as 100 times larger than previously believed (Martin-Pintado et al. 1990).

5. Conclusions

There is warm (T$\gtrsim$100 K) gas present in many types of molecular clouds. The theoretical understanding of the thermal balance in the warm gas is, in general, quite good, but there are several problems : In dense cores, the large sizes and high column densities of the warm regions implied by the multi-transition studies are inconsistent with the likely heating by dust-gas collisions. In photon-dominated regions, the large columns of ~100 K gas implied by the ^{13}CO 6→5 results are not compatible with current models for such regions. The fundamental problem in both cases is that the observed column densities are too large for primary stellar radiation to be effective. There is a strong possibility that we have failed to include some significant heating mechanisms in models of the regions. An additional but somewhat different puzzle arises when trying to explain the source of the widespread warm, dense, high velocity gas observed in many regions.

This work was supported in part by NASA grant NAG2-419, and by a David and Lucile Packard Foundation Fellowship.

References

Black, J.H., and van Dishoeck, E.F. 1987, *Ap.J.*, **322**, 412.

Brand, P.W.J., Moorhouse, A., Burton, M.G., Geballe, T.R., Bird, M., and Wade, R. 1988, *Ap.J.*, **334**, L103.

Burton, M.G., Hollenbach, D.J., and Tielens, A.G.G.M. 1990, *Ap. J.*, **365**, 620.

Butner, H.M., Evans, N.J. II, Harvey, P.M., Mundy, L.G., Natta, A., and Randich, M.S. 1990, *Ap.J.*, **364**, 164.

Carr, J.S. 1989, *Ap. J.*, **345**, 522.

Cesaroni, R., Walmsley, C.M., Kömpe, C., and Churchwell, E. 1991, *Astr. and Ap.*, in press.

Churchwell, E.B., Walmsley, C.M., and Wood, D.O.S. 1991, *Astr. and Ap.*, in press.

Draine, B.T., Roberge, W.G., and Dalgarno, A. 1983, *Ap.J.*, **264**, 485.

Egan, M.P., Leung, C.M., and Spagna, J.F. Jr. 1988, *Comp. Phys. Comm.*, **48**, 271.

Geballe, T.R., and Persson, S.E. 1987, *Ap.J.*, **312**, 297.

Graf, U.U., Genzel, R., Harris, A.I., Hills, R.E., Russell, A.P.G., and Stutzki, J. 1990, *Ap.J. (Letters)*, **358**, L49.

Güsten, R., and Fiebig, D. 1988, *Astr. and Ap.*, **204**, 253.

Hollenbach, D.J., Chernoff, D.F., and McKee, C.F. 1988, in *Proc. 22d ESLAB Symposium, Infrared Spectroscopy*, ed. B.H. Kaldeich (Noordwijk : ESA), p..

Jaffe, D.T., Genzel, R., Harris, A.I., Lugten, J.B., Stacey, G.J., and Stutzki, J. 1989, *Ap.J.*, **344**, 265.

Mangum, J.G. 1991, personal communication.

Martin–Pintado, J., Rodriguez-Franco, A., and Bachiller, R. 1990, *Ap.J.(Letters)*, **357**, L49.

Mauersberger, R., Henkel, C., Wilson, C., and Walmsley, C.M. 1986, *Astr. and Ap.*, **162**, 199.

Parmar, P.S., Lacy, J.H., and Actermann, J.M. 1991, *Ap.J. (Letters)*, **372**, L25.

Plume, R. 1991, personal communication.

Scoville, N.Z., Hall, D.N.B., Kleinmann, S.G., and Ridgway, S.T. 1979, *Ap.J.*, **232**, L121.

Scoville, N.Z., Kleinmann, S.G., Hall, D.N.B., and Ridgway, S.T. 1983, *Ap. J.*, **275**, 201.

Smith, M.D., Brand, P.W.J.L., and Moorhouse, A. 1991, *MNRAS*, **248**, 451.

Sternberg, A. and Dalgarno, A. 1989, *Ap.J.*, **338**, 197.

Tielens, A.G.G.M., and Hollenbach, D.J. 1985, *Ap.J.*, **291**, 722.

Watson, D.M., Genzel, R., Townes, C.H., and Storey, J.W.V. 1985, *Ap.J.*, **298**, 316.

HOT CORE CHEMISTRY: GAS PHASE MOLECULE FORMATION *IN SITU*.

S.B. CHARNLEY[†] and A.G.G.M. TIELENS
Space Science Division, NASA Ames Research Center, California 94035.

ABSTRACT. The possibility that the observed abundances of several molecules in the Orion Hot Core are due to gas phase neutral chemistry has been examined.

The Orion Hot Core is a high temperature (200-300K), dense ($n(H_2) \sim 10^7 cm^{-3}$) clump of gas. It may represent a state of interstellar gas which is common to star-forming regions. The chemistry of the Hot Core is markedly different from that observed cold molecular clouds and also from that of nearby sources within OMC-1 itself (Blake *et al.* 1987; Lacy *et al.* 1989; Turner 1991). It is believed that the Hot Core is a region in which molecular ice mantles have recently been evaporated from warm dust grains. Brown, Charnley & Millar (1988) presented a simple hot core theory which qualitatively reproduced the observed composition. In this picture the presence of a given molecule in a hot core can be accounted for in one of three ways. It may either 1) be formed by cold chemistry during isothermal collapse, accreted at its abundance at the free-fall time, and subsequently released unaltered from the mantle in the hot core, or 2) be formed by grain surface chemistry between accreted atoms and molecules, or 3) be formed *in situ* by neutral-neutral chemistry in the hot gas by reactions involving radicals (present due to a 'burst' of ion-molecule chemistry following mantle evaporation) such as OH, CN, NH_2 and CH_3, and mantle molecules. Starting with the simplest mantle compositions, we have attempted to constrain what reactions may have occurred on grain surfaces, and also the original chemical composition of the cold phase, by first examining the plausibility of the thesis that several molecules are produced *in situ*. We have modelled the neutral chemistry which ensues following the evaporation of ice mantles containing various ratios of the composition $H_2O:NH_3:CH_4:CO:HCN:HNC:C_2H_2:H_2CO:C_2H_4$ into gas with physical conditions similar to that found in Orion. Figure 1 shows the chemical evolution in a representative model.

The original model of Brown *et al.* produced too much HNC and insufficient HCN due to the accretion of gas with a cold cloud HCN/HNC ratio (~ 1). It was postulated that the conversion of HNC to HCN by H atom reaction may occur on the dust. This reaction probably has a large barrier in the solid state; we have included it as a gas phase process with a barrier of 100K as suggested by Pineau des Forêts *et al.* (1990) and find that this, not unexpectedly, leads to HCN/HNC ratios more in accord with observation. The calculated abundance of CN in the models is always below the observed upper limit ($< 5 \times 10^{-10}$).

The reaction of CN with acetylene to form HC_3N will proceed rapidly hot core temperatures (Lichtin & Lin 1986). The observed range of estimates for the abundance of acteylene (Evans *et al.* 1991) are sufficient to produce large abundances of this molecule. Brown *et al.* showed that the hot core abundance of HC_3N could be explained by cold phase chemistry. It is likely that the presence of other cyanopolyynes in the Orion Hot Core (e.g. HC_5N and HC_7N, Turner 1991), which exhibit similar cold phase evolution, is also due to freeze-out of cold phase abundances. Herbst & Leung (1990) have postulated that the reaction of CN with ethene has a significant channel leading to the formation of vinyl cyanide (CH_2CHCN), if so, it will be rapid at hot core temperatures (Lichtin & Lin 1986). Reproduction of the observed vinyl cyanide abundance requires ethene

†NAS/NRC Resident Research Associate

to be present in the mantles at an abundance similar to that of acetylene. Hydrogenation of accreted acetylene to ethene and ethane may be important on grain surfaces (Tielens 1991). We predict that C_2H_4 is present in the Hot Core and that C_2H_6 and should also be present. Reaction of CN with C_2H_6 proceeds rapidly at 300K but the products probably do not include CH_3CH_2CN (Lichtin & Lin 1985); we have been unable to discover a viable neutral chemistry for this molecule. The abundances of HC_3N ($\sim 2 \times 10^{-9}$) produced by cold phase chemistry (Brown et al.) are too low to provide a viable source of CH_2CHCN and CH_3CH_2CN by grain surface hydrogenation of HC_3N, as has been suggested (Tielens 1991). Acetylene, CH_3CN and H_2CO cannot be efficiently synthesised in the hot gas from the simplest mantle compositions, however, the observed abundances of these molecules can be adequately reproduced by formation in the cold phase. It is not certain whether the NO observed in Orion is located in the Hot Core, in the Compact Ridge, or in both (Ziurys et al. 1991). NO cannot be efficiently formed at the temperature of the Orion Hot Core but it can be at lower temperatures ($\sim$ 100K) suggesting that, if formed in situ, it is present in the Compact Ridge. If NO is indeed present in the Hot Core then perhaps the accreted NO is hydrogenated to HNO on grains and, following evaporation, recovered as NO in the hot gas by reaction with H atoms (Westley 1980).

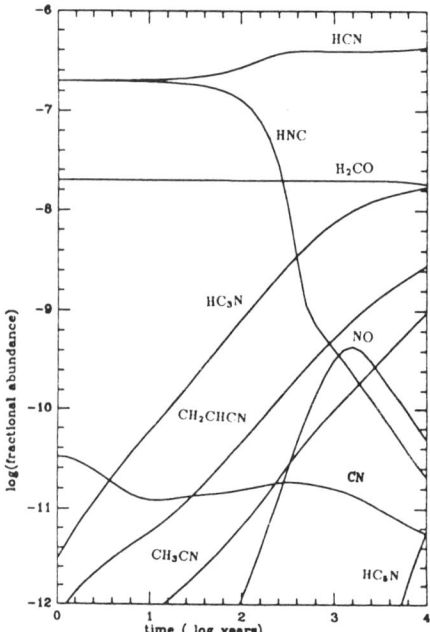

A complex hot core chemistry may be driven by relatively simple mantle compositions. The actual chemistry which occurs depends upon the presence, or otherwise, of certain mantle molecules and the density and temperature of the core gas. This can lead to strong chemical differentiation between cores (Charnley 1991; Charnley, Tielens & Millar 1991).

Figure 1. Evolution of some neutral molecules in a hot core following mantle evaporation ($n(H_2) = 10^7 cm^{-3}$; T=250K). The mantle composition was that inferred from observation (see references in text) for water, ammonia, methane, formaldehyde and CO. HCN and HNC were assumed to have equal abundances. The largest estimate for the acetylene abundance was used ($\sim 10^{-6}$). Ethene was included at an abundance equivalent to that of acetylene.

References

Blake, G.A., Sutton, E.C., Masson, C.R. & Phillips, T.G. 1987, Ap.J., **315**, 621.
Brown, P.D., Charnley, S.B. & Millar, T.J. 1988, M.N.R.A.S., **231**, 409.
Charnley, S.B. 1991, in preparation.
Charnley, S.B., Tielens, A.G.G.M. & Millar, T.J. 1991, in preparation.
Evans, N.J., Lacy, J.H. & Carr, J.S. 1991, preprint.
Herbst, E. & Leung, C.M. 1990, Astr.Ap., **177**, 180.
Lichtin, D.A. & Lin, M.C. 1985, Chem.Phys. **473**, 482.
Lichtin, D.A. & Lin, M.C. 1986, J.Chem.Phys. **325**, 330.
Pineau des Forêts, Roueff, E & Flower, D. 1990, M.N.R.A.S., **244**, 668.
Tielens, A.G.G.M. 1991, in Chemistry and Spectroscopy of Interstellar Molecules, ed. N. Kaifu, University of Tokyo Press
Turner, B. 1991, Ap.J.Suppl., **76**, 617.
Westley, F. 1980, NSRDS-NBS 67; Washington DC.
Ziurys, L., McGonagle, D., Minh, Y. & Irvine, W.M. 1991, Ap.J., **373**, 535.

THE PHYSICAL AND CHEMICAL STRUCTURE OF WARM, DENSE REGIONS : IC 63 AND IC 443

D.J. Jansen[1], E.F. van Dishoeck[1], J.H. Black[2] and T.G. Phillips[3]

[1] *Leiden Observatory, P.O. Box 9513, 2300 RA Leiden, The Netherlands*
[2] *Steward Observatory, Univ. of Arizona, Tucson, AZ 85721, USA*
[3] *Div. of Physics, Mathematics & Astronomy, Caltech 320-47, Pasadena, CA 91125, USA*

1. Introduction

Elevated temperatures in molecular clouds can result either from heating by ultraviolet photons or from the passage of shock waves. The effect that these processes have on the chemical abundances is not well established observationally, but is of great importance for the interpretation of molecular line observations not only in our own Galaxy, but also in external galaxies. We present here initial results from our study of two "template" regions: the photon–dominated region IC 63 and the shocked region IC 443.

In order to derive molecular abundances, it is necessary to have accurate constraints on the temperature and density in the sources. These have been obtained from observations of (sub)millimeter lines of CO ($2\to1$, $3\to2$), HCN, HCO$^+$ ($1\to0$, $3\to2$, $4\to3$), CS ($2\to1$, $5\to4$, $7\to6$) and H$_2$CO (various lines), combined with detailed statistical equilibrium calculations.

2. Photon–dominated region : IC 63

IC 63 is a reflection nebula located close to the star γ Cas (type B0.5 IV, $D \approx 200$ pc), in which fluorescent ultraviolet H$_2$ emission has been detected (Witt *et al.* 1989). The cloud is exposed to several hundred times the general Galactic UV background radiation. Our CO $2\to1$ observations show a small ($1'\times2' \approx 0.05\times0.1$ pc), isolated molecular cloud coincident with the brightest nebulosity. The line widths are of order 2 km s^{-1}, so that there cannot be an important shocked component. From the relative strengths of the CO, CS, HCO$^+$ and HCN lines, we find $T \approx 35$ K and $n \approx 3 \times 10^4$ cm^{-3}. The derived abundances at this temperature and density are listed in Table 1.

3. Shocked region : IC 443

IC 443 is a well-known supernova remnant, which has been studied before by White *et al.* (1987) and Ziurys *et al.* (1989). Its interaction with the molecular cloud shows up as strong and broad ($\Delta V \gtrsim 20$ km s^{-1}) molecular lines in small clumps. Observations of shocked H$_2$ reveal a ring–like structure (Burton *et al.* 1988). Comparison of CO $1\to0$, $2\to1$ and $3\to2$ line strengths along the ring shows that the low–J CO lines are optically thin. We have concentrated our searches for other molecules on three positions: one in clump B, and two in clump G (nomenclature from Huang *et al.* 1986).

From the modeling of the relative strengths of the CS, HCN, HCO$^+$ and H$_2$CO lines (see Figure 1), we find temperatures of about 100 K and densities around 5×10^5 cm^{-3}, with little variation between the clumps. Our column densities for CS, HCN and HCO$^+$ (see Table 1) agree with those of Ziurys *et al.* within a factor 2 at the same positions.

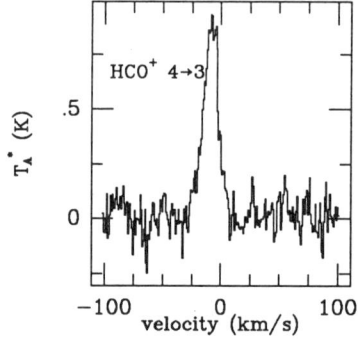 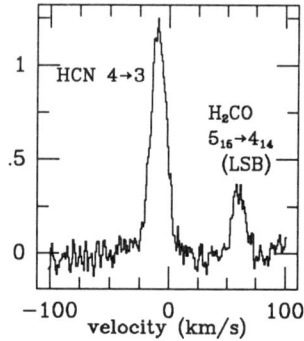

Fig. 1. Spectra of IC 443 G obtained at the Caltech Submillimeter Observatory

4. Conclusions

The preliminary results presented in Table 1 show that within factors of a few, there are no significant differences between the abundances in IC 63 and IC 443, and dark clouds like L134N and TMC-1, for the molecules studied in this work. In IC 63, the abundances may be slightly lower due to the enhanced photodissociation rates. In IC 443, the shock has clearly compressed and heated the surrounding molecular cloud, but has hardly affected the chemistry. In particular, the HCO^+ abundance is not significantly enhanced, contrary to earlier speculations. Volatile molecules such as CH_3OH, and sulfur–bearing molecules such as SO and SO_2 have low abundances in both clouds.

Table 1. Column densities N (in cm^{-2}) and relative abundances X (CO / H_2 = 1×10^{-4}).

	IC 63		IC 443 B		IC 443 G		L134N	TMC-1
	N	X	N	X	N	X	X	X
H_2	3 (21)	1.0	5 (21)	1.0	7 (21)	1.0	1.0	1.0
CO	3 (17)	1 (-4)	5 (17)	1 (-4)	7 (17)	1 (-4)	1 (-4)	1 (-4)
^{13}CO	2 (15)	7 (-7)	8 (15)	2 (-6)	2 (16)	3 (-6)	4 (-6)	4 (-6)
CS	1 (13)	3 (-9)	5 (13)	1 (-8)	3 (13)	4 (-9)	7 (-10)	1 (-8)
HCN	4 (12)	2 (-9)	4 (13)	8 (-9)	1 (14)	1 (-8)	3 (-9)	2 (-8)
HCO^+	5 (12)	2 (-9)	4 (13)	8 (-9)	8 (13)	1 (-8)	6 (-9)	3 (-9)
H_2CO	8 (12)	3 (-9)	3 (13)	6 (-9)	6 (13)	9 (-9)	2 (-8)	2 (-8)
CN	3 (13)	1 (-8)	...	...	<3 (13)	<4 (-9)	<3 (-9)	3 (-8)
SO	<1 (12)	<3 (-10)	...	...	...	...	5 (-9)	5 (-9)
SO_2	<1 (13)	<3 (-9)	...	...	<1 (13)	<1 (-9)	1 (-9)	<1 (-9)
CH_3OH	<4 (12)	<2 (-9)	...	...	...	...	2 (-9)	2 (-9)

Burton, M.G., Geballe, T.R., Brand, P.W.J.L., Webster, A.S. 1988, MNRAS **231**, 617.
Huang, Y.-L., Dickman, R.L., and Snell, R.L. 1986, ApJ **302**, L63.
White, G.J., Rainey, R., Hayashi, S.S., and Kaifu, N. 1987, A&A **173**, 335.
Witt, A.N., Stecher, T.P., Boronson, T.A., and Bohlin, R.C. 1989, ApJ **336**, L21.
Ziurys, L.M., Snell, R.L., and Dickman, R.L. 1989, ApJ **341**, 857.

Preliminary Observations of the Galaxy with a 7° Beam by the Cosmic Background Explorer (COBE)

EDWARD L. WRIGHT

UCLA Astronomy Dept., Los Angeles CA 90024-1562

September 6, 1991

Abstract. The FIRAS instrument on COBE has mapped almost the entire sky with a 7° beam and a spectral resolution of about 1 cm^{-1} over the 1-100 cm^{-1} range. Maps showing the strengths of the three main components: dust continuum, [C II] and [N II] are presented. A mean spectrum of the galactic emission is found. These results are used to compare the Milky Way to other galaxies.

1. Introduction

This paper will summarize results that are more extensively discussed in Wright et al. (1991). The maps presented here include data collected during the entire ten month lifetime of the helium cryogen on COBE, while Wright et al. only considered the first half of the mission. However, since only six months are required to cover the sky, the increased sky coverage is from 72% to 89%. Data taken in the high spectral resolution long scan mode will fill in some of the remaining holes, leaving the total sky coverage from the FIRAS instrument at 98%. Maps of the weaker lines such as [C I] and CO only show significant signals in the inner galactic plane, while the maps of the stronger components presented here show structure over much of the sky.

2. Approach

The details of the calculation can be found in Wright et al., but the basic approach is to approximate the emission from the galaxy as the product of a function of position and a function of frequency: $I_\nu(l,b) = G(l,b)g(\nu)$. The continuum map in Figure 1 is the function $G(l,b)$. The spectrum $g(\nu)$ has been modeled as a dust continuum plus lines by Wright et al.

The dust continuum was fit using two populations of dust with ν^2 emissivity laws: a warm component at 20.4 K and a cold component at 4.77 K with 6.7 times more opacity than the warm component.

3. CO Fitting

The four measured CO lines and the limit on the 1-0 line can be used to make a model of the CO distribution. The model that I have used is one that assumes warm and cold clouds, both optically thin and in LTE. The column density and temperature in each type of cloud give four parameters,

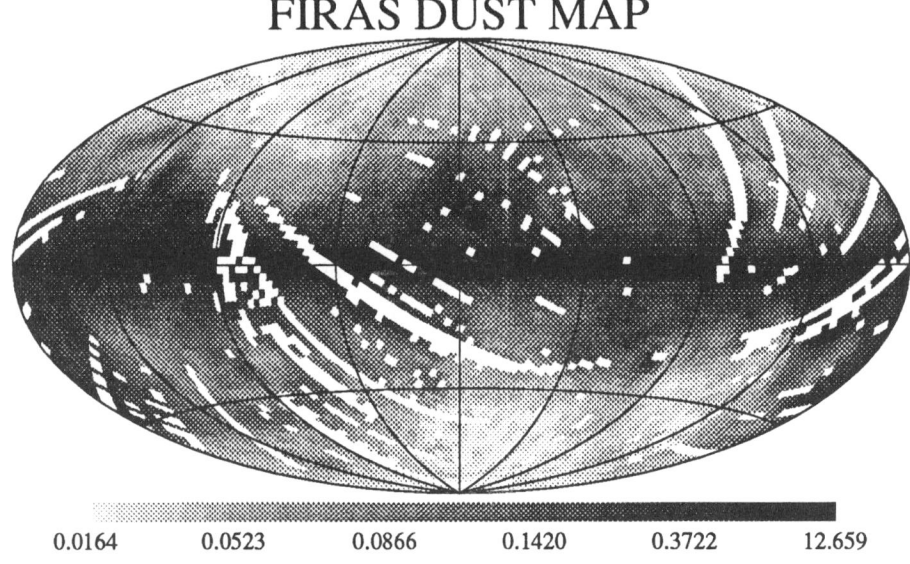

Fig. 1. A continuum map based on data from 23-50 wavenumbers.

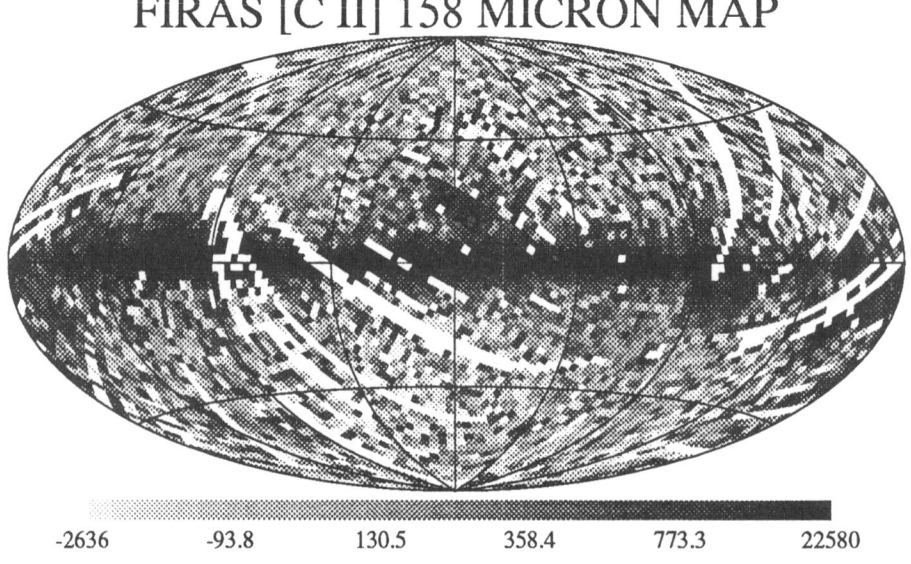

Fig. 2. Map of the [C II] 158 μm line.

FIRAS [N II] 205 MICRON MAP

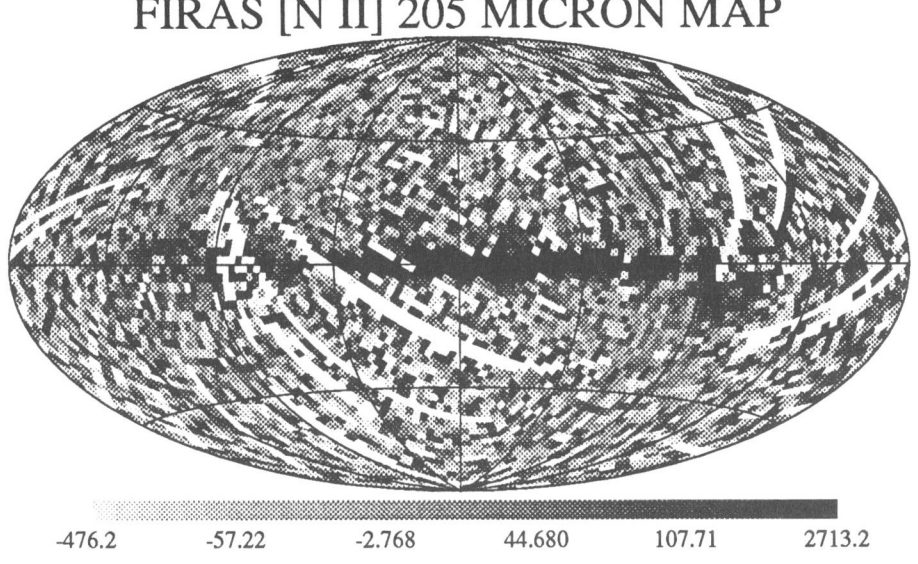

Fig. 3. Map of the [N II] 205 μm line.

and given only four measured lines, a good fit is expected even though the model is not very realistic or complete. The model has 88% of the CO at 5 K and 12% at 19 K. Figure 4 shows the line strengths and the predicted strengths from the model. This model predicts $0.5\ 10^{-8}$ erg/cm^2/sec/sr for the 1-0 line, while the CO-C$^+$ correlation from Wolfire, Hollenbach and Tielens (1989) predicts $0.4\ 10^{-8}$ erg/cm^2/sec/sr, compared to the observed $(0.6 \pm 0.7)\ 10^{-8}$ erg/cm^2/sec/sr.

The total CO flux from this model is $16\ 10^{-8}$ erg/cm^2/sec/sr, or 0.9% of the 158 μm flux. Note however, that the FIRAS sensitivity at short wavelengths is much worse than the sensitivity at 1 mm, so a fairly strong "hot" CO component could be present without being detected. Such a component could dominate the total CO luminosity. For example, if one includes a hot component with 1% of the total CO at T = 70 K, then the total CO luminosity doubles while χ^2 of the fit increases by only four units, even after including reasonable upper limits for the CO lines from 7-6 to 16-15.

4. [N II] Lines

The [N II] lines seen be FIRAS are quite strong relative to the [C II] lines, and cast some doubt on the idea that the [C II] line emission of galaxies is dominated by PDRs which produce no [N II] emission. Gry et al. (1991) have computed the [N II] and [C II] emissivities in the local ISM using

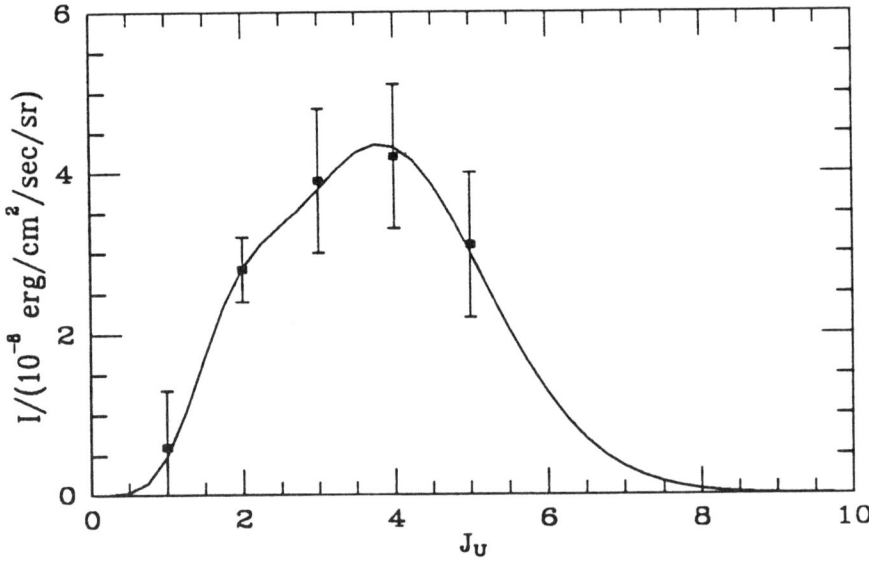

Fig. 4. Two temperature fit to the CO data.

Copernicus observations of the excited state column densities in these ions. I have taken their ratio of I(205)/I(158) for the ionized part of the local ISM and plotted it as the steeper line in Figure 5, where the points are from Wright et al., while the line with the smaller slope is the ratio of the median I(205)/N_H from Gry et al. to the average I(158)/N_H. Note that the latter line fits the low intensity, high galactic latitude points well, while the ionized ISM value fits the higher intensity galactic plane points. This shows that a combination of the local diffuse ISM with low density H II regions can explain the observed [N II] to [C II] relation without invoking much PDR emission.

However, for a mean density of 1 H cm^{-3}, the diffuse ISM can have an opacity in the [C II] line of $\approx 1/\sin(2l)$ if all carbon is in the form of C$^+$ and the velocity gradient follows the galactic rotation curve. If high spectral resolution observations of distant sources such as W49 show a number of sharp absorption dips superimposed on the emission line from the source, then the true [C II] emission from the galaxy would be higher than flux measured by FIRAS, allowing for a greater contribution from PDRs to the [C II] emission.

5. Discussion

The FIRAS observations of the Milky Way show that its ratio of [C II] 158 μm line flux to the total far-infrared flux is typical of other galaxies

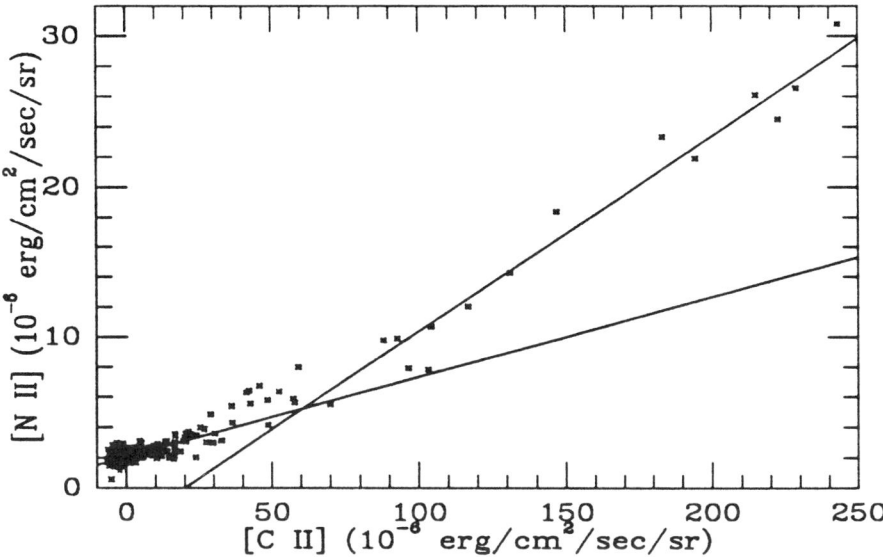

Fig. 5. N$^+$ vs. C$^+$. Lines show ratio for the median of 7 local lines-of-sight and the average of the local ionized ISM. Data are from Wright et al..

discussed by Stacey et al. (1991). The species CO and [C I] each contribute about 1% as much flux as [C II], while the [N II] line at 205 μm contributes a surprisingly large 10% of the power from [C II].

Acknowledgements

The National Aeronautics and Space Administration/Goddard Space Flight Center (NASA/GSFC) is responsible for the design, development, and operation of the Cosmic Background Explorer (COBE). Scientific guidance is provided the the COBE Science Working Group. GSFC is also responsible for the development of the analysis software and for the production of the mission data sets.

References

Gry, C., Lequeux, J., and Boulanger, F. 1991: 'Fine Structure Lines of C$^+$ and N$^+$ in the Galaxy', *AA* **TBD**, TBD

Stacey, G. J., Geis, N., Genzel, R., Lugten, J. B., Poglitsch, A., Sternberg, A., and Townes, C. H. 1991: 'The 158 Micron [C II] Line: A Measure of Global Star Formation Activity in Galaxies', *ApJ* **373**, 423-444

Wolfire, M. G., Hollenbach, D. J., and Tielens, A. G. G. M. 1989: 'The Correlation of C$^+$ and CO', *ApJ* **344**, 770-778

Wright, E. L. et al. 1991 : 'Preliminary Spectral Observations of the Galaxy with a 7° Beam by the Cosmic Background Explorer (COBE)' , *ApJ* **TBD**, TBD

QUESTIONS AND ANSWERS

V.V.Burdyzha: What is the possibility for observation of the distortion CMB in range 100 - 150 μ by COBE, which creats in the moment of the recombination of the universe?

E.L.Wright: The smallest galactic flux seen by COBE at 100 - 150 μm has about 10^{-1} times the CMB power. This is much larger than the predicted distortion in the CMB caused by the Lyman α line at recombination.

S.P.Tarafdar: You gave two temperatures for CO. Are they excitation temperature or kinetic temperature? These temperatures are very similar to two dust temperatures you obtained. Can you comment?

E.L.Wright: The temperatures are excitation temperatures. The similarity to the dust temperatures is probably coincidental.

E.van Dishoeck: Are you worried that COBE's lack of detection 557 GHz water emission will mean that the submillimeter Wave Astronomy Satellite (SWAS) will not see water?

G.Melnick: No - for three reasons. First, water emission is expected to arise predominantly in warm ($T > 30\,K$), dense ($n_{H_2} > 10^4\,cm^{-3}$) gas. Such gas comprises a very small fraction of COBE's 7 degree beam. SWAS will have a solid angle 10^4 times smaller than COBE's, which means the filling factor will favor SWAS. Second, COBE's spectral resolution, $\lambda/\Delta\lambda$, was about 100. SWAS will have a spectral resolution of 5×10^5, or 5000 times higher than COBE's. This increased spectral resolving power will aid SWAS immensely in distinguishing line emission from the (spectrally) neighboring dust continuum emission. Finally, the upper limit COBE set on the 557 GHz water emission is less than a factor of 10 lower than COBE's measurements of the lower-J CO line strengths. The known antenna temperatures produced by these CO lines for fields-of-views comparable to that of SWAS ($\sim$ 4 arcminutes) are very strong by SWAS's sensitivity expectations. Even if the 557 GHz water line strength is 10 times below COBE's upper limit, SWAS should have no problem detecting this emission.

CHEMICAL COMPOSITION VARIATIONS IN THE GALAXY

Ju. L. FRANTSMAN
Radioastrophysical observatory
Latvian Academy of Sciences
Turgeneva 19, 226524, Riga, Latvia

ABSTRACT. Comparing the data about the Cepheid periods and the star evolution calculations, the conclusion is made that inside 5 kpc distance from the Sun heavy element abundance Z changes between 0.01 and 0.04.

Most Cepheids are believed to be in the phase of core helium burning. Theoretical frequency-period distributions were obtained [1] by taking into account where, during its helium burning phase, each model of a series of models of different composition crosses the instability strip and how long it spends pulsating there. The results show the extreme sensitivity of the distribution to variations in the composition. These theoretical results agree with the observations of Cepheid metallicity [2] (Figure 1).

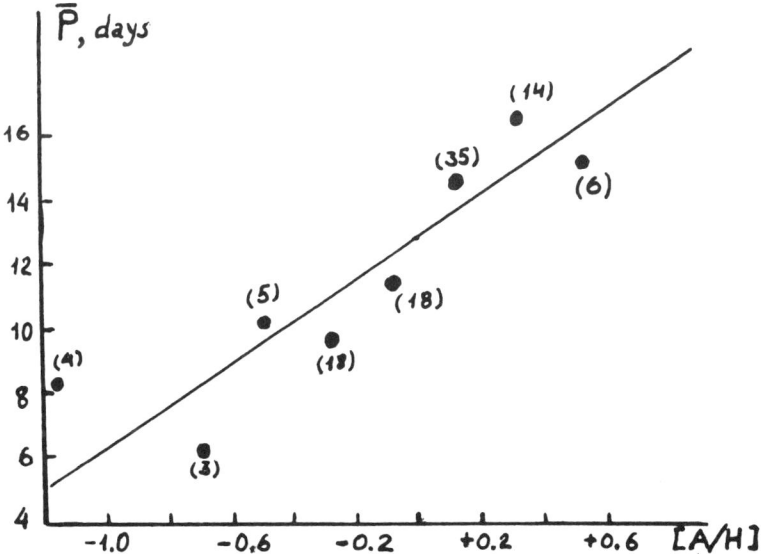

Figure 1. The observational mean period – metal abundance A/H diagram. Numbers – the amount of Cepheids.

In Figure 2 the mean period distributions are shown in rings on the different distances from the Sun. It is possible to distinguish regions with considerable amount of long period Cepheids: in the direction of the center of the Galaxy beginning with 2 kpc (apparantly due to the gradient of the heavy elements in the Galaxy), the regions with l=70-90 within 2-3 kpc of the Sun and l=225-270 within 3-4 kpc of the Sun. Comparing these observational data about Cepheid periods and the results obtained by the star evolution calculations (Figure 2) the conclusion is made that within 5 kpc from the Sun the heavy element abundance Z of matter changes between 0.01 and 0.04.

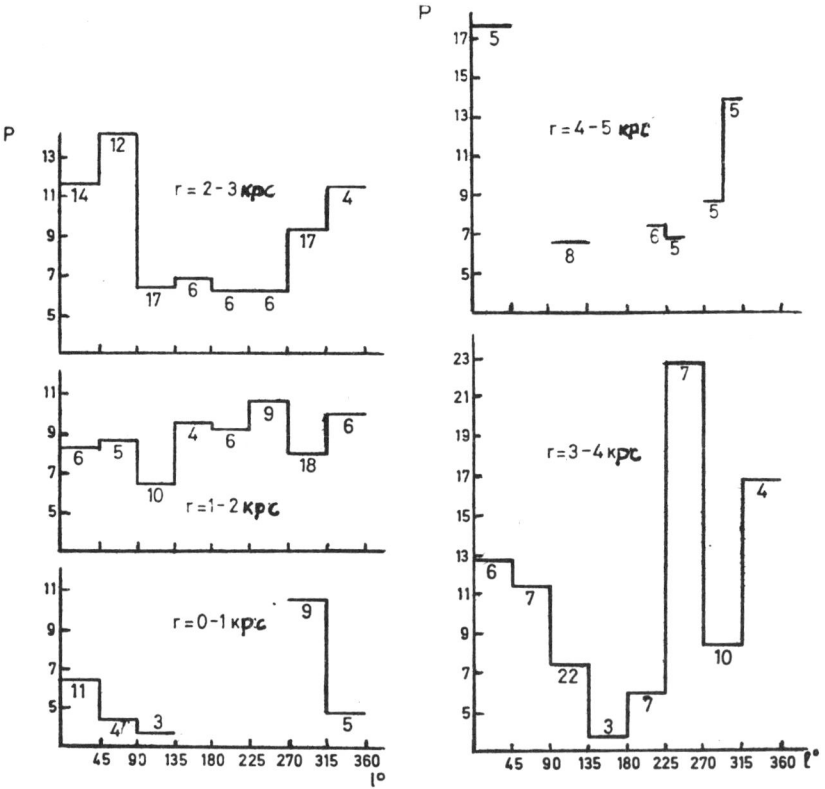

Figure 2. The mean period of Cepheids in the Galaxy distributions along the longitude in the rings on the different distances from the Sun.

References.
1. Becker S.A., Iben I.,Jr., Tuggle R.S.(1977) 'On the frequency-period distributions of Cepheid variables in galaxies in the local group', Ap. J., 218, 633.
2. Harris H.C. (1981) 'Photometric abundances of classical cepheids and the gradient in the Galactic disk', A. J., 86, 707.

The Chemistry of Photon-Dominated Regions

AMIEL STERNBERG

School of Physics and Astronomy, Tel Aviv University, Ramat Aviv, Israel

September 5, 1991

Abstract. Theoretical models of photon-dominated regions (PDRs) of molecular cloud surfaces are described. Key aspects of the chemistry of dense PDRs are presented.

1. Introduction

The surface layers of the clumpy molecular clouds in star forming regions are commonly exposed to intense fluxes of far-ultraviolet (6-13.6 eV) photons emitted by external or embedded OB stars. The physical and chemical structures of these neutral hydrogen surface layers are controlled by the FUV photons and are therefore called photon-dominated regions (PDRs). Many atomic and molecular emission lines are produced in PDRs. Recent observational results are summarized elsewhere in this volume (see articles by White, Stutzki, and Jaffe). Here I discuss theoretical models of the chemistry of PDRs. Other reviews of PDRs have been presented by Genzel, Harris and Stutzki (1989) and by Hollenbach (1990).

Most of the PDR models in the literature (*e.g.* De Jong, Boland and Dalgarno 1980, Tielens and Hollenbach 1985, van Dishoeck and Black 1988, Sternberg and Dalgarno 1989) are one dimensional and assume steady state conditions. The models of Tielens and Hollenbach (1985) and Sternberg and Dalgarno (1989) explicitly compute the gas temperature and the gas-phase chemical abundances as functions of cloud depth.

The hydrogen particle density, n (cm^{-3}), and the UV intensity, χ (in units of the average interstellar FUV intensity) are crucial parameters which govern PDR structure. In star forming regions n ranges from 10^3 to 10^7 cm^{-3} and χ ranges from 10^2-10^6. In many low density ($n < 10^5$ cm^{-3}) PDRs the dominant gas heating mechanism in the outer layers is probably dust photoelectric emission. In low density PDRs the cooling is dominated by fine structure emission of atoms and ions (*e.g.* CII 157 μm and OI 63 μm). In high density regions ($n > 10^5$ cm^{-3}) collisional deexcitation of UV-pumped H_2 molecules becomes an effective heating mechanism. Molecular line cooling (H_2, CO, OH) becomes significant in high density PDRs.

The model calculations show that in dense PDRs the outer atomic hydrogen surface layers become hot (T> 1000K). The gas temperature falls rapidly in the transition region where the hydrogen becomes molecular. The extent of the hot atomic hydrogen layer depend on the ratio χ/n (Sternberg and Dalgarno 1989). When $\chi/n > 0.01$ cm^3 the extent of the atomic layer is limited by dust absorption and is of order one visual extinction, A_v. When

$\chi/n < 0.01$ cm^3 the extent of the atomic layer is limited by the opacity of the H_2 FUV absorption lines and is therefore much smaller. Thus, for fixed χ H_2 self-shielding reduces the sizes of the hot layers as n increases.

2. Chemistry

The chemical structure of dense PDRs is illustrated in Fig. 1 (adapted from Sternberg and Dalgarno 1991) which diplays the fractional abundances of various atomic and molecular species as functions of A_v for a cloud with $n = 10^6$ cm^{-3} and $\chi = 10^5$. The total abundances of carbon and oxygen relative to hydrogen are 3×10^{-4} and 6×10^{-4}. At the cloud surface grain surface H_2

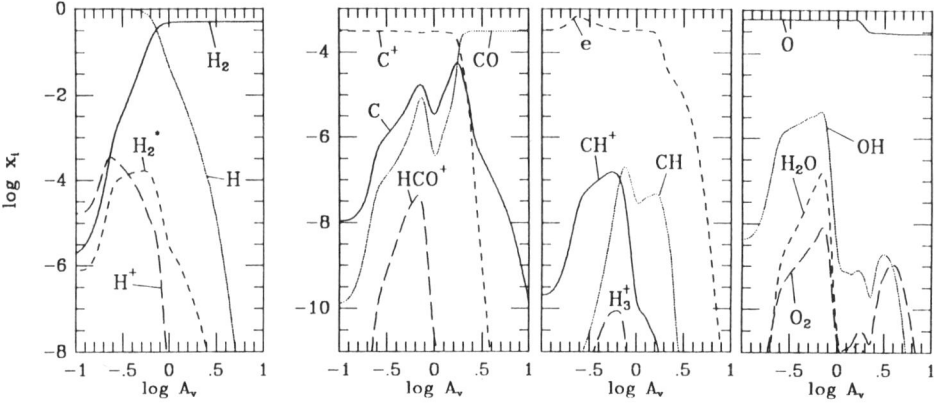

Fig. 1.

formation is slow compared to molecular photodissociation and the hydrogen is atomic. At large cloud depths the FUV radiation is attenuated and the hydrogen becomes molecular. Similarly, at the cloud surface all of the carbon is in the form of C^+ and at large cloud depths it is fully incorporated into CO. At intermediate depths the neutral atomic carbon fraction becomes large. Note that the atomic carbon abundance profile has two peaks. At the inner peak ($A_v = 1.6$) the gas temperature is less than 100 K, the hydrogen is fully molecular, and the carbon chemistry is governed by ion-molecule and neutral reactions which proceed at low temperature. The atomic carbon abundance is set by the balance between photoionization and recombination

$$C + \nu \leftrightarrow C^+ + e \quad . \tag{1}$$

CH^+ and CH are formed by the slow sequence initiated by radiative association

$$C^+ + H_2 \to CH_2^+ + \nu \tag{2}$$

$$CH_2^+ + e \to CH + H \tag{3}$$

$$CH + \nu \rightarrow CH^+ + e \qquad (4)$$

$$CH + C^+ \rightarrow CH^+ + C \qquad (5)$$

CO and OH are formed by the neutral reactions

$$CH + O \rightarrow CO + H \qquad (6)$$

$$CH + O \rightarrow OH + C \qquad (7)$$

and are removed by photodissociation.

At the outer atomic carbon peak ($A_v = 0.6$) the gas temperature is greater than 1000 K, the hydrogen is primarily atomic and the chemistry is dominated by reactions with moderate endothermicities or activation energies. Thus, CH^+ is formed rapidly by the abstraction reaction

$$C^+ + H_2 \rightarrow CH^+ + H \qquad (8)$$

which leads to the rapid production of CH via another abstraction reaction

$$CH^+ + H_2 \rightarrow CH_2^+ + H \qquad (9)$$

followed by reaction (3). The resulting CH abundance is sufficiently large that the reactions

$$CH + \nu \rightarrow C + H \qquad (10)$$

$$CH + H \rightarrow C + H_2 \qquad (11)$$

dominate the production of atomic carbon rather than reaction (1). A large proton abundance is maintained in the hot gas via the photodissociation reaction

$$CH^+ + \nu \rightarrow C + H^+ \quad . \qquad (12)$$

OH and H_2O are formed rapidly via the neutral reactions

$$O + H_2 \rightarrow OH + H \qquad (13)$$

$$OH + H_2 \rightarrow H_2O + H \qquad (14)$$

which become rapid at high temperatures. OH is a crucial intermediary in hot gas chemistry (*cf.* Neufeld and Dalgarno 1989) and its large abundance generates the CO and HCO^+ and O_2 peaks in Fig. 1 via the reactions

$$C^+ + OH \rightarrow CO + H^+ \qquad (15)$$

$$C^+ + OH \rightarrow CO^+ + H \qquad (16)$$

$$CO^+ + H_2 \rightarrow HCO^+ + H \qquad (17)$$

$$O + OH \rightarrow O_2 + H \quad . \qquad (18)$$

The large OH abundances produced in dense PDRs suggests that they may be the sites of the OH maser spots that surround compact HII regions (Hartquist and Sternberg, 1991).

The molecular abundance peaks that form in the hot regions are sources of intense collisionally excited molecular line emission. Near infrared emission from vibrationally excited hydrogen molecules (H_2^*) and far-infrared lines of rotationally excited CO are produced in the hot gas (Sternberg 1986, Hollenbach 1988, Sternberg and Dalgarno 1989, Burton, Hollenbach and Tielens 1990). The double lobed H_2 emission source recently discovered in the narrow line region of the Seyfert galaxy NGC 1068 may be produced in dense PDRs (Rotaciuc et al. 1991).

3. Future Work

The theoretical models have been successful in interpreting and stimulating many observations. Detailed comparison with the observations now require that the models be improved to include the effects of time dependence, clumpy cloud structure, finite cloud size, and non-stellar radiation fields.

Acknowledgements

This work was supported in part by the Max-Planck Institut für extraterrestrische Phyisk, Garching, Germany.

References

Burton M., Hollenbach D.J. and Tielens A.G.G.M., 1990, *Ap.J.* **365**,620.

deJong T., Dalgarno A. and Boland W., 1980, *Astr.Ap.* **91**,68.

Genzel R., Harris A.I. and Stutzki J., 1989, in *Infrared Spectroscopy in Astronmy* ed. M. Kessler (ESA SP series).

Hartquist T.W. and Sternberg A., 1991, *M.N.R.A.S.* **248**,48.

Hollenbach D.J., 1988, *Astr. Lett. and Comm.* **26**, 191.

Hollenbach D.J., 1990 in *Evolution of the Interstellar Medium* ed. L. Blitz

Neufeld D.A. and Dalgarno A., 1989, *Ap. J.* **340**,869.

Rotatiuc V., Krabbe A., Cameron M., Drapatz S., Genzel R., Sternberg A. and Storey J.W.V., 1991, *Ap.J.Letters* **370**,L23.

Sternberg A., 1986, *Ph.D. thesis* Columbia University

Sternberg A., 1988, *Ap.J.* **322**,400.

Sternberg A. and Dalgarno A., 1989, *Ap.J.* **338**,197.

Sternberg A. and Dalgarno A., 1991, in prep.

Tielens A.G.G.M. and Hollenbach D.J., 1985, *Ap.J.* **291**,722.

van Dishoeck E.F. and Black J.H., 1988, *Ap.J.* **334**,711.

NEAR INFRARED EMISSION OF NEUTRAL CARBON FROM PHOTON-DOMINATED REGIONS

V. ESCALANTE[1], A. STERNBERG[2], and A. DALGARNO[3]
[1] *Instituto de Astronomía, UNAM, Ap. Postal 70-264, Mexico, DF 04510, Mexico*
[2] *School of Physics and Astronomy, Tel Aviv University, Ramat Aviv, Israel*
[3] *Center for Astrophysics, 60 Garden St. Cambridge, MA 02138, USA*

ABSTRACT. Detailed calculations are reported of the intensities of the near infrared forbidden lines of neutral carbon atoms at $\lambda\,985.0$ nm, 982.3 nm and 872.7 nm emitted from dense clouds subjected to intense radiation fields. The metastable levels that produce the lines are excited by radiative recombination of the C^+ ions produced by photoionization. Impacts of electrons with C atoms in the heated edge zones of the clouds contribute an insignificant part to the excitation. The lines observed in M42 and NGC 2024 can be interpreted as arising in gas with densities in excess of $10^5\,{\rm cm}^{-3}$ and radiation fields with intensities between 10^3 and 10^6 times the average interstellar field intensity. Radiative recombination of C^+ ions may also be an important source of the emission lines detected in the planetary nebulae NGC 6270 and NGC 7027.

1. Excitation of the NIR Forbidden Lines of C

The near infrared forbidden lines of neutral carbon are emitted in the transitions $2p^2\ ^3P_2 - \ ^1D_2\ \lambda\,985.0$ nm, $2p^2\ ^3P_1 - \ ^1D_2\ \lambda\,982.3$ nm, and $2p^2\ ^1D_2 - \ ^1S_0\ \lambda\,872.7$ nm. The $^1D_2$ and $^1S_0$ levels can be excited by electron impacts from the C ^3P ground term or by radiative recombination of C^+ followed by cascading. Non-hydrogenic quantum-mechanical calculations of the recombination and transition rates were carried out to predict the emission of the forbidden lines of C (Escalante and Victor, 1990).

When the lines are excited by radiative recombination, their intensity depends weakly on temperature and is proportional to the C^+ and electron column densities at low densities. Furthermore the line intensity ratio of $\lambda\,985.0$ (or $\lambda\,982.3$) to $\lambda\,872.7$ is almost independent of temperature. This ratio, however, depends on density because of quenching of the upper levels. On the other hand, excitation by electron impact produces line intensities that depend strongly on temperature. If the temperature and C/C^+ ratio are sufficiently high, electron impact becomes the dominant mechanism. Thus, observations of these lines gives a diagnostic of the excitation mechanism, the radiation field and density in the region.

2. Results and Comparison with Observations

UV radiation from hot stars creates transition regions in molecular cloud boundaries where hydrogen is mostly in atomic form due to photodissociation. In these photon-dominated regions, also known as C^+ regions or photodissociation regions, the chemistry is controlled by reactions initiated by photons, and, for high UV fields, the temperature reaches 1000 K and decreases to ~ 100 K at the atomic/molecular boundary.

The calculations used models of plane-parallel regions with uniform proton density n similar to those of Sternberg and Dalgarno (1989). Because of the low temperature of the region and the low C/C^+ ratio throughout the region, the contribution of electron impact to the excitation of the lines is negligible. From the model calculations it is possible to derive minimum values for the density and radiation field intensity that are necessary to produce the observed intensities of the lines for different objects as shown in Table 1. These minimum values are consistent with the densities and radiation field intensities obtained from observations of far infrared fine-structure lines of C^+, radio recombination lines of C, and H_2 emission in M 42, NGC 2024, and NGC 6720.

Table 1

Object	$I_1(\lambda\,985.0\,\text{nm})^a$	$I_2(\lambda\,872.7\,\text{nm})$	$R = I_1/I_2$	$n_{min}(\text{cm}^{-3})$	χ_{min}^b
M42 pos. 24	5.1×10^{-5}	1.0×10^{-5}	5.1	10^6	10^3
NGC 2024	4.5×10^{-6}	...	...	3×10^4	10^2
NGC 6270	1.3×10^{-5}	$<1.2 \times 10^{-6}$	>10.8	10^5	2×10^2
NGC 7027	1.5×10^{-3}	6.0×10^{-4}	2.5	$> 10^7$	$> 10^5$

[a] Intensities are in units of $\text{erg cm}^{-2}\text{s}^{-1}\text{sr}^{-1}$. References for the intensities are given by Escalante, Sternberg, and Dalgarno (1991).
[b] $\chi_{min} = 1$ for the average interstellar field (Draine, 1978)

The line ratio $R = \lambda\,985.0/\lambda\,872.7$ is a good diagnostic for the excitation mechanism although the $\lambda\,872.7$ line intensity has more observational uncertainties. $R = 4.7$ for excitation by recombination at low electron densities. The low value of R in M 47 is consistent with excitation by recombination. The high R value in NGC 6720 is larger than expected if the excitation is by recombination, and allows the possibility that the lines are excited by electron impact in a warm neutral gas (Jewitt et al., 1983). The low R value in NGC 7027 along with the high density derived from the models suggest that the metastable levels may be quenched by electron collisions.

3. References

Draine, B.T., 1978, *Ap. J. Suppl.*, **36**, 595.
Escalante, V., Sternberg, A., and Dalgarno, A., 1991, *Ap. J.*, **375**, 630.
Escalante, V., and Victor, G.A.V., 1990, *Ap. J. Suppl.*, **73**, 513.
Jewitt, D.C., Kupferman, P.N., Danielson, G.E., and Maran, S.P., 1983, *Ap. J.*, **268**, 683.
Sternberg, A., and Dalgarno, A., 1989, *Ap. J.*, , 338, 197.

OBSERVATIONS OF WATER AND MOLECULAR OXYGEN IN THE INTERSTELLAR GAS

DAVID A. NEUFELD

The Johns Hopkins University, Baltimore, MD 21218

Abstract. Water and molecular oxygen are two simple molecules which are not easily observed in the interstellar medium. O_2 emissions have yet to be detected from interstellar space, and H_2O has been observed only in rather special environments where the interstellar gas has been warmed by embedded infrared sources or by shock waves. This sorry state of affairs does not necessarily reflect a low interstellar abundance for these species, but rather is a consequence of the very high abundance of O_2 and H_2O within our own atmosphere: strong atmospheric absorption lines make it extremely difficult to carry out observations of water and molecular oxygen emissions using ground-based or even airborne telescopes. Future observations from two *orbiting* telescopes scheduled for launch in the coming five years - the Infrared Space Observatory (ISO) and the Submillimeter Wave Astronomy Satellite (SWAS) - promise to improve radically the observational data on H_2O and O_2.

1. Introduction

Out of all the hundreds of interstellar molecules that have been detected or are potentially detectable, water and molecular oxygen are of particular interest for at least two reasons. First, the abundances of these species are of intrinsic interest in constraining our theories for the formation and destruction of interstellar molecules; simple species bearing the most abundant heavy element (oxygen) seem a good place to start in testing astrochemical models. Second, these molecules are potentially important coolants of molecular gas in the interstellar medium, both because they are potentially abundant, and - in the case of water - because of a large dipole moment. Indeed, in cold interstellar gas of density greater than $\sim 10^6 \, \mathrm{cm}^{-3}$, water may well be the *dominant* coolant (Goldsmith and Langer 1978) and thus may control the thermal balance and equilibrium temperature within cold dense cloud cores. Since gravitational collapse can occur *only* insofar as cold clouds are capable of cooling radiatively, the interstellar water abundance may prove relevant to the process of star-formation itself.

In this paper, I will review briefly attempts to observe H_2O and O_2 emission from two specific astrophysical environments: cold molecular clouds (with kinetic temperatures, $T \sim 10 - 40$ K); and hot shock-heated regions ($T \sim$ few $\times 100 -$ few $\times 10^3$ K). Warm ($T \sim 100$ K) water has been extensively observed in radiatively-heated regions - such as the Orion hot core -

which surround embedded infrared sources, but observations of H_2O and O_2 in this third astrophysical environment will not be considered here.

2. Cold Dense Clouds

2.1. OBSERVATIONAL UPPER LIMITS

Neither water nor molecular oxygen has yet been observed from the cold quiescent gas which makes up much of the molecular material within the interstellar medium. Atmospheric absorption presents a tremendous problem for observations made from suborbital platforms. In the case of O_2, two approaches have been adopted in attempting to detect emission from the interstellar gas. The first approach, suggested by Black and Smith (1984), is to search for radio emission from the isotopically substituted species $^{16}O^{18}O$, for which the atmospheric attenuation is less severe. The theoretical calculations of Langer et al. (1984) suggest that no significant chemical fractionation is expected so that $^{16}O^{18}O/O_2 = [^{18}O]/[^{16}O] \simeq 1/500$. Observations of six clouds by Goldsmith et al. (1985) have yielded (1σ) upper limits on the O_2/CO abundance ratio in the range $0.5 - 4$. Listz and vandenBout (1985) have obtained corresponding upper limits (1σ) of 0.34 for Orion A and 0.07 for ρ Ophiucus A. An alternative approach has been to search for O_2 emission in external galaxies of sufficient redshift that the astrophysical O_2 line emission would be shifted out of the corresponding atmospheric absorption feature; for the $(N, J) = (1,1) - (1,0)$ line at 119 GHz the required redshift is ~ 0.02. This method has led to (1σ) upper limits on the (galaxy-averaged) O_2/H_2 abundance ratio of 1.4×10^{-5} for NGC 7674 (Liszt 1985) and 4×10^{-6} for VIIZw31 (Goldsmith and Young 1989).

In the case of H_2O, an upper limit upon the abundance within cold quiescent clouds has also been obtained by means of a search for the isotopically substituted $H_2^{18}O$ molecule. Observations of a set of seven clouds (which included both warm and cold regions) has been carried out from the Kuiper Airborne Observatory by Wannier et al. (1991) and has yielded (1σ) upper limits on the H_2O/CO abundance ratio in the range 0.001-0.03.

2.2. FUTURE PROSPECTS

The best upper limits on the O_2 abundance are just beginning to become uncomfortable for theoretical models for the chemistry of oxygen-bearing species (e.g. Sternberg, Lepp, and Dalgarno 1987; Langer and Graedel 1989; Herbst and Leung 1989). Unfortunately, no significant improvements in sensitivity can be expected for observations carried out from within the Earth's atmosphere, and the full range of chemical model predictions will be tested only from space. The Submillimeter Wave Astronomy Satellite (SWAS) -

a small Explorer mission to be launched by NASA in 1995 - has been designed to carry out a large scale survey of radio emission from five species of crucial importance to the oxygen chemistry (Melnick *et al.* 1991). The emission lines to be observed are: the $1_{10} - 1_{01}$ line of H_2O at 557 GHz; the corresponding line of $H_2^{18}O$ at 548 GHz; the $(N, J) = (3, 3) - (1, 2)$ line of O_2 at 487 GHz; the CI $^3P_1 - ^3P_2$ fine-structure line at 492 GHz; and the ^{13}CO $J = 5 - 4$ line at 551 GHz. Vast improvements can be expected in the sensitivity to interstellar O_2 and H_2O. For example, in giant molecular clouds cores of angular diameter 4 arcminutes - and with temperature 35 K, H_2 density 5×10^5 cm^{-3} and H_2 column density 1.5×10^{23} cm^{-2} - SWAS will detect H_2O at 3σ in 1 hour of integration provided its abundance relative to H_2 exceeds 5×10^{-9}. O_2 will be similarly detected provided its abundance relative to H_2 exceeds 1.5×10^{-6}.

3. Hot Shocked Regions

Interstellar shock waves are a common phenomenon in regions of active star formation. When they propagate through molecular clouds, shock waves heat the gas, leading to substantial changes in the chemical composition.

3.1. THEORETICAL EXPECTATIONS

Detailed theoretical modeling of molecular shock waves has been carried out by several groups. Slow, non-dissociative shocks have been considered, for example, by Elitzur and de Jong (1978), by Chernoff, Hollenbach and McKee (1982), and by Draine, Roberge and Dalgarno (1983), while fast dissociative shocks have been modeled by Hollenbach and McKee (1979; 1989) and by Neufeld and Dalgarno (1989). Common to every one these calculations is the prediction that the water abundance in the hot region behind the shock will be substantially enhanced by chemical reactions with substantial activation energies that can be overcome only in the hot gas behind a shock wave. Detailed calculations show that behind a non-dissociative shock the H_2O molecule accounts for most of the gas-phase oxygen that is not tied up in CO.

3.2. OBSERVATIONS OF MASER EMISSION

Interstellar water masers are observed as intense spots of emission in the 22 GHz $6_{25} - 5_{23}$ radio line which have been widely detected within regions of active star formation. Large radial and/or proper motions suggest an association with hypersonic gas motions and thus with shock waves, and detailed calculations have shown (Elitzur, Hollenbach and McKee 1989) that the physical conditions within shocked dense clumps can be expected to produce maser emission with the enormous brightness temperatures ($> 10^{12}$ K) that are typically observed. Thus it seems that in observing the 22 GHz maser transition we are probing dense condensations within a shock-heated molecular outflow.

For twenty years following the first detection of the 22 GHz maser transition (Cheung et al. 1969), no other masing transitions were unequivocally identified. The observational picture has been radically changed in the past two years, however, with the firm detection of maser action in three additional transitions: the $10_{29} - 9_{36}$ line at 321 GHz (Menten, Melnick and Phillips 1990); the $3_{13} - 2_{20}$ line at 183 GHz (Cernicharo et al. 1990); and the $5_{15} - 4_{22}$ line at 325 GHz (Menten et al. 1990). Maser action in each of these lines is expected under much the same physical conditions that are required to produce the 22 GHz water maser (Neufeld and Melnick 1991), and the ratios of different maser lines can be used as a powerful constraint upon the nature of the emitting region. The 321 GHz/22 GHz line ratio, in particular, is a valuable probe of the temperature (Neufeld and Melnick 1990), which indicates that in at least some masing regions the gas temperature must exceed 900 K: such high H_2O temperatures can be achieved only behind non-dissociative (rather than dissociative) shocks.

3.3. FUTURE PROSPECTS

While H_2O maser line emissions can be used to probe dense condensations within a shocked molecular flow, *non-masing* far-infrared H_2O emissions account for much of the energy radiated by the lower density material. Models for the emission spectrum of slow, non-dissociative shocks - such as the shocks believed to be present in the Orion-KL region (Draine and Roberge 1982) - have shown (Neufeld and Melnick 1987) that about one-half of the emitted radiation is expected to emerge in far-infrared rotational lines of H_2O. To date, the severity of the atmospheric attenuation has prevented any far-infrared H_2O lines from being observed even at airplane altitudes. However, *several hundred* such lines are expected to show fluxes greater than 10^{-18} W cm^{-2} into a 1 arcminute beam and therefore to be readily detectable from space by the Infrared Space Observatory (ISO) - an ESA mission which is scheduled for launch in 1993 - and a few lines are predicted to show fluxes greater than 10^{-15} W cm^{-2} into a 1 arcminute beam! Within the next few years, the careful modeling of spectrophotometric observations of several far-infrared H_2O lines promises to yield reliable estimates for the H_2O abundance within hot shocked gas. Such observations will therefore provide a crucial test of one of the key untested predictions of gas-phase interstellar chemistry: that oxygen is efficiently incorporated into water within hot shock-heated regions of the interstellar medium.

Acknowledgements

I am grateful for partial travel support from the *American Astronomical Society* and from an *IAU* Young Astronomer Grant.

References

Black, J. H., and Smith, P. L. 1984, ApJ, 277, 562.
Cernicharo, J., Thum, C., Hein, H., John, D., Garcia, P., and Mattioco, F. 1990, A& A, 231, L15.
Chernoff, D. F., Hollenbach, D. J., and McKee, C. F. 1982, ApJL, 259, L97.
Cheung, A. C., Rank, D. M., Townes, C. H., Thornton, D. D., and Welch, W. J. 1969, Nature, 221, 626.
Draine, B. T., Roberge, W. G., and Dalgarno A. 1983, ApJ, 264, 485.
Draine, B. T., and Roberge, W. G. 1982, ApJL, 259, L91.
Elitzur, M., and de Jong, T. 1978, A& A, 67, 323.
Elitzur, M., Hollenbach, D. J., and McKee, C. F. 1989, ApJ, 246, 983.
Goldsmith, P. F., and Langer, W. D. 1978, ApJ, 222, 881.
Goldsmith, P. F., Snell, R. L., Erickson, N. R., Dickman, R. L., Schloerb, F. P., and Irvine, W. M. 1985, ApJ, 289, 613.
Goldsmith, P. F., and Young, J. S. 1989, ApJ, 341, 718.
Herbst, E., and Leung, C. M. 1989, ApJS, 69, 271.
Hollenbach, D. J., and McKee, C. F. 1979, ApJS, 41, 555.
Hollenbach, D. J., and McKee, C. F. 1989, ApJ, 342, 306.
Langer, W. D., Graedel, T. E., Frerking, M. A., and Armentrout, P. B. 1984, ApJ, 277, 581.
Langer, W. D., and Graedel, T. E. 1989, ApJS, 69, 241.
Liszt, H. S. 1985, ApJ, 298, 281.
Liszt, H. S., and vandenBout, P. A. 1985, ApJ, 291, 178.
Melnick, G. J. *et al.* 1991, in *Atoms, Ions and Molecules: New Results in Spectral Line Astrophysics*, ed. A. D. Haschick and P. T. Ho (San Francisco: Astronomical Soc. of the Pacific), p. 439.
Menten, K. M., Melnick, G. J., and Phillips, T. G. 1990, ApJL, 350, L41.
Menten, K. M., Melnick, G. J., Phillips, T. G., and Neufeld, D. A. 1990, ApJL, 363, L27.
Neufeld, D. A., and Dalgarno A. 1989, ApJ, 340, 869.
Neufeld, D. A., and Melnick, G. J. 1987, ApJ, 322, 266.
Neufeld, D. A., and Melnick, G. J. 1990, ApJL, 352, L9.
Neufeld, D. A., and Melnick, G. J. 1991, ApJ, 368, 215.
Sternberg, A., Lepp, S., and Dalgarno, A. 1987, ApJ, 320, 676.
Wannier, P. G. *et al.* 1991, ApJ, 377, 171.

QUESTION AND ANSWER

M.Guelin: (comment) Cernicharo et al. (1990) detected a broad smooth spectral component in the 183 GHzs line profile toward W49, that they interpreted as thermal emission of H_2O. Alternately, the broad line wings could be a large collection of maser features. If the line wings are thermal and optically thin, the water abundance derived is $\sim 10^{-5}$ of H_2 and is consistent with that deduced from $H_2^{18}O$ and HDO observaations (Jacq et al. 1990).

METHANOL MASERS IN W3(OH)

Qin Zeng
Purple Mountain Obsevatory
210008 Nanjing
P.R.China

ABSTRACT. The formation conditions of the methanol masers in W3(OH) are studied separately. The lower limits of the densities of the A- and E-type maser regions are 10^5 and 10^6 cm^{-3}. The A-type maser requires a stronger excitation from HII region than the E-type maser. The relative abundance of methanol to molecular hydrogen is more than 5×10^{-5}. The calculation indicates that there are maser series $(J_0-(J+1)_{-1})E$, $(J_1-(J+1)_0)E$ $J=1,2$ and $(J_2-(J+1)_1)A^+$ $J=7,8,9$ in W3(OH).

W3(OH) is a compact HII region $9_2-10_1 A^+$ (Wilson et al.1984), $2_1-3_0 E$ (Wilson et al.1985) and $2_0-3_{-1} E$ (Batrla et al.1987) masers were detected by single dishes and subsequently by VLA and VLBI (Menten et al.1988a,b)

The goals of this paper are studying the formation conditions of the masers separately and focusing our attention on the correlation between these masers and the radiation field.

The statistical calculations covers the lowest 203 energy levels of A-type methanol (Zeng and Lou 1990, hereinafter Paper I) below 250 cm^{-1} and 68 energy levels of E type methanol (Zeng et al.1987, hereinafter Paper II) below 231 cm^{-1} (Fig.1). Adopting a large velocity gradient model and escape probability method (Golgreich and Kwan 1974, Paper I,II) the statistical equilibrium and radiative transfer equations are solved. The compact HII region excited by stars (or stellars clusters) is charaterized as a blackbody with a radiation temperature Trd and a filling factor f. The bulk of the dust, which emits primarily in the far-infred, lies in a dense shell immediately outside the ionized zone with an average optical depth of $\tau_d=0.5$ and dust temperature Td=45K (Thronson and Harper 1979). The observed dust spectrum is fitted by $\eta_{ij} B_{ij}(Td)$, Where B_{ij} is the Planck function,
$\eta_{ij} = (\nu_{ij}/\nu_o)$ if $\nu_{ij} < \nu_o$
$\eta_{ij} = 1$ if $\nu_{ij} \gg \nu_o$. $\nu_0 = 8.565 \times 10^{12}$ Hz (Paper I).
The kinetic temperature is taken as 100K similar to the value obtained from NH$_3$ .We refer to Lees et al.(1973), Lovas et al.(1982) and Moruzzi et al.(1990) for the energy data and to Lees (1973),Lovas et al. (1982)

and Pei, Zeng and Gou (1988) for the Einstein A-values. A tentative estimation for the collision rates is made based on Paper II.

Fig.2 (a) and (b) show the maser brightness temperatures Tb's of $(2_0-3_{-1})E$ and $(2_1-3_0)E$ as the functions of density. The other parameters are $F/Vgr=10^{-6}$ km^{-1} s pc , where F is the relative abundance between methanol and molecular hydrogen .The values of Trd*f equals 300 K and 30 K for curve a and b separately . It is found that Tb's of $(2_0-3_{-1})E$ and $(2_1-3_0)E$ sensitively depend on the density. The density of more than 2×10^5 cm^{-3} is required to get a Tb of more than 10^6 K even there is an external field of Trd*f=300 k.

$Tb(9_2-10_1 A^+)$ is the sensitive functions of density and external field (Fig.3-4). Shown in Fig.3 , $Tb(9_2-10_1 A^+)$ could be higher than 10^5 K if Trd*f larger than 900K. While if Trd*f less than 250K there would be no $(9_2-10_1)A^+$ maser to display even the density is 10^5 cm^{-3}. Fig.4 shows $Tb(9_2-10_1 A^+)$ could be higher than 10^5 K if the density larger than 10^5 cm^{-3} and Trd*f=900K.

According the observation results (Menten et al.1988 (a),(b)) ,assume the ratios of the Tb's as
  $Tb(2_0-3_{-1} E)/Tb(2_1-3_0 E)=10$, $Tb(2_0-3_{-1} E)/Tb(9_2-10_1 A^+)=100$.
Suppose that $(2_0-3_{-1} E)$ and $(2_1-3_0 E)$ masers emerge in the same region which is called E region. This status can be found in Fig.4,where the density of E and A-type maser regions are around 10^6 and 10^5 cm^{-3} ,the Trd*f=300K and 900K (or Trd=3000K, the filling factors of both maser regions is 0.1 and 0.3.).Since either $(9_2-10_1)A^+$ or (2_0-3_{-1}),$(2_1-3_0)E$ masers require the pumping from HII ,and ,the masers coexist with the absoption line $10_1-9_2A^-$, the maser regions should be infront of the HII region. The schematic diagram of E,A maser and HII regions is shown in Fig.5. Where L_A and L_E are the distances between HII region and A and E-type maser regions. $L_A/L_E = \sqrt{1/3}$. The lower limits of the densities of A and E region are 10^5 and 10^6 cm^{-3}. $F/Vgr=10^{-6}$ km^{-1}s pc is necessary for fitting the high brightness temperature of the masers. According the results of VLA observation assume a lower limit of Vgr of 50 km s^{-1} pc^{-1}. Therefore the relative abundance of CH_3OH/H_2 should be larger than 5×10^{-5}. The overabundance of methanol may be caused by rich C^+ in the partly ionized region ,the interface of HII region and molecular cloud, because carbon gas phase chemistry is initiated by the radiative association of C^+ and H_2 . It also may be caused by the shock and the dust reactions.

The calculation results indicate there are maser series :$(J_0-(J+1)_{-1})E$,$(J_1-(J+1)_0)E$ J=1,2 and $(J_2-(J+1)_1)A^+$ J=7,8,9 in W3(OH). Some masers in the series have not been detected, because they are relatively weak or/and hard to be detected by based ground instruments. Only $(7_2-8_1)A^+$ at 111.289GHz could be the candidate to be detected and to prove the existence of the maser series of $(J_2-(J+1)_1)A^+$.

This work was sponsored by Purple Mountain Observatory , The Laboratory of Radioastronomy and Astronomy Commission, Academia Sinica.

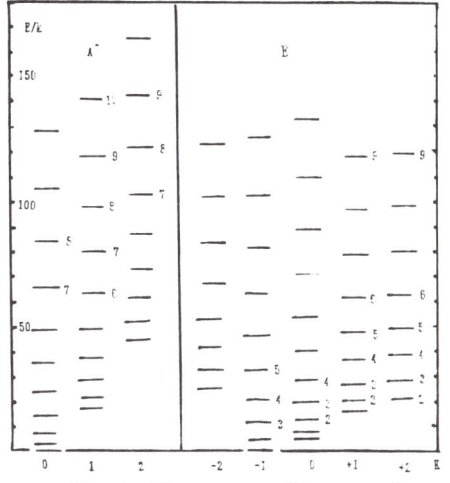

Fig.1 The energy diagram of methanol

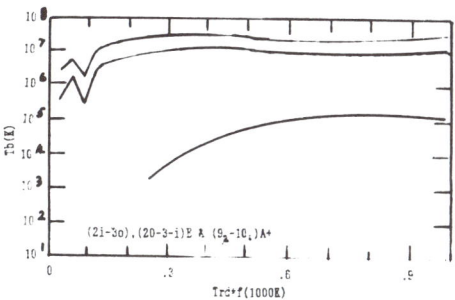

Fig.3 The Tb of (2_0-3_{-1})E (curve a), (2_1-3_0)E (curve b) and $(9_2-10_1)A^+$ (curve c) are shown as the funtions of Trd*f. Where densities are 7×10^5 cm^{-3} for curves a and b and 10^5cm^{-3} for curve c.

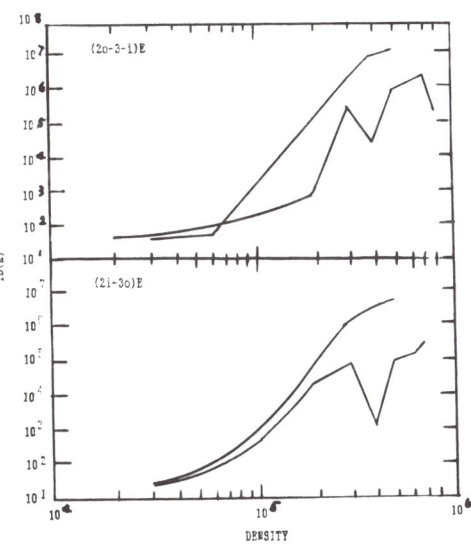

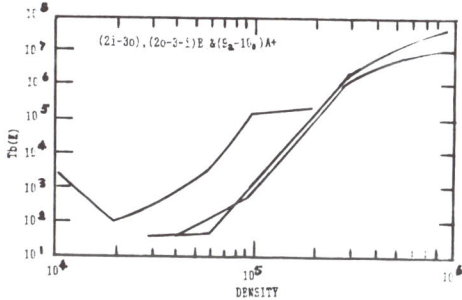

Fig.4 The Tb of (2_0-3_{-1})E (curve a), (2_1-3_0)E (curve b) and $(9_2-10_1)A^+$ (curve c) are shown as the functions of density. Where the value of Trd*f are taken as 300K for the curves a and b and 900K for the curve c.

Fig.2 (a) (b) The brightness temperatures Tb of (2_0-3_{-1})E and (2_1-3_0)E are shown as the functions of density. Other parameters are indicated in the text.

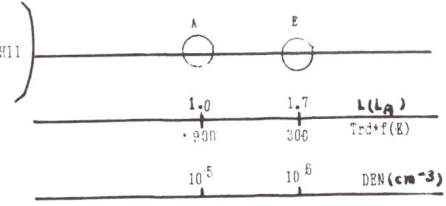

Fig.5 The schematic diagram of the regions of HII and E,A-type masers.

References

Batrla,W.,Matthews,H.E.,Menten,K.M.,Walmsley,C.M.(1987) 'Detection of strong methanol masers towards galatic HII regions',Nature 326(5),49-51

Goldreich,P.,Kwan,J.(1974) 'Molecular clouds',Astrophys.J.189,441-453

Lees,R.M.(1973)'On the E_2-E_1 labeling of energy levels and the anomalous excitation of interstellar methanol',Astrophys.J.184, 763-771

Lees,R.M.,Lovas,F.G.,Kirchhoff,W.H.,Johnson,D.R.(1973) 'Microwave spectra of molecules of astrophysical interest III.Methanol',J. Phys.Chem.Ref.Data,2,205-214

Lovas,F.J.,Suenram,R.D.,Snyder,L.e.,Hollis,J.M.,Lees,r.M.(1982) 'Detection of the torsionally excited state of methanol in Orion A',Astrophys.J.253,149-153

Menten,K.M,Johnston,K.J.,Wadiak,E.J.,Walmsley,C.M.,Wilson,T.L. (1988a) 'High resolution VLA observations of the W3(OH) methanol masers',Astrophys.J.,331,L41-L45

Menten,K.M.,Reid,M.J.,Moran,J.M.,Wilson,T.L.,Johnston,K.J.Batrla, W.(1988b) 'VLBI observations of 12 GHz methanol masers', Astrophys.J.333,L83-L86

Moruzzi,G.,Riminucci,P.,Strumia,F.,Carli,B.,Carlotti,M.,Lees,R.M., Mukhopadhyay,I.,Johns,J.W.C.,Winnewisser,B.P.,Winnewisser,M. (1990) 'The spectrum of CH_3OH between 100 and 200 cm torsional and forbidden transitions',J.Mole.Spectro.144,139-200

Pei,C.C.,Zeng,Q.,Gou,Q.Q.(1988) 'Einstein A-values of A-type methanol',Astron.Astrophys.Suppl.Ser.76,35-52

Thronson,H.A.Jr.,Harper,D.A.(1979) 'Compact HII regions in the far-infrared',Astrophys.J.230,133-148

Wilson,T.L.,Walmsley,C.M.,Snyder,L.E.,Jewell,P.R.(1984) 'Detection of a new type of methanol maser',Astron. Astrophys. 134,L7-L10

Wilson,T.L.,Walmsley,C.M.,Menten,K.M.,Hersen,W.(1985) 'The discovery of a new masering transition of interstellar methanol' ,Astron.Astrophys.147,L19-L22

Zeng,Q.,Lou,G.F.(1990) 'Interstellar A-type methanol masers', Astron.Astrophys.228,480-482(Paper I)

Zeng,Q.,Lou,G.F.,Li,S.Z.(1987) 'The masers of E-type methanol in Orion KL and Sgr B2',Astrophys. Space Sci.132,263-268(Paper II)

VLBI OBSERVATIONS OF AMMONIA (9,6) MASERS

PREETHI PRATAP, KARL M. MENTEN, MARK J. REID,
JAMES M. MORAN
Harvard-Smithsonian Center for Astrophysics, 60 Garden St.,
Cambridge, Massachusetts 02138

C. MALCOLM WALMSLEY
Max-Planck-Institut für Radioastronomie
Auf dem Hügel 69, D-5300 Bonn 1, Germany

ABSTRACT   We present the first VLBI measurement of interstellar ammonia (NH_3) masers. Two masers were found toward the ultracompact HII regions, W51-e1 and e2. The masers are unresolved in angle and smaller than 0.1 milliarcseconds. Unless these masers are highly beamed, they appear to be saturated.

Maser emission in the (J,K) = (9,6) transition of NH_3 was first detected by Madden et al. (1986) toward several star-forming regions. One of the strongest masers was found to arise from near the ultracompact HII regions W51 e1/e2 and the prominent H_2O maser sources W51S and W51M. NH_3 masers are found to be strongly variable like H_2O masers but unlike OH and CH_3OH masers. NH_3 maser emission covers the same velocity range as the low-velocity H_2O masers but in general have fewer spectral features. If the NH_3 masers are coincident with the H_2O masers, they could provide information about the physical conditions in the maser region. Accurate position determinations of these masers are necessary in order to obtain such information.

The VLBI observations were obtained using the Haystack 37m antenna, the NRAO 43m antenna, the Caltech 40m antenna and the Bonn 100m antenna. The minimum fringe spacing on the longest baseline (Bonn-OVRO) was 0.4 milliarcseconds. The 2 MHz band was centered at a velocity of 60 km s^{-1} based on a rest frequency of 18499.393 MHz for the (9,6) line of NH_3. The total power spectrum of the maser shows that there is a strong component at 55.0 km s^{-1} and a weaker component at 64 km s^{-1}. The lower limit for the 55 km s^{-1} feature is found to be 1.2×10^{13}K for a gaussian source model. The saturation temperature for this transition is found to be 10^{10}K which implies that the maser feature is saturated unless the beaming angle is less than 10^{-5} sr. The strong feature was found to be at $\alpha(1950) = 19^h21^m26\overset{s}{.}216 \pm 0.007, \delta(1950) = 14°24'42\overset{''}{.}7 \pm 0.1$. The weak feature was found to be offset by $0\overset{s}{.}0088780 \pm 0\overset{s}{.}0000027$ to the east and $-8\overset{''}{.}41700 \pm 0\overset{''}{.}00010$ to the south of the strong one. The spots are very compact and are smaller than 0.1 millarcseconds.

Figure 1 shows the positions of both the NH_3 maser features, along with the OH and H_2O masers, the thermal NH_3 peaks and the continuum source, W51 e1 and e2. It is evident from the figure that the NH_3 masers are not situated in front of the continuum source and are offset from the thermal peaks. Within the joint positional errors, the NH_3 and H_2O masers may be coincident.

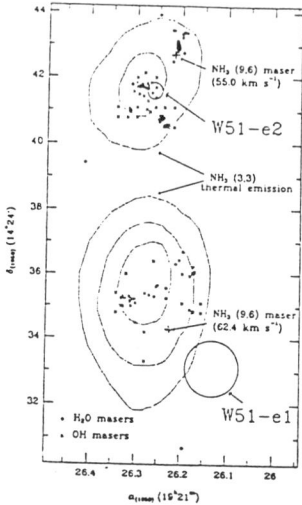

Figure 1. Map of the W51 e1/e2 region showing the positions of the two NH_3 masers with the positions of the OH masers (crosses - Gaume and Mutel 1987) and H_2O masers (dots - Genzel et al. 1978). The absolute position uncertainties of the OH and H_2O masers are 0".4 and 0".1 respectively. The two circles correspond to the sizes of the two continuum sources measured by Scott (1978). The positions of the continuum sources was taken from Ho et al. (1983). The dotted contours indicate the peaks of thermal emission as traced by the NH_3 (3,3) line (Ho et al. 1983).

There appear to be differences between the immediate molecular environments of the two compact continuum sources, W51-e1 and W51-e2. The northern OH masers are located close to the W51-e2 continuum peak while the H_2O masers are definitely offset. In contrast, the southern OH and H_2O masers overlap and are offset from W51-e1. Ho et al. (1983) suggest that this difference may be caused by outflows from the continuum sources creating adjacent density enhancements. As noted in Pratap et al. (1991), the velocity structure of the NH_3 masers appears to be strikingly different from that of the OH and H_2O masers. The centroid of the OH masers toward W51-e2 is about 58 km s^{-1}, with the masers distributed in two groups - the northeastern group having redshifted velocities and the southwestern group having blueshifted velocities (Gaume and Mutel 1987). The centroid of the OH masers toward W51-e2 is about 57 km s^{-1}. The H_2O masers in this region have velocities ranging from 41 km s^{-1} to 66 km s^{-1}. In contrast the NH_3 masers have a single component toward each source, the velocities of which correspond to the ambient molecular cloud velocity.

REFERENCES

Gaume, R. A. and Mutel, R. L. 1987, Ap. J. Suppl., 65, 193.
Genzel, R. et al. 1978, Astr. Ap., 66, 13.
Ho, P. T. P., Genzel, R., and Das, A. 1983, Ap. J., 266, 596.
Madden, S. C., Irvine, W. M., Matthews, H. E., Brown, R. D., and Godfrey, P. D. 1986, Ap. J. (Letters), 300, L79.
Pratap, P., Menten, K. M., Reid, M. J., Moran, J. M., and Walmsley, C. M. 1991, Ap. J. (Letters), 373, L13.
Scott, P. F. 1978, M.N.R.A.S., 183, 435.

A MODEL FOR THE MASER SOURCE NGC 7538 IRS 1

PREETHI PRATAP
Harvard-Smithsonian Center for Astrophysics, 60 Garden St.,
Cambridge, Massachusetts 02138

LEWIS E. SNYDER
University of Illinois, Astronomy Dept., 1002 W. Green St.,
Urbana, Illinois 61801

WOLFGANG BATRLA
Fachhochschule Coburg, Abteiling Muenchberg,
Kulmbacher Str. 76, D-8660 Muenchberg, Germany

ABSTRACT   High resolution observations of the molecular cloud around NGC 7538 IRS 1 in the J=1-0 transition of ^{13}CO show that the lower bound for density in that region is about 2×10^3 cm^{-3}. The study of various molecular transitions and the continuum measurements of the H II region indicate that the conditions required by the Boland and de Jong model for the excitation of the H$_2$CO maser are not very stringent or unique.

NGC 7538 IRS 1, IRS 2, and IRS 3 is a group of infrared sources situated in a dense molecular cloud. Several molecular masers have been observed toward IRS 1, an ultracompact H II region (Campbell 1984). IRS 1 is also one of two Galactic sources where formaldehyde (H$_2$CO) masers have been observed. The few H$_2$CO masers detected (Forster et al. 1985; Gardner et al. 1986) could be due to stringent conditions required for the excitation of the H$_2$CO maser or because the maser emission is hidden by the more common wide absorption features of Galactic H$_2$CO. Boland and de Jong (1981) have modeled the H$_2$CO masers in NGC 7538 utilizing pumping by free-free emission from the ultracompact H II region. In order to examine the conditions required by the Boland and de Jong (1981) model, high resolution maps have been made of the J=1-0 transitions of HCN (Pratap et al. 1989) and HCO$^+$ (Pratap et al. 1990 - Paper I). The molecular transitions trace high density material around IRS 1 since the HCN line thermalizes at H$_2$ densities of about 10^6 cm^{-3} and the HCO$^+$ line thermalizes at H$_2$ densities of about 10^5 cm^{-3} (assuming that the lines are optically thin). The HCN and HCO$^+$ maps show the presence of a cavity in the material around the H II region (Paper I). The cavity is interpreted as being caused by lower density material in that region and thus the molecular emission surrounding the cavity implies a density enhancement in the molecular cloud. The high density tracers provide an upper bound for the H$_2$ density in the cavity of about 10^5 cm^{-3}. High resolution observations of ^{13}CO indicate that the lower bound for the density in the cavity is 2×10^3 cm^{-3}. The millimeter continuum fluxes from IRS 1 indicate that the H II region can be represented by an cool, less dense, extended component and a hot, dense, compact component.

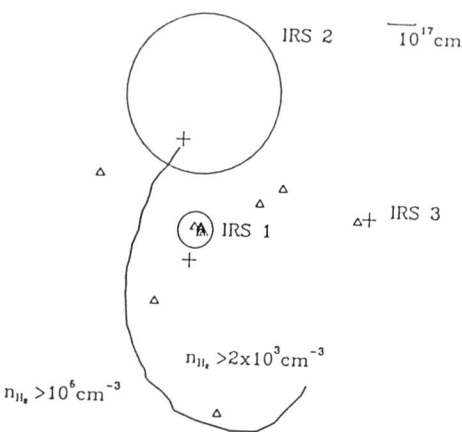

Figure 1. Front view of the molecular cloud around IRS 1, IRS 2 and IRS 3. The circles around IRS 1 and IRS 2 indicate the relative sizes of the radio continuum sources associated with them (Campbell 1984; Henkel et al. 1984). The crosses are the infrared positions of the three sources. The solid contour shows the edge of the high density material surrounding the low density cavity. The open triangles are positions of the H_2O masers (Kameya et al. 1990). All the other masers are directly in front of IRS 1.

Figure 1 shows a front view of the NGC 7538 IRS 1 region. All the masers are seen projected against the low density cavity surrounding IRS 1 and IRS 2. Calculations of the Boland and de Jong (1981) model for the H_2CO maser with the parameters obtained above show that the masers can be excited by either component of the HII region. The masering gas should be situated between 0.011 and 0.017 pc from the exciting source. The resulting fractional abundance of H_2CO with respect to H_2 should range between 2×10^{-7} and 5×10^{-8} in order to produce the observed brightness temperature for the maser. These values are consistent with the prediction of 8×10^{-7} made from chemical models (de Jong et al. 1980). This work has been supported by the Laboratory for Astronomical Imaging with funds provided for the Berkeley-Illinois-Maryland-Array project by the University of Illinois.

REFERENCES

Boland, W., and de Jong, T. 1981, *Astr. Ap.*, **98**, 149.
Campbell, B. 1984, *Ap. J. (Letters)*, **282**, L27.
de Jong, T., Dalgarno, A., and Boland, W. 1980, *Astr. Ap.*, **91**, 68.
Forster, J. R., Goss, W. M., Gardner, F. F., and Stewart, R. T. 1985, *M.N.R.A.S.*, **216**, 35P.
Gardner, F. F., Whiteoak, J. B., Forster, J. R., and Pankonin, V. 1986, *M.N.R.A.S.*, **218**, 385.
Henkel, C., Wilson, T. L., and Johnston, K. J. 1984, *Ap. J. (Letters)*, **282**, L93.
Kameya, O., Morita, K., Kawabe, R., and Ishiguro, M. 1990, *Ap. J.*, **355,**, 562
Pratap, P., Batrla, W., and Snyder, L. E. 1989, *Ap. J.*, **341**, 832.
Pratap, P., Batrla, W., and Snyder, L. E. 1990, *Ap. J.*, **351**, 530.
Scoville et al. 1986, *Ap. J.*, **303**, 416.

RESULTS OF THE MONITORING OF A VERY STRONG WATER MASER EVENT IN W49N

E. SCALISE JR., G.M. PACHECO, A.M. GÓMEZ BALBOA[1] AND Z. ABRAHAM[2]
SCT - INPE, CP 515, 12201 S.J.Campos, São Paulo;
[1]CNPq - ON, CP 23002, 20921 Rio de Janeiro;
[2]USP - IAG, CP 9638, 04301 São Paulo. BRASIL

ABSTRACT. Since the discovery of the 6_{16}–5_{23} rotational transition of interstellar water in maser emission, several hundred sources have been discovered. Their profiles are very variable, presenting correlated variability (Gammon, 1976) or anticorrelated variability (Cesaroni, 1990). Several models have been proposed to explain this variability. Strong Galactic masers, such as W49 and W51 (Kylafis et al. 1991), and also an extremely strong outburst detected by Abraham et al. (1981) in Orion, could not be fully explained based on current models. We have monitored from October 1989 to July 1991 a very strong water vapour eruption in the +28 km s^{-1} feature, originated from W49N. Here we present the results and discuss the possible correlated variability of this feature with the one that appeared at +62 km s^{-1}.

Equipment and observations

The observations were carried out with the Itapetinga 13.7m radiotelescope. At 22 GHz the HPBW is 4.2 arc min. The system temperature is of about 1000K. The signal is detected by an acousto-optical spectrometer with 1000 channels of 40kHz each.

The spectra were initially corrected to account for the change of atmosphere attenuation. In order to eliminate any gain fluctuation the spectra were normalized, one against the other, using the average antenna temperature of an "undisturbed" region situated in the velocity range from -20 to -30 km s^{-1}. We have calculated the uncertainties in the antenna temperature to be of the order of 10%.

Discussion and results

In Figure 1 we can see that the +28 km s^{-1} feature was observed for the first time, above the minimum detectable temperature, on October 1989, reaching a relative maximum by April 1990. Superimposed to this activity a very strong event happened in the region causing the emission to rise, almost linearly, three folds in 50 days. This level of emission remained high during the next 200 days, presenting some fluctuations. Polarization measurements carried out at this phase showed no percentage of linear polarization larger than 10%.

The decay phase, to the pre-flare level, was slower than the rise time and lasted for 90 days. The return to the zero level is hapenning in a similar way as of its rise but it has not yet been achieved.

Assuming the radiation to be isotropic, the line width to be of the order of our spectrometer resolution, and the source to be placed at 14 kpc, at its maxima, this event reached the intensity of 0.073 solar luminosities, becoming one of the strongest water vapour eruptions ever recorded in the Galaxy, but still 4 orders of magnitude weaker than extragalactic megamasers.

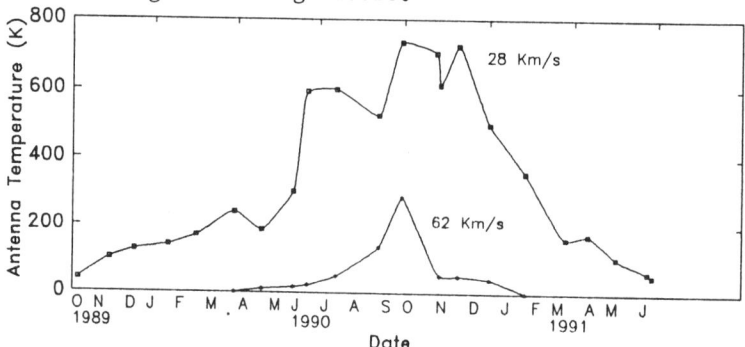

Figure 1 - Time evolution of the +28 km s^{-1} and +62 km s^{-1} features of W49N covering the period from October 1989 to July 1991.

The other feature at +62 km s^{-1} rised slowly during 145 days, decreased in only 30 days to a quiescent level and disappeared completely after 80 days. The overall shape of this event disagrees with the diffusive model of Burke et al. (1978).

From the analysis of the VLBI map of this region (Walker et al 1982), it looks possible that these two features correspond to two spots lying 0.230 milliarcsecond apart. If the excitation source of both features is spatially coincident with the +28 km s^{-1}, its front of excitation has to propagate at a velocity larger than 2000 km s^{-1} to reach the +62 km s^{-1} spot, what will cause the ionization of all surrounding molecules. But if it lies somewhere between the spots, the velocity of the front would be such to interact successively with the two features without destroying the molecules of the medium.

Acknowledgement: The authors wish to express their gratitude to the Itapetinga Radio Observatory staff. This work was partially supported by the Brazilian agencies FAPESP, CAPES and CNPq.

References:
Abraham, Z., Cohen, N.L., Opher, R., Raffaelli, J.C. & Zisk, S.H. (1981) Astron. Astrophys., **100**, L10.
Burke, B.F., Giufrida, T.S. & Haschick, A.D. (1978), Ap.J., **226**, L21.
Cesaroni, R. (1990), Astron. Astrophys., **233**, 513.
Gammon, R.H. (1976), Astron. Astrophys., **50**, 71.
Kylafis, N.D. & Norman, C.A. (1991), Ap. J. , in press.
Walker, R.C., Matsakis, D.N. & Garcia-Barreto, J.A. (1982), Ap.J., **255**, 128.

H ATOM OBSERVATIONS IN NEAR-STELLAR ENVIRONMENTS

LUIS F. RODRIGUEZ
Instituto de Astronomía, UNAM
Apdo. Postal 70-264
México, DF 04510, México

ABSTRACT. Interferometric observations of the 21-cm line of atomic hydrogen resolve out the emission from extended, line-of-sight clouds and allow the detailed study of compact H I structures in the surroundings of some stars. These atomic hydrogen components most probably are the result of photodissociation of gas that originally was in molecular form. They have been observed in H II regions, reflection nebulae, and planetary nebulae. The study of this atomic hydrogen component is important to determine the mass and physical conditions of gas in the environment of luminous stars and to test our theoretical knowledge of photodissociation regions.

1. INTRODUCTION

If a star of sufficiently high temperature is embedded in a molecular environment, it is expected that photodissociation of this environment will take place. A very important tracer of this photodissociated gas is the presence of atomic hydrogen, that could in principle be observed via its 21-cm hyperfine transition. However, the actual observation of H I in near-stellar environments is a difficult observational problem.

Consider, for example, observations made with a single dish with a beam size of 30 arc min, pointing in a direction that goes through 10 kpc of H I with a mean density of 1 cm^{-3}. In the volume observed by the beam we will have of the order of 5×10^5 M$_\odot$ of atomic hydrogen. Since the mass of the H I associated with the object of interest could be as small as a fraction of a solar mass, it is obvious that the observer may have a hard time disentangling the origin of the observed emission.

The successful observation of H I in the nearby environment of a star requires of several conditions: i) an interferometer is needed to obtain high angular resolution and to establish the positions of the sources of emission, ii) the radial velocity of the H I associated with the object studied should differ from that of large diffuse clouds in the same line of sight, and iii) preferably the source should be located away from the galactic plane where most of the H I exists. Under these favorable conditions it has been possible to study photodissociated H I in association with planetary nebulae, H II regions, and reflection nebulae.

2. PLANETARY NEBULAE

The first detection of H I in planetary nebulae was made in NGC 6302 (Rodríguez and Moran 1982). Since then, H I has been found in association with six other planetary nebulae (see Taylor et al. 1990). This atomic hydrogen is usually detected in absorption against the thermal continuum produced by the inner, ionized part of the planetary nebula. In most of the sources, the mass in H I is comparable with that present in the envelope in the form of ionized and/or molecular hydrogen, implying then an important correction to the mass of the planetary nebula. It is believed that the planetary nebula was originally molecular and that as the stellar nucleus becomes progressively hotter, photodissociation and ionization produced the H I and H II zones.

Taylor et al. (1989) have been able to observe the H I associated with IC 418 both in absorption and in emission. From these observations they derived a total mass of circumnebular neutral hydrogen of 0.35 $M_\odot$ and a kinetic temperature in the range of 150 to 350 K.

3. H II REGIONS AND REFLECTION NEBULAE

Interferometric observations made mainly during the last decade have determined that there is H I associated with several H II regions (Roger and Pedlar 1981; Joncas et al. 1985). The H I is often located in broad intermediary zones between the ionized and molecular components. This atomic component is almost certainly due to photodissociation of H_2 by the ultraviolet radiation from the exciting star(s).

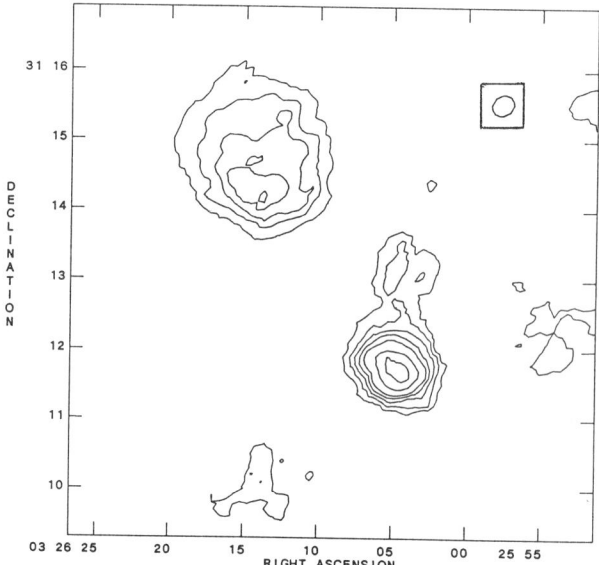

Figure 1. H I emission from the NGC 1331 region. The H I emission zones are approximately centered on BD+30°549 (northeast) and SVS 3 (southeast), the illuminating stars of the reflection nebula NGC 1333.

Using the Very Large Array, Rodríguez et al. (1990) detected H I in association with the reflection nebula NGC 1333. Two H I regions were detected, one associated with the star BD+30°549, and the other with the star SVS 3 (Figure 1). Rodríguez et al. (1992) have observed these regions with greater angular and velocity resolutions and have been able to determine their mass and radius, which are given in Table 1.

Table 1
Parameters of Photodissociated H I Regions

Source	M_{HI} ($M_\odot$)	Radius (pc)	$\dot{N}_{UV}$ ($s^{-1}\,Hz^{-1}$)	Sp (H I)	Sp (Other)
BD+30°549	0.077	0.084	1.3×10^{29}	B8	B9
SVS 3	0.057	0.042	1.3×10^{30}	B6	B6
S187	70	1.2	2.0×10^{32}	B1	B0
IC5146	450	3.5	1.9×10^{32}	B1	B0
LkHα101	85	1.2	4.9×10^{32}	B0	B0

In Figure 2 we show an H I spectrum taken at the center of the region associated with BD+30°549. From the width of this line it is possible to set an upper limit to the kinetic temperature of the gas, $T_K \leq 240$ K, while from the peak line brightness temperature it is possible to set a lower limit, $T_K \geq 66$ K. Then, the gas in this photodissociated region has $66\,K \leq T_K \leq 240\,K$. This range is in agreement with the values expected for a photodissociated region around a B-type star (Hollenbach et al. 1991).

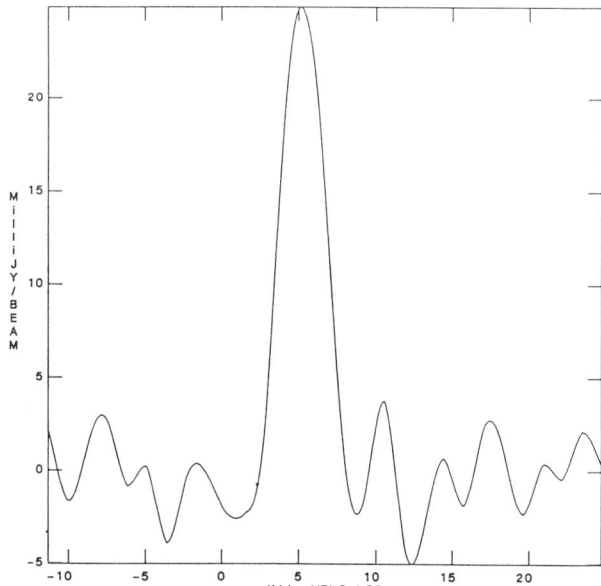

Figure 2. Spectrum of the H I emission from the region associated with BD+30°549.

Knowing the mass and radius of the H I region, it is possible to estimate the rate of dissociating photons from numerical models of the photodissociated region around B-type stars (Escalante et al. 1992). These models have spherical symmetry, uniform density and gas temperature, and use the same parameters of the plane-parallel models of Sternberg and Dalgarno (1989). The results of the numerical models can be fitted to an expression of the form

$$\left[\frac{\dot{N}_{UV}}{10^{30} s^{-1} Hz^{-1}}\right] = 0.73 \left[\frac{M_{HI}}{0.01 M_\odot}\right]^2 \left[\frac{R}{0.01 pc}\right]^{-3} exp\left(4.70 \left[\frac{M_{HI}}{0.01 M_\odot}\right]\left[\frac{R}{0.01 pc}\right]^{-2}\right), \quad (1)$$

where $\dot{N}_{UV}$ is the photon rate per Hz produced by the central star at 1000 Å, M_{HI} is the hydrogen mass, and R is the radius of the region. From $\dot{N}_{UV}$, and asssuming a ZAMS star, one can determine the spectral type of the exciting star. In Table 1 we compare the spectral types derived from the H I observations with those derived by independent techniques for the H I regions associated with BD+30°549, SVS 3 (Rodríguez et al. 1992), S187 (Joncas et al. 1992), IC5146 (Roger and Irwin 1982), and LkHα101 (Dewdney and Roger 1982). As can be seen, there is good agreement within one subclass.

4. CONCLUSIONS

Photodissociated H I regions have been observed in the 21-cm line of H I in planetary nebulae, H II regions, and reflection nebulae. In the case of H II regions and reflection nebulae there is good agreement between observations and theory. Future observations of H I regions around deeply obscured, late B-type stars could provide a useful tool for the study of these stars that do not ionize significantly their surroundings, but that do produce detectable photodissociation.

5. ACKLOWLEDGEMENTS

I am indebted to J. Cantó, V. Escalante, and S. Lizano for valuable comments.

6. REFERENCES

Dewdney, P. E. and Roger, R. S. 1982, Ap. J., 255 564.
Escalante, V., Lizano, S., Cantó, J., and Rodríguez, L. F. 1992, in preparation.
Hollenbach, D. J., Takahashi, T., and Tielens, A. G. G. M. 1991, Ap. J., 377, 192.
Joncas, G., Dewdney, P. E., Higgs, L. A., and Roy, J. R. 1985, Ap. J., 298, 596.
Joncas, G., Durand, D., and Roger, R. S. 1992, in preparation.
Rodríguez, L. F. and Moran, J. M. 1982, Nature, 239, 323.
Rodríguez, L. F., Lizano, S., Cantó, J., Escalante, V., and Mirabel, I. F. 1990, Ap. J. 365, 261.
Rodríguez, L. F., Lizano, S., Cantó, J., and Escalante, V. 1992, in preparation.
Roger, R. S. and Pedlar, A. 1981, Astron. Astrophys., 94, 238.
Roger, R. S. and Irwin, J. A. 1982, Ap. J., 256, 127.
Sternberg, A. and Dalgarno, A. 1989, Ap. J., 338, 197.
Taylor, A. R., Gussie, G. T., and Goss, W. M. 1989, Ap. J., 340, 932.
Taylor, A. R., Gussie, G. T., and Pottasch, S. R. 1990, Ap. J., 351, 515.

QUESTIONS AND ANSWERS

J.C.Pecker: 1) Any Be star observed in your survey? 2) Have you detected (and measured) stellar winds around the B stars of your survey?

L.F.Rodriguez: 1) No, we do not have Be stars in our survey. Perhaps the Herbig Be stars, being associated with clouds, are good candidates to have HI regions around. 2) We do not detect any obvious effects of a wind in our data.

J.A.de Freitas Pacheco: We observe for PN an anti-correlation between M_{HII} x electron density, which can be explained by the propagation of an ionization front through the ejected envelope. Thus we would expect that the neutral H mass would decrease with time. Is there any correlation between M_{HI} and kinematical age?

L.F.Rodriguez: Yes, you are right. Indeed, Taylor et al. (1990, Ap.J.,351, 515) do find an anticorrelation between HI optical depth and radius that could point to an anticorrelation between HI mass and age.

Infall in Collapsing Protostars

J M C RAWLINGS

Department of Physics, University of Oxford, Nuclear Physics Laboratory, Keble Road, Oxford, OX1 3RH, England

T W HARTQUIST

Max Plank Institute for Extraterrestrial Physics, 8046 Garching, Germany

K M MENTEN

Harvard-Smithsonian Center for Astrophysics, 60 Garden Street, Cambridge, Massachusetts 02138, USA

and

D A WILLIAMS

Department of Mathematics, UMIST, PO Box 88, Manchester M60 1QD, England

August 28, 1991

1. Introduction

High spectral resolution observations of dense cores in lines of NH_3 have provided no clear evidence for infall, even when the cores possess embedded low mass protostars or young stellar objects (Menten et al., 1984; Menten and Walmsley, 1985; Myers and Benson, 1983). Infall is expected to continue until protostellar winds begin to affect the dynamics of the cores.

Menten et al. (1984) suggested that the freeze-out of molecules containing heavy elements on to dust grains in the higher density infalling gas prevents these high velocity regions from contributing broad wings to the NH_3 line profiles.

Though NH_3 gas phase abundances decline as freeze-out continues, the abundances of some other species, e.g. CH (Hartquist and Williams, 1989), actually increase, at least initially. Thus, CH lines may show broad wings, while NH_3 profiles remain narrow. Comparisons of such line profiles would give information about freeze-out rates, and the dynamics, and age of the collapse.

We have calculated the abundances of molecular species in a collapsing core, in which freeze-out of molecules on to dust and gas-phase chemistry are both occurring. We determine the velocity distribution of parcels of gas within the core and determine the line profiles of a variety of observable species.

2. The Model

Calculations are performed for a spherical core of 1 $M_\odot$ with initial number density $n = n(H) + 2n(H_2) = 2.8 \times 10^3$ cm^{-3} and temperature 10 K. A modified free-fall collapse to a truncated singular isothermal sphere occurs, giving a density profile of r^{-2} and an outer radius of 1.58×10^{17} cm, and a number density, n, at the boundary of 1.86×10^4 cm^{-3}.

Immediately following the establishment of the singular isothermal configuration, the self-similar collapse begins, with a collapse expansion wave (CEW) propagating outwardly in a self-similar fashion (Shu, 1977). Inside the CEW, matter approaches free-fall with density $\alpha\ r^{-\frac{3}{2}}$, and infall speed $\alpha\ r^{-\frac{1}{2}}$.

During the collapse we follow the time-dependent chemistry of 85 species involving H, He, C, N, O, S, Na, and electrons, linked in 1147 chemical reactions in a closed network. Photoprocesses stimulated by the cosmic ray induced radiation field (Prasad and Tarafdar, 1983) are included. Freeze-out rates adopted include an allowance for grain charge (Umebayashi and Nakano, 1989). We also consider the possibility of enhanced freeze-out due to the presence of many small grains (Duley and Williams, 1984).

3. Results

Nine calculations were performed for depletion rates which are D = 1,2, and 3 times the canonical rate, and for three values of the collapse age, t_c : 4×10^{12} s, 6×10^{12} s, and 8×10^{12} s. The chemical results show clearly that while NH$_3$ declines in abundance as radius decreases, other species show significant enhancements as the inflow progresses. The ten most abundant species showing significant enhancements as the collapse proceeds are OH, OCN, CH, HCO$^+$, HCO, SO, N$_2$H$^+$, HNO, H$_2$S, and SH. These species may be expected to show broad profiles.

For most parameter combinations, the calculated NH$_3$ line profiles are narrow, confirming the suggestion of Menten et al. (1984). Profiles of other species depend sensitively on the parameters D and t_c. Detailed observations of these lines towards a few of the nearest globules should provide direct evidence for the infall of protostellar envelopes, and will allow an understanding of the dynamics of protostellar collapse.

References

DULEY, W W and WILLIAMS, D A: 1984, *Interstellar Chemistry*, Academic Press, London
HARTQUIST, T W and WILLIAMS, D A: 1989, *Mon. Not. R. astr. Soc.* **241**, 625
MENTEN, K M, WALMSLEY, C M, KRUGEN, E and UNGERECHTS, H: 1984, *Astron. Astrophys* **137**, 625
MENTEN, K M and WALMSLEY, C M: 1985, *Astron. Astrophys.* **146**, 369
MYERS, P C and BENSON, P J: 1983, *Astrophys. J.* **266**, 309
PRASAD, S S and TARAFDAR, S P: 1983, *Astrophys. J.* **267**, 603
SHU, F H: 1977, *Astrophys. J.* **214**, 488
UMEBAYASHI, T and NAKANO, T: 1980, *Publ. Ast. Soc. Japan* **32**, 405

CARBON TO HELIUM RATIO IN THE WIND OF CENTRAL STARS OF PLANETARY NEBULAE WITH WC SPECTRUM

J.A. DE FREITAS PACHECO and R.D.D. COSTA
Instituto Astronômico e Geofísico, Universidade de São Paulo, Brazil

F.X. DE ARAÚJO
Observatório Nacional - CNPq, Rio de Janeiro, Brazil

and

D. PETRINI
Observatoire de la Côte D'Azur, Nice, France

Abstract. We present $\frac{C}{He}$ ratios in the wind of a sample of 5 central stars of planetary nebulae having WC spectrum. The resulting values are comparable to those observed in population I WR stars.

Key words: Planetary nebulae, Wolf-Rayet stars, Chemical abundances

1. Introduction

UV spectra of the central stars of planetary nebulae (CS) indicate that most of these objects have a fast wind. The study by Cerruti-Sola and Perinotto (1985) led to the conclusion that CS's with a Wolf-Rayet spectrum (WC type) have a detected wind in the UV range. CS's are believed to be remnants of red giants which have thrown off their outer envelope. In this work we give a preliminary estimate of the $\frac{C}{He}$ in the wind of 5 CS's having a WC spectrum. Our analysis supports the view that the $\frac{C}{He}$ ratios in the wind and in the nebula are different, indicating that mass loss rates estimated with the usual assumption of equal chemical composition are incorrect.

2. Observations and Data

The data were obtained using the facilities of the National Laboratory for Astrophysics (Brazopolis - Brazil). Cassegrain+Reticon and Coudé+CCD observations were performed for all the stars of our sample.

Most of the lines in the wind of the observed objects display PCyg profiles, indicating that the medium is not transparent to the considered photons. Optical depth effects in an expanding envelope can be treated using the "first moment" of the line (Castor, Lutz and Seaton 1981). A detailed calculation based on such an approach will be presented elsewhere (de Freitas Pacheco et al. 1991). Here we give only a first estimate of the $\frac{C}{He}$ ratio, neglecting self-absorption effects and assuming that the lines are formed by recombination processes.

The table below gives the measured equivalent widths (in Å) of lines formed in the wind, which were used in our analysis. These lines were selected because they are probably formed mainly by recombination processes (Clegg 1989).

	BD+30	SwSt-1	N5315	Hen 2-99	Hen 2-113
$HeI\lambda5876$	$21.0^{(a)}$	nebular	nebular	33.0	$9.7^{(a)}$
$HeII\lambda4686$	$7.8^{(b)}$	$5.0^{(a)}$	70	$4.0^{(a)}$	-
$CII\lambda4267$	$12.4^{(c)}$	$0.95^{(c)}$	-	$21.3^{(c)}$	$11.6^{(d)}$
$CIII\lambda4650$	$49.0^{(a)}$	} $22.1^{(a)}$	280.0	$49.0^{(a)}$	$3.5^{(a)}$
$CIV\lambda4659$	$32.3^{(b)}$				
$CIV\lambda5806$	$29.7^{(a)}$	$4.9^{(a)}$	672.0	73.9	$0.4^{(a)}$

Notes:(a)line displaying a PCyg profile. Data refer to the emission component only; (b)line with probably a PCyg profile; (c)from Cassegrain observations; (d)data from Kaler et al. (1989)

3. Carbon-to-Helium Ratio

Under the discussed conditions, we found the following results for the relative $\frac{C}{He}$ ratios:

	BD+30	SwSt-1	N5315	Hen 2-99	Hen 2-113
$\frac{C}{He}$	0.33	0.67	≥ 1.4	0.37	0.48

4. Conclusions

In spite of the simplicity of our analysis, the conclusion that the $\frac{C}{He}$ ratio in the wind of CS's with WC spectrum differs from that in the surrouding nebula cannot be avoided. Our procedure, similar to that used by Torres (1988) and by Kaler et al. (1989) gave a $\frac{C}{He}$ ratio for Hen 2-99 in agreement with the latter authors. These $\frac{C}{He}$ ratios, comparable to those found in the wind of population I WC stars (de Freitas Pacheco and Machado 1988), indicate contamination of the expanding stellar envelope by the core material. Another open possibility is to consider that CS's having WC type are post-AGB helium-burning objects.

References

Castor,J., Lutz,J.H., Seaton,M.J. 1981, Mon. Not. R. astr. Soc. **194**,547
Cerruti-Sola,M., Perinotto,M. 1985, Ap.J. **291**, 237
Clegg,R.E.S. 1989 in *Planetary Nebulae* ed. S. Torres-Peimbert - Reidel, Dordrecht, pg. 139
De Freitas Pacheco,J.A., Araujo,F.X., Costa,R.D.D.,Petrini,D. 1991 - in preparation
De Freitas Pacheco,J.A., Machado, M.A.D. 1988, Astron.J. **96**, 365
Kaler,J.B., Shaw,R.A., Feibelmann,W.A., Lutz,J.H. 1989, Ap.J.Supp. **70**,213
Torres,A.V. 1988, Ap.J. **325**,759

CHEMICAL COMPOSITION OF A SOUTHERN PLANETARY NEBULAE SAMPLE

J.A. DE FREITAS PACHECO and R.D.D. COSTA
Instituto Astronômico e Geofísico, Universidade de São Paulo, Brazil

Abstract. We report the results of our analysis of a well observed sample of southern planetary nebulae. The average $\frac{S}{O}$ and $\frac{Ar}{O}$ ratios are comparable to the solar value and to those observed in galactic H II regions. The He abundance correlates with the $\frac{N}{O}$ ratio, confirming the trend found by previous studies, indicating the surface contamination of the progenitors by dredge-up episodes.

Key words: Planetary nebulae - Chemical abundances

1. Introduction

Planetary nebulae are the end of the evolutionary path of intermediate mass stars ($1 < \frac{M}{M_\odot} < 9$). Planetaries with low mass progenitors ($M < 3M_\odot$) or type II nebulae have their envelopes contaminated by products of CN cycle. They are nitrogen enriched due to carbon convertion. These objects may contribute significantly to the nitrogen enrichment of our galaxy. Type I nebulae are associated with massive progenitors ($3 < \frac{M}{M_\odot} < 9$). During their AGB phase, several mixing episodes occur following each He-shell flash, dredging-up helium and carbon as well as some s-process elements. Type I planetaries may be another secondary source of the carbon in the galaxy. Therefore the study of the abundances of these objects is quite relevant for comparison with theoretical calculations of advanced stages of intermediate mass stars. In this work, we report the results of the analysis of a southern sample of planetary nebulae. The observations were performed at the National Laboratory for Astrophysics (Brazopolis - Brazil). The data were already partially published in a series of papers and will not be discussed here.

2. Physical conditions

The interstellar extinction was estimated from Balmer decrement. The ionic concentrations to be determinated require a previous knowledge of the electron temperature and density. From optical spectra these parameters can be estimated from the line intensity ratios

$$R(OIII) = \frac{\lambda 4363}{\lambda 5007} \; ; \; R(NII) = \frac{\lambda 5754}{\lambda 6584} \; ; \; R(SII) = \frac{\lambda 6717}{\lambda 6730}$$

In our computations we considered a three-level atom model, including collisional excitation and de-excitation and radiative transitions in the statistical equilibrium equations. The relevant atomic data used in our calculations are those compiled by Mendoza (1983). Collisional effects in the He^{+2} concentration were taking into account using the formulae by Clegg (1987).

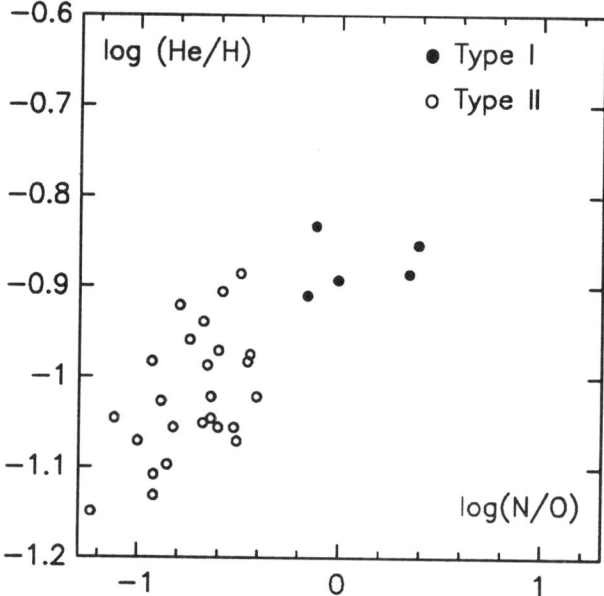

Fig. 1. $\frac{He}{H}$ versus $\frac{N}{O}$ for our sample of southern planetary nebulae. The correlation observed, in the sense that nebulae with high helium content have higher $\frac{N}{O}$ ratio, confirms some earlier results and theoretical expectations.

3. Conclusions

The $\frac{Ne}{O}$, $\frac{S}{O}$ and $\frac{Ar}{O}$ ratios are not expected to be affected by dredge-up episodes. In fact, the average values for our sample are

$$\frac{Ne}{O} = 0.23 \ ; \ \frac{S}{O} = 0.017 \ ; \ \frac{Ar}{O} = 0.0050$$

which are comparable to the solar values and to those observed in galactic H II regions.

The average helium abundance of type II planetaries (25 objects) present in our sample is $\frac{He}{H} = 0.097$, while that of type I (5 objects) is $\frac{He}{H} = 0.133$, indicating the effects of the third dredge-up. Moreover a correlation between the $\frac{He}{H}$ and $\frac{N}{O}$ ratios are expected. Figure 1 shows such a data. In spite of the observed scatter in the data points, a clear trend is observed in sense that nebulae with high helium content also have high $\frac{N}{O}$ ratio. This confirms some earlier results and theoretical expectations (Renzini and Voli 1981).

References

Clegg,R.E.S. 1987, Mon. Not. R. astr. Soc. **229**,31p
Mendoza,C. 1983, in IAU Symp. 103 - Planetary Nebulae - ed. D. Flower, Reidel, Dordrecht
Pottasch,S. 1984, *Planetary Nebulae*. Reidel, Dordrecht
Renzini,A. and Voli,M. 1981, Astron. Astrophys. **94**, 175

Chemistry in Protoplanetary Nebulae

D A HOWE, T J MILLAR and D A WILLIAMS
Department of Mathematics, UMIST, PO Box 88, Manchester M60 1QD, UK

September 12, 1991

Abstract. We have investigated gas-phase chemistry in a remnant red giant wind, during transition to a planetary nebula, using the interacting stellar winds model. Rapid destruction by UV of most existing molecules is predicted, within ~ 100 yrs of the core star heating up, suggesting that the large molecules in CRL 618 may be destroyed within decades. However, significant abundances of some hydrogenated molecules and ions (eg. CH^+, CH_2^+, CH_3^+, CH, CH_2, NH) may form behind the shock predicted by the interacting stellar winds model. Also, survival and/or formation of observable amounts of some molecules (eg. HCN, CN, HC_3N) may occur in dense clumps which survive transition, and may explain the existence eg. of HCN in NGC 7027.

1. Introduction

Planetary nebulae (PNe) are thought to form from winds of red giant (RG) stars, either as a result of sudden envelope ejection, or after steady envelope depletion during a period of high mass loss. A model has been developed (eg. by Kwok, 1983; Kahn, 1983), which assumes the latter case. This model, called the interacting stellar winds (ISW) model, involves an expanding bubble of ionised gas surrounding the hot, exposed stellar core. This bubble pushes against the remnant, high density "superwind", and progressively sweeps it into a thin, dense shell, with a shock at its leading edge. Meanwhile, the UV from the central star causes some ionisation and dissociation. We have constructed models to investigate the origin and fate of observed molecules and to predict others which might be observable in transition objects.

2. Destruction of molecules in the precursor wind

High mass loss RG winds (eg. IRC+10216) show a variety of molecular species, evidence of a rich chemistry. Our models predict that, during transition to PN, most existing molecules (except H_2 and possibly CO, which self-shield) will be destroyed, within about 100 yrs of the star becoming hot.

3. Chemistry in the shocked neutral wind

We have constructed a chemistry appropriate for a carbon-rich object. We find that significant amounts (column densities $\sim 10^{12}$ - 10^{14} cm^{-2}) of CH^+, CH_2^+, CH_3^+, CH, CH_2 and NH may be sustained for the first few hundred years.

4. Chemistry in a dense clump

Dyson et al. (1989) suggest that dense clumps in the Helix nebula (NGC 7293) may have originated as SiO maser spots in the wind of its RG precursor, and have survived transition to PN. We find that the UV extinction inside such a clump may allow the survival and/or formation of small but observable amounts of some molecules (eg. HCN, CN, HC_3N in a carbon-rich nebula) well into PN evolution.

5. Discussion and conclusions

Some proposed proto-PNe, eg. CRL 2688 and CRL 618 have large abundances of heavy molecules (eg. HC_5N, HC_7N), whereas most PNe have few or no molecules. Our results suggest that the large molecules formed in the RG wind will be destroyed by the UV flux from the exposed stellar core, within about 100 yrs of the beginning of transition (assuming the star becomes hot before this). CRL 618 is thought to have ceased producing its "superwind" about 200 years ago, and its central star now has a temperature of about 25000 K, so we expect that if the large molecules observed in this object are relics of the RG era, they will be destroyed within decades (we have not included molecular sources such as grain shattering, see Jura and Kroto, 1990).

Significant abundances of some hydrogenated molecules and molecular ions may be formed behind the shock predicted by the interacting stellar winds model, at least during the first few hundred years.

Survival and/or formation of small amounts of some molecules may occur in dense clumps, and may explain eg. the existence of HCN in the young PN, NGC 7027 (see eg. Sopka et al., 1989).

The above results will be presented in detail in *Monthly Notices*.

References

DYSON, J E, HARTQUIST, T W, PETTINI, M and SMITH, L J: 1989, *Mon. Not. R. astr. Soc.* **241**, 625-630
JURA, M and KROTO, H: 1990, *Astrophys. J.* **351**, 222-229
KAHN, F D: 1983, in Flower, D R, ed(s)., *Planetary Nebulae, IAU Symp.No.103*, Reidel, 305-316
KWOK, S: 1983, in Flower, D R, ed(s)., *Planetary Nebulae, IAU Symp.No.103*, Reidel, 293-303
SOPKA, R J, OLOFSSON, H, JOHANSSON, L E B, RIEU, N-Q, ZUCKERMAN, B: 1989, *Astr. Astrophys.* **210**, 78-92

MOLECULES IN NOVAE AND SUPERNOVAE

J.M.C. RAWLINGS
Department of Physics
Nuclear Physics Laboratory
Keble Road, Oxford OX1 3RH, UK

ABSTRACT. Molecular observations and models of the chemical processes in the ejecta of novae and supernovae are reviewed. Although only a few molecular species have been identified, the information that they give has provided great insight into the physical and chemical conditions. We now have quite a detailed understanding of the processes at work in both novae and supernovae.

1. Introduction

Until about ten years ago the only molecular feature seen in novae was the "5μm excess" attributed to CO v=1→0 (or C_3^+). Before SN1987a there was no evidence for molecule or dust formation in the ejecta of supernovae. The situation now is quite different; CO, CN, SiO, SiO_2, SiC and PAH features (as well as H_2 at late times) have been seen in novae. In SN1987a CO, (CO^+), SiO and H_3^+ have been detected. Dust has been observed in the ejecta of SN1987a and about a third of all novae produce optically thick dust shells.

Novae and supernovae are chemically and physically very dissimilar. The essential characteristics that they share are the presence of high densities (in excess of 10^{10} cm^{-3}), high temperatures (2000-10000 K), harsh, intense and strongly time-dependent radiation fields and complex physical conditions (eg. the ionization structure).

2. Novae

2.1. OBSERVATIONS

Novae are of particular interest at IR wavelengths in that they are often seen to produce very optically thick dust shells (completely obscuring the nova) at a time some 10-100 days after the outburst (eg. see Ney and Hatfield 1978). They are therefore useful laboratories in which to examine the processes of dust grain nucleation and growth. Typical parameters for a classical nova are given below:

Outburst energy	= 10^{45} ergs
Mass loss	= $10^{-5} - 10^{-4}$ M$_\odot$
Bolometric luminosity	= 2×10^4 L$_\odot$ (constant)
Ejecta velocity	= 300-10,000 Kms^{-1}
Ejecta density	= 5×10^{11} cm^{-3} (at t=5 days)
Composition:-	Metals (eg. C,N,O) enhanced over cosmic by 10-1000X

In general two physical nova types have been identified (according to the white dwarf composition):

1. C-N-O types; typically associated with duller, slow ($\dot{m}_v \simeq 0\overset{m}{.}01$ day^{-1}), dusty novae and *tend* to produce an optically thick carbon dust shell (eg. FH Ser, NQ Vul). In most cases CO (at 4.8 and 2.3 μm) is the only molecule that has been detected and in the pre-dust formation epoch only. (eg. Ferland *et al.* 1979)

2. He-Mg-Al types; typically associated with brighter, faster novae ($\dot{m}_v \simeq 0\overset{m}{.}1$ day^{-1}) and tending to produce an optically thin silicate type shell if any dust is formed at all (eg. V1370 Aql, QU Vul). In these novae, the predominant molecular lines seen are of SiC, SiO and other silicate features sitting on top of the dust continuum. (eg. Gehrz *et al.* 1984)

This is, however, a very general categorization (Gehrz 1990) and there are several recent novae which do not easily fit into these categories. A notable example is V842 Cen (1986) (Gehrz 1990): At least three different dust/molecular types seem to have formed in the ejecta; PAH features at 3.28 and 3.4 μm were detected in the ejecta shortly after the transition. The relative strength of these features changed substantially with time which could indicate a changing structural nature (Hyland and McGregor 1988). The origin of the PAH features is likely to be the destruction of dust grains by shocks associated with the progress of the ionization fronts (Rawlings and Evans 1991). The subsequent exposure to the intense UV and ions such as C$^+$ will lead to the eventual destruction of the PAHs.

In addition, the presence of both carbon and silicate dust is required so as to account for the spectrum between 7 and 13 μm. A feature at 11.3 μm could be due to PAHs or annealed olivine. Amorphous olivine smokes also gives very good fits to the 8-13 μm spectra of several "silicate" type novae although the absence of the 20μm O-Si-O bend feature in many novae is puzzling (Roche *et al.* 1984).

2.2. MODELS

To date, only carbon-rich novae have been studied in any detail.

The chemical modelling has concentrated on two epochs of the nova evolution:

- The pre-dust formation epoch, when the CO 5μm feature is seen (Rawlings 1986,1988).

- The dust nucleation epoch (Rawlings and Williams 1989). This is the first attempt at detailed chemical modelling of the kinetics of the formation of nucleation sites.

A 'typical' nova may maintain a constant bolometric luminosity for several hundreds of days after the outburst. Thus, as the ejecta expands, the effective photosphere contracts and the radiation field hardens. As a result, the expanding ejecta is overtaken by a series of ionization fronts. In the studies of the chemistry at early times (Rawlings 1986) it was found that simple molecules such as H$_2$ and CO could have appreciable abundances only in the region where the carbon is neutral (CI). In these conditions the carbon continuum ($\lambda \leq 1100$Å) and the H$_2$/CO self-mutual shielding protect the molecules from the radiation field. As a result of the intensity of the radiation field and the high densities, the chemistry is in steady-state (at T$\simeq$3500 K). The main H$_2$ formation routes are by three-body and H$^-$ reactions. The main loss route is collisional dissociation by atomic hydrogen. In the CII region (T$\simeq$6000 K) the H$_2$ can not build up appreciable column densities due to the presence of the unshielded Lyman flux. In a model of the CI region limited to H,C and O chemistry, the only other molecules to achieve abundances greater than about 10^{-10} are

CH, OH, C_2 and O_2. The chemistry is also *very* temperature sensitive: temperatures of less than 3500 K are required for H_2 to be optically thick. The CO formation route is typical for hot circumstellar environments:

$$O + H_2/H \longrightarrow OH, \quad O^- + H_2 \longrightarrow OH$$
$$C + H_2/H \longrightarrow CH, \quad C^- + H_2 \longrightarrow CH$$
$$OH + C \longrightarrow CO + H, \quad CH + O \longrightarrow CO + H, \quad C + O \longrightarrow CO + h\nu$$

The main CO loss routes are photodissociation and collisional dissociation by H atoms. Simple ionization models suggest that the neutral zone ceases to exist within a few days of the outburst. This is contradicted by the presence of CO at later times. If, however, a thin ($\Delta r/r_{ej} \simeq 0.01$), cool, neutral shell of ejecta of enhanced density could survive, then the observationally deduced column density of $> 10^{18}$ cm^{-2} could be sustained.

When modelling the chemical formation of dust nucleation sites we should note that the ejectae of novae are very far from LTE and a microscopic approach is required. The intensity of the radiation field together with the extreme inefficiency of molecule formation in ionized regions limits nucleation to the CI ionization zone (Rawlings and Williams 1989). At the temperatures and densities appropriate to the nucleation epoch, $X(H_2) \simeq 10^{-4} - 10^{-2}$ and CO saturates in this region. This is an important point since C>O in the ejecta. The only viable nucleation mechanism is based on an extended hydrocarbon chemistry. This chemistry is limited to molecules containing 8-10 carbon atoms (ring closure rapidly stabilizes molecules with greater than about 10 carbon atoms or less) and incorporates all data that is available on large molecule chemistry. The saturation of CO prevents oxygen attack on the hydrocarbons which would inhibit the formation of nucleation sites.

It is found that small species (such as C_2) are more important 'building blocks' to nucleation sites than larger molecules. This is due to the radiation field which keeps the abundance of the larger molecules low. In addition the chemistry is *extremely* sensitive to the density, fractional ionization and H_2 abundance - an ionization of less than 10^{-3} and $X(H_2) > 10^{-4}$ are required for nucleation sites to be formed in sufficient abundance. This in turn requires the gas temperature to be low (1000-1500 K) and the shell density to be enhanced over the spherical mean by a factor of 50 or more.

3. Supernovae

3.1. OBSERVATIONS

SN1987a was the first supernova in which molecules and dust were seen to form. The extensive coverage has revealed the presence of several molecular species. CO 2.3 μm and 4.6 μm were detected at times of 112 and 117 days post-outburst onwards respectively (Meikle et al. 1989). Fitting to the 2.3 μm line shows an apparent rise of the CO mass with time (from Spyromilio et al. 1988):

t/days	T/K	v(Kms^{-1})	$M_{co}/M_\odot$
192	3000	2000	1.7×10^{-5}
255	1800	1200	4.7×10^{-5}
284	1600	1200	1.2×10^{-4}

Note that the implied CO:C ratio (10^{-3}) is very high. The same authors have suggested that CO$^+$ may have been present in the ejecta at 255 days. The identification is however weak and could quite easily be explained by ArII fine structure.

SiO v=1→0 emission at 8.1 μm was first seen at 160 days and was clearly visible in the period 450-578 days post-outburst. Roche et al. (1991) have fitted the emission spectrum with an SiO temperature of 1500K and a total mass of $4\pm2\times10^{-6}$ M$_\odot$ (corresponding to about 15% of the total dust mass).

Miller et al. (1991) have identified and fitted H_3^+ to the features at about 3.4 and 3.5 μm as seen in the day 192 spectra of Meikle et al. (1989). They find the H_3^+ mass to be 1.1×10^{-7} M$_\odot$ with an excitation temperature of 2050 K.

3.2. MODELS

The features of the supernova ejecta that distinguish it are that the ejecta is both chemically and physically *highly* stratified and the main radiation field is an *intrinsic* source function derived from the radioactive decay of ^{56}Co.

Of the various models that have been developed so as to describe the behaviour of SN1987a, one of the most successful has been the partially mixed model 10HM of Pinto and Woosley (1988). Some characteristics of this model (at t=1 year) are given below:

	CORE	MANTLE
v(kms^{-1})	400-1000	1500-3500
n(cm^{-3})	1.1×10^9	6.7×10^8
T(K)	2000-3000	6000-7000
X_H	≤ 0.001	0.45
X_{He}	0.54	0.54

Note that hydrogen is highly deficient in the core. This is of great interest as the velocity and temperature of the CO emission lines are consistent with the core being the origin of emission. We may therefore be seeing evidence of a non-hydrogen based chemistry.

The situation is further complicated by the radiation field. The He*(2^1S) metastable state of helium is indirectly excited by the decay of ^{56}Co ($\lambda^{-1} \simeq 112$ days). The consequent decay to the ground state results is a strongly time- and density-dependent radiation field. An estimate of the strength of this field has been made by Petuchowski et al. (1989) and Rawlings and Williams (1990) on the basis of Fe ionization ratios. In addition there are contributions from the photospheric radiation field (of constant temperature 5500K) and the decay of ^{57}Co ($\lambda^{-1} \simeq 392$ days). Some 80% of all the fast electron energy is deposited in the core region (McCray 1990).

In addition to the radiation field, the fast electron flux has direct consequences on the chemistry; collisional ionizations lead to a high He$^+$ abundance which then attacks any molecular species (such as CO) which may be present. Collisional detachments of negative ions are also significant.

In these conditions the most effective molecular formation routes will be simple one-step reactions such as the radiative association:

$$C + O \longrightarrow CO + h\nu$$

The rate for this reaction has recently been calculated by Dalgarno et al. (1989). Other important reaction types are destruction by He$^+$, $(CO + He^+ \longrightarrow He + C^+ + O)$ and photons, negative ion reactions, charge exchanges, three-body reactions (with He as the third body), dissociative recombinations and collisional dissociations.

Petuchowski et al. 1989 [1], Rawlings and Williams 1990 [2], and Lepp et al. 1990) [3], have developed models of the chemistry in which it was found that the CO formation occurs via several channels. The direct radiative association referred to above is dominant but formation via C^-, C_2, and O_2 are all significant. The CO formation rate is therefore robust to parameter changes and the CO abundance is essentially controlled by the efficacy of the destruction mechanisms (photoionization/dissociation and reaction with He^+). The models differ in the relative importance that they ascribe to CO destruction by the radiation field and He^+. Different hydrogen abundances are also assumed (hydrogen chemistry is only significant in the model of [3]). A much greater CO^+ abundance is predicted by the model of [3] than those of [1] or [2] but both [2] and [3] predict CO abundances that are 100X too small as compared to the observations. Lepp et al. [2] point out that this could be improved if the charge exchanges of He^+ with low IP metals have rate coefficients of 3.3×10^{-9} cm^{3}s^{-1}. What is more probable is that the supernova ejecta is *poorly* mixed. Rayleigh-Taylor instabilities may then form at the boundaries of the abundance discontinuities so that fingers of mantle material are pushed back into the core region. The CO emission could then originate from this low velocity mantle material. If this is the case then both models predict a CO mass that could rise to as high as $10^{-2} M_\odot$ at late times. The only other molecular species predicted to be present are SiO, C_2, O_2, H_3^+ and HeH^+.

4. The Future

The direct associative ionization reaction of H(n=2) with H to form H_2^+ has been shown to be highly significant in dense protostellar outflows (Rawlings et al. 1991). Miller et al. (1991) have suggested that (as a result of the reaction of H_2^+ with H_2) this may be the main formation route of H_3^+ in SN1987a. It is likely that reactions involving H(n=3) may be even more important. In any case, as data on these newly studied reaction types emerges we can expect major alterations in our understanding of the chemistry in these environments.

References

Dalgarno, A., Du, M.L. and You, J.H. 1990, *Astrophys. J.*, **349** 675
Ferland, G.J., et al. 1979, *Astrophys. J.*, **227** 489
Gehrz, R.D., et al. 1984, *Astrophys. J.*, **281** 303
Gehrz, R.D. 1990, In: *"Physics of Classical Novae"* (Springer-Verlag), Eds. A. Cassatella and R. Viotti
Hyland, A.R. and McGregor, P.J. 1988, In *"Interstellar Dust"*, *IAU Symposium No. 135*
Lepp, S., Dalgarno, A. and McCray, R. 1990, *Astrophys. J.*, **358** 262
McCray, R. 1990, In: *"Molecular Astrophysics"* (CUP), Ed. T. Hartquist
Meikle, W.P.S., et al. 1989, *Mon. Not. R. ast. Soc.*, **238** 193
Miller, S., Tennyson, J., Lepp, S. and Dalgarno, A. 1991 *Submitted to Nature*
Ney, E.P. and Hatfield, B.F. 1978, *Astrophys. J.*, **219** L111
Petuchowski, S.J., Dwek, E., Allen, J.E. Jr. and Nuth, J.A. 1989, *Astrophys. J.*, **342** 406
Pinto, P.E. and Woosley, S.F. 1988, *Astrophys. J.*, **329** 820
Rawlings, J.M.C. 1986, *Ph.D. Thesis, UMIST*
Rawlings, J.M.C. 1988, *Mon. Not. R. ast. Soc.*, **232** 507
Rawlings, J.M.C. and Williams, D.A. 1989, *Mon. Not. R. ast. Soc.*, **240** 729
Rawlings, J.M.C. and Williams, D.A. 1990, *Mon. Not. R. ast. Soc.*, **246** 208
Rawlings, J.M.C., Drew, J.E. and Barlow, M.J. 1991, *These proceedings*
Rawlings, J.M.C. and Evans, A. 1991, *In preparation*
Roche, P.F., Aitken, D.K. and Whitmore, B. 1984, *Mon. Not. R. ast. Soc.*, **211** 535
Roche, P.F., Aitken, D.K. and Smith, C.H. 1991, *Mon. Not. R. ast. Soc.*, **252** 39P
Spyromilio, J., Meikle, W.P.S., Learner, R.C.M. and Allen, D.A. 1988, *Nature*, **334** 327

QUESTIONS AND ANSWERS

V.Escalante: At which stage of the nova outburst did you carry out the chemistry calculations? We know that red giant stars produce dust. How do we know that the dust observed in a nova outburst comes from the outburst itself and not from the red giant companion?

J.M.C.Rawlings: Calculations are performed at times that are consistent with observations. Thus the pre-dust formation chemistry is studied over the period 3 - 20 days post-outburst (when CO is seen) and the dust nucleation chemistry is studied from $\sim$ 30 to 80 days post-outburst (typically 50 days) ie. immediately prior to dust formation. There are many reasons why the dust cannot originate from the red star. I list a few below: (i) dust formation can result in an optically thick dust shell which completely covers the nova 'sky'. Dust picked up from the red star would be localized into one area; (ii) the radius of the IR pseudo-photo sphere and the dust temperature are consistent with rapid condensation - in any case the blackbody radius of the dust shell is very much larger than the binary separation; (iii) it is highly unlikely that any pre-existent dust will survive a wind moving at between 500 and 10,000 km/s.

J.P.Maillard: Providing the detection of H_3^+ is correct - which is not completely convincing at the resolution of the spectrum - what would be the rate of production of the HeH^+? And did you look at it because the fundamental band of HeH^+ is located in the same spectral range 3 to 4 μm?

J.M.C.Rawlings: This work was done by Miller et al. As we have two of the authors here I will pass this question to one of them:

S.Lepp: Our models produce between 10 and 100 times less HeH^+ than H_3^+. Still we have tentatively identified two lines as being from vibrationaly excited HeH^+.

J.A.de Freitas Pacheco: Concerning your dust calculations for novae, what CNO enhancement have you assumed?

J.M.C.Rawlings: We have considered various enhancements in the range 10 times to 100 times (being compatible with the observations). In the results that I have presented here we assume enhancement of about 50 times. Note that, of course, we require $C > O$ in this model.

M.Guelin: Did anybody detect $C^{18}O$ in the ejecta of SN1987A? Would it be possible to derive the $C^{18}O/C^{16}O$ abundance ratio? ^{18}O could be comparable to ^{16}O in these ejecta?

J.M.C.Rawlings: To my knowledge $C^{18}O$ has not been detected in SN1987a, but I would not like to make a definitive statement on that. The problem is that the near infra-red spectrum is heavily crowded with Ar II and Nickel lines. I would also think it unlikely that an abundance ratio could be determined as the CO emission is almost certainly very optically thick.

S.Lepp: I just wanted to comment that in a recent work by Liu, Dalgarno and Lepp, we have analysed the CO spectra and found that much larger CO masses may be fit, when optical depth and non-LTE populations are accounted for.

J.M.C.Rawlings: Yes, I think our models have predicted this to be in the core: either conditions are unfavorable in which case CO has a very low abundance or, in favorable situation CO is very optically thick. The condition required so as to predict the CO mass deduced in the optically thin approximation would have to be somewhat contrived.

SILICON MONOXIDE IN SUPERNOVA SN1987A

Craig H. Smith, David K. Aitken
Dept. of Physics, University College ADFA, Campbell, ACT, 2601, Australia

Patrick F. Roche
Dept. of Astrophysics, University of Oxford, Oxford, OX1 3RH, U.K.

ABSTRACT. The 8.1 μm $\Delta v = 1$ emission band of silicon monoxide detected in SN 1987A is modelled. Near day 500 the SiO mass is $4 \pm 2 \; 10^{-6} \, M_\odot$ and the excitation temperature is $\sim$1500 K. The mass of SiO is about 10 percent of the mass of dust inferred from the mid infrared emission near day 600, while the temperature is close to the condensation temperature of silicate grains. The SiO molecules may have been precursors to dust grain formation.

1. Introduction

- Silicon Monoxide was clearly present in the infra-red spectrum of SN 1987a from about day 160 to day 517, but could have been present at both earlier and later times.
- Spectra taken with the UCL spectrometer on the Anglo-Australian Telescope on days 465 and 517 show the SiO emission clearly between 8 and 9.5 μm, while the data between 9.5 and 13 μm allow the level and slope of the continuum emission to be established.
- We have extracted the SiO band emission by adopting a continuum of slope $F_\lambda \propto \lambda^{-3.7}$ on day 465 and $\lambda^{-3.5}$ on day 517 (Fig 1).

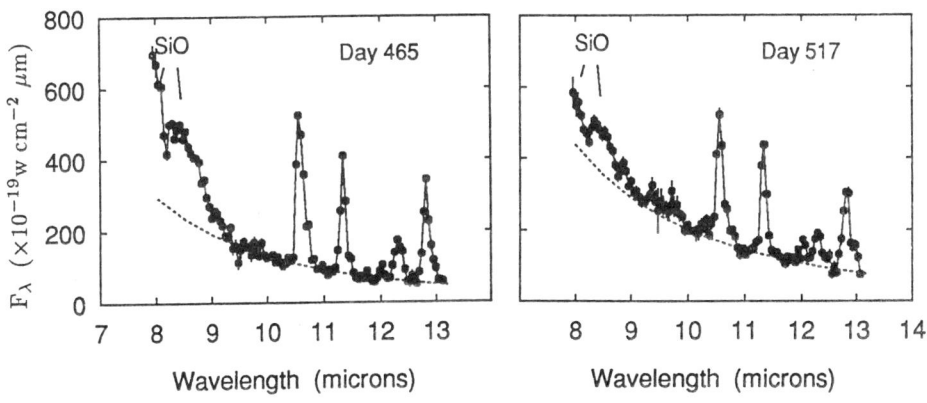

Fig 1. Spectra at 8-13 μm of SN 1987A 465 and 517 days after the explosion.

2. Modelling the SiO emission

- The emission from the supernova is attributed to the fundamental $\Delta v = 1$ SiO vibration-rotation band at 8.1 μm.
- The emission from each rotational component of the vibrational level is calculated and then summed to give the emission from the band at the adopted temperature.
- Fig 2 shows the continuum-subtracted SiO emission on Day 465 together with the 1500 and 2000 K SiO emission spectra convolved to the 0.09 μm resolution of the UCL spectrometer, (similar fits are also available for day 517). Temperatures of 1000 and 2500 K produce emission bands that are respectively narrower and broader than the observed spectra.

3. Mass of SiO

- The mass of SiO is estimated from the observed intensity of the emission band and the model fits. With a temperature of 1500 K, we obtain a mass of SiO of $5.6 \; 10^{-6}$ $M_\odot$ on day 465 and $2.8 \; 10^{-6}$ $M_\odot$ on day 517; a temperature of 2000 K would decrease the mass by 20 percent while a temperature of 1000 K would increase it by 50 percent.
- It appears that the mass of SiO decreased between days 465 and 517 and by day 578 the SiO emission was no longer detectable. This coincides with the onset of emission from dust from the supernova, which started near day 450 and increased beyond day 578.
- The SiO molecules may be the precursors of silicate grains, in which case the mass of SiO, at about 10 percent of the mass of dust, is close to that required if the SiO provides the seed for dust condensation to occur, while the excitation temperature is close to the condensation temperature of silicate materials.

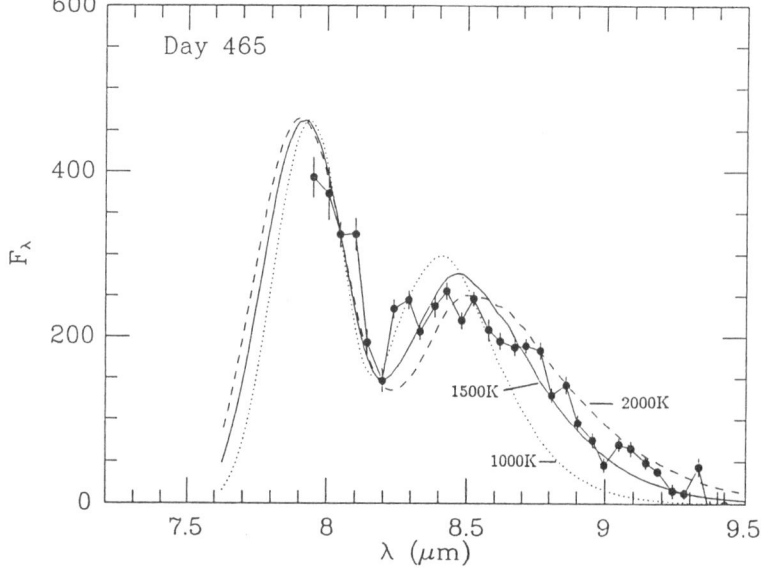

Fig 2. Comparison of observed SiO emission bands with model fits for day 465.

MOLECULAR STUDIES OF HERBIG-HARO OBJECTS

Salvador Curiel
Harvard–Smithsonian
Center for Astrophysics
60 Garden Street (MS 10)
Cambridge, MA 02138

ABSTRACT. Recent ultraviolet, infrared, millimeter and centimeter-wavelength observations have revealed that HH objects are also sources of molecular line emission. Infrared observations have shown that H_2 line emission in HH objects presents a wide and complex variety of morphologies, being in most of the cases similar to that observed at optical wavelengths. New high–angular resolution radio observations of molecular lines have revealed that small high–density condensations are frequently found in association with HH objects. Below, a brief review of molecular emission in HH objects is presented.

H_2 EMISSION ASSOCIATED WITH HH OBJECTS

The near-infrared spectral region contains a number of emission lines of molecular hydrogen, which have been proven to be important diagnostics for the molecular component of the gas in HH objects. The strongest and most commonly studied infrared line is the v=1–0 S(1) H_2 line at 2.12 μm. This line was first detected and mapped in HH objects using single beam techniques, and recently, using two–dimensional array detectors. H_2 emission has been searched and detected in many Herbig-Haro objects, including HH1-2 (Harvey et al.1986; Hartigan et al.1991), HH 7-11 (Zealey et al.1984; Lightfoot and Glencross 1986; Hartigan et al.1989; Burton et al.1989; Garden et al.1990; Stapelfeldt et al.1991), HH32 (Zealey et al.1986), HH43 (Schwartz et al.1988b), HH52/53/54 (Sandell et al.1987), HH 6 and HH 12 (Lane and Bally 1986; Stapelfeldt et al.1991), Cep A/GGD 37 (Bally, and Lane 1991), and several additional sources in surveys by Elias (1980), Schwartz et al.(1987), and Wilking et al.(1990); see also the review by Lane (1989). High-resolution infrared spectra of H_2 lines, from several HH objects, have been obtained by Doyon and Nadeau (1988), Zinnecker et al.(1989), Brand et al.(1989) and Carr (1990). A general result of these observations is that H_2 lines are much more narrow than optical lines. In particular, the width of the S(1) line is typically only half of the width of the Hα line in the same objects, and the observed central velocities of the H_2 lines are also lower than those of the optical lines (Zinnecker et al.1989).

The vibrational and rotational transitions of the H_2 molecule can be excited either by shocks (via collisional excitation in gas heated by the shocks to a few thousand degrees) or by absorption of ultraviolet radiation (in the Lyman and Werner bands with subsequent cascade to lower levels). Because the line strengths produced by the collisional and the fluorescent processes are significantly different, it is possible to distinguish between these mechanisms by comparing the line strength of lines from different levels such as the v=1–0 and v=2–1 S(1) transitions at 2.122 μm and 2.247 μm, respectively. A general description of the characteristics expected for H_2 emission excited in several different situations is given by Wolfire and Königl (1991). Numerical simulations predict that the expected line ratio of these lines from shocked gas is of the order of 10, whereas in photo excited gas it is ~ 2 (e.g., Black and van Dishoeck 1987). Molecular line ratios measured in several HH objects are ~ 10 (e.g., Schwarts et al.1987, 1988a), which are consistent with the predictions of nondissociative, low velocity shock wave models (e.g., Shull and Hollenbach 1978). In general, the H_2 emission seems to trace low velocity shock waves (with velocities around 10 to 40 km s^{-1}), while the optical emission traces much faster shocks (with typical velocities of about 100 km s^{-1}, or more). However, there is some evidence that other molecular excitation mechanisms may be at work in some of the objects. For instance, ultraviolet lines identified as H_2 Lyman band emission have been observed in some low-excitation HH objects such as HH 43 and HH 47 (Schwartz 1983). Likewise, the UV continuum that has been detected in several HH objects (such as HH 1 and 2) and previously attributed to atomic hydrogen two-photon emission (e.g., Dopita et al.1982; Brugel et al.1982) may arise, at least in part, from H_2 photodissociation (Böhm et al.1987). A detailed review of UV observation of HH objects has been recently presented by Brugel (1989).

COMPARISON OF OPTICAL AND IR EMISSION

Since H_2 lines and optical lines trace different components of the shocked gas, a comparison of the spatial distribution of optical and infrared emission in HH objects may provide essential clues to their flow structure. With the new infrared array detectors, it is now possible to study the H_2 emission distribution at resolutions comparable to optical images. Such studies have been performed by Schwartz et al.(1988), Hartigan et al.(1989), Garden et al.(1990), Stapelfeldt et al.(1991), and Lane et al.(1991). The technique and new results are discussed in a recent review by Lane (1989). These studies have shown that the H_2 line emission in HH objects presents a wide and complex variety of morphologies. Although the overall distribution of the H_2 emission is similar to that observed at optical wavelengths, they have subtle but important differences. To illustrate these differences, a general description of the morphology of three regions is presented; HH 1-2, Cepheus A, and HH 7-11.

   a) **HH1-2.**   Large proper motions and wide optical lines have been observed in this region indicating that high velocity (up to $\sim$200 km s^{-1}; Hartmann and Raymond 1984) shock waves are taking place. However, H_2 line profiles exhibit narrow linewidths (of about 40 km s^{-1}; Zinnecker et al.1989), suggesting that H_2 molecules are excited by weaker shock waves. Although HH1 and HH2 share a common energy source, they exhibit different morphologies at both, infrared and optical wavelengths (see Figure 1). HH1 is a clear example of a well defined bow shaped optical object where the H_2 lines are emitted from the wings (away from the apex) of the bow shock. This morphology is consistent with a jet model (or a bullet model) with a bow shock at its end. In this model, the external H_2 molecules entering the bow shock will be dissociated near its apex (where the shock velocity is generally greater than 100 km s^{-1}) and H_2 line emission will therefore predominantly arise from the wings of the bow shock (where the

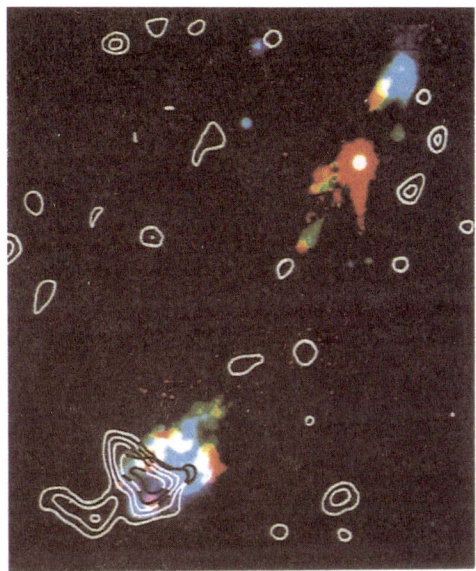 

Figure 1. Comparison of an infrared H_2 line image with optical H_α and [SII] images of the HH1-2 (left) and HH7-11 (rigth) regions obtained by Hartigan et al.(1991) and Hartigan et al.(1989). This figure also shows the spatial distribution of NH_3 in HH1-2 and HCO^+ in HH7-11 with respect to the optical and infrared emission.

 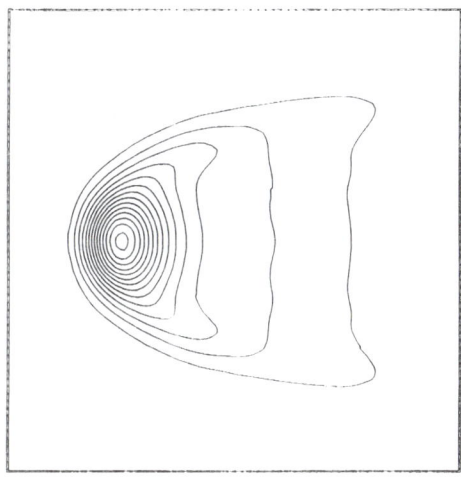

Figure 2. Comparison of the integrated 1-0S(1) H_2 flux map of HH7 (left) with that calculated for a bow-shaped shock wave with a magnetic precursor (rigth), using a shock velocity of 100 km s^{-1}, and an inclination angle of 40^0 with respect to the line-of-sight.

shocks are oblique and thus, weaker), having small radial velocities (compared with optical lines). The fact that the H_2 emission is observed mainly along the edge of one side of the optical object may be due to a geometrical projection (being favored the regions in the wings of the shock wave where, viewed from an angle, the molecular column density along the line-of-sight is greater) or to inhomogeneities in the ambient gas (e.g., if the flow is hitting the edge of a molecular condensation such as that detected in this region by Davis et al.[1990]). On the other hand, HH 2 has a more complicated optical structure, with several knots and extended emission connecting them. A similar morphology is observed at infrared wavelengths but with less evident extended emission. When infrared images are superimposed to optical images (see Figure 1), the H_2 emission seems to be produced predominantly upstream with respect to the H_α and [SII] lines (i.e., toward the energy source of the system). However, most of the optical knots seem to have an H_2 counterpart, suggesting a common origin for the H_2 and optical lines.

b) **Cepheus A/GGD 37.** A particularly intriguing case is the Cepheus A region (e.g., Hartigan et al.1986, Lenzen 1988), which has been mapped at arcsecond resolution (Lane 1989, Bally and Lane 1991). This HH object presents large proper motions (110–250 km s^{-1}) and extremely broad optical lines indicating shock velocities up to almost 500 km s^{-1}. On the other hand, the H_2 lines are much narrower with linewidths of 25–45 km s^{-1} (Doyond and Nadeau 1988), suggesting that the H_2 molecules are excited by much less energetic shock waves. The GGD 37 HH object has an overall bow shock morphology, with complex bow shaped substructures (Bally and Lane 1991; Lane et al.1991). A similar morphology is seen in molecular hydrogen, but in this case the arcs are softer and less protruding. As in the case of HH1, it appears that the H_2 lines are emitted mainly from the wings of the bow shocks, where the shocks are oblique and thus much weaker (Bally and Lane 1991). The morphology and spectroscopic characteristics of this object seem to be consistent with a model of a bow shock formed around an interstellar "bullet" moving through a molecular medium and viewed from an angle (Hartigan et al.1986; Lane 1989; Bally and Lane 1991).

c) **HH7-11.** HH7-11 is a nearly collinear system of optical emission knots in the L1450 molecular cloud near NGC 1333 (see Strom et al.1986). These objects have optical radial velocities (–200 to –40 km s^{-1}) and line profiles indicative of shock velocities $\sim$ 100 km s^{-1} (Solf and Böhm 1987), but the spectra show strong [OI] and [SII] lines characteristic of a low-velocity ($\sim$ 30 km s^{-1}) shock. The H_2 1–0S(1) emission line is centered at much lower radial velocities of –60 to 0 km s^{-1} (Zinnecker et al.1989), an important fact to shock models of the region. Intense H_2 emission is observed in three of the HH objects (HH7, HH8, and HH10), while weak emission is observed in HH9 and no emission is associated with HH11 (Hartigan et al.1989; Figure 1). HH7 is a particularly interesting object which has a bow-shaped morphology at both, optical and infrared-wavelengths, with the H_2 and [SII] images coinciding within an arcsecond. Recent results indicate that in the case of HH7 the H_2 emission seems to arise from a "magnetic precursor" in front of a J-shock (Hartigan et al.1989; Carr 1990; and Stapelfeldt et al.1991). In HH8 and HH10, the H_2 emission seems to be produced upstream with respect to the H_α and [SII] lines (i.e., toward the energy source of the system). In these two objects, the spatial and velocity characteristics of the emitting gas reveal a geometry more complex than that of a simple bow shock or planar shock. HH11 remains an enigma, and could represent a second ejection from SVS 13 (the presumed energy source of the system). The low–excitation spectrum observed in this object is particularly difficult to explain since HH 11 has no visible H_2 emission.

Recently, detailed H_2 velocity mapping with spatial resolution of the order of $1''$ has been obtained combining two-dimensional infrared detector arrays, such as IRCAM at UKIRT, with Fabry-Perot spectrophotometers (e.g., Carr 1990). This new type of observations will be very useful to establish the shock type and geometry of individual HH objects and optical jets. Likewise, further information can be obtained by calculating infrared line intensities and profiles from bow-shaped shock waves with magnetic precursors, similar to those calculated for optical lines by Hartigan et al.(1987). This type of model has been used by Carr (1990) and Curiel (1991) to calculate the H_2 1–0S(1) line profile for a bow shock geometry, for which they obtained a profile very similar to the spatially integrated line profile for HH7. Figure 2 shows a comparison of an integrated H_2 flux map obtained for HH7 by Hartigan et al.(1989) and those calculated for a bow-shaped shock wave with a magnetic precursor, using a shock velocity of 100 km s^{-1}, and an inclination angle of 40° with respect to the line-of-sight (Curiel 1991). The adopted shock is a very simplified jet model in which both the temperature and velocity increase linearly from an initial value to their maximum values in the precursor. It is important to notice how the global morphology predicted by this simple model resemble that observed. This type of shock wave seems to be a very promising way to explain the spatial coincidence of optical and infrared emission, and the line profiles observed in HH7 and other HH objects.

HCO$^+$ AND NH$_3$ EMISSION ASSOCIATED WITH HH OBJECTS

Recent high-angular resolution observations of molecular lines, carried out with the VLA, Hat Creek, NMA and JCMT radio telescopes, have shown that small high-density condensations are frequently found at the border of HH objects and aligned with the outflow direction. The most commonly studied lines are those of HCO$^+$(J=1→0) and (J=3→2) at millimeter wavelengths, and NH$_3$(1,1) and (2,2) at centimeter wavelengths. HCO$^+$ clumps have been detected in association with HH 7-11 (Rudolph and Welch 1988; Figure 1), and with HH 1-2 (Davis et al.1990). Ammonia condensations have been observed toward HH 25-26 (Torrelles et al.1989), and HH 2 (Torrelles et al.1991, private communication; Figure 1); see also the review by Torrelles (1990). These NH$_3$ and HCO$^+$ condensations or clumps have typical densities of $n(H_2) \simeq 10^5$ cm^{-3}, radial velocities similar to those of the ambient molecular cloud and typical sizes of a few arcseconds. Although the number of reported cases is still comparatively small, the high detection rate suggests that this emission may be a common feature of HH objects.

The association of HH objects with molecular condensations suggests that at least these HH objects could be ambient dense gas shocked by a wind. If this is the case, the observed high-density clumps associated with the HH objects would be ambient cloudlets (perhaps uniformly distributed in the molecular cloud) that have been compressed by a shock wave, enhancing the density by at least a factor of $\sim$ 100 in order to be detectable with molecular tracers of high-density gas such as NH$_3$ and HCO$^+$ (e.g., Rudolph and Welch 1988, and Torrelles 1990). However, the compression of the ambient molecular gas would also imply an acceleration of the molecular gas to velocities different to the ambient molecular cloud velocity. At present, such acceleration has not been observed in the high-density condensations associated with HH objects. The molecular lines observed typically have radial velocities of only $\sim$ 1 km s^{-1} and widths of $\leq$ 1 km s^{-1}. Another possible explanation to this association is that the uv radiation emitted by the shock waves excite and heat ambient high-density cloudlets nearby. In this case, the cloudlets would have radial velocities similar to that of the molecular cloud velocity, and line widths given by thermal broadening. However, calculations will be needed to establish if the radiation emitted by shock waves is capable of exciting HCO^+ and NH_3 in molecular condensations to the observed levels.

—I am grateful to A.P. Lane and B. Reipurth who kindly provided some excelent images which I used during the presentation of this work. I am also grateful to J.M. Torrelles for comments on an early version of this manuscript and for providing the unpublished NH_3 contour map used in Fig. 1.

REFERENCES

Bally, J., and Lane, A.P. 1991, in Astrophysics with Infrared arrays, ASP vol. 14, ed. R. Elston, p. 273-278.
Black, J.H., and van Dishoeck 1987, *Ap. J.*, **322**, 412.
Böhm, K., Bührke, Th., Raga, A., Brugel, E., Witt, A., and Mundt, R. 1987, *Ap. J.*, **316**, 349.
Brand, P.W.J.L., Toner, M.P., Geballe, T.R., and Webster, A.S. 1989, *MNRAS*, **237**, 1009.
Brugel, E.W. 1989, in ESO-Workshop on Low Mass Star Formation and Pre-Main Sequence Objects, ed. Bo Reipurth, p. 311-329.
Bugel, E.W., Shull, J.M., and Seab, C.G. 1982, *Ap. J.*, **262**, L35.
Burton, M.G., Brand, P.W.J.L., Geballe, T.R., and Webster, A.S. 1989, *MNRAS*, **236**, 409.
Carr, J.S. 1990, *Bull. AAS*, **21**, 1086.
Curiel, S. 1991, *PhD Thesis, Universidad Nacional Autónoma de México*.
Davis, C.J., Dent, W.R.F., and Bell Burnell, S.J. 1990, *MNRAS*, **244**, 173.
Dopita, M.A., Binette, L., and Schwartz, R.D. 1982, *Ap. J.*, **261**, 183.
Doyon, R., and Nadeau, D. 1988, *Ap. J.*, **334**, 883.
Elias, J.H. 1980, *Ap. J.*, **241**, 728.
Garden, R.P., Russell, A.P.G., and Burton, M.G. 1990, *Ap. J.*, **354**, 232.
Hartigan, P., Curiel, S., and Raymond, J. 1989, *Ap. J. (Letters)*, **347**, L31.
Hartigan, P., Curiel, S., and Raymond, J. 1991, in preparation.
Hartigan, P., Lada, C.J., Stocke, J., and Tapia, S. 1986, *A. J.*, **92**, 1155.
Hartigan, P., Raymond, J., and Hartmann, L. 1987, *Ap. J.*, **316**, 323.
Hartmann, L., and Raymond, J. 1984, *Ap. J.*, **276**, 560.
Harvey, P.M., Joy, M., Lester, D.F., and Wilking, B.A. 1986, *Ap. J.*, **301**, 346.
Lane, A.P. 1989, in ESO-Workshop on Low Mass Star Formation and Pre-Main Sequence Objects, ed. Bo Reipurth, p. 331-348.
Lane, A.P., and Bally, J. 1986, *Ap. J.*, **310**, 820.
Lane, A.P., Bally, J., and Hartigan, P. 1991, in preparation.
Lenzen, R. 1988, *Astr. Ap.*, **190**, 269.
Lightfoot, J.F., and Glencross, W.M. 1986, *MNRAS*, **221**, 993.
Rudolph, A., and Welch, W.J. 1988, *Ap. J. (Letters)*, **326**, L31.
Sandell, G., Zealey, W.J., Williams, P.M., Taylor, K., and Strey, J.V. 1987, *Astr. Ap.*, **182**, 237.
Schwartz, R.D. 1983, *Ann. Rev. Astr. Ap.*, **21**, 209.
Schwartz, R.D., Cohen, M., and Williams, P.M. 1987, *Ap. J.*, **322**, 403.
Schwartz, R.D., Cohen, M., and Williams, P.M. 1988a, *Ap. J.*, **333**, 1035.
Schwartz, R.D., Williams, P.M., Cohen, M., and Jennings, D.G. 1988b, *Ap. J. (Letters)*, **334**, L99.
Shull, J.M., and Hollenbach, D.J. 1978, *Ap. J.*, **220**, 525.
Solf, J., and Böhm, K.H. 1987, *A. J.*, **93**, 1172.
Stapelfeldt, K.R., Beichman, C.A., Hester, J.J., Scoville, N.Z., and Gautier III, T.N. 1991, *Ap. J. (Letters)*, , in press.
Strom, K.M., Strom, S.E., Wolff, S.C., Morgan, J., and Wenz, M. 1986, *Ap. J. Suppl.*, **62**, 39.
Torrelles, J.M. 1990, in Atoms, Ions, and Molecules: New Results in Spectral Line Astrophysics, ASP vol. 16, eds. A.D. Haschick and P.T.P. Ho, p. 257-268.
Torrelles, J.M., Ho, P.T.P., Rodríguez, L.F., Cantó, J., and Verdes, L. 1989, *Ap. J.*, **346**, 756.
Wilking, B.A., Schwartz, R.D., Mundy, L.G., and Schultz, A.S.B. 1990, *A. J.*, **99**, 344.
Wolfire, M.G., and Königl, A. 1991, *Ap. J.*, in press.
Zealey, W.J., Williams, P.M., and Sandell, G. 1984, *Astr. Ap.*, **140**, L31.
Zealey, W.J., Williams, P.M., Taylor, K.N., Storey, J.W., and Sandell, G. 1986, *Astr. Ap.*, **158**, L9.
Zinnecker, H., Mundt, R., Geballe, T.R., and Zealey, W.J. 1989, *Ap. J.*, **342**, 337.

MOLECULES IN STELLAR WINDS
Alfred E. Glassgold
Department of Physics, New York University
2 Washington Place, New York, NY 10003, USA

ABSTRACT. High mass-loss molecular winds are characterstic of two important phases of stellar evolution, very young protostars and stars on the asymptotic branch. Novel chemical processes in these winds suggest a variety of new astronomical observations.

1. PHYSICAL CONDITIONS IN STELLAR WINDS

Molecules occur in stellar winds if the physical conditions are right: the density must be high and the temperature low. It also helps if the wind can shield itself from the stellar radiation. High density winds are generated by stars at both the beginning and the end of a star's life, i.e. by protostars and evolved red giants. The rate and the duration of the mass loss of these objects are such that a significant fraction of a solar mass (or more) is ejected, large enough to affect the evolution of the stars. How these stars generate their winds is an important, open problem in astrophysics.

Table 1 gives the physical conditions for two, nearby, representative outflows. IRC +10216 is a very bright, evolved (AGB) carbon star with a rich circumstellar chemistry characterized by the formation of complex hydrocarbon and small refractory molecules. The physical conditions in the photochemical region of IRC +10216 are similar to the photodissociation transition region of a dense interstellar cloud. SVS 13 is the prototype of the extremely high velocity (EHV) outflows recently discovered in highly embedded (and therefore very young) protostellar sources. The EHVs are believed to power the moderate velocity, bipolar molecular outflows that are ubiquitous near young stellar objects (Lada 1985). The conditions in the inner regions of a fast, protostellar wind are extreme by the standards of interstellar chemistry, and define a new branch of astrochemistry. In addition to the short dynamical timescale (about one day), the photorates in the inner wind are extremely large, as illustrated by the photoionization rate for neutral carbon atom. However the wind also has a tremendous capacity for self- and mutual-shielding, as witnessed by the maximum value of the far UV optical depth of neutral carbon.

Table 1. Physical Conditions for Wind Chemistry

Quantity	IRC +10216	SVS 13
R_* (cm)	10^{14}	7×10^{11}
T_* (K)	2200	5000
r (cm) [1]	5×10^{16}	10^{12}
u_W (km s^{-1}) [2]	15	150
τ_{dy} (s)	3.3×10^{10}	6.7×10^{4}
$\dot{M}$ ($M_\odot$ yr s^{-1})	3×10^{-5}	3×10^{-6}
n (cm^{-3})	2.5×10^{4}	10^{15}
T (K)	~ 20	~ 2000
$G_o(C)$ (s^{-1}) [3]	3×10^{-10}	5×10^{-4}
$\tau(C)$ [4]	10	8×10^{6}

1. Location of intense, *in situ* chemical activity.
2. Terminal speed.
3. Unshielded photoionization rate of neutral carbon.
4. UV optical depth, assuming all carbon is in CI.

2. PROTOSTELLAR WINDS

It is important to distinguish between phenomena associated with the protostellar wind itself and those produced by interaction with the environment, which we refer to as primary and secondary (without any value judgement about their importance). In the former category are the formation and destruction of molecules close to the source of the wind; in the latter are chemical changes induced by the interaction of the wind with infalling, accreting, or ambient gas. One should also distinguish between regions of low- and high-mass star formation. A paradigm for the formation of single, low-mass stars has been developed by Shu and his collaborators (e.g., Shu et al. 1987, 1988), and we focus here on the chemical properties of their primary winds. The Manchester group has discussed some of the chemical phenomena associated with wind interactions (Hartquist et al. 1986, Charnley et al. 1988a, 1988b, 1990).

Typical parameters for a very young protostar are given in Table 1. At a later stage of protostellar evolution, the object becomes visible as a T-Tauri star and the mass-loss rate can be 100 or 1000 times smaller. This case was analyzed by Rawlings, Williams, and Cantó (1988), who found that molecule formation does not occur in such low-density winds. Stimulated by the discovery by Lizano et al. (1988) of both EHV HI and CO, my colleagues and I independently pointed out that the efficiency of molecular synthesis was a sensitive function of density and that CO was likely to be present in the primary wind of SVS 13 (Glassgold et al. 1989). The simultaneous presence of atomic and molecular gas can be understood in terms of the difficulty in forming and preserving H_2 in comparison with CO.

We recently completed a much more detailed study of a quasi-spherical wind (Glassgold, Mamon, & Huggins 1991, henceforth GMH), inspired by the X-celerator model of Shu et al. (1988). In order to capture the essential elements of the model, Ruden, Glassgold, and Shu (1989, henceforth RGS) approximated the velocity u and density n by

$$u(r) = \mathcal{U}(r)\, v_W, \quad n(r) = \frac{C}{r^2} \frac{1}{\mathcal{U}\mathcal{A}}, \tag{1}$$

where $\mathcal{U}$ and $\mathcal{A}$ describe the acceleration and collimation of the wind and $C = \dot{M}/4\pi m u_W$; for SVS 13 as described in Table 1, $C = 6 \times 10^{35}$ cm^{-1}. RGS used the simple form, $\mathcal{U} = \mathcal{A} = 1 - R^*/r$ and, after considering wide range of heating and cooling mechanisms, and concluded that:
1. The dominant cooling process is adiabatic expansion.
2. The dominant heating processes are
   a. three-body formation of H_2 (in the inner envelope)
   b. ambipolar diffusion heating (in the outer envelope).

As a result, protostellar winds start out warm but quickly cool down. Note that the inverse process to three-body formation *cools* the gas and helps limit the maximum temperature. Because ambipolar diffusion heating depends on the electron fraction, both heating mechanisms couple strongly to the chemistry.

To understand protostellar wind chemistry, we start with the results of dynamical and thermal modeling, i.e., both n and T decrease rapidly with distance beyond several protostellar radii. The wind can become molecular by a two-stage process whose details depend sensitively on the density, temperature, and the ultraviolet radiation field (GMH):
1. Formation of H_2
   a. Radiative attachment

$$e + H \rightarrow H^- + h\nu, \quad H^- + H \rightarrow H_2 + e$$

$$H^+ + H \rightarrow H_2^+ + h\nu, \quad H_2^+ + H \rightarrow H_2 + H^+$$

   b. Three-body formation

$$H + H + H \rightarrow H_2 + H, \quad H + H_2 + H \rightarrow H_2 + H_2$$

The weak radiative attachment reactions are effective for low mass-loss rates and three-body formation at high densities.

2. Hydride & Heavy Molecule Formation

$$A + H_2 \rightarrow AH + H, \quad AH + B \rightarrow AB + H$$

The most important channel for hydride formation involves A = O because of the high abundance of O, the low reaction threshold, and large rate coefficient.

The GMH calculations show that the acceleration and collimation factors considerably enhance molecular synthesis. The density and the dynamical timescale close to the protostar are about 400 and 20 times larger than for the simple case where $\mathcal{U} = \mathcal{A} = 1$ (spherically symmetric, impulsively started wind). GMH also find that, except for H_2, molecular synthesis is almost complete in that all of the available carbon and silicon are in CO and SiO and almost all of the residual oxygen is in H_2O. Even the abundance of molecular hydrogen is quite large, ~ 0.15. A novel aspect of the chemistry is $CO/C_{tot} \gg H_2/H_{tot}$. Thus, there is generally a substantial abundance of atomic H capable of emitting the 21 cm line.

3. THE PROTOTYPICAL C-RICH AGB STAR IRC +10216

The circumstellar envelopes of evolved stars are important for both stellar evolution and the interstellar medium. Understanding how evolved stars lose mass means solving the twin problems of the dynamical origins of the winds and the formation of dust. The latter is surely one of the most fundamental problems in astrochemistry, and is likely to involve the close coupling of dynamical, thermal, chemical, and radiative transfer effects. The nearby carbon star IRC +10216 is the prime target for both observational and theoretical studies of circumstellar chemistry and here we focus on the outer envelope. A broader review of circumstellar chemistry has recently been published by Omont (1991). Over forty molecules have been detected in IRC +10216, mainly in emission at mm wavelengths, but thirteen have been detected with near infrared absorption spectroscopy. The latter technique is indispensible for detecting symmetric molecules; it also provides important spatial information from analyses of the measured rotational excitation and the line profiles (Keady & Ridgway 1991).

The molecules observed in the circumstellar envelope of IRC +10216 are formed mainly at three locations:

1. The upper atmosphere of the star $(1 - 2R_*)$, characterized by thermal equilibrium and freeze-out of well-bound molecules.

2. The "transition" region $(2 - 30R_*)$, where dust is formed and surface reactions occur.

3. The "outer" envelope (beyond $30R_*$), which is dominated by photochemistry. These regions may not be all that distinct; they have been introduced to emphasize that different types of chemistry occur in the flow.

Because thermal equilibrium should hold near the photosphere, it might seem reasonable to suppose that the situation there is straightforward. However, the freeze-out depends on the run of density and temperature in the upper atmosphere, where the conditions for cool stars with unstable atmospheres (i.e., with strong winds) are not all that well known. Still, careful use of the equilibrium abundance calculations can guide the choice of initial chemical abundances to use for the transition region.

The transition region has long been regarded as nearly intractable because of its complicated and unknown dynamics and poorly understood dust-formation chemistry.

Among the dynamical effects that have to be considered are the inhomogeneities in the upper atmosphere, shocks produced by stellar pulsations, and the wind generation process itself. The chemistry involves non-equilibrium, inhomogeneous, nucleation and simultaneous treatment of gas-phase and surface chemistry. Omont (1991) gives a critical review of the problems facing the the theory, which he emphasizes must account for the *partial* condensation of the wind. Despite the formidable difficulties, there seems to be a consensus that a chemical-kinetic approach can provide some insights into the *precursors* of the solid phases (e.g., Frenklach & Feigelson 1989).

Equally important are observations that probe the region of dust formation. For example, Danchi et al. (1989) have made new 10 μm interferometric measurements of IRC +10216 which indicate that dust forms close to the star, $\sim 0.05''$ or within 2-3 R_*, and most efficiently at minimum luminosity. The dust continues to evolve physically and chemically as it moves into the outer envelope. Evidence for depletion of the gas onto dust is provided by interferometric measurements of the mm lines of SiS (Bieging & Rieu 1989), which peak strongly right on IRC +10216 and then decrease rapidly in a few arcseconds ($10^{15} - 10^{16}$ cm). Evidence for molecule formation in this same region has been deduced from the lineshapes of several molecules in the 10 μm window (Keady & Ridgway 1991). In particular, NH_3, CH_4, and SiH_4 all appear to form in the region from $10-30$ R_* (or $1-3 \times 10^{15}$ cm). This is in accord with thermal equilibrium calculations for the upper atmosphere of the star, which give very *low* abundances for these fully hydrogenated species. The presumption is that they are formed on grains.

The photochemical chains introduced by Huggins & Glassgold (1982) form the basis of the theory of the outer envelope. Two of the earliest photochains considered start with CO and C_2H_2 and lead to chemically reactive ions and radicals. The theory predicts variations on a spatial scale that can now be measured with existing facilities like the IRAM 30-m and the Nobeyama 45-m telescopes and several mm arrays. Pioneering work by Bieging & Rieu (1988,1989) demonstrate the technique and confirm the qualitative predictions of the photochemical model of the outer envelope. The following general comments on the connections between the theory and observation are relevant in this connection.

1. The molecules in the outer envelope are either (a) progenitors or (b) synthesis products. A progenitor may have survived the transition region intact (e.g., CO), particpated in dust fromation (e.g., C_2H_2), or have been synthesized in the transition region (e.g., NH_3, CH_4, and SiH_4). Progenitors and synthesis products have essentially different spatial distributions, respectively, (a) uniform, with a cut-off and (b) shell. Interferometric measurements with angular resolution better than $10''$ can distinguish between them. This is obviously important in the absence of a theory of the inner regions.

2. Photodissociation by the interstellar radiation field eventually destroys all

simple molecules, leaving all atoms except H and O singly ionized. The finite size of a molecular distribution is determined by the unshielded photorate *and* by shielding. In addition to CO (discussed above), measurements of molecular cutoffs in IRC +10216 have been published for HCN (Bieging, Chapman, & Welch 1984), C_2H, HNC, HC_3N (Bieging & Rieu 1988), and SiS (Bieging & Rieu 1989). The far UV dust extinction needed to fit the observations is greater than that of interstellar dust, consistent with a smaller size for newly formed grains (Huggins, Morris, & Glassgold 1984, Truong-Bach et al. 1987). In principle, measurements of the decline in abundances at large distances for a well selected set molecules with known photodissociation rates could determine the far UV optical properties of the circumstellar dust of IRC +10216.

A new direction for circumstellar chemistry involves silicon: seven silicon molecules have already been identified in IRC +10216 (SiH_4 SiC, SiC_2, SiC_4, SiO, SiN, and SiS). Glassgold, Lucas, & Omont (1987) first proposed that SiC_2 was synthesized by the condensation reactions

$$Si^+ + C_2H_2 \rightarrow SiC_2H^+ + H$$

$$Si^+ + C_2H \rightarrow SiC_2^+ + H, \qquad SiC_2^+ + H_2 \rightarrow SiC_2H^+ + H,$$

on the assumption that Si^+ comes from photoionization of Si; they later suggested that the Si comes from the photodissociation of SiS (observed in the outer envelope with an abundance of $\sim 3 \times 10^{-7}$, e.g., Bieging & Rieu 1989) and proposed that ion-molecule reactions of SiS could produce the entire silicon carbide family. The products of the SiS chain display the customary shell distribution (Glassgold & Mamon 1990), with an unusually broad SiC_2 distribution. Howe & Millar (1990) proposed a similar model but assumed that SiC_2 was a progenitor molecule with an abundance of 7.5×10^{-8}. Several ongoing projects to map SiC_2 in IRC +10216, e.g., with the IRAM 30 m telescope and interferometer (Lucas 1991) and the BIMA array (Gensheimer et al. 1991), show that SiC_2 is distributed in a shell, in qualitative agreement with the theory of Glassgold & Mamon (1990).

The recent detection of SiN (Turner 1991) links circumstellar silicon and nitrogen chemistry. Turner suggests that is formed by

$$Si^+ + NH_3 \rightarrow HSiNH^+ + H.$$

This route should work (if the recombination of $HSiNH^+$ produces SiN) because NH_3 is observed in the IR at the beginning of the outer envelope with an abundance $\sim 7.5 \times 10^{-8}$ (Keady & Ridgway 1991). Turner also suggests that the Si^+ comes from the silane photochain, again based on the fact that SiH_4 is observed in the dust formation region with an abundance of $\sim 10^{-7}$. However, silane is just one of several sources of Si and Si^+: SiO ($x \sim 4 \times 10^{-7}$), SiS ($x \sim 3 \times 10^{-7}$), SiH_4 ($x \sim 10^{-7}$), and possibly SiC_2 and Si_2C. It is interesting that, assuming that its photospheric abundance of is solar, more than 2% of the silicon survives the dust-formation process in IRC +10216.

4. CONCLUSION

The above discussion of circumstellar chemistry illustrates the close links between dynamical, thermal, chemical, and radiative transfer effects. A similar coupling occurs in interstellar clouds. The advantage of circumstellar chemistry is that realistic models of the flow can be obtained so that basic thermal and chemical processes can be tested in well-defined environments.

5. REFERENCES

Bieging, J., Chapman, B., & Welch, W.J. 1984, ApJ, 285, 256
Bieging, J. & Rieu, N.-Q., 1988, ApJ (Letters), 329, L107
Bieging, J. & Rieu, N.-Q., 1989, ApJ (Letters), 343, L25
Charnley, S.B., Dyson, J.E., Hartquist, T.W., Williams., D.A., 1988a, MNRAS, 231, 169; 1988b, MNRAS, 235, 1257; 1990, MNRAS, 243, 405
Danchi, W.C., Bester, M., Degiacomi, C.G. McCullough, P.R., Townes, C.H. 1990, ApJ (Letters), 359, L59
Frenklach, M. & Feigelson, E.D. 1989, ApJ 341, 372
Gensheimer, P., Likkel, L., Snyder, L.E., 1991, BAAS, 23, 911
Glassgold, A.E., Lucas, R., Omont, A. 1986, A&A, 157, 35
Glassgold, A.E. & Mamon, G.A., 1990, *Chemistry and Spectroscopy of Interstellar Molecules*, ed. N. Kaifu, (Tokyo), in press
Glassgold, A.E., Mamon, G.A., Huggins, P.J., 1989, ApJ (Letters). 336, L29
Glassgold, A.E., Mamon, G.A., Huggins, P.J., 1991, ApJ 373, 254 (GMH)
Hartquist, T.W., Dyson, J.E., Pettini, M., Smith, L.J., 1986, MNRAS 221, 715
Howe, D.A. & Millar, T.J., 1990, MNRAS, 244, 444
Huggins, P.J. & Glassgold, A.E., 1982, ApJ, 252, 201
Huggins, P.J., Glassgold, A.E., & Morris, M., 1984 ApJ, 279, 284
Keady, J.J. & Ridgway, S., 1991, preprint
Lizano, S., Heiles, C., Rodriguez, L.F., Koo, B.-C., Hasegawa. T., Hayashi, S., Mirabel, I.F., 1988, ApJ, 328, 763
Lucas, R. 1991, *Astrochemistry of Cosmic Phenomena*, in press
Nejad, L.A.M., Millar, T.J., Freeman, A., 1984, MNRAS, 134, 129
Omont, A. 1991, *Chemistry in Space*, eds. J.M. Greenberg & V. Pirrinello, (Kluwer:Netherlands), p.171
Rawlings, J.M.C., Williams, D.A., Cánto, J., 1988, MNRAS, 230, 695
Ruden, S.P., Glassgold, A.E., Shu, F.H., 1990, ApJ, 361, 546 (RGS)
Shu, F.H., Adams, F.C., Lizano, S., 1987, ARAA, 25, 23
Shu, F.H., Lizano, S., Ruden, S.P., Najita, J., 1988, ApJ (Letters), 328, L19
Truong-Bach, Rieu, N.-Q., Omont, A., Oloffson, H., Johansson, L.E.B. 1987, A&A 176, 285
Turner, B.E., 1991, BAAS, 23, 933

QUESTIONS AND ANSWERS

L.d'Hendecourt: Apparently, 2% of the silicon survives dust formation. Is this a measured value? How does this compares with the strong depletion of Si in the diffuse ISM?

A.E.Glassgold: Infrared measurements of IRC+10216, summarized by Keady & Redgway in the 1991 Ap.J. are the basis of the approximate 2% figure. In the diffuse interstellar medium, it is about the same, but silicon is much more depleted in dense regions.

D.A.Williams: There is a remark on the requiremente for dust with high FUV extinction in IRC+10216. I think this indeed likely. There is no reason to expect that dust in IRC+10216 should have the same extintion character as the mean interstellar dust. In IRC+10216 dust is likely to be carbonaceous and hydrogen-rich. Films made in the laboratory of hydrogen-rich amorphous carbon have a high band gap and will automatically provide high far UV extinction. Such grains, if they survive to the ISM, will lose hydrogen and become more typical of general ISM, with stronger visible extinction, at the expense of the far UV extinction.

A.E.Glassgold: Neat!

J.M.C.Rawlings: (1) What is the source of your radiation field? A black-body at T~2000 - 5000 K (a 2000 K black-body is *fainter* in the UV than the I-S radiation field) surely does not emit a significant amount of UV photons bearing in mind the high densities in the outflow. Is Ly α trapping included in the model? (2) In the three cases for the temperature profiles that you presented, you included a $(1 - R_*/r)$ factor. What is the justification for this?

A.E.Glassgold: (1) The protostellar radiation field was usually represented ed by a blackbody spectrum of 5000 K.Depending on the spectral properties of the absorber, the photorates near the protostar are typically 6-10 orders of magni- tude larger than the interstellar medium. We also investigated protostellar spectra ctra with large UV excesses and found that the abundance of CO is quite robust. We have never considered using a photospheric temperature as small as 2000 K for a low mass protostar.It is true that the proper measure of the effects of the radiation field is the parameter G/n, where G is a photorate and n is the density.In our variational studies, this parameter well exceeded the value for the diffuse interstellar medium. Lyman alpha trapping was included in the ionization theory of the H atom.

(2) The modulation factors for velocity and collimation were chosen arbitrarily to go to unity in a few stellar radii, as suggested by the dynamical wind model of Shu et al.(1988). We have since verified that the choice made by RGS gives a good representation of the flow obtained by exact integration of of the appropriate MHD equations.

J.P.Maillard: If I understand the reaction proposed to produce SiN even if it is very unabundant from thermo-equilibrium reactions in carbon-rich environment, it supposes that it forms at the edge of the circumstellar envelope. But it is where it should be destroyed by photoionization. Are the models able to reconcile this contradiction?

A.E.Glasssgold: SiN should have a "shell" distribution due to the combined effects of production by Si^+ and radiative destruction. There is little SiN in the inner envelope because the abundance of Si^+ goes to zero, and photodissociation limits the ammount of SiN in the outer envelope.

EXCITED HYDROGEN CHEMISTRY IN PROTOSTELLAR OUTFLOWS

J.M.C. RAWLINGS, J.E. DREW
Department of Physics
Nuclear Physics Laboratory
Keble Road, Oxford OX1 3RH, UK

M.J. BARLOW
Department of Physics and Astronomy
University College London
Gower street, London WC1E 6BT, UK

ABSTRACT. Chemical models of protostellar and other outflows have been reassessed in the light of new chemical data. In particular, reactions involving excited hydrogen (2s,p) are shown to be important in hot, dense outflows. The $H(n=2) + H \rightarrow H_2 + h\nu$ reaction is much less of a contributor to the H_2 formation rate than the recently measured $H(n=2) + H \rightarrow H_2^+ + e^-$ reaction, providing conditions allow the 0.75eV endothermicity of this reaction to be overcome.

1. Chemistry

We report here for the first time the significance of the associative ionization reaction:

$$H + H(n = 2) \longrightarrow H_2^+ + e^-. \tag{1}$$

Cross-sections for this reaction have been measured by Urbain et al. (1991). Using these we calculate rate coefficients (in cm^3 s^{-1}) of: 1.7×10^{-13} at 3000K, 1.1×10^{-12} at 5000K, and 6.0×10^{-12} at 10000K. A good fit to the rate coefficient at T $\gtrsim$ 2500 K is given by:

$$k = 3.5 \times 10^{-14} + 3.1 \times 10^{-11} e^{-16355/T} \text{ cm}^3 \text{s}^{-1} \tag{2}$$

The reaction is endothermic by 0.75eV and the cross-section shows a secondary barrier of $\simeq$2eV. H_2 formation then occurs as a result of charge exchange:

$$H_2^+ + H \longrightarrow H_2 + H^+ \tag{3}$$

We expect this path is particularly important in the chemistry of protostellar outflows. In all cases of astrophysical interest it is faster than the radiative association reaction proposed by Latter and Black (1991):

$$H + H(n = 2) \longrightarrow H_2 + h\nu \tag{4}$$

Furthermore, the negative-ion H_2 formation route is ineffective in outflows from cool as well as hot stars because of photo-detachment of H^- in the IR radiation field. In short, reaction (1) is a major contributor to the H_2 formation rate in both (i) high-temperature, strongly irradiated regions where the n=2 level is well populated, (ii) cooler (3000 - 5000 K), denser regions. In (ii), lower ionization and large departures from LTE (eg. at 3000K, $H(n=2)/H^+ > 2\times10^{-4}$) are necessary for reaction (1) to be more important than direct radiative recombination of H^+ with H.

2. Models

We have tested the importance of the excited hydrogen chemistry in four situations: T-Tauri winds, cool neutral outflows, BN winds and the ejecta of supernovae.

In the case of T-Tauri winds and cool neutral outflows we have adapted the models of Rawlings et al. 1988 and a model of Glassgold et al. 1989 to accommodate the new chemistry. In both cases the initial ionization was calculated by balancing recombination against collisional ionization. The results show that within a few stellar radii H_2 is significantly enhanced on including reaction (1). However, at later times the temperature and density drop whilst, in the case of the T-Tauri wind, the ionization level remains quite high. At this stage radiative recombination of H^+ with H becomes relatively more efficient (see also Latter and Black's discussion of $H(n=2)$ chemistry in the early universe). Note these calculations do not include Lyα trapping and so set a *lower* limit to the efficiency of the $H(n=2)$ chemistry.

The winds associated with BN-type objects are hot, dense and fairly well ionized – the most promising conditions for the excited hydrogen reactions. In our model $R_* = 4\times10^{11}$ cm, $\dot{M}= 1.0\times10^{-6}$ $M_\odot yr^{-1}$, $v_\infty = 150$ km s^{-1}, and the underlying radiation field is a Kurucz model atmosphere for T_{eff}=25000K, log(g)=4.0. We adopt a slowly accelerating velocity law and a temperature profile consistent with the wind models of Drew (1989). Self-shielding for H_2, CO and the carbon continuum have been included. To begin with, we use a two-level H^o atom + continuum: transfer in Lyα is accounted for using escape probabilities based on the Sobolev approximation, and the effect of Balmer continuum opacity is included. For the initial conditions we assume that $H(n=1)/H(n=2)$ are in detailed balance and that photoionization of $H(n=2)$ balances case B recombination. Preliminary calculations confirm H_2 formation via reaction (1) is *very* important in hot, partly ionized winds.

In the case of supernovae we have looked at both the "core" and the "mantle" of the ejecta. Despite the presence of a strong, very non-thermal high energy electron flux in the core region, NLTE models of the ejecta of SN1987a (eg. Schmutz et al. 1990) show the departure coefficients for H^o n=1,2 and 3 are less than 10. Hence, as the temperature is also low (2-3000 K), there is unlikely to be a significant enhancement to the chemistry in the core. In the "mantle" region, observations indicate that the n=2 level may be markedly overpopulated. However, even here, the importance of excited hydrogen chemistry in SN1987a appears to be marginal.

In some situations, the reaction: $H(n=3) + H \longrightarrow H_2^+ + e^-$ may also be significant. It is exothermic by 1.14eV but no cross-sections are yet available.

References

Drew, J.E. 1989, *Astrophys. J. Suppl.*, **71** 267
Glassgold, A.E., Mamon, G.A. & Huggins, P.J. 1989, *Astrophys. J.*, **336** L29
Latter, W.B. & Black, J.H. 1991, *Astrophys. J.*, **373** 161
Rawlings, J.M.C., Williams, D.A. & Canto, J. 1988, *Mon. Not. R. ast. Soc.*, **230** 695
Schmutz, W., et al. 1990, *Astrophys. J.*, **355** 255
Urbain, X., Cornet, A., Brouillard, F. & Giusti-Suzor, A. 1991, *Phys. Rev. Lett.*, **66** 1685

MOLECULES IN THE ENVELOPES OF LATE-TYPE STARS

R. LUCAS
IRAM
300 rue de la Piscine
38406 Saint-Martin-d'Hères
FRANCE

1. Introduction

With physical conditions very similar to those of molecular clouds, circumstellar envelopes (CSEs) are the sites for a rich chemistry. Hovever their simpler geometries, the variations in elemental composition, and the short time scales of the outflows, make them objects of particular interest to the astrochemist. For a recent and general review of circumstellar chemistry the reader may refer to Omont (1990). Here we give only a very short overview (Section 2), then we report on some of the very recent observational advances in this field (Section 3).

2. Overview of circumstellar chemistry

LTE is expected to be valid in the photosphere and in the very inner layers, where the temperature and the density are high. Calculations by Tsuji (1987) and others, show most radicals such as CN, HNC, NH_3, cyanopolyynes (except HC_3N) cannot be formed in LTE. Outside of the LTE region, radical reactions are dominated by reactions by H, the abundance of which is highly uncertain. Grain processes are also important in this region. Ice has been found on grains in some massive CSEs. Diffuse IR bands carriers (PAHs ?) are abundant in PPNs, but whether they are also important in the earlier phases is unknown (Frenklach and Feigelson 1989). Shocks are also probably present, related to pulsations and high velocity winds, which could dissociate abundant molecules, and result in the synthesis of otherwise unexpected products (HCN, CS in O-rich CSEs).

The chemistry in the outer layers is dominated by photon induced processes. CO photodissociation is fast, but self shielding occurs. It is expected to be effective between $3\ 10^{16}$ and 10^{18} cm, depending on the mass loss rate (Mamon et al. 1987). It leads to C and C^+. Similar photodissociation chains starts from C_2H_2 and HCN, in C-rich objects. Ionization is initiated by cosmic rays and by interstellar UV. The most abundant ions are $C_2H_2^+$, H_3^+, H_3O^+, HCO^+, depending on the C/O ratio. Chromospheric UV may also play a role in some objects, such as α Ori, or PNs, where HCO^+ is found in abundance. Ion-molecule reactions were proposed by Glassgold and Huggins (1985). They lead to the synthesis of

C_3H, C_3, C_4H, HC_3N, HNC, SiC_2, C_2S, C_3S, ... Radical reactions are also important due to abundance of radicals produced by photodissociation. They could synthesize HC_3N, C_4H, C_6H, SiC_4, SO, SO_2 ....

3. Recent observations

3.1. NEW MOLECULES

Since the review by Lucas and Guélin (1990), several new molecules have been found in the last few years, all of them in the C-rich object IRC+10216:
— CP (Guélin et al. 1990). Its column density is $\simeq 10^{13}$ cm^{-2}. Thermodynamics would imply that CP should be less abundant than PN, and much less than HCP, which is not found, while photochemistry would produce CP in the outer envelope. However from the profile shape, this molecule has an extent of $\simeq 15''$, which is smaller than that of photodissociation products.
— H_2CCC (propadienylidene) was tentatively detected in IRC+10216 with a column density of $\simeq 2.6 \; 10^{12}$ cm^{-2} (Cernicharo et al. 1991a). It is the linear isomer of the C_3H_2 ring.
— H_2CCCC (butatrienylidene), isomer of diacetylene, has been identified in IRC+10216 (Cernicharo et al. 1991b). The U-shaped lines imply a spatial distribution similar to that of C_4H. Formation by $C_2H_2^+ + C_2H_2 \to C_4H_3^+ + H$, followed by dissociative recombination, is proposed (Glassgold et al. 1987).
— The HCCN radical was found in the outer envelope with a column density $\simeq$ 1.2 10^{13} cm^{-2} (Guélin and Cernicharo 1991a). Its formation mechanism remains unclear.
— SiN (see Turner 1991)

3.2. HIGH-RESOLUTION OBSERVATIONS OF IRC+10216

The nearby object IRC+10216 is by far the best studied of carbon-rich CSEs. Its chemical structure has been studied with the Hat Creek interferometer (Bieging and Rieu 1989). With more sensitive instruments it is now possible to obtain very detailed information on this prototype object.

The 30 m telescope has been used by Kahane et al. (1991) to study the distribution of SiO, SiS, C_4H, HC_3N and SiC_2, in several transitions. While SiO and SiS are clearly present in the central regions ($r < 15''$), C_4H is distributed in a hollow shell ($14'' < r < 26''$). The case of HC_3N is less clear, since its 2 mm emission seems to fill a sphere, while at 3mm Bieging and Rieu (1988) found it in a hollow shell. Finally the distribution of SiC_2 emission is asymmetric and stronger in the southern part of the source.

With the Plateau de Bure interferometer, Guélin et al. (1991b) have observed C_2H with high resolution and sensitivity. The C_2H shell appears asymmetric (fig. 1). They simultaneously mapped emission from C_5H and C_4H in the ν_7 vibrational state, which are both distributed in hollow shells. With the same instrument, Cernicharo et al. (1991c) mapped the emission of NaCl and of SiC_2 (presented in a poster paper).

3.3. INTERFEROMETRIC OBSERVATIONS OF THERMAL SiO IN EVOLVED STARS

Lucas et al. 1991, using the IRAM interferometer on Plateau de Bure, have observed the SiO $v=0$ $J=2-1$ emission from 12 evolved stars, mostly O-rich. Simple source models were

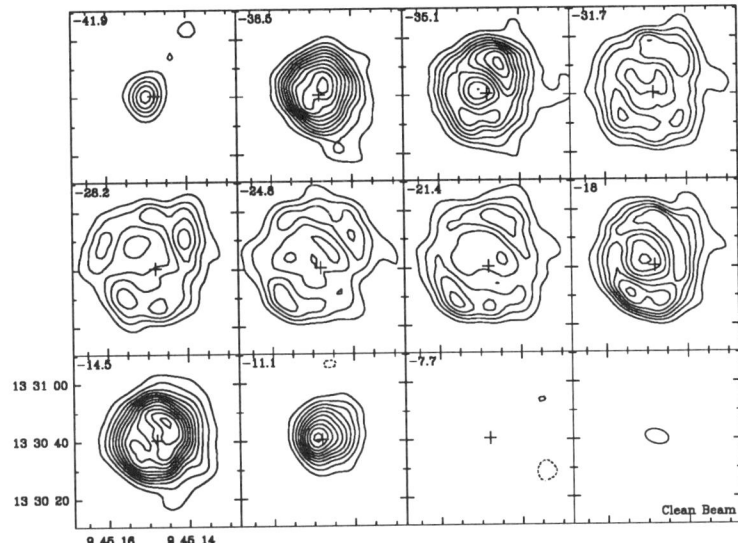

Figure 1: Combined 30-m and Plateau de Bure observations of CCH (N,J)=(1,3/2)–(0,1/2) (87.3 GHz) emission from IRC+10216. Channel maps labelled by LSR velocity in km s^{-1}. The resolution is 8.6" by 5.1", the contour spacing is 0.1 Jy (Guélin et al. 1991b).

fitted in the visibility curves, thus determining fluxes and sizes. In some stars, the u, v coverage was good enough to obtain maps.

In RX Boo, R Cas, χ Cyg, and IK Tau, the emission is circularly symmetric and centered on the star's optical position. For the other objects the results are also compatible with that geometry. The half intensity sizes are 0.9" to 2.4". They are larger than those expected from previous work, though much smaller than the extents of CO (Bujarrabal and Alcolea 1991). SiO is widespread within the envelopes up to $\sim$ 1-7 10^{15} cm, and even further for the supergiants IRC+10420 and NML Cyg. The profile shapes are mostly gaussian, which indicates the emitting regions do not have the expected high and constant expansion velocity. This is confirmed by the distribution of size as a function of velocity, which shows no clear increase at the line center. Grain formation is proposed to continue in O-rich and S-type Miras as far as 5 10^{15} cm from the central star, which would explain the large SiO abundance and extended acceleration region that are probably at the origin of the non-standard line profiles.

3.4. NITROGEN CHEMISTRY IN O-RICH ENVELOPES

Olofsson et al. (1991) observed CO, ^{13}CO, CN, HCN and CS in TX Cam with the 30-m telescope. HC$_3$N, SiS, and HNC had been found by Lindqvist et al. 1988, but neither OH nor H$_2$O masers are found. For CN, and CS, the observed abundances agree with theoretical models (Nejad and Millar 1988, Nercessian et al. 1989), but there is a strong disagreement for HCN. The low CN/HCN ratio seems particular to O-rich environments.

S. Guilloteau et al. (1991) have used the IRAM interferometer on Plateau de Bure, to observe the HCN v=0 J=1–0 emission from O-rich evolved stars. They found that the extent of HCN emission, a few 10^{16} cm, is comparable to that of the 1612 MHz OH masers. These sizes are in agreement with the assumption that HCN is formed by gas-phase reactions initiated by the photodissociation of CH_4, as proposed by Nejad and Millar (1988) and Nercessian et al. (1989).

3.5. Observations of Sulfur-bearing Species in Oxygen-rich Envelopes

H_2S was first detected in OH231.8+4.2 by Ukita and Morris (1983), while SO_2 and SO observed were found with the 30m (Lucas et al. 1986, Guilloteau et al. 1986). A systematic study of SO_2, SO, H_2S has been undertaken with the 30m (Omont et al. 1991). H_2S was detected in 15 O-rich stars in its ortho and para states. The line profiles of H_2S are much narrower than those of SO_2, which indicates that H_2S is present in the inner regions (where acceleration probably occurs), while SO_2 (and probably SO) are in an outer shell. Modelization of the H_2S emission shows it comes from a region of radius $\simeq 10^{16}$ cm. SO_2 and SO are formed by reactions of S with OH, in the H_2O photodissociation region (a few 10^{16} cm).

3.6. Studies of CO Emission

CO, detected in hundreths of CSEs, has been extensively used to derive mass losses, following Knapp and Morris (1985). Margulis et al. (1990) observed all M giants, S stars, and O stars in the 2 μm Sky Survey, with $\delta > 10°$. The line profiles are simple for C ans S stars, but for M stars they indicate complex envelope structures. Heske et al. (1990) found a deficiency of CO in 13 OH/IR stars, particularly in the most massive ones. The (2–1)/(1–0) ratio increases drastically with increasing optical depth of the dust (9.7 μm feature turning into absorption). This may be due to low kinetic temperature, or to lower mass loss in the outer shells.

Olofsson et al. (1990) haved mapped carbon stars with detached envelopes (R Scl, U Ant, S Sct, TT Cyg). These are thin envelopes, detached from the stars (inner radii $> 10^{17}$ cm), which probably present episodic mass loss, triggered by helium shell flashes. Bujarrabal and Alcolea (1991) have mapped with the 30m telescope, seven O-rich stars (RR Aql, RX Boo, R Cas, S CrB, R Leo R LMi, and IK Tau), 1 S star (χ Cyg), and 2 C stars (S Cep and V Cyg). Most objects are resolved, roughly circular in shape, and much smaller than the sizes predicted by envelope models, such as Knapp and Morris (1985).

Sahai (1990) developed a self-consistent model for the CO emission, including thermal equilibrium, and showed that mass loss derivation is very sensitive to the kinetic temperature law. In particular the gas may be cooled to temperatures below 2.7 K by adiabatic expansion, producing CO J=1–0 profiles with reduced intensity and width. This model was applied to the C-rich object U Cam. Truong-Bach et al. (1990) have observed and modelled the CO emission of CRL2688. In their two-shell model of IRC+10216 (1991), the mass loss was higher in the outer envelope (4 10^{-5} $M_\odot$/yr) than in the inner 4" (2.5 10^{-5} $M_\odot$/yr).

3.7. The 200 km s^{-1} Bipolar Outflow in CRL618

This outflow, discovered by Cernicharo et al. (1989) has been observed in HCN J=1–0 with the Plateau de Bure interferometer (Neri et al. 1991). The angular resolution was 2.4". The circumstellar envelope itself is extended, and centered on the radio continuum source.

The red-shifted outflow emission comes from a an unresolved region 1" to the west of the HII region, while the blue-shifted high-velocity gas arises from a source af size $\simeq$ 2", partly in front of the HII region. The high velocity HCN is proposed to be formed behind shocks at the wind-envelope interface surface.

References

Bieging, J.H., N-Q-Rieu 1988, ApJ 329, L107
Bieging, J.H., Nguyen-Q.-Rieu 1990, in "From Miras to Planetary Nebulae: which Path for Stellar Evolution?", Eds M.O. Mennessier and A. Omont, p164
Bujarrabal, V., Alcolea, J. 1991, A&A, to be published
Cernicharo, J., Guélin, M., Martin-Pintado, J., Peñalver, J., Mauersberger, R. 1989, A&A 222, L1
Cernicharo, J., Gottlieb, C.A., Guélin, M., Killian, T.C., Paubert, G., Thaddeus, P., Vrtilek, J.M. 1991a, ApJ 368, L39
Cernicharo, J., Gottlieb, C.A., Guélin, M., Killian, T.C., Thaddeus, P., Vrtilek, J.M. 1991b, ApJ 368, L43
Cernicharo, J., Guélin, M., Lucas, R., 1991c, in preparation
Frenklach, M., Feigelson, E.D. 1989, ApJ 341, 372
Glassgold, A.E., Mamon, G.A., Omont, A., Lucas, R. 1987, A&A 180, 183
Guélin, M., Cernicharo, J., Paubert, G., Turner, B.E. 1990, A&A 230, L9
Guélin, M, Cernicharo, J. 1991a, A&A 244, L21
Guélin, M., Cernicharo, J., Guilloteau, S., Lucas, R., 1991b, in preparation
Guilloteau., S., Lucas, R., N-Q-Rieu, Omont, A. 1986 A&A 165, L1
Guilloteau, S., Forveille, T., Lucas, R., Morris, M., Radford, S.J.E., Fuente, A. 1991, in preparation
Glassgold, A.E., Huggins, P.J. 1985, M-Type stars, ed. by H.R. Johnson and F. Querci (NSF-CNRS 1988), p291
Heske, A., Forveille, T., Omont, A., van der Veen, W.E.C.J., Habing, H. 1990, A&A 239, 173
Jackson, J.M., N-Q-Rieu 1988, ApJ 335, L83
Kahane, C. et al. 1991, paper presented to the IAU.
Knapp, G.R., Morris, M. 1985, ApJ 292, 640
Lindqvist, M., Nyman, L.-Å., Olofsson, H., Winnberg, A., 1988, A&A 205, L15
Lucas, R., Omont, A., Guilloteau, S., N-Q-Rieu, 1986, A&A 154, L12
Lucas, R., and Guélin, M. 1990, in "Submillimetre Astronomy", G.D. Watt and A.S.Webster eds., p97
Lucas R., Bujarrabal, V., Guilloteau, S., Bachiller, R., Baudry, A., Cernicharo, J., Delannoy, J., Forveille, T., Guélin, M., Radford, S.J.E., 1991, submitted to A&A
Mamon, G.A., Glassgold, A.E., Huggins, P.J. 1988, ApJ 328, 797
Margulis, M., Van Blerkom, D.J., Snell, R.L., Kleinmann, S.G. 1990, ApJ 361, 673
Nejad, L.A.M., Millar, T.J. 1988, MNRAS 230, 79
Nercessian, E., Guilloteau, S., Omont, A., Benayoun, J.J. 1989, A&A 210, 225
Neri, R., Garcia-Burillo, S., Cernicharo, J., Guélin, M., Guilloteau, S., and Lucas, R. 1991, in preparation
Olofsson, H., Carlström, U., Eriksson, K., Gustafsson, B., Willson, L.A. 1990, A&A 230, L13
Olofsson, H., Linsqvist, M., Nyman, L.-Å., Winnberg, A., N-Q-Rieu, 1991, A&A 245, 611
Omont, A. 1990, in Chemistry in Space, Eds. Greenberg and Pironello, p171
Omont, A., Morris, M., Lucas, R., Guilloteau, S. 1991, in preparation
Sahai, R. 1990, ApJ 362, 652
Truong-Bach, Morris, D., Nguyen-Q.-Rieu 1990, A&A 230, 431
Truong-Bach, Morris, D., Nguyen-Q.-Rieu 1991, A&A, to be published
Tsuji, T. 1987, in Astrochemistry, IAU Symp 120
Turner, B.E., preprint
Ukita, N., Morris, M. 1983, A&A 121, L5

QUESTIONS AND ANSWERS

B.Khare: Larger molecules are found in the outer envelopes. How large are these molecules?

R.Lucas: $HC_{11}N$ is the largest one.

L.F.Rodriguez: Your HCN results on VY CMA show that the source is very elongated. Do you have an explanation for that?

R.Lucas: Not at this time.

LITHIUM ABUNDANCE AND SPACIAL DISTRIBUTION OF T TAURI STARS

Jane Gregorio-Hetem and Jacques R.D. Lépine
Instituto Astronômico e Geofísico, Universidade de São Paulo
CP 9638, 01065 São Paulo, Brasil

ABSTRACT. We determined temperatures and Lithium 6707 A resonance line equivalent width of a sample of 62 T Tauri stars. Lithium abundances were then estimated by using a grid of curves of growth. The lithium abundance is shown to decrease with the distance of the stars to the nearest dense core of a molecular cloud. This effect is interpreted as being due to the ages of the stars, the youngest ones being closer to still active star formation regions.

1. Introduction

We report the preliminary results of an investigation of the relation of the Li abundance in T Tauri stars with their distance to the nearest star-forming region. The lithium abundance is potentially an age indicator; it is believed that the youngest T Tauri stars present the same Li abundance of the interstellar medium, and that the surface abundance gradually decreases with time, due to the combined effect of convection and destruction in the stellar interior (Spite and Spite, 1982; Duncan and Jones, 1983). It would be of great interest to be able to estimate the age of T Tauri stars, in order to study their evolution as well as the evolution of the star-forming regions.

2. Observational data

Our program stars have been observed in a recent survey for new T Tauri stars (TTS) based on the IRAS point source catalog (Gregorio-Hetem et al.,1991). In the part already completed of the survey, we obtained Coudé spectra of 62 TTS (including previously known TTS), with the 1.6 m telescope of the Laboratorio Nacional de Astrofísica, Brazópolis, Minas Gerais. The observed spectral region is 6550-6750 A, which contains Hα and the Li resonance line at 6707 A; the resolution is 0.1 A. We also made UBVRI photometric measurements of most of the program stars with a 0.6m telescope at the same observatory.

3. Temperature and Li abundance determinations

Since the surface lithium abundance is known to be a function of both age and mass (or temperature), we determined the temperatures of the stars of our sample by fitting, with a model, our photometric data and the IRAS data. The model consists of a central star, represented by a blackbody, surrounded by a spherically symmetric circumstellar dust shell (CDS). Good fits are obtained by taking CDS density decreasing like $r^{-1.5}$ and temperature decreasing like $r^{-0.4}$, and an extinction law like λ^{-1}. The internal radius of the CDS is fixed by considering that the radiative equilibrium temperature of the dust grains cannot exceed typical evaporation temperature of about 400 K. In some cases an optically thick, geometrically thin disk with temperature decreasing outwards like $r^{-0.75}$ is needed to obtain a good fit. The stellar temperatures obtained in this way are in general in good agreement with the spectral types, when they are known.

Lithium abundances were estimated by comparing the stellar temperature and the equivalent width of the Li I λ6707 line with a grid of curves of growth presented by Duncan (1991), from Kurucz and Bell and Gustafsson model atmospheres, for log g = 4.0 and log g = 3.75.

4. Results and discussion

We present in Figure 1 log Li abundance as a function of log T. The well known depletion of Li in low temperature TTS can be observed. We remark that we obtain, for many TTS with T > 5000 K, abundances of the order of log N(Li)= 4, on the log N(H) = 12 scale. This is much larger than the accepted primordial abundance of Li, log N(Li) < 3 (Boesgard and Steigman, 1985). One possibility would be to suspect that the model atmospheres are not correct. Another possibility would be a segregation mechanism to act during the last stages of formation of stars, if for instance Li is tied to dust grains submitted to radiation pressure, so that its accretion on the star is delayed, resulting in a relative enrichment of the photosphere.

We present in Figure 2 log of Li abundance versus distance of the TTS to the nearest dense core of a star forming molecular cloud. The distances are projected distances measured in degrees; the reason for the use of angular units is that the distances of most of the relatively isolated TTS are not known. At 150 pc, which is the approximate distance of nearby star-forming cloud complexes like ρ Oph and Cham I, one degree correspond to about 3 pc. Most TTS are situated within 2 degrees of a molecular cloud core; up to such separation they can be considered as belonging to the same association. The relative position, in galactic coordinates, of the TTSs studied here and of the cloud complexes nearest to them are illustrated in Figure 3. For the stars situated at $d > 2°$, we observe in Figure 2 a tendency of Li abundance to decrease with d.

The obvious interpretation is that this is an effect of age, the more isolated TTS being the oldest ones. Since we do not observe any systematic variation of temperature with distance, we can eliminate the possibility of the more distant TTS being less massive stars. This could be the case if the more distant stars were objects ejected from star-forming regions with larger velocities. Such a mass-selection effect can probably only be observed if stars of a same association (within 2°) are compared.

From the log N(Li) versus T and versus age relation presented by Basri, Martin and Bertout (1991), we conclude that Hen 1, the TTS with highest galactic latitude discovered in our survey, is $5*10^7$ years old, and that the group of TTS situated around the "isolated" TTS TW Hya, is about $3*10^7$ years old.

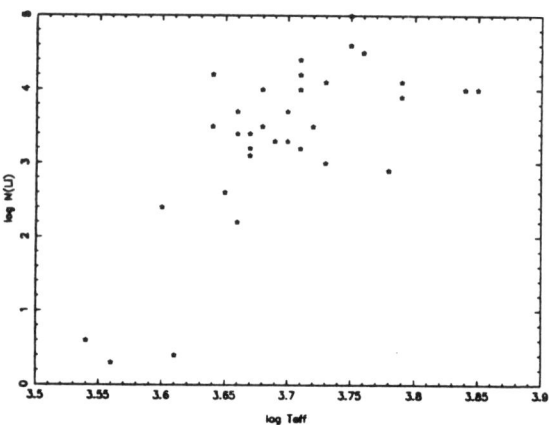

Figure 1: log of Lithium abundance versus log of effective temperature of T Tauri stars.

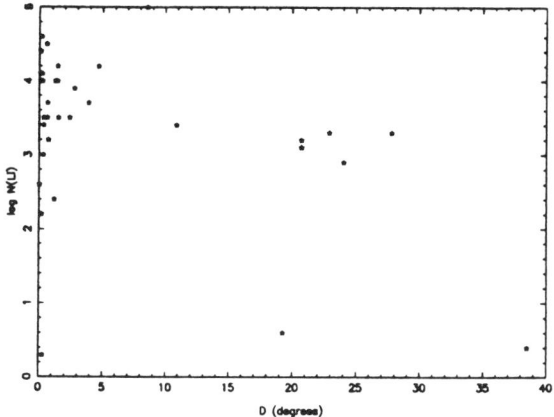

Figure 2: log of Lithium abundance versus projected distances to the nearest dense core of a star forming molecular cloud.

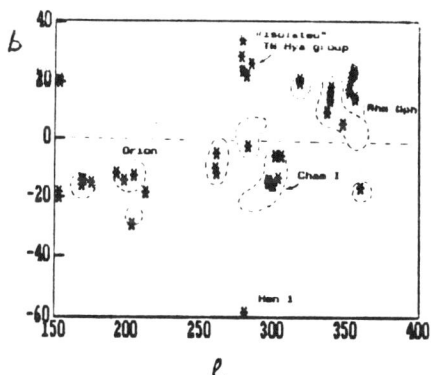

Figure 3: The spatial distribution of the T Tauri stars of our sample, in galactic coordinates. Some of the most important star-formation regions are indicated.

References

Basri,G., Martin,E.L., Bertout,C, 1991: A & A, in press

Boesgard, A.M., Steigman,G.: 1985, ARA & A 23,319

Duncan, D.K.: 1991, Ap.J, in press

Duncan, D.K., Jones,B.F., 1983, Astrophys. J. 271:66

Gregorio-Hetem,J., Lépine, J.R.D., Torres, C.A.O, Quast, G.R.,
    de la Reza, R.: 1992, A J., in press

Spite,F., Spite, M.: 1982, A & A 115, 357

SUB-MILLIMETRE OBSERVATIONS OF SiCC IN IRC 10216

L.W. AVERY[1], T. AMANO[1], M.B. BELL[1], P.A. FELDMAN[1], J.W.C. JOHNS[1],
J.M. MACLEOD[1], H.E. MATTHEWS[2], D.C. MORTON[1], J.K.G. WATSON[1],
B.E.TURNER[3], S.S. HAYASHI[4], G.D. WATT[2], and A.S. WEBSTER[5]
[1]Herzberg Inst. of Astrophysics, NRC, Ottawa, K1A0R6, Canada
[2]Joint Astronomy Center, 665 Komohana St., Hilo, HI 96720, U.S.A.
[3]National Radio Astronomy Obs., Charlottesville, VA 22901, U.S.A.
[4]National Astronomical Obs., Osawa 2-chome, Mitaka, Tokyo 181, Japan
[5]Royal Observatory, Blackford Hill, Edinburgh, EH9 3HJ, Scotland

ABSTRACT. We have used the James Clerk Maxwell Telescope to observe 11 lines of SiCC with excitation energies up to 330 K above the ground state in IRC 10216. An excitation analysis indicates T_{rot} = 43 ± 5 K within a given Ka ladder and T_{exc} = 160 ± 50 K between Ka ladders. The column density of SiCC is estimated to be 3 x 10^{15} cm^{-2}, more than an order of magnitude higher than previous estimates.

1. Introduction

In 1984 Thaddeus et al. assigned nine unidentified millimetre-wavelength lines to the ring molecule SiCC in the circumstellar envelope of the carbon star IRC 10216. For transitions between levels, J_{KaKc}, characterized by the same value of Ka, they found T_{rot} = 10 K, while lines arising in different Ka ladders yielded T_{exc} = 140 K. Transitions within a Ka ladder can occur by both collisional and rapid radiative deexcitation. However, the "cross-Ka ladder" radiative transitions are only weakly permitted, so that the relative population distributions between the Ka ladders depends more strongly upon collisions. For these reasons Thaddeus et al. (1984) argued that the high cross-Ka ladder excitation temperature should reflect the kinetic temperature in the star's atmosphere.

2. Data

We report observations with the James Clerk Maxwell Telescope (JCMT) of 11 high-frequency SiCC lines in IRC 10216. The range of excitation energies covered by these new detections is from 70 K to 330 K above the ground state, considerably higher than the transitions discussed by Thaddeus et al. (1984). The data are summarized in Table 1 where S is the quantum mechanical line strength and E/k the excitation energy of the line in K (Gottlieb et al. 1989). T_R^* is the line radiation temperature corrected for both atmospheric and telescope losses, and also forward spillover and scattering. The JCMT beamwidth (FWHM) is 16 arcsec at 345 GHz and 21 arcsec at 240 GHz.

3. Analysis

We have combined our observations with the lower excitation data used by Thaddeus et al. (1984) to carry out an extended excitation analysis, using a source diameter of 20 arcsecs (Cernicharo et al. 1989). Our results confirm that a markedly different value of excitation temperature is found within a given Ka ladder compared to that between Ka ladders. For our JCMT lines we find average values of T_{rot} = 43 ± 5 K within a given Ka ladder, and 160 ±

50 K across Ka ladders. Evaluation of the total column density, as described by Thaddeus et al. (1984), using these temperatures leads to $N[SiCC] \approx 3 \times 10^{15}$ cm^{-2}.

TABLE 1. SiCC TRANSITIONS DETECTED

Freq (MHz)	T_R^* (K)	Transition	S	E/k (K)
232534	1.0	$10_{2,9}$-$9_{2,8}$	9.6	70
234534	0.4	$10_{8,3}$-$9_{8,2}$	3.6	187
		$10_{8,2}$-$9_{8,1}$	3.6	187
235713	0.7	$10_{6,5}$-$9_{6,4}$	6.4	132
		$10_{6,4}$-$9_{6,3}$	6.4	132
259433	0.5	$11_{6,6}$-$10_{6,5}$	7.7	145
		$11_{6,5}$-$10_{6,4}$	7.7	145
342805	0.6	$15_{2,14}$-$14_{2,13}$	14.7	141
344906	0.9	$16_{0,16}$-$15_{0,15}$	15.9	144
346110	0.75	$14_{2,12}$-$13_{2,11}$	13.7	132
350280	0.25	$15_{10,5}$-$14_{10,4}$	8.3	328
		$15_{10,6}$-$14_{10,5}$	8.3	328
352437	0.4	$15_{8,8}$-$14_{8,7}$	10.7	260
		$15_{8,7}$-$14_{8,6}$	10.7	260
354790	0.6	$15_{6,10}$-$14_{6,9}$	12.6	206
354798	0.6	$15_{6,9}$-$14_{6,8}$	12.6	206

4. Conclusions

The 11 high frequency SiCC lines we have observed in IRC 10216 permit an extended excitation analysis to be carried out. We find that the excitation temperature within a given Ka ladder is 43±5 K, considerably higher than the result of 10 K (Thaddeus et al. 1984) based on wider-beam, lower-frequency data. Population distributions between different Ka ladders correspond to T_{exc} = 160±50 K between ladders. Such values seem too high to be characteristic of kinetic temperatures, and may reflect infrared excitation of the molecule. The column density of SiCC is estimated to be 3×10^{15}cm^{-2}, more than an order of magnitude higher than previous estimates.

5. References

Cernicharo, J., Gottlieb, C.A., Guélin, M., Thaddeus, P. and Vrtflek, J.M. 1989, Ap.J. **341**, L25.
Gottlieb, C.A., Vrtflek, J.M. and Thaddeus, P. 1989, Ap.J. **343**, L29.
Thaddeus, P., Cummins, S.E., and Linke, R.A. 1984, Ap.J. **283**, L45.

A MOLECULAR LINE SURVEY OF THE CARBON STAR IRAS 15194-5115

L.-Å. Nyman
ESO, Casilla 19001, Santiago 19, Chile
R.S. Booth, U. Carlström, L.E.B. Johansson, H.Olofsson
Onsala Space Observatory, S-439 00 Onsala, Sweden
R. Wolstencroft
Royal Observatory, Blackford Hill, Edinburgh EH9 3HJ, UK

IRAS 15194-5115 was discovered by IRAS and identified as a carbon star by Meadows et al. (1987). It is the third brightest carbon star at 12 μm, the brighter ones are IRC+10216 and CIT6. Its infrared properties are similar to those of IRC+10216. Le Bertre and Epchtein (1990) have monitored the star in the near–infrared and derived a period of 578 days. The distance to IRAS 15194-5115 is estimated to about 1 kpc from infrared observations, and also by comparing its bolometric luminosity with that of IRC+10216.

We have used the 15m SEST (Swedish–ESO Submm Telescope) to make a survey of molecular lines toward IRAS 15194-5115 in the 3 and 1.3 mm bands. The idea was to derive molecular abundances and compare them with those of IRC+10216.

The observed lines are listed in Table 1, and a few spectra are shown in Fig. 1. In total 23 transitions of 14 molecular species and their isotopes have been detected. Abundances were calculated using the optically thin lines and the expression given by Olofsson et al. (1990). An excitation temperature of 10K was assumed for all species, and their spatial extents were estimated from photodissociation calculations or in some cases scaled from interferometric observations of IRC+10216. The same species were also observed toward IRC+10216 with the SEST, and the abundances were calculated in the same way. The abundances for the two sources are similar to within a factor of 2 to 3. From our CO data we estimate a mass loss rate of $5 \cdot 10^{-5}$ $M_\odot yr^{-1}$ and an expansion velocity of 22 km s^{-1} for IRAS 15194-5115.

The narrow, blueshifted component seen in the CO spectra in Fig. 1 emanates from an interstellar cloud in the line of sight. The CS(2-1) spectrum (Fig. 1f) shows a narrow, redshifted component whose intensity changes with time. It has a maximum near the infrared maximum. The intensity variations may be due to maser action.

REFERENCES
Le Bertre and Epchtein, 1990, "The infrared spectral region of stars", Montpellier
Meadows et al., 1987, MNRAS 225, 43p
Olofsson et al., 1990, A&A 230, 405

TABLE 1. Observed lines, intensities, and abundances

Transition	Frequency (MHz)	T_{mb} (K)	$\int T_{mb}dv$ (K kms^{-1})	Abundance [X][/H$_2$]
C$_3$H$_2$(2(1,2)-1(0,1))	85338.9	0.02	0.95	$1.1 \cdot 10^{-7}$
H^{13}CN(1-0)	86340.2	0.57	20.4	$7.4 \cdot 10^{-7}$
SiO(2-1 $v=0$)	86847.0	0.15	5.4	$1.6 \cdot 10^{-7}$
HN^{13}C(1-0)	87090.9	0.03	1.0	$4.1 \cdot 10^{-8}$
C$_2$H(1-0)	87316.9	0.12	12.5	$9.9 \cdot 10^{-6}$
HCN(1-0)	88631.8	0.53	17.2	Opt. thick
HNC(1-0)	90663.5	0.09	3.4	$9.6 \cdot 10^{-8}$
SiS(5-4)	90771.5	0.06	2.4	$5.0 \cdot 10^{-7}$
HC$_3$N(10-9)	90979.0	0.10	3.7	$3.8 \cdot 10^{-7}$
C$_4$H(10-9)	95150.3	0.04	3.5	$1.4 \cdot 10^{-5}$
CS(2-1)	97981.0	0.43	16.4	$1.1 \cdot 10^{-6}$
C$_3$H($^2\Pi_{3/2}$J=9/2-7/2)	97995.5	0.01	1.1	$5.8 \cdot 10^{-8}$
C$_3$N(11-10)	108834.3	0.03	2.3	$3.8 \cdot 10^{-7}$
SiS(6-5)	108924.3	0.06	2.3	$3.8 \cdot 10^{-7}$
HC$_3$N(12-11)	109173.6	0.07	3.1	$3.9 \cdot 10^{-7}$
^{13}CO(1-0)	110201.4	0.35	15.3	Opt. thick
CN(1-0)	113491.0	0.15	16.4	$1.1 \cdot 10^{-6}$
CO(1-0)	115271.2	1.3	49.0	Opt. thick
SiC$_2$(5(0,5)-4(0,4))	115383.0	0.08	2.8	$3.5 \cdot 10^{-7}$
^{13}CO(2-1)	220398.7	0.78	32.0	Opt. thick
CO(2-1)	230538.0	3.5	121.5	Opt. thick
CS(5-4)	244935.6	0.57	18.4	$3.5 \cdot 10^{-7}$
HCN(3-2)	265886.4	3.4	94.7	Opt. thick

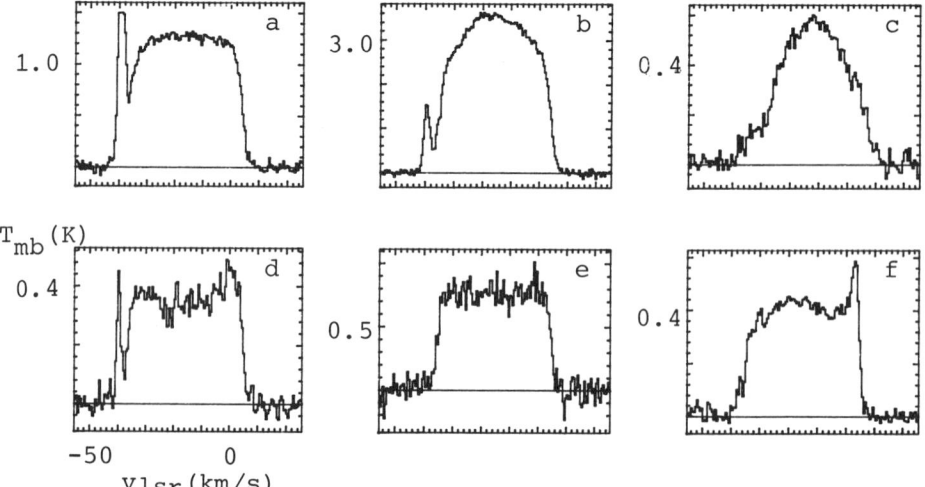

Figure 1a. Spectrum of CO(1-0), b. CO(2-1), c. HCN(1-0), d. ^{13}CO(1-0), e. ^{13}CO(2-1), f. CS(2-1).

CARBON STARS WITH OXYGEN-RICH CIRCUMSTELLAR ENVELOPES

Silvia Lorenz Martins and Sayd José Codina Landaberry
Observatório Nacional - CNPQ
Rua José Cristino, 77 - São Cristóvão
CEP : 20921 -Rio de Janeiro - Brasil

Abstract. We observed two carbon stars with silicate emission, C1003 and BM GEM, and we calculated the abundances ratios. The results are compared with them of RO stars(Dominy, 1984).

Introduction

Typical spectra of carbon stars present 11.2μm SiC emission. From a study based on IRAS observations Little-Marenin(1986) and Willems and de Jong(1986, 1988) identified independently nine carbon stars surrounded by circumstellar shells with silicate dust grains.
Little-Marenin(1986) and Willems and de Jong(1986) presented some suggestions to explain this anomaly:
1) These stars could be components of binary systems with an M and a C star;
2) We may be observing an unusual chemical equilibrium in the circumstellar shell (Little-Marenin, 1986);
3) The transit time of material through the circumstellar shell could be longer than the time scale of the transformation M-S-C, allowing the existence of an oxygen-rich envelope around the carbon-rich star.
Recently Lloyd Evans(1990) and Lambert et al.(1990) proved that these anomalous carbon stars with 9.8μm silicate emission are J-type carbon stars and questioned the considerate evolutive sequence for the carbon stars. They suggested that early R stars(R0, R1...) are J-type carbon stars progenitors.
In an earlier work we studied, and rejected, the first hypothesis (Silvia Lorenz Martins, Master's Thesis, 1990) and this paper presents the abundance ratios C^{12}/C^{13}, C/O and the s process abundances for two of the special carbon stars (C1003 and BM GEM) and R0-type carbon stars. There are great difficulties in the determination of these abundances in cool carbon stars. One is due to the spectral density of rotational molecular lines of CN molecule.

The Method

The method applied is the spectral synthesis in small regions (20 Å), as suggested by Kilston(1973, 1975). The synthesis program includes ionized elements and molecular chemical equilibrium calculations, (Tsuji, 1973). The Hönl-London factors in the gf values for the molecules are normalized by the sum rule (Whitting et al., 1980); the gf values for the atomic lines are taken from the Kurucs' dates. The regions were chosen because of the physical parameters and elemental abundances. The spectra were obtained at Laboratório Nacional de Astrofísica, Brasópolis, Brasil.

Results

The results of the synthesis for the C1003 and BM GEM are compared in the table with of Dominy(1984) for early R stars.

COMPARISON TABLE

	C1003	BM GEM	Early R stars*
[<s>/<Fe>]	+0.95	+0.68	+0.14
C/O**	2	3	2
C^{12}/C^{13}	3	3	9

* HD 113801, R0; BD +33°2399, R2; HD 156074, R0 (Dominy, 1984);
** C/O = $(C^{12}+C^{13})/O$
[X] = $\log X_* - \log X_\odot$

Conclusions

The s abundances and C^{12}/C^{13} ratios for C1003, BM GEM are expected for J-type stars (see Utsumi, 1985) and such abundances are relatively similar to the early R stars. Abundance analysis (Dominy, 1984) showed that these R stars are C^{13}-rich with no detectable enrichments of the s-process elements. The temperature of R0 stars are about 4200-5000K and the J-type carbon stars are about 2500-3000K. As these R stars evolve to become cooler and more luminous, they can appear as J-type cool carbon stars.

References

DOMINY, J. F. (1984) Ap. J. Supp. Ser. **55**: 27-43;
KILSTON, S. D. (1973) "On the nature of the carbon stars" - University of California, Los Angeles, PhD.;
KILSTON, S. D. (1975) Publ. Astr. Soc. Pacific **87**: 189;
LAMBERT, D. L.; HINKLE, K. H. and SMITH, V. V. (1990) Ap. J. vol **99** no. 5 1612-1620;
LITTLE-MARENIN, I. R. (1986) Ap. J. **307**: L15-L19;
LLOYD EVANS, T. (1990) M. N. R. A. S. **243**: 336-348;
LORENZ MARTINS, S. (1990) "Estrelas Carbonadas Peculiares", Master's Thesis;
TSUJI, T. (1973) A & A **23**: 411-431;
UTSUMI, K. (1985) "Cool stars with Excesses of Heavy Elements"; M. Jaschek and P. C. Keenan (eds.), p 243-247;
WHITING, E. F.; SCHADEE, A.; TATUM, J. B.; HOUGEN, J. T. and NICHOLLS, R. W. (1980) Journal of Molecular Spectroscopy **80**: 249-256;
WILLEMS, F. J. and DE JONG, T. (1986) Ap. J. **309**: L39-L42;
WILLEMS, F. J. and DE JONG, T. (1988) A & A **196**: 173-184.

G305.8 - 0.2: A YOUNG OBJECT WITH A DUST AND GAS ENVELOPE

J.W.S. Vilas Boas, E. Scalise Jr.[1], G.G. Sanzovo[2],
G. Mendes Pacheco[1], A.M. Gómez Balboa[3]
CRAAE - EPUSP/PTR, CP 8174, 05508 S.Paulo, SP;
[1]SCT - INPE - DAS, CP 515, 12201 S.J.Campos, SP;
[2]Univ. Est. de Londrina, Depart. de Física, 86020 Londrina, PR;
[3]SCT - CNPq - ON, CP 23002, 20921 Rio de Janeiro, RJ, BRASIL.

Abstract. The H_2O maser source in G305.8 - 0.2 was first detected by Haynes et al. (1984). Inspection of the Southern Hemisphere radio surveys did not show any strong compact HII region or water maser placed nearby this source. The IRAS point source catalog (2.0) shows a strong source in the direction of the maser. We analysed the temporal variation and line profile of the H_2O maser emission and discussed the nature of its associated IRAS source. These source characteristics seem compatible with the hypothesis that the maser is associated with a protostellar disk around an early-type star of spectral type O7 to O9 surrounded by a spherical shell of gas and dust.

Equipment and observations

Radio observations were made using the 13.7 meter antenna of the Itapetinga Radio Observatory. In the first period (Jun-Dec84) weekly observations were made using a filter bank consisting of 46 channels of 100 kHz. During the second period (Sep90-Jun91) monthly observations were made using an acousto-optical spectrometer with 1000 channels of 40 kHz. A 1000K receiver was used in both periods.

Results and Interpretation

During the first period the flux density of the -26 kms^{-1} line doubled, reaching a maximum in 8 days (1500 Jy), returning to pre-burst level after 20 days. Assuming a kinematic distance of 2 kpc Vilas Boas et al. (1991) computed that the isotropic luminosity of the maser was 10^{30} erg.s^{-1}. Its temporal evolution could be explained by a pulse of energy E_0 produced by the heating source radiating diffusivelly through the gas cloud. This maser line is saturated. The pumping changes as radiation diffuses through the cloud. To fit the model proposed by Burke et al.(1978), a pulse with energy of 10^{41} erg should be injected in the cloud. If we assume that this energy is supplied to the cloud in a time scale of a day and the emission is anisotropic (Alcock and Ross, 1985) it implies a heating source with maximum luminosity of 10^{39} erg.s^{-1}. The Plank's fit to the IRAS fluxes suggests the existence of two shells, one of 130 K with 3×10^{-12} sr and other of 50K with 10^{-9} sr. The nfrared integrated luminosity of 2×10^4 L$_o$, together with the maser luminosity, suggests the presence of an O7 to O9 ZAMS star. If the temporal evolution of the maser indicates the heating source activity, then the 20 solar mass star we are dealing with is younger than 10^5 yr (Panagia,1973). The absence of radio continuum implies that the circumstellar ionized gas is extremely thick at radio wavelengths or that the circumstellar dust is optically thick within the Lyman continuum, suppressing the formation of an HII region. Specially during the second period

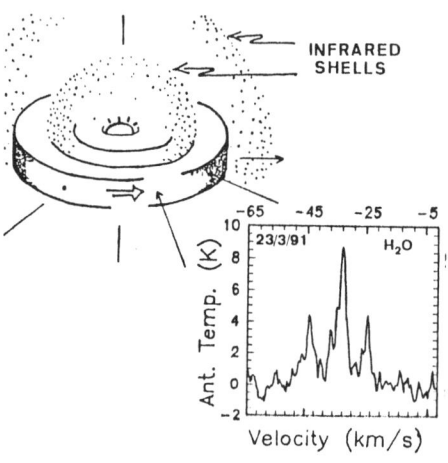

Fig. 1 - Schematic representation of the source.

the profiles observed were in good agreement with line profiles from protostellar disks (Grining and Gregor'ev, 1980). Figure 1 shows the H$_2$O maser emission observed in Mars, 91. If we assume that the disk is in gravitational equilibrium with the central star, the velocity of the lateral lines suggests that the emission is originated in a region of the disk placed 10^{15} to 4×10^{15} cm from the central star, what is in agreement with Elmegreen and Morris (1979) for maser emission from a 20 solar mass star surrounded by a one solar mass disk. Only between June and September, 84 we have observed a weak line with a velocity of -56 kms^{-1}. Its intensity changed in a time scale of days and its microwave emission rate was 0.5 s^{-1}. This suggests that the emission is unsaturated. Acceleration of cloudlets by the wind of the central star (Elmegreen and Morris, 1979) could explain the high velocity of this emission. Radio continuum and CO line observation of this region at millimeter wavelengths are of extreme importance to study the envelope of this star looking for bipolar outflow structures and continuum emission from cold dust.

References
Alcock, C. & Ross, R.R. (1985) Ap. J., **290**, 433.
Burke, B.F., Giufrida, T.S. & Haschick, A.D. (1978) Ap. J., **266**, L21.
Elmegreen, B.G. & Morris, M. (1979) Ap. J., **299**, 593.
Grinin, V.P., & Gregor'ev, S.A. (1983) Sov. Astron. Lett., 9, 244.
Haynes, R.F., Roger, R.S., Forster, B.J., Robinson, R., Batchelor, A. & Wellington, K.J. (1984), IAU Circular nº 3943.
Vilas Boas, J.W.S., Scalise Jr., E. & Sanzovo, G.C. (1991) Astr. J., in press.

BIPOLAR OUTFLOW AND TURBULENCE IN MOLECULAR CLOUDS DUE TO PROTOSTELLAR ALFVÉN WAVES

V. Jatenco-Pereira and R. Opher
Instituto Astronômico e Geofísico da
Universidade de São Paulo
C.P. 9638, 01065 São Paulo, SP
Brazil

ABSTRACT. There is strong evidence for magnetic fields in molecular clouds to be oriented parallel to the direction of bipolar outflows. We have evidence for Alfvén waves in molecular clouds from the facts that: (i) the velocity width of CO lines, ΔV, is approximately twice the gravitational energy, and (ii) polarization maps indicate well-defined magnetic fields. These facts indicate that ΔV is wave-like and not eddy-like. We study the Alfvén wave protostellar model of Jatenco-Pereira and Opher (1989) to explain the observed bipolar outflows and turbulence in molecular clouds. We assume that the Alfvén waves are primarily dissipated by non-linear damping and examine the physical parameters necessary to produce the observed massive low-velocity bipolar outflows ($u_\infty \sim 10 - 50$ km s^{-1} and $\dot{M} \sim 10^{-3}$ $M_\odot$ yr^{-1}) and turbulence in molecular clouds.

1. INTRODUCTION

The outflows from young stellar objects are highly energetic with kinetic energies of up to 10^{47} erg and exhibit a wide variety of observable phenomena. Some level of collimation usually exists for the outflows, the most notable examples being massive flows of cold molecular gas (e.g. Snell, Loren and Plambeck 1980). In general, these molecular outflows have a mass loss rate $\dot{M} \sim 10^{-3} - 10^{-4} M_\odot$ yr^{-1}, and terminal velocity $u_\infty \sim 10-50$ km s^{-1}, lower than the escape velocity (v_{eo}). Highly collimated optical jets have also been observed in H_α images (e.g. Mundt and Fried 1983). A disklike structure perpendicular to the directions of the outflows has been found in redor of the central objects (e.g. Kaifu, et. al. 1984).

2. THE MODEL

We study the Alfvén wave protostellar model of Jatenco-Pereira and Opher (1989), to explain the observed bipolar outflows and turbulence in molecular clouds. The magnetic structure studied was similar to the coronal hole geometry of the sun of an initial nonradial diverging magnetic field of area $A(r)=A(r_0)(r/r_0)^S$, where $A(r)$ is the cross-sectional area of the geometry at a radial distance r, $S > 2$, and r_{max} for the nonradial divergence being determined by $(A(r)/r^2)_{max}/(A(r_0)/r_0^2) = 10$. For long periods ($10^6$ s $> P > 2 \times 10^4$ s) collision absorption is small, and we assume that the dominant mechanism for the absorption of the Alfvén waves is nonlinear damping, with the damping rate (e.g. Lagage and Cesarsky, 1983) being given by: $\Gamma_{NL} \propto \omega (V_S/V_A)(\rho<\delta v^2>/(B^2/8\pi))$ and the damping length $L_{NL} = V_A/\Gamma_{NL}$, where V_S is the sound velocity, V_A is the Alfvén velocity, $\rho<\delta v^2>$ is the energy density of the Alfvén waves and ω is a characteristic Alfvén frequency.

3. RESULTS

In Figure 1, we show the preliminary results of this study.

Using the initial conditions of Jatenco-Pereira and Opher (1989), with a central object of $M = 1M_\odot$, radius $r_0 = 10^{13}$ cm, surface mass density $\rho_0 = 0.257 \times 10^{-7}$ g cm^{-3}, a jet temperature $T = 100$ K, surface magnetic field $B \sim 40G$, surface flux of Alfvén waves $\Phi \sim 10^9$ ergs cm^{-2} s^{-1}, and varying the opening angle by varying, S, and the initial damping length, L_{NLO}, we obtain:

a) mass loss rate $\dot{M} \sim 10^{-3} - 10^{-4}$ $M_\odot$ yr^{-1} and $u_\infty \sim 10 - 50$ km s^{-1} lower than the escape velocity (v_{eo}), as observed; and
b) a non-zero Alfvén waves flux, far from the star ($r \sim 300\, r_0$), that could explain the turbulence in molecular clouds (Fig. 1).

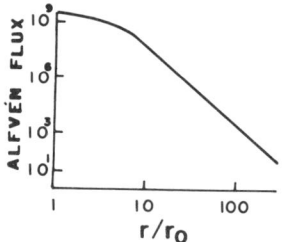

Fig. 1. Alfvén wave flux versus (r/r_0).

Acknowledgements

The authors would like to thank the brazilian agency CNPq for partial support.

4. REFERENCES

Jatenco-Pereira, V., Opher, R. (1989) Mon. Not. R. astr. Soc. 236, 1.
Kaifu, S.S., Hasegawa, T., Morimoto, M., Inatami, J. et al. (1984) Astron. and Astrophys. 134, 7.
Lagage, P.O., Cesarsky, C.J. (1983) Astron. and Astrophys. 125, 249.
Mundt, R., Fried, J.W. (1983) Astrophys. J. 275, L83.
Snell, R.L., Loren, R.B. and Plambeck, R.L. (1980) Astrophys. J. 239, L17.

Spectral Synthesis of Cool Components of Symbiotic Stars

Claudio B. Pereira and Sayd Codina Landaberry
Observatório Nacional
Rua José Cristino, 77 São Cristóvão - CEP 20921
Rio de Janeiro - Brasil

ABSTRACT. We analyzed the optical spectra of cool component of symbiotic star SY Mus by means of spectral synthesis technique in order to derive the atomic abundances using the Minneart formulae for computing the atmospheric opacities. We obtained a satisfactory agreement between the observed and computed spectra and the resulting abundances were consistent to the solar abundances.

INTRODUCTION. The optical spectra of symbiotic stars contain various emission lines superimposed on the red continuum of the cool component. In these continuous spectra we observed molecular bands of TiO at $\lambda 5850 Å$, $\lambda 6200 Å$ and $\lambda 7200 Å$ which characterize the spectra of the cool red giants. Sometimes is difficult to analyze the spectra of these cool components. This difficulty arises due the large strengths of Tio bands blanketing which strike out the atomic absorption lines or by the presence of the emission lines that blends it.

THE MODEL. We used as first approximation an isothermal model with a linear variation of the source function with the optical depth and the absorptions which are the only interactions between the photons and atoms and molecules. This model can be written by Minneart formulae as

$$\frac{1}{R_\nu} = \frac{1}{R_c} + \frac{1}{\alpha_\nu}$$

where

R_ν, is the residual intensity;

R_c, is the central intensity and

α_ν, is the absorption coefficient per particle in the lower level of the transition.

OBSERVATIONS. Our study is based on spectra obtained at Laboratório Nacional de Astrofísica. The data were obtained with a coudé spectrograph (first order) attached to the 1.6m telescope. The detector used was a CCD with 576 pixels with a grating of 18 Å/mm. We analyzed two regions of SY Mus; one between 5400Å and 5425Å and the other between 5430Å and 5445Å. The raw data were corrected for flatfield and wavelength scale was established through the spectrum of a thorium lamp.

RESULTS AND DISCUSSION. Despite of this simple model with one slab for the formation of absorption lines (atomic and molecular) the agreement between the computed and observed spectra was good. From our synthesis we found solar abundances in the region analyzed. Very little components of symbiotic stars had their spectra been analyzed in detail until now. The studies in this directions are still in the beginning. **ER Del**, (Johnson, H.R. and Ake, T.B.,1989) ; **BD-21.3873**, (Jorisson, A.,1989); **R Aqr** and **UV Aur**, (Little, Little-Marenin and Bauer, 1987) are good candidates to analyze if these stars have enhanced s-elements lines such us Sr, Y, Ba, Nd and Sm in their spectra. Techenetium lines could also be investigated in cold giants of symbiotic stars. We plan to observe in the region between 4100Å and 4400Å where technetium lines of 4238Å, 4262Å and 4297Å are visible despite of low levelof spectral energy in the blue continuum. Some of them have period longer than 300 days (Whitelock,1987) and mass of 3M☉ (Luud and Leédyarv, 1986) so we have to look for if some peculiar red giants could belongs to one of these symbiotic systems.

References :
Johnson, H. R. and Ake, T.(1989) 'BD-21.3873: An Heavy Element-Rich Symbiotic?', Evolution of Peculiar Red Giants Stars. IAU Colloquium, 106.
Jorissen, A.(1989) 'ER Del : A True Symbiotic S Star?', Evolution of Peculiar Red Giants Stars. IAU Colloquium, 106.
Little, S. J., Little-Marenin, I. R. and Bauer, W.(1987). 'Additional Late-Type Stars with Techenetium', Astron. J. **94**,981-995
Luud, L and Leédyarv, L.(1986).'Properties of the Cold Components Of Symbiotic Stars', Astrophysics, **24**, 154-162.
Whitelock, P. A.(1987),'Symbiotic Miras', Publ. Ast. Soc. Pac.,**99**, 573-591.

RADIOACTIVE ISOTOPE ^{26}Al IN THE INTERSTELLAR MATTER
(resulting from a mass loss by AGB stars)

Ju. L. FRANTSMAN
Radioastrophysical observatory
Latvian Academy of Sciences
Turgeneva 19, 226524, Riga, Latvia

ABSTRACT. Several calculations of the Asymptotic Giant Branch star evolution have been performed with the aim of explaining the synthesis of interstellar ^{26}Al in these stars. The agreement of theoretically calculated mass of interstellar ^{26}Al and observations is rather satisfactory, the best for the abrupt tenfold jump in the mass-loss rate for the stars reaching the luminosity $\log(L/L_\odot) \simeq 4.0$.

Although possible sources of the measured interstellar ^{26}Al (about $3M_\odot$ [1-2]) have been investigated, they appear inadequate to account for the observed ^{26}Al gamma-ray flux [3]. The origin of the ^{26}Al in the interstellar medium has remained a mystery. It has been suggested that ^{26}Al, synthesized in the AGB stars, are carried to the surface by a process of convective dredging and that, in the result of rapid mass loss from the surface, this isotope could contribute to the enrichment of the interstellar medium. Several different cases have been considered for the value of the mass-loss law. The main processing in the He shell follows the sequence:
$^{14}N(\alpha,\gamma)^{18}F(\beta^+)^{18}O(\alpha,\gamma)^{22}Ne(\alpha,n)^{25}Mg$.
The ^{25}Mg brought into the envelope could be efficiently transformed into the unstable isotope ^{26}Al: $^{25}Mg(p,\gamma)^{26}Al$. The star is presumed to shed mass by stellar wind; the loss rate will then be expressed by Reimer's low: $\dot{M}=-4\times10^{-13}\alpha L/gR$, were $\dot{M}$ is in units of $M_\odot$ yr^{-1}, L, g, R denote the star's luminosity, surface gravity, and radius in solar units. The coefficient α is undetermined but is usually taken to be of order unity. Many observations, however, suggest that apart from the conventional stellar wind and planetary nebulae ejection, some other mechanism also ought to operate during the AGB phase, substantially raising the mass-loss rate [4-5]. The results of our calculations are presented in Table 1.

Table 1. The mass of ^{26}Al in the interstellar medium in the Galaxy (for different α)

Nr	α	$M_{26_{Al}}(M_\odot)$
1	1.0	39.4
2	3.0	19.2
3	$\alpha=1$ if $\log(L/L_\odot)<4.1$	
	$\alpha=10$ if $\log(L/L_\odot)>4.1$	6.4
4	$\alpha=1$ if $\log(L/L_\odot)<4.0$	
	$\alpha=10$ if $\log(L/L_\odot)>4.0$	5.0

The contribution of different initial mass stars on the amount of ^{26}Al in the interstellar medium is illustrated in Table 2. Successive columns contain the range of masses M, the number of stars in this range N (the assumed total number of stars between 1 and 100 $M_\odot$ is 10^3), and the contribution of ^{26}Al in the interstellar medium of stars in successive mass ranges (for cases 2 and 4 of mass loss laws, see Table 1).

Table 2. The contribution of different mass stars on the interstellar ^{26}Al

ΔM, $M_\odot$	N	$\Delta M_{26_{Al}}/M_{26_{Al}}$, %	
		Nr 1	Nr 2
1 - 2	646	0.4	9
2 - 3	162	4	10
3 - 4	67	7	7
4 - 5	36	12	9
5 - 6	21	13	16
6 - 7	14	27	20
7 - 8	10	38	30

References
1. Mahoney W.A., Ling J.C., Jakobson A.S. (1982) Ap.J.,262, 742.
2. Clayton D.D. (1984) Ap.J.,280, 144.
3. Prantzos N. (1991) Institut D'Astrophysique de Paris, Pre-publ. Nr 344.
4. Frantsman Ju.L. (1986) Astrofizika, 24, 131 (Astrophysics, 1986, 24-25, Nr1).
5. Frantsman Ju.L. (1988) Astrophysics and Space Sci., 145, 251.

ARE MOLECULES RESPONSIBLE FOR ORIGIN OF COLD GIANTS MASS LOSS?

I. K. SHMELD
Radioastrophysical observatory
Latvian Academy of Sciences
Turgeneva 19, 226524 Riga, Latvia

V. S. STRELNITSKIJ, A. V. FEDOROVA O. V. FEDOROVA
Institut for Astronomy Ac. Sci. USSR,
Pjatnickaja 48, 109017 Moscow USSR

ABSTRACT. It is shown that radiative pressure in vibronic transitions of molecules may play an important role in origin of mass outflow of cold giants and supergiants.

The mechanism of mass loss of cold giants is of great interest during last years. It seems that it is hard to explain the mass loss from these stars only as the dust driven winds (A. G. Hearn, 1989), see however(C. Dominik et. al., 1990). Usually some additional mechanisms explaining the initial gas acceleration near the star surface are considered. Most papers deal with different kinds of waves to explaine that (see A. G. Hearn, 1989 for details).

One of the possible additional mechanism may be light pressure on the circumstellar molecules. In the layers near the photosphere the main source of absorption are molecular bands and the integral is reduced to a sum upon the most important molecular bands. According to calculations of chemical equilibrium in the upper atmospheres M-class giants and supergiants all the carbon is tied in the CO molecules and the most of remaining O is in the H_2O molecules. These molecules are characterized by strong rotational-vibrational bands near the peak of photosphere emission and may contribute the major part into radiative acceleration of the gas. Pure rotational transitions also may be important as quasi-continuum absorbers in the $10 - 40\mu$ region (T. Tsuji, 1966). CO vibrational bands have an important role in absorption.

If we assume that inward radiation flux is negligible, then the distribution of radiation intensity is isotropic ($F_\nu = \pi I$) and constant within the i-th vibrational band ($I_{\nu i}$ = const), the most molecules are in the lower state of the vibrational transition, then, if $a_{rad} \simeq a_{grav}$:

$$\sum_i \frac{n_i}{n} I_{\nu i} f_i + \frac{37.7\, n_{H_2O}}{n} \int k^r_\nu I_\nu d\nu \simeq 4.0 \cdot 10^{-20} \frac{M}{r^2}, \quad (1)$$

where n_i - the number density of molecules in the lower level i, k^r_ν(cm) - the mean extinction coefficient in the quasi-continuum rotational spectrum and f_i - the oscilator strength. Assuming that the most of molecules are in the ground vibrational state and taking into account only the strongest bands from this state (4-6)μ for CO, with $f \simeq 1.1 \cdot 10^{-5}$, 6.3$\mu$ and 2.7μ for H_2O, with $f \simeq 1.0 \cdot 10^{-5}$ and $0.9 \cdot 10^{-5}$, in accordance with (1) follows:

$$\frac{n_{H_2O}}{n} \left\{ I_{2.7} f_{2.7} + I_{6.3} f_{6.3} + 37.7 \int k^r_\nu I_\nu d\nu + \right.$$

$$+ \frac{n_{CO}}{n} I_{4.6} f_{4.6} \simeq 4.0 \cdot 10^{-20} \frac{M}{r^2} \quad (2)$$

We assume for a typical supergiant star $R = 10^{14}$ cm, T_{eff} =3000K, $M = 10 M_\odot$ and for Mira variable $R = 7 \cdot 10^{13}$ cm, T_{eff} = 2000K, $M = 1 M_\odot$. If for I_ν we take the Plank appromaximation, instead of (2) we have:

$$2.5 \frac{n_{H_2O}}{n} + \frac{n_{CO}}{n} \simeq 3.3 \cdot 10^{-3}, \quad 2.4 \frac{n_{H_2O}}{n} + \frac{n_{CO}}{n} \simeq 1.3 \cdot 10^{-3} \quad (3),$$

for supergiant and Mira variable correspondingly.

If usual abundances He/H = 0.8, O/H = $0.8 \cdot 10^{-3}$, O/C = $0.3 \cdot 10^{-3}$ are assumtd and if most of the carbon is in the CO and remaining O in the H_2O, then for oxygen stars radiation pressure on molecules may be important only for stars with $M \simeq 10 M_\odot$.

For carbon stars all the oxygen is tied in CO one obtains from (3), for $a_{rad} \simeq a_{grav}$ O/H $\simeq 4 \cdot 10^{-3}$ for a supergiant, O/H $\simeq 2 \cdot 10^{-3}$ for a Mira variable) and C/H $\simeq 3 \cdot 10^{-4} (M/M_\odot)$ for both.

The radiative pressure on the carbon bearing molecules may initiate the gas outflow from the massive carbon stars.

REFERENCES

C. Dominik et. al., (1990) Astron. and Astroph., **240**., 365.
A. G. Hearn. (1989) in: "From Miras to Planetary Nebulae: which Path for Stellar Evolution" - Proc. of International Colloqium. Sept. 4-7. Montpellier France.
T. Tsuji (1966) Publ. Astron. Soc. Jap., **18**, N3, 127.

Cometary Chemistry

MICHAEL F. A'HEARN

Department of Astronomy, University of Maryland

September 16, 1991

Abstract.
The key problem in cometary chemistry is to observe abundances of species in the coma and to reassemble those species into the species that are present in the nucleus. The limitations are primarily due to poorly constrained models and the lack of uniqueness in reassembling parent species from the fragments when not all of the fragments are observable. These problems will be illustrated with several examples.

Key words: comets sulfur ammonia water grains

1. Introduction

Snyder(this volume) will provide a comprehensive survey of the molecules found in comets and Despois (this volume) will relate the composition of comets to that of the interstellar medium. I will therefore concentrate on determining the composition of the cometary nucleus from observations of the coma. This includes reconstruction of parent molecules that were frozen as ices in the nucleus and reconstruction of relatively refractory grains.

The standard paradigm is the dirty snowball model of Whipple (1950) in which ices evaporate from the nucleus, the resultant gas flows into the vacuum of space, and the gas drags sufficiently small solid grains with it. The outflow velocity is of order 1 km/s at r=1 AU and the density in the coma is sufficiently high for chemical reactions to occur only in the first few hundred to few thousand km. Beyond that distance, only photochemistry and reactions with the solar wind have significant rates. At a few thousand km from the nucleus of P/Halley, the total density of ions was $10^{-3} \times$ the total density of neutrals so the $10^5 \times$ greater cross-section of reactions involving ions means that they can be significant at somewhat larger distances from the nucleus than neutral-neutral reactions. The rapid outflow yields only a short time during which a given particle can undergo an n-body reaction, whereas the time during which photoprocesses or solar wind reactions can take place is orders of magnitude longer.

Most measurements of cometary comae are made *via* remote sensing so the region sampled is dominated by the region in which photoprocesses dominate. Even *in situ* measurements, such as those made by the ICE, Suisei, Vega, and Giotto spacecraft, were made primarily in the region where photoprocesses dominate. Only Giotto clearly penetrated the region in which chemical reactions should occur. Thus I will emphasize the fragments that are seen in the outer coma and use of those to infer the composition that was released from the nucleus.

2. Reassembling Parent Molecules

2.1. H_2O

In order to evaluate our ability to put the pieces together, I will first consider the relatively well understood case of H_2O and its frequently observed products – OH, H, $O(^1D)$, and H_2O^+. Direct measurements of H_2O are rare. There have been no measurements of the other dissociation product, H_2. There is no doubt that all the expected fragments of H_2O are present and the only question is whether we can reliably put the fragments together quantitatively to say that H_2O is the sole parent for all the fragments or conversely to derive the amount of H_2O from measurements of one or more of the fragments.

Remote measurements of OH (Festou et al., 1986) showed roughly 35% more OH than expected from the amount of H_2O deduced almost simultaneously from the neutral mass spectrometer (NMS) on Giotto (Krankowsky et al., 1986). The NMS also showed that H_2O was roughly 80% of the total gas. No subsequent study has clearly eliminated the discrepancy although much of the discrepancy is undoubtedly in the use of somewhat different models for the interpretation of the data. Thus an accuracy of 10% in reassembling the pieces is overoptimistic with our present knowledge. Two weeks later, infrared measurements of H_2O yielded a production rate 2× higher than nearly simultaneous measurements of OH (Weaver et al., 1987). The lack of an adequate time-dependent model of the coma may play a role in the discrepancy as may uncertainties in the excitation of the two species. The uncertainties are highlighted by comparing measurements of the same species in different spectral domains as illustrated by Festou (1990, Figure 3). He shows the large discrepancy between ultraviolet and radiowave measurements of OH as well as the discrepancy between those results and the ultraviolet measurements of H. Taken as a whole, one must conclude that H_2O is the dominant parent of OH and H and, in most comets, of $O(^1D)$ but the uncertainties in the data and in the models prohibit any firm conclusion about additional minor parents of these species.

2.2. NH_3

NH_3 presents a particularly interesting example because there is only one direct measurement, a radio measurement for comet I-A-A which did not yield a reliable abundance (Altenhoff et al., 1983). While in comet P/Halley, two mutually exclusive sets of products of NH_3 led to estimates of the NH_3 abundance which differ by an order of magnitude.

Allen et al. (1987) used the *in situ* measurements from the ion mass spectrometer (IMS) on Giotto in conjunction with a model for the ion-chemistry of the inner coma to interpret the radial profiles of the mass peaks

TABLE I
Sulfur-Bearing Species in Comets

Species	Technique	Comets	Abundance	Source	Comments
S	uv	all	?	various	
CS	uv	all	0.2%	CS_2	
S_2	uv	I-A-A	0.01-0.1%	grains?	only comet
H_2S	radio	Austin+Levy	0.2%	parent?	all comets?
H_3S^+	Giotto IMS	Halley	0.2% H_2S	H_2S	all comets?
SO	uv	several	< 0.01%	parent, SO_2	no comets?
SO_2	uv	several	< 0.01%	parent	no comets?
SH	optical	P/Bro-Met	< 1%	H_2S	
OCS	radio	Levy	< 0.2%	parent	
H_2CS	radio	Levy	< 0.1%?	parent	

associated with H_2O^+, NH_3^+, and their associated compounds. Fitting these profiles with various assumed abundances of the corresponding neutrals, they concluded that an NH_3 abundance of 1-2% that of water provided the best fit to the data. It should be pointed out, however, that the resultant fit was still not a good one.

Remote observations of NH and NH_2, on the other hand, imply much lower abundances of NH_3. Wyckoff et al. (1991) have found from observations of NH_2 that the abundance of NH_3 is only 0.2% in P/Halley. They find somewhat lower but comparable values for three other comets. Analysis of narrow-band photometry of NH for comet P/Halley also suggests an NH_3 abundance of a few times 0.1% (Schleicher, A'Hearn, and Samarasinha, unpublished), comparable to the values found for other comets. Although the errors on all the results are large, the discrepancy between the interpretations of the ions and of the neutrals is beginning to strain the quoted uncertainties.

Since the primary dissociation channel of NH_3 is into NH_2 and NH, while direct ionization and protonization to NH_4^+ are minor channels for the destruction of NH_3, it seems likely that the results based on the neutrals are correct. Since all investigators agree on the primary destruction channel, the NH_3 abundance implied by the ions would require an unknown process to destroy NH_3 without producing NH_2 and NH. Thus this is a case in which one must make a judicious choice of which pieces to reassemble into the parent.

2.3. SULFUR COMPOUNDS

A quite different example of reassembly uses atomic column densitites to determine whether or not all molecules containing a given atom have been identified, as in the case of sulfur shown in Table I. It has been known for many years that in some comets the observed production of CS is insufficient to explain the observed column density of S if the CS comes from CS_2. With the discovery of H_2S in comets (Bockelee-Morvan et al., 1991), Roettger (1991) has found that all the observations of S can be explained if one assumes amounts of H_2S less than or equal to that observed in comets Austin and Levy. At first glance, this seems to explain everything very well but there are several complications. One significant limitation is that S is always observed with a relatively small field of view. Thus the observed column density of sulfur is insensitive to any sulfur-bearing molecules with long lifetimes, say $\geq 10^5$ sec. An additional complication is indicated by the fact that some observations of very active comets yield a production of CS_2 which should produce more S than is observed. One way out of this is to assume that CS comes from a different (unknown) parent, thereby reducing the source of S atoms by a factor two. Another way out is suggested by the fact that in some spectra the sulfur triplet at 1812Å exhibits a profile inconsistent with the known intensity ratios of the three components implying that optical depth effects are important. These effects have not been modelled although in the simplest approximation the finite optical depth merely redistributes the flux among the three components without changing the net flux. Finally, the recently discovered lines of S at 1425/1474Å (Roettger et al., 1989) yield abundances that are a factor of five greater, implying an error either in the model of the excitation or in the atomic parameters. Thus the possibility remains of a substantial amount of other sulfur bearing molecules.

3. Grains as Parents

The *in situ* observations of CO in comet P/Halley (Eberhardt et al., 1987) were the first to provide solid evidence for an extended, non-gaseous source for an observed gaseous species. At least half the CO was released from the extended source. Although doubts have been expressed about the interpretation, suggesting that the spatial profile observed by Giotto was an artifact induced by passage through a CO-rich jet, the extended source interpretation is still more convincing. Since the lifetimes of icy grains would be much too short to explain the spatial distribution, and since the sum of all heavier gases was no more than the abundance of CO, it is generally hypothesized that the CO was released, either thermally or by photoprocesses, from refractory grains, presumably the CHON grains also detected *in situ*. Although no other observations provide such a convincing case, there are many observations that are most easily interpreted under this scenario. *E.g.*, Feldman *et al.* (1991) have used the extended source to explain the different production rates of CO in comet Levy deduced from instruments with different fields of view. Similarly Snyder (this volume) has pointed out that the radio observations of H_2CO require an extended source (either gaseous or refractory) in order to avoid a ridiculously high abundance while Krankowsky (1991) has pointed out that the radial profile observed with the Giotto NMS also implies an extended source with the same spatial characteristics as that of the CO source.

Additional evidence for refractory sources of gas comes from observations of CN. A'Hearn et al. (1986) originally suggested that the jets observed in CN images of P/Halley were so well collimated that a grain source (collisionless even very close to the nucleus) was required. More recent processing of those same images by Klavetter (private communication) shows a region of excess production of CN which is semicircular and at the same distance from the nucleus as was the extended source of CO. It therefore appears that, at least in some comets, very many different gaseous species are produced from refractory grains.

I suggest that this may also be the source of S_2 in comet I-A-A. The absence of SO and SO_2 from comets (Kim and A'Hearn, 1991) rules out our original hypothesis that S_2 was a residue of interstellar processing of sulfur compounds in the icy mantles of interstellar grains. An alternative procedure for producing S_2 is by sputtering of solid sulfur-bearing grains. Russell et al. (1987) pointed out that the interplanetary magnetic field had an unusual radial alignment when the S_2 was observed so that solar wind particles could penetrate unusually deeply into the coma. Boring et al. (1985) and Chrisey et al. (1987, 1988) showed that S_2 was readily produced by sputtering solid sulfur or solid H_2S. Sputtering from the nucleus requires a quantum yield at least an order of magnitude too high but spreading the material into grains reduces the required quantum yield. Grains of H_2S are unlikely because they are too volatile and no mechanism is know to produce solid S_8, but H_2S is readily converted to FeS in the presence of Fe and liquid water. Whether the process also works in the presence of ice is unclear but I suggest that the S_2 was produced by sputtering of either grains of FeS.

In summary, refractory grains are playing an increasingly important role in our understanding of the gases observed in comets. Unfortunately, the physical processes involving grains are less well understood than are gaseous processes so it is very difficult to reassemble grains from their fragments.

References

A'Hearn, M. F., S. Hoban, P. V. Birch, C. Bowers, R. Martin, and D. A. Klinglesmith, III: 1986, *Nature* **324**, 649

Allen, M., M. Delitsky, W. Huntress, Y. Yung, W.-H. Ip, R. Schwenn, H. Rosenbauer, E. Shelley, H. Balsiger, and J. Geiss: 1987, *Astron.Astrophys.* **187**, 502

Altenhoff, W. J., W. Batria, W. K. Huchtmeier, J. Schmidt, P. Stumpff, and M. Walmsley: 1983, *Astron.Astrophys.* **125**, L19

Bockelee-Morvan, D., P. Colom, J. Crovisier, D. Despois, and G. Paubert: 1991, *Nature* **350**, 318

Boring, J. W., Z. Nansheng, D. B. Chrisey, D. J. O'Shaugnessy, J. A. Phipps, and R. E. Johnson: 1985, in C.-I. Lagerkvist, B. A. Linblad, H. Lundstedt, and H. A. Rickman, ed(s)., *Asteroids, Comets, Meteors II*, Uppsala Univ., Uppsala, 229

Chrisey, D. B., R. E. Johnson, J. W. Boring, and J. A. Phipps: 1988, *Icarus* **75**, 233

Chrisey, D. B., R. E. Johnson, J. A. Phipps, M. A. McGrath, and J. W. Boring: 1987, *Icarus* **70**, 111

Feldman, P. D., A. F. Davidsen, W. P. Blair, C. W. Bowers, W. V. Dixon, S. T. Durrance, H. C. Ferguson, R. C. Henry, R. A. Kimble, G. A. Kriss, J. Kruk, K. S. Long, H. W. Moos, O. Vancura, and T. R. Gull: 1991, *Astrophys.J.*, in press

Festou, M. C.: 1990, in J. Mason, ed(s)., *Comet Halley Investigations, Results, Interpretations*, Ellis Horwood, Chichester, 245

Festou, M.C., P. D. Feldman, M. F. A'Hearn, C. Arpigny, C. B. Cosmovici, A. C. Danks, L. A. McFadden, R. Gilmozzi, P. Patriarchi, G. P. Tozzi, M. K. Wallis, and H. A. Weaver: 1986, *Nature* **321**, 361

Kim, S. J. and M. F. A'Hearn: 1991, *Icarus* **90**, 79

Krankowsky, D.: 1991, in R. L. Newburn, M. Neugebauer, and J. Rahe, ed(s)., *Comets in the Post-Halley Era*, Kluwer, Dordrecht, 855

Krankowsky, D., P. Lämmerzahl, I. Herrwerth, J. Woweries, P. Eberhardt, U. Dolder, U. Herrmann, W. Schulte, J. J. Berthelier, J. M. Illiano, R. R. Hodges, and J. H. Hoffman: 1986, *Nature* **321**, 326

Roettger, E. E.: 1991, *Ph.D. Thesis*, The Johns Hopkins University, Baltimore

Roettger, E. E., P. D. Feldman, M. F. A'Hearn, M. C. Festou, L. A. McFadden, and R. Gilmozzi: 1989, *Icarus* **80**, 303

Whipple, F.: 1950, *Astrophys.J.* **111**, 375

Wyckoff, S., S. C. Tegler, and L. Engel: 1991, *Astrophys.J.* **368**, 279

QUESTIONS AND ANSWERS

S.Sandford: Given that methanol (CH_3OH) is now known to exist in interstellar ices and that it's been recently detected in comets, could its presence bear on the question of the extended sources of CO and H_2CO? (We know from lab experiments that photolysis of CH_3OH in H_2O-rich ices produces CO, H_2CO, CO_2, CH_4,...)

M.F.A'Hearn: If the CH_3OH leaves the comet in grains it may very well help explain the extended source of CO and H_2CO. If it evaporates directly from the nucleus it cannot explain the observations.

W.F.Huebner: The idea of sputtering sulfer off FeS grains is very interesting. Are FeS grains consistent with PIA and Puma experiments on Giotto and Vega?

M.F.A'Hearn: I have not yet checked this point.

L.d'Hendecourt: How is the oxygen distributed in the comet? 1) mostly in H_2O? 2) why not in molecular oxygen O_2? Some question for the nitrogen: 1) in NH_3, CN... 2) why not in molecular nitrogen N_2?

M.F.A'Hearn: Oxygen is observed as [OI] at $\lambda 6300, 6363$ and brightness and spatial distribution are both consistent with prompt emission following dissociation of 5 - 10% of water via $H_2O + h\nu \rightarrow H_2 + O(^1D)$. It is also observed in the $\lambda 1304$ resonance line of OI which has been observed with small fields of view using IUE for many comets. The ensemble of $\lambda 1304$ data on all comets has not, to my knowledge, been checked for consistency with H_2O dissociation but I think that in some comets the results are consistent, within the usual factors of two, with production from H_2O. A.Delsemme will comment on the nitrogen and the abundance of N_2.

COMETARY ORIGIN OF CARBON AND WATER ON THE TERRESTRIAL PLANETS

A. H. DELSEMME
DEPARTMENT OF PHYSICS & ASTRONOMY
THE UNIVERSITY OF TOLEDO
TOLEDO, OH 43606, U.S.A.

The origin of carbon and water on the terrestrial planets is not trivial, because their presence seems to be excluded by the high temperatures requested by the accretion disk models (Morfill 1988) as well as by Lewis' (1973) adiabat that can explain the different densities of the planets. These high temperatures imply that all dust was outgassed and dehydrated and most carbon stored in CO, that is in the gas phase, before the separation of dust from gas, that occurred by dust sedimentation to the mid-plane of the accretion disk. Thermochemical equilibrium was easily reached, since chemical kinetics (Lewis & Prinn 1980) had time constants shorter than the time needed to agglomerate the first planetesimals (Weidenschilling 1988).

Lewis et al. (1979) do not succeed in explaining the large retention of carbon by the Earth, even if they choose an adiabat in the accretion disk that brings the Earth's zone near the peak of graphite activity. But this peak is needed, not at one astronomical unit (AU) but near or beyond 2.6 AU, in order to explain the rich carbon chemistry of the carbonaceous chondrites -at least partially- by Fischer-Tropsch-Type reactions (Anders 1986). This alone strongly implies that the bulk of the observed carbon on the terrestrial planets has an exogenous origin and was brought about after the bulk of planetary accretion had ended.

The paradigm that has emerged to describe the origin of the Solar System implies that, during the final stages of planetary accumulation, the orbital evolution of the planetesimals, which is inevitable, automatically provides this source of carbon and water in the form of objects made at cooler temperatures because further away in the Solar System. At steady state, the mean relative velocities in the swarm of smaller bodies grow in proportion to the escape velocity of the largest body (Safronov 1969) inducing larger and larger orbital changes that widen the zone swept by the minor bodies. Because of the growth of Jupiter's embryo, the proto-Earth will be hit first by chondrites coming from the asteroid belt, then by comets coming from Jupiter's zone and eventually from Saturn's zone.

A simple model predicts on the Earth a veneer of 3-4 km of chondritic silicates, some 5 km of water, 1.2 km of organic compounds and an atmosphere of 240 bars. Comets bring 75% of the silicates, whereas they bring 98% of water and organic compounds and more than 99% of the atmosphere. A major erosion of this veneer is to be expected from late giant impacts predicted by the accumulation models and also used to explain the formation of the Moon.

Some cosmochemical models of the terrestrial planets have independently converged to an inhomogeneous accretion, for instance the "chondrite model" (Larimer and Anders 1970). A recent version (Anders and Owen 1977) concludes that volatile elements were added at least on the Earth and Mars. Prinn and Fegley (1989) and Dreibus & Wänke (1989) also recognize the need of an inhomogeneous accretion of two components whose origin they do not try to explain.

Since observational evidence has recently established the ubiquity of accretion disks around very young stars, and since we have just shown that there seems to be a general mechanism in accretion disks to make terrestrial planets and to bring them a veneer of water and organic compounds, it is tempting to conclude that the number of terrestrial planets ready to develop life-as-we-know-it, from water and carbon compounds, is extremely large in the Universe.

Figure 1. Mid-plane temperature of the accretion disk as a function of radius. The crosses are Lewis 1974 aggregation temperatures of planets and satellites. The solid line is the adjustment of the disk model (Morfill 1988). The two dotted lines correspond to accretion rates 10 times smaller or larger.

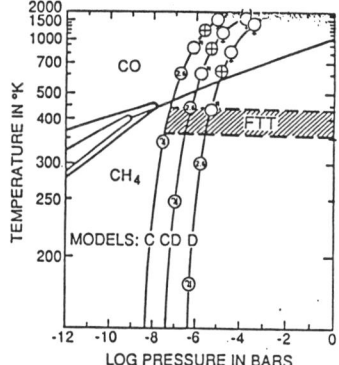

Figure 2. - Thermochemical equilibrium for carbon in a gas of solar composition. The quasi vertical curves are Cameron's 1985 adiabats. These adiabats can be interpreted as a cooling sequence. Adiabat CD brings the distance 2.6 AU at the high end of the FTT temperature zone, implying that dust sedimentation took place at that time. The same adiabat shows that, in the Earth's zone (symbol ⊕) all carbon was then in gaseous CO.

REFERENCES

Anders, E. 1986, p. 31-39 in Comets Nucleus Sample Return, ESA SP 249.
Anders, E. and Grevesse, N. 1989 Geochim. Cosmochim. Acta, 53, 197-214.
Cameron, A.G.W. 1985, p. 1073-1099 in Protostars and Planets II, Black and Matthews (eds), Univ. of Arizona Press, Tucson.
Chyba, C. F. 1987, Nature 330, 632-635.
Delsemme, A. H. 1981, p. 141-159 in Comets and the Origins of Life, Ponnamperuma (ed.), Reidel Publ., Dordrecht.
Delsemme, A. H. 1984, Origins of Life 14, 51-60.
Delsemme, A. H. 1990a, in Comets in the Post-Halley Era, Newburn et al. (eds.) Kluwer Acad. Press, Dordrecht (in press).
Delsemme, A. H. 1990b, in Proc. COSPAR XXVIII Plenary Meeting, The Hague (in press).
Hayashi, C., Nakazawa, K., and Nakagawa, Y. 1985, p. 1100-1153 in Protostars and Planets II, Black and Matthews (eds.), Univ. of Arizona Press, Tucson.
Ip, W. H., and Fernandez, J. A. 1988, Icarus 74, 47-62.
Lewis, J. S. 1974, Science 186, 440-443.
Lewis, J. S., Barshay, S. S., Noyes, B. 1979, Icarus 37, 190-206.
Lewis, J. S. and Prinn, R. B. 1980, Astrophys. J. 238, 357-364.
Lin, D.N.C., Papaloisou, J. 1985, p. 981-1072 in Protostars and Planets II, Black and Matthews (eds.), Univ. of Arizona Press, Tucson.
Lynden-Bell, D., and Pringle, J. E. 1974, M.N. Roy. Astron. Soc. London 168, 603-637.
Matsui, T., and Abe, Y. 1986, Nature 322, 526-528.
Morfill, G. E. 1988, Icarus 75, 371-379.
Morfill, G. E., Tscharnuter, W., Völk, H. J. 1985, p. 493-533 in Protostars and Planets II, Black and Matthews (eds.), Univ. of Arizona Press, Tucson.
Morfill, G. E., and Wood, J. A. 1989, Icarus 82, 225-243.
Smith, B. A. and Terrile, R. J. 1984, Science 226, 1421-1424.
Weidenschilling, S. J. 1988, p. 348-371 in "Meteorites and the Early Solar System", J. F. Kerridge and M. S. Matthews, eds., Univ. of Arizona Press, Tucson.
Wood, J. A., and Morfill, G. E. 1988, p. 329-347 in "Meteorites and the Early Solar System", J. F. Kerridge and M. S. Matthews, eds., Univ. of Arizona Press, Tucson.

INTERSTELLAR AND METEORITIC ORGANIC MATTER AT 3.4 μm

P. EHRENFREUND [1,2], F. ROBERT [3], L.d'HENDECOURT[2], F. BEHAR[4]
[1] *Huygens Laboratory Astrophysics, Leiden, The Netherlands*
[2] *Groupe de Physique des Solides, Univ. Paris 7, Paris,France*
[3] *Lab. Geochimie des Isotopes Stables, Paris, France*
[4] *Institute Francais du Petrole, Rueil-Malmaison, France*

ABSTRACT. The 3 micron spectrum of the galactic center source IRS 7 is compared to a spectrum obtained on the deuterium-rich organic polymer extracted from the Orgueil carbonaceous meteorite. The almost perfect match between the two spectra in the 3.4 μm region suggests that the chemical composition of the interstellar organic matter resembles that of the meteoritic macromolecule.

1. Introduction

The IR spectrum of the galactic center source IRS 7 indicates, that aliphatic hydrocarbons are coating the surface of interstellar grains in the line of sight (Butchart, 1986).
   There has been considerable discussion on the 3.4 μm interstellar absorption. Several attempts have been made to identify this matter by fitting the IRS 7 spectrum with known carbonaceous polymers (Sagan & Khare, 1979, Ogmen & Duley, 1988, Hoyle et al., 1982, Greenberg, 1982, Sandford et al.,1991). We have analyzed recently a series of organic samples extracted from terrestrial and extra-terrestrial rocks to study the relation between various organic samples at 3.4 μm. Such a comparison was considered important because any type of organic macromolecule would exhibit a signature at this wavelength, that characterizes aliphatic chains branched to aromatic cycles.

2. Results

   We have compared the 3.4 micron spectrum of Orgueil to the spectrum of the galactic center source IRS 7 in Fig. 1., obtained by Butchart et al. (1986). This spectrum represents actually the best available fit of IRS 7 and suggests that the chemical composition of the interstellar organic matter and that of the meteoritic polymer are similar.
   Because thermal events at the surface of the parent body meteorite or in the solar nebula can yield such an organic structure, the meteoritic polymer was compared to various terrestrial samples (kerogens) of known thermal history. Kerogens can be considered as a mixture of complex macromolecules deriving from organic remains. Similar macromolecules, with a highly aromatic cross linked 3-dimensional network have been described in meteorites by Hayatsu & Anders (1981).
   In Fig. 2 we show the signature at 3.4 μm of Type II kerogens with increasing C/H ratio. The C/H ratio acts as a parameter of maturation of the sample. However, the details of the interstellar organic structure cannot be found in the vast majority of natural or synthetic kerogens.

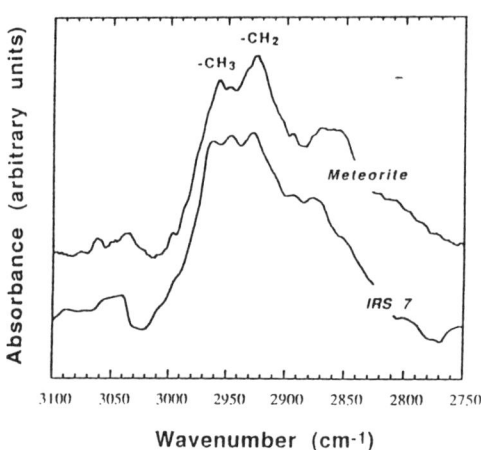
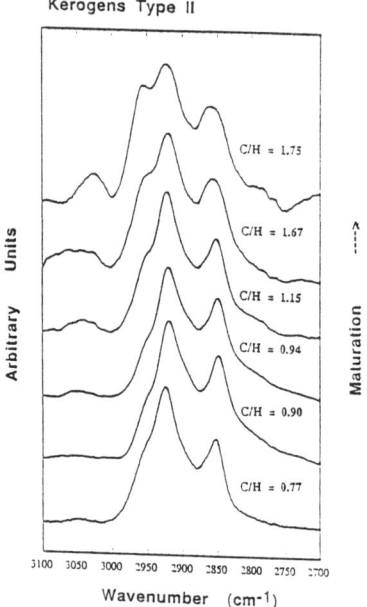

Figure 1. Comparison of the spectrum observed towards IRS 7 (Butchart et al., 1986) with a meteoritic macromolecule of Orgueil in the range 3100-2750 cm^{-1}. Ordinate is given in absorbance units (ln I_0/I).

Figure 2. Kerogen Type II (marine) at 3.4 µm as a function of their C/H ratio.

3. Discussion

Several processes may lead to this organic structure and thus severe transformations of the original interstellar organic carbon may have taken place during or after the formation of the solar system. However, a thermal degradation of the meteoritic macro-molecule remains unlikely, since labile organic structures are still bounded with the aromatic cycles. Furthermore the meteoritic sample is enriched in deuterium D/H (3.5 10^{-4}) and no other possibility than ion-molecule reactions in molecular clouds can account for this enrichment. The present comparison studies of the near infrared feature in organic matter indicate the existence of large molecules in the interstellar medium, at least on grain mantles.

An interesting implication of the match between the 3.4 µm spectra of Orgueil and IRS 7 is the possibility to estimate with some accuracy the fraction of the cosmic carbon locked up in complex interstellar organic molecules. Compared to the carbon density towards IRS 7 (6.8 10^{19}), one finds that about 12 % of carbon is locked up in aliphatic CH bonds whereas 30 % participates to other carbon bonds.

References
Butchart, I., et al. (1986), *Astron. Astrophys.* **154**, L 5
Greenberg, J.M. (1982), *Comets*, eds. Wilkening L.L., Univ. of Arizona Press, Tucson, 131
Hayatsu, R., Anders E. (1981), *Topics of Current Chemistry* **99**, ed. Springer Verlag, 1
Hoyle, F., Wickramasighe, N.C., Al Mufti, S., Olavesen, A.H. (1982), *Astrophys. Space Sci.* **83**, 405
Ogmen, M., Duley, W.W. (1988), *Astrophys. J. Lett.* **334**, L 117
Sagan, C., Khare, B.N. (1979), *Nature* **277**, 102
Sandford, S.A., et al. (1991), *Astrophys J.* **371**, 607

Diamagnetic Abundance Differentiation in the Solar System

Raphael Steinitz and Estelle Kunoff
Physics Dept. Ben-Gurion University
Be'er Sheva, Israel

ABSTRACT Chemical abundances in the solar corona or solar wind compared to those in the photosphere differentiate according to first ionization potential (FIP). We suggest that the effect is the result of diamagnetic diffusion pumps operating in the presence of gravity and diverging magnetic structures. We then comment briefly on implications concerning abundances in the solar system and chemically peculiar stars.

The relative elemental abundances in the solar corona (SC), solar wind (SW) and solar energetic particles (SEP) are essentially similar [1]. These abundances, however, differ from those in the photosphere (see fig. 1 in Breneman and Stone [2]). On comparison to those in the solar photosphere, they divide into two distinct groups: elements with FIP < 10 eV are 3-4 times more abundant than higher FIP elements. Similar results were obtained by Cook et. al.[3], Meyer [4], Veck and Parkinson [5] and Schmidt et. al. [6]. According to Meyer [1], *this division is strictly according to FIP and is independent of both mass and charge*. Since the SC is the origin of both the SEP and SW particles, it is sufficient to account for FIP < 10 eV particle enrichment in the SC. We are thus looking for a mechanism operating on FIP < 10 eV particles as they pass from the chromosphere to the SC.

We note that temperatures in the chromosphere, being about 10^4 K, are hot enough to ionize the low FIP elements but not those with FIP > 10 eV [7]. It appears that ions in the chromosphere are thus fed into the SC more easily than neutrals. We therefore explore, in the next section, the possibility that the enrichment is due to diamagnetic pumping.

The mirroring force, invoked by Fermi [8] to explain cosmic ray acceleration is (Spitzer [11]):

$$m\frac{dv_{//}}{dt} = -\frac{\varepsilon_\perp}{B}\nabla_{//}B \qquad (1)$$

where $v_{//}$ is the velocity along the magnetic field lines, $\varepsilon_\perp = mv_\perp^2/2$ is the portion of the ion's kinetic energy orthogonal to the field lines, B is the magnetic field and we take the component of the gradient along B. This equation describes the acceleration of the gyrating center of a charged particle along diverging magnetic field lines. Since gravity acts on both neutral and charged particles while the mirroring force on the charged particles only, we have a magneto-gravitational mechanism that separates between the two species of particles. In the folowing, we thus develop a model of the diamagnetic diffusion pump, enabling us to estimate the enrichment of ions relative to neutrals.

Now consider the ratio of gyrofrequency to collision frequency; as this ratio increases, so does the relative enrichment. For collisions,

$$v_{coll} = \frac{n\pi e^4}{\sqrt{m/2}(kT)^{3/2}} \qquad (2)$$

so that the ratio of gyration to collision frequencies is:

$$R = \frac{v_{gyr}}{v_{coll}} = \left(\frac{\sqrt{k}}{e}\right)^3 \frac{1}{\pi^2\sqrt{m}}\left(\frac{BT^{3/2}}{n}\right). \qquad (3)$$

Substituting typical chromosphere values, $T \sim 10^4$ K, $m \sim 20 m_p$ and for $10^9 < n < 10^{10}$, we find that $2.9 \times 10^9 B < R < 2.9 \times 10^{10} B$ and conclude that even for very small fields, a typical collision time includes many gyration periods. Therefore, we can incorporate the effect of collisions in a dynamic friction term in the equation of motion.

The motion of a charged particle's gyration center can be described by:

$$m\dot{v}_z = -\alpha v_z + \frac{\varepsilon_\perp}{L} - mg. \qquad (4)$$

Here $L = [\partial(\ln B)/\partial z]^{-1}$ and we assume that **B** is essentially vertical; α is the dynamical friction. To eliminate gravity, we take the time derivative and use $\varepsilon_\perp = \mu B$, where μ is the magnetic momoent (adiabatic invariant) and assume that $\partial B/\partial T = 0$. Since μ is conserved, substitute for μ its value at t=0, $\mu = \varepsilon_{\perp 0}/B_0$ and obtain:

$$\ddot{v}_z + \frac{\alpha}{m}\dot{v}_z - \frac{\mu}{mL^2}v_z = 0. \qquad (5)$$

In the chromosphere, $\varepsilon_\perp \sim 1$ eV $\sim 10^4$ K and $L \sim 10^5$ km, so that the last term is negligible compared to friction. The assymptotic velocity is thus:

$$-\alpha v_z + \frac{\varepsilon_\perp}{L} - mg = 0. \qquad (6)$$

which we then substitute for the average velocity in the definition of the (material) current density $J_z = n\langle v_z \rangle$ in the diffusion equation and obtain:

$$-D\nabla n_I - \frac{1}{\alpha}\left(\frac{kT}{L} - mg\right) = 0. \qquad (7)$$

Since the diffusion constant $D = kT/\alpha$ and the neutrals follow a barometric decay, we finally obtain for the abundance ratio of ions to neutrals:

$$\frac{n_I}{n_N} = \frac{n_{I0}}{n_{N0}} \exp(z/L) \qquad (8)$$

If large, vertical, magnetic field intensity gradients exist over typical distances $z \sim L$, an enrichment factor of 2 - 4 ensues. We can now consider the effect of this result on Solar System bodies and chemically peculiar (CP) stars. Solar System bodies, not shielded by their own magnetic fields, may acquire the abundance patterns carried by the SW. Their original surface chemical composition could therefore be modified. We also note that the rare earth elements have among the lowest FIP values in the periodic table of elements. We can therefore speculate that the Diamagnetic Effect may play an important role in the CP stars.

We have considered the problem of the enrichment of elemental abundances in the solar corona according to FIP, which has as yet received no satisfactory explanation. Noting that only elements with FIP<10 eV are ionized in the chromosphere, we have constructed a model using the diamagnetic effect as a mechanism to selectively pump ions against gravity into the corona. This mechanism must operate, irrespective of other processes. Finally, we have shown how this result can be applied to the solar system and chemically peculiar stars.

1. J.-P. Meyer, Ap. J. Sup. **57**, 173 (1985).
2. H. H. Breneman and E. C. Stone, Ap. J. (Letters) **299**, L57 (1985).
3. W. R. Cook, E. C. Stone and R. E. Vogt, Ap. J. **279**, 827 (1984).
4. J.-P. Meyer, Ap. J. Sup. **57**, 151 (1985).
5. N. J. Veck and J. H. Parkinson, MNRAS, **197**, 41 (1981).
6. J. Schmidt, P. Bochsler and J. Geiss, Ap. J. **329**, 956 (1988).
7. C. Jordan, MNRAS **142**, 501 (1969).
9. E. Fermi, Phys. Rev. **75**, 1169 (1949); Ap. J. **119**, 1 (1954).
10. L. Spitzer Jr., *Physics of Fully Ionized Gases*, (Wiley and Sons, NY,1962).

COMETARY MOLECULES

Lewis E. Snyder
Astronomy Dept., University of Illinois
Urbana, Illinois U.S.A.

ABSTRACT. Which molecular species are firmly identified as cometary species and which are also interstellar? How may excitation effects bias the data interpretation?

1. Introduction

Because space is limited, only those molecules with observed and identified gas-phase spectra will be discussed. Thus, there can be no discussion of either the interesting mass spectra acquired by the Comet Halley spacecraft, or current work on the proposed polyoxymethylene identification (Huebner 1987). Modern theory links the composition of comets to that of the interstellar medium, with many uncertainties. Table 1 lists the known interstellar and circumstellar species by number of atoms, except for the polycyclic aromatic

TABLE 1. The 89 reported interstellar and circumstellar molecules as of July, 1991
----
2 AlF AlCl C_2 CH CH^+ CN CO CP CS CSi HCl H_2
KCl NH NO NS NaCl OH PN SO SO^+ SiN SiO SiS
3 C_2H C_2S HCN HCO HCO^+ HCS^+ H_3^+ H_2O H_2S HNC
HNO N_2H^+ OCS SO_2 c-SiC_2 4 c-C_3H l-C_3H C_3N
C_3O C_3S C_2H_2 $HCNH^+$ H_2CO H_2CS H_3O^+ HNCO HNCS
$HOCO^+$ NH_3 5 C_4H C_4Si c-C_3H_2 CH_2CN CH_4 HC_3N
HCOOH H_2C_2O H_2CHN H_2NCN SiH_4 6 C_5H C_5O C_2H_4
CH_3CN CH_3NC CH_3OH CH_3SH $HCONH_2$ 7 C_6H CH_2CHCN
CH_3C_2H HC_5N $HCOCH_3$ NH_2CH_3 8 CH_3C_3N $HCOOCH_3$
9 CH_3C_4H CH_3CH_2CN $(CH_3)_2O$ CH_3CH_2OH HC_7N
10 CH_3C_5N $(CH_3)_2CO$ 11 HC_9N 13 $HC_{11}N$

hydrocarbon (PAH) compounds. Ring structures are preceded by c-, but where the structure is also linear an l- is used. Mann and Williams (1980) and Lovas (1986; 1987) have prepared extensive frequency lists.

2. Cometary Molecules

The cometary molecules with observed and identified gas-phase spectra are listed in Table 2. The known interstellar/

TABLE 2. The 27 reported cometary molecules as of July 1991

<u>2</u> C_2 CH CH^+ CN CN^+ CO CO^+ CS N_2^+ NH OH OH^+ S_2
<u>3</u> HCN H_2O H_2O^+ H_2S C_3 CO_2 CO_2^+ NH_2 OCS
<u>4</u> H_2CO NH_3 <u>5</u> CH_4 <u>6</u> CH_3CN CH_3OH

circumstellar molecules are more chemically complicated than these; along with the Halley mass spectra and meteoritic chemistry, this suggests that comets are more chemically evolved than their gas-phase spectra indicate.

Two important questions are: how secure is a given cometary molecular identification; and what molecules are identified in multiple spectral regions? Several reviews have addressed these questions (e.g., A'Hearn 1983; Feldman 1983; Wyckoff 1983; Snyder 1982), so this will be a brief update. The ultraviolet region will be defined as $\lambda < 3200$ Å, optical as 3200-8000 Å, infrared as 8000 Å to 0.5 mm (20 cm^{-1}), and radio as $\lambda > 0.5$ mm wavelength. Table 3 lists ultraviolet molecules, approximate band centers, and assignments. All are

TABLE 3. Cometary molecules with ultraviolet spectra

C_2 (2313 Å, $D^1\Sigma^+_u - X^1\Sigma^+_g$) CN^+ (2181 Å, $f^1\Sigma - a^1\Sigma$; 3185 Å, $c^1\Sigma - a^1\Sigma$) CO (1510 Å, $A^1\Pi - X^1\Sigma^+$) CO^+ (2190 Å, $B^2\Sigma^+ - X^2\Sigma^+$) CS (2576 Å, $A^1\Pi - X^1\Sigma^+$) OH (3090 Å, $A^2\Sigma^+ - X^2\Pi_i$) S_2 (2957 Å, $B^3\Sigma^-_u - X^3\Sigma^-_g$) CO_2^+ (2890 Å, $B^2\Sigma^+_u - X^2\Pi_g$)

commonly found except for S_2, which was detected only in Comet IRAS-Araki-Alcock or I-A-A (A'Hearn 1983; Feldman 1983; 1991; Wyckoff 1983). Table 4 lists optical molecules, approximate band centers, and assignments (A'Hearn 1983; Wyckoff 1983).

TABLE 4. Cometary molecules with optical spectra

C_2 (5165 Å, $d^3\Pi_g - a^3\Pi_u$; 7715 Å, $A^1\Pi_u - X^1\Sigma^+_g$)
$^{12}C^{13}C$ (4745 Å, $d^3\Pi_g - a^3\Pi_u$) CH (3889 Å, $B^2\Sigma^- - X^2\Pi_r$; 4315 Å, $A^2\Delta - X^2\Pi_r$) CH^+ (4225 Å, $A^1\Pi - X^1\Sigma^+$)
CN (3883 Å, $B^2\Sigma^+ - X^2\Sigma^+$; 7873 Å, $A^2\Pi_i - X^2\Sigma^+$)
CO^+ (3954 Å, $B^2\Sigma^+ - A^2\Pi_i$; 4273 Å, $A^2\Pi_i - X^2\Sigma^+$)
N_2^+ (3914 Å, $B^2\Sigma^+_u - X^2\Sigma^+_g$) NH (3360 Å, $A^3\Pi_i - X^3\Sigma^-$)
OH^+ (3565 Å, $A^3\Pi_i - X^3\Sigma^-$) H_2O^+ (6198 Å, $\tilde{A}^2A_1 - \tilde{X}^2B_1$)
C_3 (4040 Å, $\tilde{A}^1\Pi_u - \tilde{X}^1\Sigma^+_g$) CO_2^+ (3509 Å, $\tilde{A}^2\Pi_u - \tilde{X}^2\Pi_g$; 3674 Å) NH_2 (5007 Å, $\tilde{A}^2A_1 - \tilde{X}^2B_1$)

Table 5 lists cometary molecules observed in the infrared,

TABLE 5. Cometary molecules with infrared spectra

CO (4.7 μm/2143 cm^{-1}, 1-0)
OH (1.5-1.9 μm/6666-5263 cm^{-1}, 3-1, 4-2, 5-3, 12-9,
  6-4, 7-5; 119.23 μm/83.9 cm^{-1}, $^2\Pi_{3/2}$ J=5/2-3/2,+ -)
H_2O (1.38 μm/7249.8 cm^{-1}, $\nu_1+\nu_3$; 1.88 μm/5331.3 cm^{-1},
  $\nu_2+\nu_3$; 2.44 μm/4100 cm^{-1}, $\nu_1+\nu_3-2\nu_2$ (tent. assignment);
  2.64 μm/3782.2 cm^{-1}, $\nu_2+\nu_3-\nu_2$; 2.66 μm/3755.9 cm^{-1}, ν_3)
CO_2 (4.3 μm/2349 cm^{-1}, ν_3) OCS (4.85 μm/2062 cm^{-1}, ν_3)
H_2CO (3.5 μm/2843 cm^{-1}, ν_5; 3.6 μm/2783 cm^{-1}, ν_1)
CH_4 (3.3 μm/3019 cm^{-1}, ν_3)

wavelengths (in μm/cm^{-1}), and assignments. Encrenaz and Knacke (1991) and Weaver et al. (1991) have given recent reviews. The 1.38 μm H_2O band was detected in Halley by Knacke et al. (1986) and Krasnopolsky et al. (1986), and the 1.88 μm band by Knacke et al. (1986). The 2.44 μm band is a tentative H_2O hot band assignment in Halley (Maillard et al. 1987), but the 2.64 μm assignment is definite (Weaver et al. 1986). The 2.66 μm assignment (ν_3) is well confirmed (Mumma et al. 1986; Weaver et al. 1986; Combes et al. 1988). Larson et al. (1989) also detected 10 H_2O emission lines from ν_3 and a hot band line from $\nu_2+\nu_3-\nu_2$ in Comet Wilson. CO, CO_2, OCS, and H_2CO were identified in Halley by Combes et al. (1988), but CO and OCS were weak. Stacey et al. (1987) detected OH in Halley at 119.23 μm and Krasnopolsky et al. (1986) detected the 1.5-1.9 μm OH bands. CH_4 in Halley was reported by Kawara et al. (1988) and in Wilson by Larson et al. (1989). Table 6 lists the molecules observed in the radio region, wavelengths/frequencies, and assignments. The radio CH line has been reported only in Comet Kohoutek (Black et al. 1974). Radio OH lines are routinely observed in comets. HCN detections in Kohoutek and Halley were followed by detections in Brorsen-Metcalf, Austin, and Levy (cf., Crovisier and Schloerb 1991; Colom et al. 1990; Schloerb and Ge 1990a;b). The 1.348 cm H_2O line was reported in comets Bradfield and I-A-A (Jackson et al. 1976; Altenhoff et al. 1983). H_2S and CH_3OH (3.099 and 2.066 mm) were detected in Austin (Bockelée-Morvan et al. 1991). CH_3OH (1.240 mm) also was detected in Levy (Schloerb and Ge 1990a). The 6.207 cm H_2CO line was detected in comets Halley and Machholz by Snyder et al. (1989; 1990), the 1.3283 mm line in Brorsen-Metcalf, Austin, and Levy by Colom et al. (1991), and the 0.8522 mm line in Levy by Schloerb and Ge (1990b). The 1.256 cm line of NH_3 was detected only in I-A-A (Altenhoff et al. 1983). Vibrationally excited CH_3CN (2.7079 mm) was observed only in Kohoutek (Ulich and Conklin 1974).

TABLE 6. Cometary molecules with radio spectra

CH (8.988 cm/3,335.481 MHz, $^2\Pi_{1/2}$ J=1/2,F=1-1)
OH (18.595 cm/1,612.2310 MHz $^2\Pi_{3/2}$ J=3/2,F=1-2;
18.001 cm/1,665.4018 MHz $^2\Pi_{3/2}$ J=3/2,F=1-1;
17.980 cm/1,667.3590 MHz $^2\Pi_{3/2}$ J=3/2,F=2-2;
17.424 cm/1,720.5300 MHz $^2\Pi_{3/2}$ J=3/2,F=2-1)
HCN(3.3825 mm/88,630.4157 MHz J=1-0,F=1-1;
3.3824 mm/88,631.8473 MHz J=1-0,F=2-1; 3.3824 mm/
88,633.9360 MHz J=1-0,F=0-1; 1.1285 mm/265,886.432
MHz J=3-2; 0.8457 mm/354,505.472 MHz J=4-3)
H_2O (1.348 cm/22,235.120 MHz $J_{KaKc}=6_{16}-5_{23}$,F=5-4)
H_2S (1.7764 mm/168,762.762 MHz $J_{KaKc}=1_{10}-1_{01}$)
H_2CO (6.207 cm/4829.6639 MHz $J_{KaKc}=1_{11}-1_{10}$,F=2-1;
1.3283 mm/225,697.775 MHz $J_{KaKc}=3_{12}-2_{11}$;
0.8522 mm/351,768.639 MHz $J_{KaKc}=5_{15}-4_{14}$)
NH_3 (1.256 cm/23870.1296 MHz J(K)=3(3),F=4-4)
CH_3CN (2.7079 mm/110,709.55 MHz $v_8=1$,J(K)=6(3)-5(3);
2.7079 mm/110,712.22 MHz $v_8=1$, J(K)= 6(0)-5(0))
CH_3OH (3.0990 mm/96,739.39 MHz J(K)=2(-1)-1(-1)E;
3.0989 mm/96,741.42 MHz J(K)=2(0)-1(0)A+; 2.0662 mm/
145,093.75 MHz J(K)=3(0)-2(0)E; 2.0661 mm/145,097.47
MHz J(K)=3(-1)-2(-1)E; 2.0661 mm/145,103.23 MHz
J(K)=3(0)-2(0)A+; 2.0657 mm/145,131.88 MHz J(K)=
3(1)-2(1)E; 1.2400 mm/241,767.224 MHz J(K)=5(-1)-
4(-1)E; 1.2399 mm/241,791.431 MHz J(K)=5(0)-4(0)A+)

3. Excitation Effects in Comets

In the ultraviolet and optical spectral regions, the dominant excitation mechanism for electronic transitions is resonance fluorescence. The Doppler-shifted resonant transitions of cometary molecules absorb the solar spectrum and reradiate it via various electronic transitions. Typically, collisions play only a minor role in establishing the molecular population distributions in these spectra (cf., Arpigny 1976). In the longer wavelength spectral regions, a major excitation mechanism is believed to be infrared fluorescence, but radio excitation may be more complicated.

Because ultraviolet excitation mechanisms are fairly well understood, the detection of S_2 in I-A-A, as discussed by M. A'Hearn (this book), can be attributed both to the existence of an outburst at closest approach and to the remarkable 25 km/arcsec scale of the comet.

The intensities of the radio OH lines are determined by ultraviolet pumping called the Swings effect (cf., Schloerb and Gerard 1985; Crovisier and Schloerb 1991; de Pater et al. 1991). Briefly, solar ultraviolet Fraunhofer radiation pumps the cometary OH ultraviolet bands listed in Table 3 (3090 Å, $A^2\Sigma^+-X^2\Pi_i$). The OH cascades to the ground state, establishing the intensities of the 18 cm lines in Table 6. This effect determines whether radio OH will be observed in maser emission, absorption, or not at all! For example, when the projected densities of the unsplit upper and lower levels of the ground-state Λ doublet are almost equal, the cometary OH signal may vanish, even though the OH has not. Another point is that single radio antennas can make crude maps of cometary OH with a resolution of a few arcminutes, but radio interferometers can produce detailed imaging which can greatly affect the excitation modeling (cf., de Pater et al. 1991).

The evidence for an extended cometary dust coma that began to emerge with the measurements of Comet Halley is:

A. The in situ measurements made with the Giotto neutral gas mass spectrometer by Eberhardt et al. (1987) can be explained if CO is produced both directly from the nucleus and from extended dust in the inner coma.

B. The >60,000 km CN and C_2 jets observed by A'Hearn et al. (1986a,b) can be explained by dissociation directly from CHON particles in an extended dust coma.

C. The VLA detected the 6.207 cm line of H_2CO in emission from Halley and Machholz (Snyder et al. 1989; 1990). Interpreting this line requires an extended coma model.

D. The cometary H_2CO model of Bockelée-Morvan and Crovisier (1991) shows how H_2CO will behave if produced only by the cometary nucleus, with no further gas production in an extended gas/dust coma. This is an excellent model, but it does not explain (a) the VLA detections of the 6.207 cm H_2CO line; (b) the IRAM detections of the 1.3283 mm H_2CO line at a nuclear distance of 3500 km from Austin and 6300 km from Levy by Colom et al. (1991); or (c) the CSO detection of the 0.8522 mm H_2CO line at a 30" nuclear distance offset from Comet Levy by Schloerb (1991, private communication). These observations suggest that either specious comet lines are extremely common at different frequencies and times, or there is indeed an extended cometary dust coma which needs to be taken into account in future models.

4. Summary

Newly detected species include H_2O, H_2CO, CH_4 (Halley), CH_3OH (Austin and Levy), and H_2S (Austin). Cometary nuclei appear to be more chemically evolved than the current gas-phase spectroscopy suggests. Some of the excitation conditions are exemplified by the sudden appearance of S_2 in I-A-A, the ultraviolet pumping of radio OH lines, the extended dust coma, and the unexplained radio lines of H_2O and excited CH_3CN.

Support from NASA grants NAGW-1131 and NAGW-2299 to the University of Illinois is gratefully acknowledged.

5. References

A'Hearn, M. F. (1983) 'Spectrophotometry of comets at optical wavelengths', in Wilkening and Matthews (eds.), Comets, University of Arizona Press, Tucson, pp. 433-60.

A'Hearn, M. F. et al. (1986a) 'Cyanogen jets in Comet Halley', Nature 324, 649-51.

A'Hearn, M. F. et al. (1986b) 'Gaseous jets in Comet p/Halley', in 20th ESLAB Symp. Expl. Halley's Comet, SP-250, pp. 483-6.

Altenhoff, W.J. et al. (1983) 'Radio observations of Comet 1983d', A&A 125, L19-L22.

Arpigny, C. (1976) 'Interpretation of comet spectra', in Donn et al. (eds.), The Study of Comets, NASA SP-393, pp. 797-839.

Black, J. H. et al. (1974) 'Radiofrequency emission from CH in Comet Kohoutek (1973f)', ApJ 191, L45-7.

Bockelée-Morvan, D. et al. (1991) 'Microwave detection of hydrogen sulfide and methanol in Comet Austin (1989c1)', Nature 350, 318-20.

Bockelée-Morvan, D., and Crovisier, J. (1991) 'Formaldehyde in comets: excitation of the rotational lines', A&A, in press.

Colom, P. et al. (1990) 'Millimetre observations of Comets P/Brorsen/Metcalf (1989o) and Austin (1989c1) with the IRAM 30-m radio telescope' in Huebner et al. (eds.), Workshop on Observations of Recent Comets (1990), Southwest Research Inst., San Antonio, pp. 80-5.

Colom, P. et al. (1991) 'Formaldehyde in comets: microwave observations of P/Brorsen-Metcalf (1989X), Austin (1989c1), and Levy (1990c)', A&A, in press.

Combes, M. et al. (1988) 'The 2.2-12 μm spectrum of Comet Halley from the IKS-VEGA experiment', Icarus 76, 404—36.

Crovisier, J. and Schloerb, F. P. (1991) 'Study of comets at radio wavelengths', in Newburn et al. (eds.), Comets in the Post-Halley Era, Kluwer, Dordrecht, pp. 149-73.

de Pater, I. et al. (1991) 'A review of radio interferometric imaging of comets', Ibid., pp. 175-207.

Eberhardt, P., et al. (1987) 'The CO and N_2 abundance in Comet p/Halley', A&A 187, 481-4.

Encrenaz, T., and Knacke, R. (1991) 'Carbonaceous compounds in comets: infrared observations', in Newburn et al. (eds.), Comets in the Post-Halley Era, Kluwer, Dordrecht, pp. 107-37.

Feldman, P. D. (1983) 'Ultraviolet spectroscopy of comae', in Wilkening and Matthews (eds.), Comets, University of Arizona Press, Tucson, pp. 461-79.

Feldman, P. D. (1991) 'Ultraviolet spectroscopy of cometary comae', in Newburn et al. (eds.), Comets in the Post-Halley Era, Kluwer, Dordrecht, pp. 139-48.

Huebner, W. F. (1987) 'First polymer in space identified

in Comet Halley', Science 237, 628-30.
Jackson, W. M., Clark, T., and Donn, B. (1976) 'Radio detection of H_2O in Comet Bradfield' in Donn et al. (eds.), The Study of Comets, NASA SP-393, pp. 272-80.
Kawara et al. (1988) 'Infrared spectroscopic observation of methane in Comet P/Halley' A&A 207, 174-81.
Knacke, R. F. et al. (1986) 'Ground-based detection of water in Comet Halley' in Battrick et al. (eds.), 20th ESLAB Symp. Expl. Halley's Comet, SP-250, pp. 99-101.
Krasnopolsky, V. A. et al. (1986) 'Spectroscopic study of Comet Halley by the Vega 2 three-channel spectrometer', Nature 321, 269-71.
Larson, H. P. et al. (1989) 'Airbourne infrared spectroscopy of Comet Wilson (1986l) and comparisons with Comet Halley', ApJ 338, 1106-14.
Lovas, F. J. (1986) 'Recommended rest frequencies for observed interstellar molecular microwave transitions-1985 revision', J. Phys. Chem. Ref. Data 15, 251-303.
_____, (1987) Ibid. 16, 153-4.
Maillard, J. P. et al. (1987) 'Spectrum of Comet P/Halley between 0.9 and 2.5 microns', A&A 187, 398-404.
Mann, A.P.C., and Williams, D.A. (1980) 'A list of interstellar molecules', Nature 283, 721-25.
Mumma, M. J. et al. (1986) 'Detection of water vapor in Halley's Comet', Science 232, 1523-28.
Schloerb, F. P., and Ge, W. (1991a,b) 'Comet Levy (1990c)', IAU Circs. No. 5081 and 5086.
Schloerb, F. P., and Gerard, E. (1985) 'Models of cometary emission in the 18-cm OH transitions: the predicted behavior of Comet Halley', AJ 90, 1117-35.
Snyder, L.E. (1982) 'A review of radio observations of comets', Icarus, 51, 1-24.
Snyder, L. E., Palmer, P., and de Pater, I. (1989) 'Radio detection of formaldehyde emission from Comet Halley', AJ 97, 246-53.
_____, (1990) 'Observations of formaldehyde in Comet Machholz (1988j)', Icarus 86, 289-98.
Stacey, G. J., Lugten, J. B., and Genzel, R. (1987) 'Detection of OH rotational emission from Comet P/Halley in the far-infrared', A&A 187, 451-54.
Ulich, B. L., and Conklin, E. K. (1974) 'Detection of methyl cyanide in Comet Kohoutek', Nature 248, 121-2.
Weaver, H. A., Mumma, M. J., and Larson, H. P. (1991) 'Infrared spectroscopy of parent molecules', in Newburn et al. (eds.), Comets in the Post-Halley Era, Kluwer, Dordrecht, pp. 93-106.
Weaver, H. A. et al. (1986) 'Post-perihelion observations of water in Comet Halley', Nature 324, 441-4.
Wyckoff, S. (1983) 'Overview of comet observations', in Wilkening and Matthews (eds.), Comets, University of Arizona Press, Tucson, pp. 3-55.

QUESTIONS AND ANSWERS

J.P.Maillard: You listed CO as a detected molecule in the IR (at 4.7 μm). As far I know it has not been detected. It has been looked for, of course, but only upper limit has been obtained.

L.E.Snyder: CO was identified in Halley by Combes et al (1988), but it was week.

T.J.Millar: If cometary particles are related to interstellar particles, one might expect them to contain significant fractionation in deuterium. Have there been sensitive searches for D-bearing molecules in comets?

L.E.Snyder: This question was answered by Mike A'Hearn.

T.J.Millar: What is the abundance of D?

M.F.A'Hearn: The Giotto NMS was used in its ion mode by Eberhardt to derive separate upper & lower limits on HDO/H_2O in Halley at something like 5×10^{-5} to 5×10^{-4} (I don't remember the exact numbers). Spectra from IUE have been searched for emission by OD and upper limits have been derived for 5 - 10 comets ranging from 10^{-2} to 4×10^{-4}, depending on signal-to-noise.

LABORATORY STUDIES OF PLANETARY MOLECULES AND ICES: THE CASE OF IO

F. SALAMA, S.A. SANDFORD, and L.J. ALLAMANDOLA
NASA-Ames Research Center. Space Science Division, MS: 245-6
Moffett Field, CA 94035, U.S.A

ABSTRACT. The techniques of low temperature spectroscopy are applied here to analyze infrared observational data of Io in the 2.0-5.0 µm range. The presence of solid H_2S and traces of H_2O in the SO_2-dominant surface ices are derived from this analysis and it is suggested that CO_2 clusters may as well be present near the surface of Io.

1. Introduction

Io, a satellite of Jupiter, is the reddest object in the solar system and the first body (beyond Earth) where active volcanism has been observed. Io spectra show several features in the IR. While SO_2 frost responsible for the prominent absorption features in this range, is one of the major surface components, the other components which are responsible for the weaker spectral features had been unidentified. We have focused our attention on the 2.0-2.5 and 2.9-5.3 µm regions. Originally unidentified bands falling at 2.97 µm, 3.15 µm, 3.85 µm, and 3.91 µm [1] and 2.1253 µm [2, 3], were found in Io spectra. All the features, except the 2.1253 µm band, show strong temporal and longitudinal variations. In this paper, we present the result of comparative studies between observational data on Io [1, 2] and detailed laboratory studies of plausible surface ices [1, 3].

2. Results

Among the unidentified features of the Io spectrum, the 3.91 µm and 3.85 µm bands fall close to *but not at* the position expected for the S-H stretching vibrations of pure H_2S frost; the 3.15 and 2.97 µm pair fall close to those expected for the O-H stretching vibrational modes of H_2O, and the 2.1253 µm band falls close to, *but not at* the position of an overtone of the asymmetric stretching vibration of solid CO_2. Laboratory simulations [1,3] were then performed on pure H_2S, pure H_2O, pure CO_2 and pure SO_2 frosts and their mixtures (mixed molecular ices and layered ices) in order to determine how the solid state interactions as well as temperature variations (from 9 K to 130 K) and UV irradiation would affect the spectra. These comparative studies of spectra of Io with laboratory absorption spectra allow us to draw *several main conclusions* about the composition and physical nature of the surface material on Io.
1- The good match between the laboratory spectra and the spectra of Io strongly suggests that *hydrogen sulfide is mixed in the surface material of the satellite* [1]. An upper limit of about 3% for the amount of H_2S relative to SO_2 in the surface material is indicated at the cold patches (~ 100 K). The 3.91 µm and 3.85 µm bands in the spectra of Io (Fig.1) indicate that

H_2S is present as clusters and isolated molecules in a matrix of SO_2. The laboratory experiments show that the infrared spectrum of H_2S embedded in an SO_2 ice is largely unaffected by thermal variations below 100 K and prolonged VUV irradiation, implying that H_2S can survive the harsh conditions on Io which is in Jupiter's radiation belt. Also, only mixed molecular ices (i.e, SO_2 matrices containing H_2S) can explain the observed band shifts and splitting and can account for the fact that solid H_2S is observed in the surface material of Io under temperature and pressure conditions well above the sublimation point of pure H_2S.

2- The relatively more complicated case of the variable bands of Io at 2.97 μm and 3.15 μm indicates that *traces of water are suspended in the mixtures of H_2S and SO_2*. An upper limit of about 0.1% for the amount of H_2O relative to SO_2 is implied [1]. The spatial and temporal variability of the H_2O bands appears correlated with the volcanic activity on Io.

3- The experiments also suggest that the newly discovered band at 2.1253 μm [2] may be due to *the formation of CO_2 clusters in the atmosphere of Io*. An upper limit of about 1% for the amount of CO_2 relative to SO_2 is derived and much of the CO_2 is estimated to be contained in the cold (~ 100 K) polar regions of the satellite [3].

This study stresses the importance of studying mixed ices (matrices) in carrying out laboratory simulations of the "dirty" ices covering the surfaces of satellites and planets in the solar system. Previous studies have only considered pure materials and neglected the importance of molecular interactions.

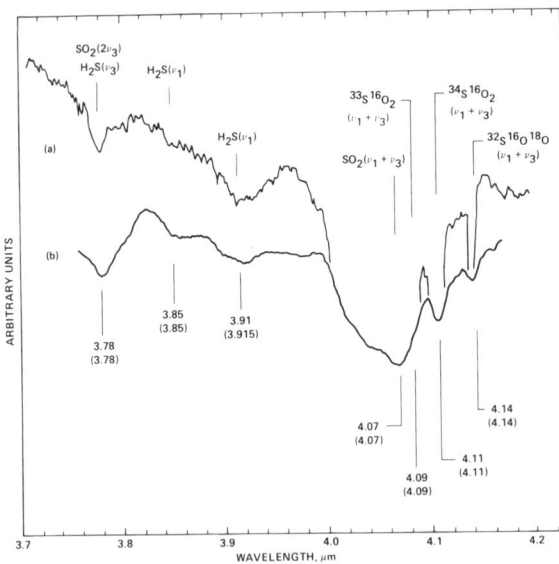

Figure 1: Comparison between (a) the IR absorption spectrum of an $H_2S:SO_2$ (3:100) ice grown on a 100K surface [1] and (b) the reflectivity of Io at $\Phi = 70°$ measured by Howell et al (Icarus 78, 27-37, 1989).

References:
1- Salama, F., Allamandola, L. J., Witteborn, F. C., Cruikshank, D. P., Sandford, S. A., and Bregman, J. D. (1990) 'The 2.5 - 5.0 μm Spectra of Io: Evidence for H_2S and H_2O Frozen in SO_2', Icarus 83, 66-82.
2- Trafton, L. M., Lester, D. F., Ramseyer, T. F., Salama, F., Sandford, S. A., and Allamandola, L. J. (1991) 'A New Class of Absorption Feature in Io's Near-Infrared Spectrum', Icarus 89, 264-276.
3- Sandford, S. A., Salama, F., Allamandola, L. J., Trafton, L. M., Lester, D. F.,and Ramseyer, T. F. (1991) 'Laboratory Studies of the Newly Discovered Infrared Band at 4705.2 cm^{-1} (2.1253 μm) in the Spectrum of Io', Icarus 91, 125-144.

TYPE II CLATHRATE HYDRATE FORMATION IN COMETARY ICE ANALOGS IN VACUO

D.F. BLAKE, L. ALLAMANDOLA,
S. SANDFORD, D. HUDGINS
MS 239-4
NASA/Ames Research Center
Moffett Field, CA 94035- USA

F. FREUND
Dept. of Physics
San Jose State University
San Jose, CA 95192
USA

ABSTRACT. Clathrate Hydrates can be formed under high vacuum conditions by annealing vapor-deposited amorphous ices of the appropriate composition. When astrophysically significant $H_2O:CH_3OH$ ices are deposited and annealed, Type II Clathrate Hydrates are formed which can hold up to 6 mole % large guest molecules such as methanol and 12 mole % small guest molecules such as CO_2 and CO. The solid state transformation of amorphous mixed molecular ice into crystalline clathrate hydrate and its sublimation at higher temperatures may serve to explain heretofore anomalous mechanical and gas release properties observed in cometary ices and laboratory ice analog experiments.

1. Introduction

Many researchers have called upon clathrate hydrates to explain the anomalous properties of mixed molecular ices in various astrophysical environments, in particular comets.[1] However, there has never been a satisfactory explanation as to how clathrate hydrates could form under the pressure/temperature conditions extant in small icy bodies or in the vacuum of space. In fact, evidence has been presented for the formation of clathrate hydrates from vapor-deposited amorphous ices by indirect methods (i.e., IR spectroscopy) by Devlin and co-authors.[2,3] More recently it has been shown by electron diffraction and direct imaging in an electron microscope that type II clathrate hydrates can also form under high vacuum conditions from astrophysically relevant amorphous ice precursors.[4]

2. Experimental Results

In the experiments described in (4), either $H_2O:CH_3OH$ (2:1) or $H_2O:CH_3OH$ (20:1) was vapor-deposited under high vacuum onto a thin carbon substrate at 85 K inside a Transmission Electron Microscope (TEM). The deposit remained amorphous until about 120 K when a solid state phase separation occurred, producing a type II clathrate hydrate of

CH$_3$OH and a second amorphous phase containing the molecules which would not fit into the clathrate structure. At higher temperatures (~145 K) the amorphous grain boundary phase sublimed leaving a porous network of clathrate. At ~150 K, the clathrate hydrate decomposed leaving porous hexagonal ice which sublimed at slightly higher temperatures in the vacuum of the TEM. When similar ice compositions were vapor-deposited and annealed in a cryogenic infrared (IR) spectrometer, only small shifts in absorption maxima were seen at and above the clathrate crystallization temperature. However, when 1% CO$_2$ was added as a local probe, the ν_3 asymmetrical stretching vibration of CO$_2$ was seen to shift from 2340 cm^{-1} to 2346 cm^{-1} during clathrate crystallization. This shift is identical to that reported for CO$_2$ contained within the small cages of type II clathrate hydrates.[3]

3. Discussion and Conclusions

The above solid state reactions have important ramifications with regard to the mechanical and gas release properties of cometary ices. First, the initial crystallization event would enclathrate some of the guest species present in the precursor amorphous ice, and sweep the remaining neutrals and radicals to grain boundaries where they would be free to recombine. The energy released by crystallization and by the recombination of radicals could cause the anomalous evolution of gases at low temperatures. If enough foreign molecules were present in the original amorphous ice, a microporosity could develop which would allow escape of gases from deep within the nucleus, independent of solid state diffusion processes. Second, type II clathrate hydrates, once formed, can hold up to 12 mole % small molecules in the two small cage sites present for each large cage site in the structure. Thus, type II clathrate hydrates, once formed, could enclathrate small molecules which percolated up from the interior during warming and annealing of deeper ice strata. All of these molecules would then be released upon the decomposition of the clathrate hydrate. Third, the mechanical properties of the clathrate hydrate will undoubtedly be different than the precursor amorphous ice. It is quite possible that surface exfoliation and mass shedding may be influenced by the presence / absence of clathrate hydrate. Last, the observed shift in the ν_3 asymmetrical stretching vibration of CO$_2$ opens the possibility to remotely observe clathrate hydrates using the IR signatures of their guest species.

4. References

1. A. H. Delsemme and P. Swings, *Ann. d'Ap.* **15**, 1 (1952). A. H. Delsemme and D. C. Miller, *Planet. Space. Sci.* **18**, 709 (1970). A. H. Delsemme and D. C. Miller, *Planet Space Sci.* **18**, 717 (1970). S. L. Miller, in *Physics and Chemistry of Ice*, E. Whalley et al., Eds. (Royal Society of Canada, Ottawa, 1973), pp. 42-50. J. I. Lunine and D. J. Stevenson, *Astrophys. J. Suppl.* Ser. **58**, 493 (1985).
2. J.E. Bertie and J. P. Devlin, *J. Chem. Phys.* **78**(10), 6340 (1983). H. H. Richardson, P. J. Wooldridge, J. P. Devlin, *J. Chem. Phys.* **83**(9), 4387 (1985).
3. F. Fleyfel and J. P. Devlin, *J. Phys. Chem.* **95**, 3811 (1991).
4. D. Blake, L. Allamandola, S. Sandford, D. Hudgins and F. Freund, *Science*, in press.

OBSERVATIONS OF PARENT MOLECULES IN COMETS AT RADIO WAVELENGTHS: HCN, H_2S, H_2CO AND CH_3OH

P. COLOM[1], D. BOCKELEE-MORVAN[1], J. CROVISIER[1],
D. DESPOIS[2], G. PAUBERT[3]
[1]Observatoire de Paris, Section de Meudon, F-92195 Meudon, France
[2]Observatoire de Bordeaux, BP 89, Avenue Pierre Sémirot, F-33270 Floirac, France
[3]IRAM, Avenida Divina Pastora, 7, N.C., E-18012 Granada, Spain

ABSTRACT. We present observations of cometary parent molecules at the IRAM radio telescope which led to the first detections of H_2S and CH_3OH in comets, and confirmed the presence of H_2CO and HCN. Production rates and abundances relative to H_2O are given.

Comets P/Brorsen-Metcalf (1989 X), Austin (1989c1) and Levy (1990c) were observed on September 2-7 1989, May 21-25 and August 26-31 1990, respectively. The observations [1] were performed with the IRAM (Institut de Radio Astronomie Millimétrique) 30-m telescope at Pico Veleta (Spain). The J(1-0) 89 GHz and J(3-2) 266 GHz transitions of HCN were marginally detected in P/Brorsen-Metcalf, whereas clear detections were obtained in comets Austin and Levy. The 3_{12}-2_{11} 226 GHz line of H_2CO was searched for and detected in the three comets [2]. The observation of H_2S 1_{10}-1_{01} at 169 GHz in comet Austin led to the first detection of hydrogen sulfide in a comet [3]. Cometary H_2S was confirmed in Levy, through both the 169 GHz and the 2_{20}-2_{11} 217 GHz lines [4, 5]. Methanol was identified in comet Austin through several J(2-1) ΔK = 0 (97 GHz) and J(3-2) ΔK = 0 (145 GHz) rotational transitions [1,3]. A dozen of CH_3OH lines around 97, 145, 165 and 218 GHz were detected in comet Levy [1].

HCN, H_2CO and H_2S production rates were derived from the observed line intensities using models treating the evolution of the excitation conditions from the collision-dominated region (inner coma: collisions with H_2O, $\sigma = 10^{-14}$ cm^2, T_{kin} = 50 K) to the radiation-dominated region (outer coma: IR excitation of the vibrational bands by the Sun). For CH_3OH, we assume LTE and use a rotational temperature of 30 K, in agreement with the observed relative line intensities. For the density distribution we assume isotropic outflow from the nucleus at constant velocity of 0.8 km s^{-1} and take into account the molecular lifetimes against photodissociation. The results are summarized in the table.

HCN seems to be more abundant by at least a factor of two in periodic comets (P/Halley, P/Brorsen-Metcalf) than in non-periodic comets (Wilson, Austin, Levy). This suggests a chemical difference between periodic and new comets. Production rates inferred with the assumption of release from the nucleus show that formaldehyde is a minor component of the nucleus with an abundance relative to water which ranges from 4×10^{-4} (Levy) to 4×10^{-3} (P/Brorsen-Metcalf) [2]. These abundances are at least an order of magnitude less than the IKS Vega value (4%) [6]. H_2S is a minor component, with a relative abundance of 2×10^{-3}. The other sulfur-bearing molecules observed (SO_2, OCS, H_2CS) are less abundant than hydrogen sulfide [4, 5]. Implications for Solar System formation are given in [5, 7]. Methanol is a substantial component of the nucleus, with a relative abundance of the order of 1% in comets Austin and Levy.

TABLE : Production rates and abundances.

Molecule	Line	Frequency GHz	Date	Q^a [s^{-1}]	$Q/Q[H_2O]^b$
P/Brorsen-Metcalf (1989 X)					
HCN	J(1-0)	88.6	89/09/04-07	4.5×10^{26}	1.8×10^{-3}
H_2CO	3_{12}-2_{11}	225.7	89/09/04-07	1.1×10^{27}	4.4×10^{-3}
Austin (1989c1)					
HCN	J(1-0)	88.6	90/05/23	2.0×10^{25}	5.0×10^{-4}
H_2CO	3_{12}-2_{11}	225.7	90/05/21-25	4.6×10^{25}	1.1×10^{-3}
H_2S	1_{10}-1_{01}	168.8	90/05/24-25	1.1×10^{26}	2.7×10^{-3}
CH_3OH	(3,0)-(2,0)A	145.1	90/05/25	4.9×10^{26}	1.2×10^{-2}
Levy (1990c)					
HCN	J(1-0)	88.6	90/08/29	6.6×10^{25}	2.6×10^{-4}
H_2CO	3_{12}-2_{11}	225.7	90/08/26-30	1.0×10^{26}	4.0×10^{-4}
H_2S	1_{10}-1_{01}	168.8	90/08/30-31	5.0×10^{26}	2.0×10^{-3}
CH_3OH	(3,0)-(2,0)A	145.1	90/08/27	1.8×10^{27}	7.2×10^{-3}
HC_3N	J(24-23)	218.3	90/08/27	$< 1.2 \times 10^{25}$	$< 5.0 \times 10^{-5}$
SO_2	7_{17}-6_{06}	165.2	90/08/29	$< 6.0 \times 10^{26}$	$< 2.5 \times 10^{-3}$
OCS	J(18-17)	218.9	90/08/28	$< 5.0 \times 10^{26}$	$< 2.0 \times 10^{-3}$
H_2CS	4_{14}-3_{13}	135.3	90/08/28	$< 2.5 \times 10^{26}$	$< 1.0 \times 10^{-3}$

a. Assuming a parent distribution; b. $Q[H_2O]$ from OH-18 cm observations.

References

[1] Bockelée-Morvan D., Crovisier J., Colom P., Despois D., Paubert G. (1990), *Observations of parent molecules in comets P/Brorsen-Metcalf (1989o), Austin (1989c1) and Levy (1990c) at millimetre wavelengths: HCN, H_2S, H_2CO and CH_3OH*, Proceedings of the 24th ESLAB Symp. "The Formation of Stars and Planets, and the Evolution of the Solar System", Friedrichshafen, 17-19 Sept. 1990, ESA **SP-315**, 243-248
[2] Colom P., Crovisier J., Bockelée-Morvan D., Despois D., Paubert G. (1992), *Formaldehyde in comets: microwave observations in P/Brorsen-Metcalf (1989 X), Austin (1989c1) and Levy (1990c)*, Astron. and Astrophys. (in press)
[3] Bockelée-Morvan D., Colom P., Crovisier J., Despois D., Paubert G. (1991), *Microwave detection of hydrogen sulfide and methanol in comet Austin (1989c1)*, Nature **350**, 318-320
[4] Crovisier J., Despois D., Bockelée-Morvan D., Colom P., Paubert G. (1991), *Microwave observations of hydrogen sulfide and searches for other sulfur compounds in comets Austin (1989c1) and Levy (1990c)*, Icarus (in press)
[5] Despois D., Crovisier J., Bockelée-Morvan D., Colom P. (1991), *Formation of comets: constraints from the abundance of hydrogen sulfide and other sulfur-bearing molecules*, IAU Symposium 150, this volume
[6] Combes et al. (1988), *The 2.5-12 µm spectrum of comet Halley from the IKS-VEGA Experiment*, Icarus **76**, 404-436
[7] Despois D. (1991), *Solar System - Interstellar Medium, a chemical memory of the origins*, IAU Symposium 150, this volume

RADIO INTERFEROMETRIC OBSERVATIONS OF COMETARY MOLECULES

Patrick Palmer[1], Lewis E. Snyder[2], and Imke de Pater[3]

[1]Department of Astronomy & Astrophysics, University of Chicago; [2]Department of Astronomy, University of Illinois; [3]Department of Astronomy, University of California, Berkeley

ABSTRACT. A useful method for extracting cometary signals is demonstrated using VLA observations of Comet Brorsen-Metcalf (1989o).

1. Introduction

The VLA has detected the 1_{11}-1_{10} transition of H_2CO at 4829.659 MHz (6cm) from comets Halley, Machholz (1988j), and possibly Brorsen-Metcalf (1989o) (Snyder, Palmer, and de Pater, 1989; 1990). Using the Brorsen-Metcalf observations, we will demonstrate one of the advantages of interferometers over single-element radio telescopes for extracting weak spectral lines.

2. Observations and Discussion

The VLA D configuration was used to search Brorsen-Metcalf for H_2CO on 1989 Sept. 2, 3, and 7, with the spectrometer arrangement described by Snyder et al. (1989). The untapered beam was 20"x20", and the tapered 40"x40". Figure 1 depicts a data cube of H_2CO images for each velocity channel. By cutting through the cube along a pixel (which is at a fixed position relative to the comet), a spectrum of intensity versus cometocentric velocity for that position is obtained. Averaging over the appropriate pixels simulates the synthesized VLA beam. A pixel (shaded square) is 10"x10" in tapered, and 5"x5" in untapered data. Blocks of 3x3, 5x5, and 9x9 pixels were used for the data reductions. Within the field of view, different pixel clusters can be sampled to optimize the coupling of the synthesized beam to the gas distribution. Thus small pointing errors caused by ephemeris inaccuracies can be corrected after the observations have been made; this would be impossible to do with single-element telescope data. Gérard's (1987) SYMCOMET routine was applied to different pixel "cuts" through the Brorsen-Metcalf data cube, and the resulting symmetrical spectra are in Figure 2. Both the 5x5 tapered and untapered spectra show the best coupling of the beam to the H_2CO cloud, with a peak at channel 13 (cometocentric radial velocity = +0.76 km/s). This suggests that the boundaries of the coma gas best match the 5x5 pixel block.

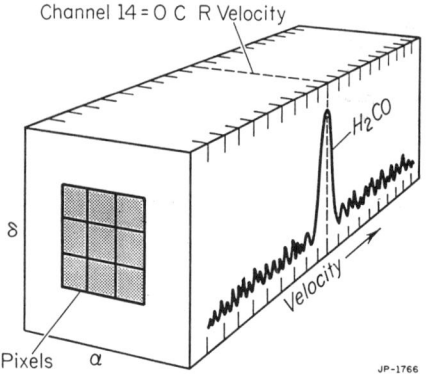

Figure 1. Representation of a data cube (images for each velocity channel as a function of α and δ). Here channel 14 is at 0 km/s cometocentric radial velocity.

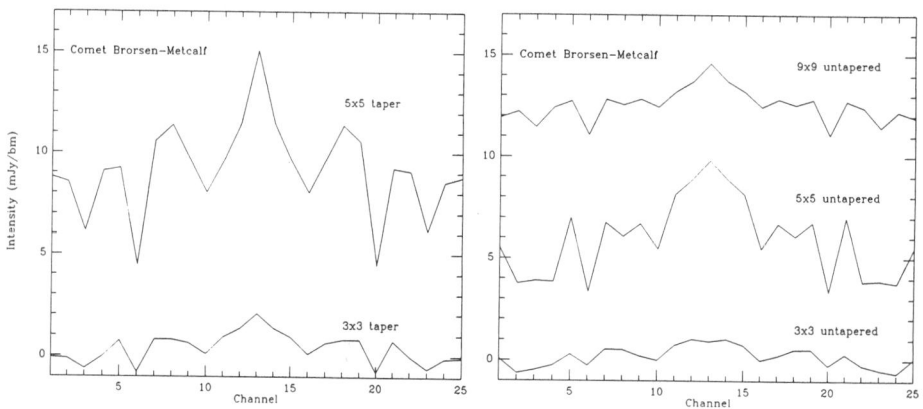

Figure 2. Spectra of the 6 cm H_2CO line obtained from applying SYMCOMET to different "cuts" through the Brorsen-Metcalf data cube. Ordinate: intensity in mJy/bm; abscissae: channel number (direction opposite to velocity). Channel 14 is 0 km/s and the channel width is 0.76 km/s.

This work was supported by NASA NAGW-1131 and NSF AST-8900156.

5. References

Gérard, E. (1987) 'The OH Radio Lines in Comets: A Review' in W. M. Irvine, F. P. Schloerb, and L. E. Tacconi-Garman (eds.), Cometary Radio Astronomy, NRAO Workshop 17, Green Bank, WV, pp. 91-9.

Snyder, L. E., Palmer, P., and de Pater, I. (1989) 'Radio Detection of Formaldehyde Emission from Comet Halley', AJ 97, 246-53.

Snyder, L. E., Palmer, P., and de Pater, I. (1990) 'Observations of Formaldehyde in Comet Machholz (1988j)', Icarus 86, 289-98.

GAS PRODUCTION RATES IN COMETS

A.A. DE ALMEIDA
Department of Astronomy, Inst. of Astronomy and Geophysics,
University of São Paulo
C.P. 9638, CEP 01065, São Paulo - Brazil

ABSTRACT. *Emission fluxes of CN, C_2 and C_3 carbon- bearing molecular species observed in the coma of comets Bennett (1969i $\equiv$ 1970II), West (1975n $\equiv$ 1976VI), P/Halley (1982i), Hartley-Good (1985ℓ) and Bradfield (1987s) are analysed in the framework of Haser model. CN, C_2 and C_3 production rates are determined using recently derived fluorescence efficiencies. The dependence of CN, C_2 and C_3 production rates on the heliocentric distance and the possible correlations among these radicals is studied and briefly discussed.*

INTRODUCTION

As a part of the systematic programme - Determination of Gas and Dust Production Rates in Comets (GDPC) - going on in this laboratory (Almeida et al., 1989; Almeida, 1991), the emission band fluxes of cometary CN, C_2 and C_3 carbon bearing molecular species observed in the coma of comets Bennett (1969i $\equiv$ 1970II) (Babu and Saxena, 1972) West (1975n $\equiv$ 1976 VI) (Sivaraman et al., 1979), P/Halley (1982i) (Goraya et al., 1989a; 1989b), Hartley-Good (1985ℓ) (Rautela et al., 1989) and Bradfield (1987s) (Rautela and Sanwal, 1988; Ojha and Joshi, 1989) are analysed. These comets are studied in the framework of the Haser model (1957) by using recently derived fluorescence efficiencies (g-factors) at $r_h = 1$ AU (Almeida et al., 1989) and appropriate numerical parameters which are known for the more prominent species CN, C_2 and C_3 (Cochran, 1985). Haser model analysis of CN, C_2 and C_3 production rates are developed for these comets and the possible logarithimic correlations among these molecular species and comets graphically analysed.

DISCUSSION

The derived CN, C_2 and C_3 production rates using recently derived fluorescence efficiencies (Almeida et al., 1989) may have a systematic error amounting to about ± 20 percent and where possible these results were compared with the ones found in the literature. Equations 1 of Almeida et al. (1989) and Konno and Wyckoff (1989), derived independently, are equivalent within this uncertainty.

In the particular case of West (1975n $\equiv$ 1976VI) since the comet broke up in four different parts (A $\longrightarrow$ D: 1976 Feb. 19.1 ± 0.2; A $\longrightarrow$ B: 1976 Feb. 27.7 ± 0.2;

A ⟶ C: 1976 Mar. 6.5 ± 0.3 (where A is the principal fragment and B, C, D are secondary fragments (Sekanina, 1982)), the expected errors should be at least twice as much since the data used (Sivaraman et al., 1979) corresponds to observations performed after the primary nucleus (A) of the comet had splited in these four fragments close to perihelion passage (February 25.22, 1976 (UT)). Surprisingly the results for comet West (1975n ≡ 1976VI) correlates very well with the ones obtained for the other four comets (see for instance Figure 2), and particularly with comet P/Halley (1982i) for CN and C_2. This might suggest that Sivaraman et al. (1979) tracked mainly the principal fragment (A) of the nucleus during their observations.

From Figures 1 to 4 one can easily conclude that as far as the global production rates are concerned it decreases for CN and C_2 and increases for C_3 with the heliocentric distance. Hartley-Good (1985 ℓ) is the comet that shows the least C_3 production rate, compared to CN and C_2 among the five comets considered in this study.

Acknowledgments: The author is thankful to Mr. Julio C. Klafke for efficient assistance in plotting the graphs and to CNPq for financial support under grant No. 306304/88-0.

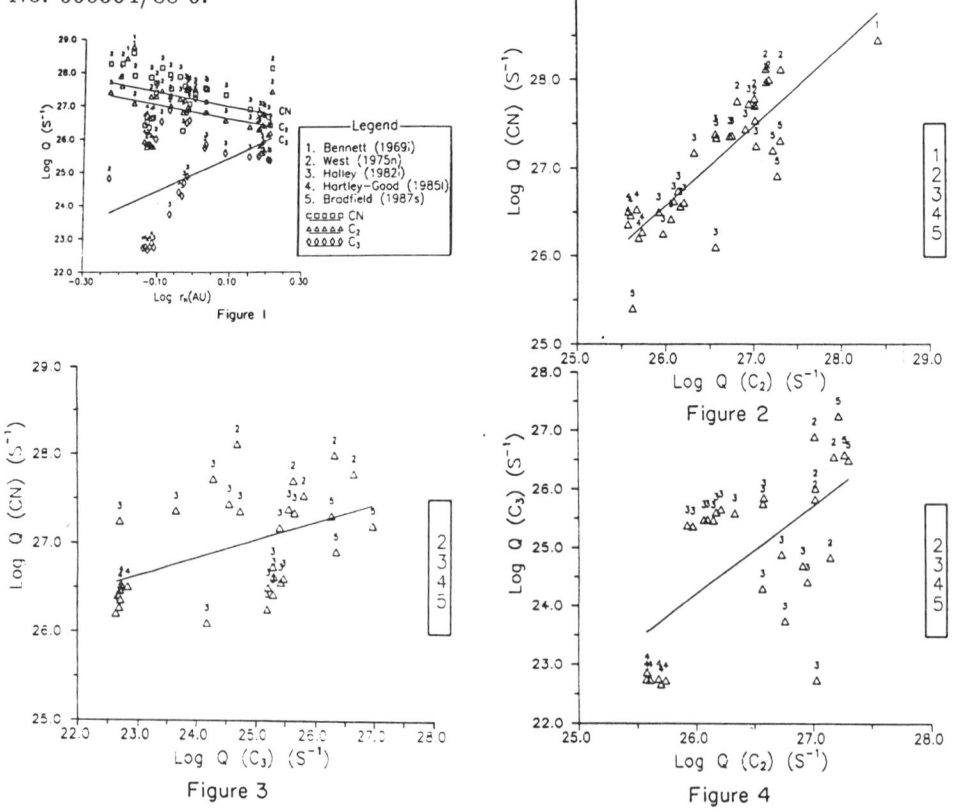

Figure 1

Figure 2

Figure 3

Figure 4

Figure 1 - Global production rates of CN, C_2 and C_3 molecules as a function of heliocentric distance for comets, Bennett (1969i $\equiv$ 1970II), West (1975n $\equiv$ 1976VI), P/Halley (1982i), Hartley-Good (1985ℓ) and Bradfield (1987s). The straight lines represent the logarithimic correlations.

Figure 2 - Logarithimic correlation between production rates of CN and C_2 molecules for comets Bennett (1969i $\equiv$ 1970II), West (1975n $\equiv$ 1976VI), P/Halley (1982i), Hartley-Good (1985ℓ) and Bradfield (1987s).

Figure 3 - Logarithimic correlation between production rates of CN and C_3 molecules for comets Bennett (1969i $\equiv$ 1970II), West (1975n $\equiv$ 1976VI), P/Halley (1982i), Hartley-Good (1958ℓ) and Bradfield (1987s).

Figure 4 - Logarithimic correlation between production rates of C_2 and C_3 molecules for comets Bennett (1969i $\equiv$ 1970II), West (1975n $\equiv$ 1976VI), P/Halley (1982i), Hartley-Good (1985ℓ) and Bradfield (1987s).

References

Almeida, A.A., Singh, P.D. and Burgoyne, C.M. (1989) 'Haser Model CN, C_2 and C_3 Production Rates in Some Comets', Earth, Moon and and Planets **47**, 15-31.

Almeida, A.A. (1991), 'An Analysis of the Spectrophotometric and Photometric Observations of Comets 1967II, 1968I, 1968V, and 1968VI', Earth, Moon and Planets (in press).

Babu, G.S.D. and Saxena, P.P. (1972), 'Spectrophotometry of Comet Bennett', Bull. Astron. Inst. Czech **23**, 346-349.

Cochran, A.L. (1985), 'A Re-Evaluation of the Haser Model Scale Lengths for Comets', Astron. J. **90**, 2609-2614.

Goraya, P.S., Sanwal, B.B., Rautela, B.S., Duggal, H.K. and Malhi, J.S. (1989a), 'Spectrophotometry of P/Halley (1982i)', Earth, Moon and Planets **44**, 243-249.

Goraya, P.S., Rautela, B.S., Sanwal, B.B., Gupta, S.K., Duggal, H.K., and Malhi, J.S. (1989b), Spectrophotometric Study of Periodic Comet Halley (1982i)'Earth, Moon and Planets **45**, 17-27.

Konno, I. and Wyckoff, S. (1989), 'Atomic and Molecular Abundances in Comet Giacobini-Zinner', Adv. Space Res. **9**, 163-168.

Ojha, D.K. and Joshi, S.C. (1989), 'Spectrophotometry of Comet Bradfield (1987s)', Earth, Moon and Planets **44**, 1-5.

Rautela, B.S. and Sanwal, B.B. (1988), 'Spectrophotometric Study of Comet Bradfield (1987s)', Earth, Moon and Planets **43**, 221-225.

Rautela, B.S., Goraya, P.S., Sanwal, B.B., and Gupta, S.K. (1989), 'Spectrophotometry of Comet Hartley-Good (1985ℓ)', Earth, Moon and Planets **44**, 233-242.

Sekanina, Z. (1982), 'The Problem of Split Comets in Review'in L.L. Wilkening (ed.), Comets, The University of Arizona Press, Tucson, pp. 251-287.

Sivaraman, K.R., Babu, G.S.D., Bappu, M.K.V. and Parthasarathy, M. (1979), 'Emission band and continuum photometry of Comet West (1975n)-I. Heliocentric dependence of the flux in the emission bands and the continuum', Mon. Not. R. ast. Soc. **189**, 897-906.

MASS LOSS RATES OF THREE COMETS

P.D. Singh[1], W.F. Huebner[2], D.C. Boice[2], I. Konno[2], and E. Scalise, Jr.[3]
[1]*Inst. Astronômico e Geofísico, Universidade de São Paulo, CP 9638, São Paulo, Brazil*
[2]*Southwest Research Inst., 6220 Culebra Rd., San Antonio, TX 78228-0510, USA*
[3]*Dept. of Astrophysics, INPE, Av. dos Astronautas, São Jose dos Campos, SP, Brazil*

ABSTRACT. Emission features of C_2, C_3, CN, and dust in Comets Thiele (1985m), Hartley-Good (1985l), and Giacobini-Zinner (1984e) have been analyzed and their mass loss rates of about 0.5, 1.1, and 0.8 Mg s^{-1} have been determined.

Photoelectric photometry of Comets Thiele (1985m), Hartley-Good (1985l) and Giacobini-Zinner (1984e) in the system of standard IHW filters was performed at the 90-cm telescope of Lena University by Stecklum et al. (1987). They reported magnitude measurements at 387.1, 406.0, and 514.0 nm corresponding to CN, C_3, and C_2 emissions and at 365.0 and 484.5 nm continuum in CU and CB filters. The continuum brightness m_{em}^{cont} in the molecular emission filters can be interpolated from "short-" and "long-wavelength" filter magnitudes m(CU) and m(CB)

$$m_{em}^{cont}(387.1 \text{ nm}) = 0.8151\, m(CU) + 0.1849\, m(CB) + 0.494,$$
$$m_{em}^{cont}(514.0 \text{ nm}) = -0.2469\, m(CU) + 1.2469\, m(CB) + 0.244, \qquad (1)$$
$$m_{em}^{cont}(406.0 \text{ nm}) = 0.6569\, m(CU) + 0.3431\, m(CB) + 0.088.$$

The filter magnitudes m_{em} at 514.0, 406.0 and 387.1 nm were converted into emission band fluxes F_{em} [erg $cm^{-2} s^{-1}$] using the reduction formula

$$F_{em}(\lambda) = (D + E \cdot T) \cdot \left(10^{-0.4 m_{em}} - 10^{-0.4 m_{em}^{cont}}\right). \qquad (2)$$

Here T denotes the filter temperature in °C assumed to be 0, D = 5.38 · 10^{-7}, 1.38 · 10^{-6}, and 6.81 · 10^{-7} for CN, C_3, and C_2, respectively; corresponding values of E are -0.021 · 10^{-7}, 0.003 · 10^{-6}, and 0.0. For the column densities of CN, C_2, and C_3, we took g-factors at r = 1 AU from Tatum (1984), Landaberry et el. (1991), and de Almeida et al. (1989) combined with an r^{-2} power law. The magnitude measurements of Stecklum et al. (1987) at wavelength 484.5 nm for scattering by coma dust were converted into continuum fluxes by Singh et al. (1991).

Comet Hartley-Good (1985l): Since OH is a photodissociation product of H_2O, the Tacconi-Garman et al. (1990) vector model OH production rates correspond to an average H_2O production rate $Q(H_2O) \approx 3.2 \cdot 10^{28}$ s^{-1} when interpolated to r = 1 AU. However, since the vector model gives production rates about 50% higher than the Haser model (Schleicher et al., 1987), the Haser model $Q(H_2O) \approx 2.1 \cdot 10^{28}$ s^{-1} at r = 1 AU, which is in agreement with 2.8 · 10^{28} s^{-1} derived by Singh et al. (1991) considering Hartley-Good as a "normal" comet [i.e. $Q(C_2)/Q(CN) \approx 1.4$]. Assuming a gas mixture of 90% H_2O and 10% other gases of mean molecular weight 44 amu, we find that early in November, 1985, the comet was loosing $\sim 9.8 \cdot 10^5$ g s^{-1} gas and $\sim 1.4 \cdot 10^5$ g s^{-1} dust, for a total mass loss of about 1.1 Mg s^{-1} at r = 1 AU preperihelion.

Comet Giacobini-Zinner (1984e): $Q(C_2)/Q(CN) \approx 0.3$ on August 14 and 29, 1985. In this comet C_2 is depleted by a factor of 5 relative to CN [$Q(C_2)/Q(CN) \approx 1.5$ in "normal" comets (Cochran, 1987)]. For Comet Giacobini-Zinner, $Q(CN)/Q(H_2O) \approx 1.8 \cdot 10^{-3}$ (Landaberry et al., 1991). Thus the CN production rate on August 14, 1985, assuming an expansion velocity of 1 km s^{-1}, yields a Haser model $Q(H_2O) \approx 4.5 \cdot 10^{28}$ s^{-1} at r = 1.08 AU. Tacconi-Garman et al. (1990) determined $Q(OH) \approx 5.15 \cdot 10^{28}$ s^{-1} during August 24 to 26, 1985, (r $\approx$ 1.04 AU) using the vector model, which corresponds to a Haser model $Q(H_2O) \approx 4 \cdot 10^{28}$ s^{-1} at r $\approx$ 1.04 AU, in agreement with our value. Landaberry et al. (1991) derived $Q(H_2O) \approx 1.6 \cdot 10^{28}$ s^{-1}. Since H_2O is a major constituent of the coma, the gas production rate is $\sim 4.7 \cdot 10^5$ g s^{-1} at r

= 1.08 AU. The dust production rate on August 14, 1985, is $\sim 3.5 \cdot 10^5$ g s^{-1} (Singh et al., 1991). Thus the comet was loosing a total mass of ~ 0.8 Mg s^{-1} at r = 1.08 AU before perihelion.

Comet Thiele (1985m): Tacconi-Garman et al. (1990) determined a Q(OH) $\approx 1.37 \cdot 10^{28}$ s^{-1} from the November 12 and 14 to 18, 1985, observations (r $\approx$ 1.41 AU). This Q(OH) corresponds to a Haser model Q(H$_2$O) $\approx 3.2 \cdot 10^{28}$ s^{-1} at r = 1 AU when extrapolated by an r^{-2} power law. This is a factor of ~ 2 lower than the H$_2$O production rate (6 $\cdot 10^{28}$ s^{-1}) derived by Singh et al. (1991) at r = 1 AU, which was derived from a peak OH production rate of 3 $\cdot 10^{28}$ s^{-1} (December 9, 1985, r = 1.3 AU; Gérard et al., 1987) and is an upper limit. On November 4, 1985 Q(C$_2$)/Q(CN) ≈ 1.1 and shows that the comet belongs to the family of "normal" comets (Cochran, 1987). In a "normal" comet Q(CN)/Q(H$_2$O) $\approx 1.33 \cdot 10^{-3}$ (Spinrad, 1987; Newburn and Spinrad, 1989). The November 4, 1985, observations of Comet Thiele by Stecklum et al. (1987) show a mean Haser model Q(CN) $\approx 1.6 \cdot 10^{25}$ s^{-1} at r = 1.48 AU and hence the extrapolated Haser model Q(H$_2$O) $\approx 1.2 \cdot 10^{28}$ s^{-1} at r = 1.48 AU. If Q(H$_2$O) follows an r^{-2} power law, then the Haser model Q(H$_2$O) $\approx 2.6 \cdot 10^{28}$ s^{-1} at r = 1 AU. Since the vector model yields production rates about 50% higher than the Haser model, we consider our derived water production rate of $3.9 \cdot 10^{28}$ s^{-1} at r = 1 AU in good agreement with the extrapolated value of $3.2 \cdot 10^{28}$ s^{-1} obtained from the Tacconi-Garman et al. (1990) observations. Following the analyses of the above comets, our extrapolated Haser model H$_2$O production rate yields a gas production rate of $\sim 4.1 \cdot 10^5$ g s^{-1} at r $\approx$ 1.48 AU. The dust production rate on November 4, 1985, was $\sim 7.8 \cdot 10^4$ g s^{-1} (Singh et al., 1991). Thus the comet was loosing a mass of ~ 0.5 Mg s^{-1} at r = 1.48 AU.

TABLE 1. CN, C$_2$, & C$_3$ FLUXES & COLUMN DENSITIES

Date 1985	Time (UT)	r (AU)	Δ (AU)	Species	Flux (erg cm^{-2} s^{-1})	Col. density (cm^{-2})
				Comet Hartley-Good (1985l)		
Nov. 3	18.54	1.00	0.73	CN	4.77(-11)	7.966(10)
				C$_2$	4.25(-11)	7.916(10)
				C$_3$	1.42(-11)	3.720(10)
Nov. 4	18.23	0.98	0.74	CN	6.07(-11)	9.604(10)
				C$_2$	5.45(-11)	9.747(10)
				C$_3$	2.03(-11)	5.110(10)
				Comet Thiele (1985m)		
Nov. 4	00.36	1.48	0.528	CN	6.93(-12)	2.777(10)
				C$_2$	3.89(-12)	1.587(10)
				C$_3$	2.93(-12)	1.681(10)
	19.37			CN	4.22(-12)	1.691(10)
				C$_2$	4.11(-12)	1.677(10)
				C$_3$	4.04(-12)	2.317(10)
	19.48			CN	6.39(-12)	2.561(10)
				C$_2$	3.19(-12)	1.302(10)
				C$_3$	4.03(-12)	2.312(10)
				Comet Giacobini-Zinner (1984e)		
Aug 14	23.25	1.08	0.53	CN	8.66(-11)	1.741(11)
				C$_2$	1.51(-11)	3.281(10)
Aug 29	02.16	1.03	0.48	CN	2.66(-11)	6.143(10)
				C$_2$	5.33(-12)	1.053(10)
				C$_3$	6.05(-13)	1.481(09)

REFERENCES

Cochran, A.L. 1987, Astron. J. 93, 231-238.
de Almeida, A.A., Singh, P.D., and Burgoyne, C.M. 1989, Earth Moon Planets 47, 15-31.
Gérard, E., Bockelée-Morvan, D., Bourgois, G., Colom, P., and Crovisier, J. 1987, in <u>Cometary Radio Astronomy</u>, NRAO workshop, Eds., W.M. Irvine, F.P. Schloerb, and L.E.Tacconi-Garman, p. 125-133.
Landaberry, S.J.C., Singh, P.D., and de Freitas Pacheco, J.A. 1991, Astron. Astrophys., 246, 597.
Newburn Jr., R.L., and Spinrad, H. 1989, Astron. J. 97, 522-569.
Schleicher, D.G., Millis, R.L., and Birch, P.V., 1987, Astron. Astrophys. 187, 531-538.
Singh, P.D., de Almeida, A.A., and Huebner, W.F. 1991. Astron. J., in press.
Spinrad, H. 1987, Ann. Rev. Astron. Astrophys. 25, 231-269.
Stecklum, B., Pfau, W. and Hesse, M. 1987, Astron. Nachr. 308, 239-246.
Tacconi-Garman, L.E., Schloerb, F.P., and Claussen, M.J. 1990, Astrophys. J. 364, 672-686.
Tatum, J.B., 1984, Astron. Astrophys. 135, 183-187.

This research was supported by grants from NASA (Nos. NAGW-2205, NAGW-2370), FAPESP (No. 90/1384-3), CNPq (No. 304076/77), and NSF (No. INT-8901811).

A MODEL OF P/TEMPEL 2 WITH DUST AND DETAILED CHEMISTRY

W.F. HUEBNER[1], D.C. BOICE[1], I. KONNO[1], and P.D. SINGH[2]
[1] *Southwest Research Institute, San Antonio, TX 78228-0510 U.S.A.*
[2] *Instituto Astronômico e Geofísico, Universidade de São Paulo,
C.P. 9638, São Paulo, S.P. Brazil*

ABSTRACT. We apply our fluid dynamic model with chemical kinetics of dusty comet comae to P/Tempel 2. A brief summary of results concerning gas/dust dynamics and chemistry is given.

One-dimensional, multi-fluid simulations of the coma of P/Tempel 2 at perihelion have been performed. These simulations are based on our model that treats the physics and chemistry of the inner coma in great detail as summarized by Schmidt et al. (*Comp. Phys. Comm.* **49**, 17-59, 1988). Recent progress of the model includes incorporation of dust entrainment by the gas, dust size distributions, dust fragmentation, distributed coma sources of gas-phase species related to the dust, and a separate accounting of the electron energetics. An improved calculation for the sublimation rate of water from the nucleus is now performed and the chemical reaction network has been expanded to include CH_3OH, H_2S, and H_2CO_2 as possible parent molecules. The assumed volatile composition is given in Table 1 and other model parameters are listed in Table 2.

The model species profiles are appropriate for neutrals throughout the coma and for ions within the contact surface. Figure 1 illustrates that methanol is a distributed source of H_2CO upon photodissociation with a lifetime of $1.96 \cdot 10^4$ s (quiet sun). In turn, H_2CO forms a distributed source for CO with a lifetime of $5.15 \cdot 10^3$ s. Several organic species, as well as the dust, contribute to enhance the coma abundance of CH and extend its range. As shown in Fig. 2, protonated parent molecules may be major ionic species in the inner coma if the parent abundances in the nucleus are $\gtrsim 0.5\%$ with proton affinities greater than that of water (e.g., NH_3, CH_3OH, H_2CO, H_2CO_2). It can be seen that the abundance of a protonated parent species is higher than the respective parent ion in the inner coma ($R \lesssim 500$ km). For illustrative purposes, we have set the abundance of H_2CO_2 rather high (1%) in the nucleus composition which leads to its protonated ion $H_3CO_2^+$ being the most abundant ion in the inner coma. This demonstrates the sensitivity of the abundances of protonated species on the initial composition of these parents.

The dust mass distribution is approximated by a power law (exponent β) with 11 discrete sizes, logarithmically spaced between a_{min} and a_{max}. The model capabilities for dust fragmentation are not used in the present calculations. Gas and dust are rapidly accelerated upon leaving the nucleus. For standard dust densities, small particles are more efficiently entrained with the gas flow than large particles, resulting in higher terminal speeds. The acceleration zone for all particles is approximately within 10 R_{nuc}. A complete description of the model and results will be presented in a forthcoming publication.

This research was supported by grants from NASA (Nos. NAGW-2205, NAGW-2370), FAPESP (No. 90/1384-3), CNPq(No. 304076/77), and NSF (No. INT-8901811).

TABLE 1. Composition of P/Tempel 2 Model

Species	Number Abundance	Species	Number Abundance
H_2O	0.8692	C_2H_2	0.0066
CO	0.034	CS_2	0.005
CO_2	0.029	H_2S	0.005
CH_3OH	0.01	CH_3CN	0.0016
NH_3	0.01	NH_2CH_3	0.0008
H_2CO	0.01	POM_5	0.0008
H_2CO_2	0.01	HCN	0.0006
N_2	0.007	$H_2C_3H_2$	0.0004

TABLE 2. Model Parameters for P/Tempel 2

Symbol	Value	Symbol	Value
r_h	1.381 AU	Z	$2.9 \cdot 10^{17} cm^{-2} s^{-1}$
A	0.03	Q	$4.0 \cdot 10^{28} s^{-1}$
ϵ	0.97	Q_{water}	$3.5 \cdot 10^{28} s^{-1}$
R_{nuc}	1.5 km	ρ_{dust}	0.5 g cm^{-3}
A_{nuc}	14 km^2	χ	0.027
T_{nuc}	193.8 K	a_{max}	1 cm
ρ_{nuc}	0.5 g cm^{-3}	a_{min}	0.1 μm
		β	0.5

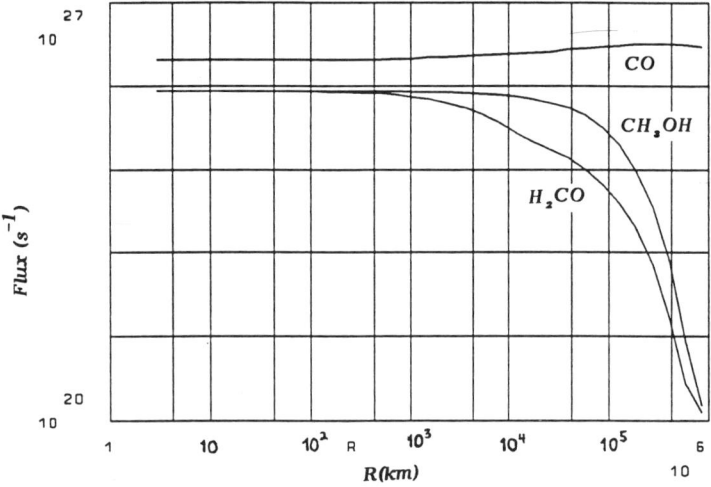

Figure 1. Profiles of the integrated flux for CO, CH_3OH, and H_2CO throughout the coma.

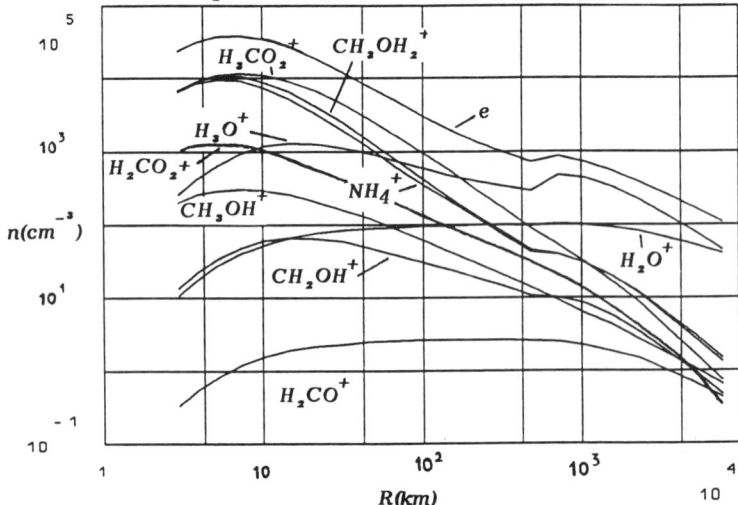

Figure 2. Number density profiles of electrons and ions related to selected parent molecules via protonization and ionization.

SOLAR SYSTEM - INTERSTELLAR MEDIUM

A Chemical Memory of the Origins

D. DESPOIS

Observatoire de Bordeaux, B.P. 89, 33270 Floirac, France

Abstract. The growing body of data on solar system objects and interstellar space provides us with new tests of the connection between the two. We emphasize here the role played by the study of comets through the properties of the dust, the chemical composition of volatiles and the elemental abundances. These data inform us on cometary matter formation, and hence on conditions in the protosolar nebula. Under the adopted scenario of formation in a cold environment, with little further processing, cometary abundances are even new constraints to interstellar (gas and solid phase) abundances. Several points specific to the chemical modelling of the collapsing cloud and of the protosolar nebula are listed.

1. Introduction

Due to the constant influx of new data from both ground-based research and spacecraft, one can trace better and better the primitive connection between the Interstellar Medium (ISM) and the Protosolar Nebula (PSN). We will focus here on the implications of recent results from cometary studies (see A'Hearn 1992 ; Snyder 1992), as comets are thought to be among the most pristine bodies in the solar system. We will further assume *as a working hypothesis* a close connection between interstellar and cometary matter, following Yamamoto and coworkers (1983,1985).

Recent reviews address related fields : meteorite properties (Meteorites and the Early Solar System, Kerridge and Matthews eds. 1988) ; giant planet atmospheres (T.Owen 1992) ; ice/rock ratio in Pluto and outer planets satellites (McKinnon and Mueller 1988). The very important issue of the D/H ratio and other isotopic ratios in the Solar System is addressed also by T.Owen (1992). Much information on cometary composition and its relation to ISM can be found in the recent works of Irvine and Knacke (1989), Yamamoto (1991), Encrenaz et al. (1991), Mumma et al. (1992), and in Comets in the Post-Halley Era (Newburn et al. eds. 1991).

Shu and Adams (1987) describe a possible scenario of the formation of solar-type stars. The four main steps are : the inside-out collapse of a gas condensation ; the formation of a central star surrounded by an accretion disk ; the emergence of a bipolar outflow perpendicular to the disk ; the final T Tauri stage where most of the gas has been blown out by the stellar wind. The history of cometary matter starts even earlier. Refractory grains form in the envelopes of evolved stars, and are eventually coated by ices, which are later processed to organic refractories when the grains are released into the diffuse interstellar medium. When the diffuse medium condenses into dense clouds, molecules begin to form a new ice mantle on the grain surface ; these mantles are believed to reevaporate partially into the gas phase due to some continuous or occasional phenomena (e.g. Tielens and Allamandola 1987, Williams 1990, Walmsley 1990).

In the collapse phase, the medium gets denser and denser, and condensation will probably dominate any reevaporation mechanism until the grain reaches the central regions. There, reheated in the accretion disk, it loses partially its mantle. The gas has now typically a density of 10^{13} cm^{-3}, and undergoes a quite specific chemistry. Some authors (e.g. Prinn and Fegley 1989) see in the gas of the protosolar nebula and the planetary subnebulae major contributors to cometary matter ; we will not discuss these models here (see Yamamoto

1991 for a presentation and references) and follow the hypothesis favoured by Yamamoto that the condensation of interstellar gas in a cold environment and the subsequent partial sublimation in the PSN are the main phenomena determining cometary composition. Grains grow also by coagulation to form cometesimals, which will aggregate to form cometary nuclei (e.g. Donn 1991). These nuclei, soon ejected by 3 body interactions with the planets toward the Oort cloud will endure only superficial alteration by UV and energetic particles from the Sun and the Galaxy during the 4 Gyr before their eventual return close to the Sun as an active comet.

To test the different models of solar system formation, comets provide us with the following tools : study of their orbits (likelihood of the formation of the Oort cloud and the Kuiper belt) ; isotopic abundances, especially D/H enrichment ; elemental abundances ; chemical abundances ; ortho/para ratios ; microstructure.

We concentrate below on the dust and volatile composition of comets. The dynamical aspects are presented by Weissman (1991) and isotopic ratios by T.Owen (1992). Species like H_2O, H_2CO, H_2S ... have two different hydrogen nuclear spin states (ortho and para) which are not easily converted one into the other and differ in energy ; the ratio of their abundances should thus inform us on the formation or last equilibration temperature of the molecule. The determination of this ratio has been attempted on H_2O in Halley (Mumma et al. 1988), but, according to the reanalysis made by Bockelée-Morvan and Crovisier (1990), there is no clear evidence of any departure from the value expected in the high temperature limit. Recent works on meteorites (see e.g. Kerridge and Matthews, eds. 1988) show that their microstructure is extremely rich in information, and a similar wealth is to be expected from the study of cometary samples.

2. Refractories : Interstellar versus cometary dust

Greenberg and Hage (1990) have proposed a detailed model of coma dust based on the interstellar dust model of Greenberg (1985). The larger interstellar grains, made of silicate cores surrounded by organic refractories, accrete molecules in a very cold region to form precometary grains. Many smaller grains, some made of carbon, others of silicates or PAHs get included in the molecular mantle. Precometary grains stick together to form a very porous structure ("bird's nest"). When the cometary matter is exposed to the sun in the active comet phase, volatiles and part of the organic refractories are released in the coma, where they can be observed ; the remaining solid particles form the cometary dust.

This model explains the low albedo observed in Halley's nucleus, the presence of many low-mass particles, the large organic component (CHON particles, e.g. Jenniskens et al. 1991), and the IR properties of coma dust. It leads to a porosity (empty space/total volume) of at least 0.6 for the nucleus.

Recently, Clairemidi et al. (1991) have proposed a tentative identification of PAHs in Halleys's coma from UV spectra ; the PAHs have been proposed as a major component of the ISM (15% of total carbon) by Léger and Puget (1984). Strong isotopic anomalies have been found by in situ analysis in individual dust grains ($^{12}C/^{13}C$ from 1 to 5000 ; Jessberger and Kissel 1991) suggesting the aggregation of unmodified solid material formed in a variety of places in the galaxy.

All these features give support to the idea of the formation of cometary matter from little processed IS material in a cold environment.

H_2O	152	SO_2	83	CH_3C_2H	65	CH_4	31
HCOOH	112	NH_3	78	H_2CO	64	CO	25
CH_3OH	99	CS_2	78	C_2H_2	57	O_2	24
HCN	95	HC_3N	74	H_2S	57	N_2	22
CH_3CN	91	CO_2	72	C_2H_4	42	H_2	5

TABLE I

Sublimation temperatures in Kelvin for major molecular species under PSN conditions (gas density 10^{13} cm^{-3}). After Yamamoto 1985.

3. Volatiles

In a series of paper Yamamoto and coworkers (1983, 1985, 1991) have studied a two stages model to explain the molecular abundances encountered in comets. First, interstellar gas freezes on grains to form ice mantles in the early stages of Solar system formation. Second, the temperature rise toward the center of the protosolar nebula evaporates partially or totally the most volatile species. In this view, cometary ices should have the same relative abundances as in the interstellar medium, with the major exception of CO and N_2, which have especially low sublimation temperatures. The abundances of CO, CO_2, N_2 are used to constrain the temperature at the final steps of cometary nuclei formation to 20-70K (for pressures corresponding to the expected typical density, 10^{13} cm^{-3}). Table 1 gives the sublimation temperature for important molecules under PSN conditions (Yamamoto 1985).

Recent advances in ground-based comet observations (cf A'Hearn 1992 ; Snyder 1992 ; Crovisier 1991a ; Weaver et al. 1991), together with Halley's results have enabled an increasing number of tests of this scenario. We assume here that gas production rates from the nucleus give a good picture of its internal composition (a noticeable departure from this is however predicted in some models, due for example to differential sublimation, Espinasse et al. 1991).

The abundance of sulfur species is a new tool made available by recent radio and visible observations (Crovisier et al. 1991, Kim and A'Hearn 1991). We summarize briefly the results of the comparison with ISM (see Despois et al. 1992). The present inventory of the sulfur volatile compounds in comets seems essentially complete (in terms of fraction of total S) ; H_2S and the parent of CS (proposed to be CS_2) dominate, with comparable amounts (0.1-0.2 % of H_2O). S is thus depleted in the volatiles, compared to solar S/O value, 2% , but not in the comet as a whole (i.e. including the dust). Unlike what is presently observed in many places in IS gas, where these species are of comparable abundances (ratio 0.1-10), H_2S strongly dominates sulfur oxides (SO and SO_2) by 2 to 3 orders of magnitude. To interpret this in the frame of the above model seems to require the conversion into H_2S of a large fraction of atomic sulfur (predicted to be the dominant S species in ISM) through grain surface reaction during the formation of cometary matter.

A summary of other cometary volatiles is given in Table 2 ; this list is rather conservative, and the species quoted are fairly securely identified. The derived abundances are however sensitive to some extent to model parameters concerning the excitation of the molecule and its spatial distribution (some species like H_2CO seem to originate at least in part from an extended source, grains or heavier parent molecule ; regarding excitation, atomic sulfur — a decay product — is one of the difficult cases) (Bockelee-Morvan and Crovisier 1992 ; Roettger 1991). On the other hand, new molecules help constraining the models ; for example, the numerous mm lines of methanol will permit to constrain rotational and kinetic

CO	40-200 (a)	NH_3	2	H_2S	2
CO_2	35	HCN	1 (a)	$CS(CS_2)$	1 (a)
CH_3OH	10-40 (c)	N_2	0.2	OCS	<2
H_2CO	0.4-40 (b)	CH_3CN	<0.1	H_2CS	<1
CH_4	<3	HC_3N	<0.05	S_2	0.2/<0.05 (b)
C_2H_2	?			SO	<0.005-0.005
				SO_2	<0.001-0.01

TABLE II
Production rates of major known or expected constituents of the nucleus, normalized to the production rate of H_2O. Unit: H_2O = 1000. See Crovisier 1991b and Mumma et al. 1992 for references. Notes: a) variable from comet to comet b) may be variable; spatial distribution not completely understood c) still under interpretation

temperature in the coma, which previously relied mainly on hydrodynamical modelling (Bockelee-Morvan et al. 1990).

Note in particular the high CH_3OH/H_2CO ratio (with the above caveat!), the low CO/H_2O, the very low abundances of N_2 and NH_3, as well as of CH_4. Important to mention also is the apparent variability from comet to comet of *some* of the species: CO (seen in many comets with different relative abundances to H_2O); S_2 seen in only one comet (but with high S/N) and absent to very good limits in others.

4. Elemental abundances

If comets are a real pristine sample of PSN material, their elemental composition should be close to solar. Recent summaries (Encrenaz et al. 1991; Wyckoff et al., 1991) using Vega and Giotto dust analyser results together with careful (but in some cases indirect) measurements of coma gas composition from ground-based observations have both concluded that nitrogen is depleted globally (by a factor ~ 6); this is mainly due to the N depletion in the volatiles (~ 75). Although not directly observable, the abundances of the two main N bearing species, N_2 and NH_3 have been carefully derived from their daughter species N_2^+, NH_2 and NH, and the reality of these depletions seems certain. Note however that some components like NH_4^+, X^- and $(HCN)_n$ have not been measured, and that another abundance summary (Delsemme 1987) is consistent with a solar value for nitrogen (mostly because of the adoption of a high N_2 abundance).

A simple explanation of this depletion is obtained in Yamamoto's model: above ~ 20 K, pure N_2 would begin to sublimate in the PSN, and N_2 is considered to be a (the ?) major nitrogen depository in the gas phase. Note however that O_2 has also a low sublimation temperature. The fact that O is not underabundant would imply that comets formed from a gas where O_2 was not a major species. This is important, as the oxygen budget of the ISM is not yet understood; it would favor atomic O, or H_2O (gas or solid), but not O_2, as main oxygen reservoir in molecular clouds. Another possibility is that of reactions destroying O_2 in the PSN (to form H_2O ?); in that case, cometary abundances are expected to depart noticeably from interstellar ones.

Other noticeable features regarding elemental abundances can be seen in Fig. 1. In P/Halley, there are, beside the expected H deficiency, a slight deficiency of C and a larger deficiency of S in the gas phase, which are roughly compensated by higher dust abundances, and a marginal indication of non solar Si/Fe. The latter, if confirmed by future spacecraft measurements with higher S/N, would be difficult to explain in the frame of a formation

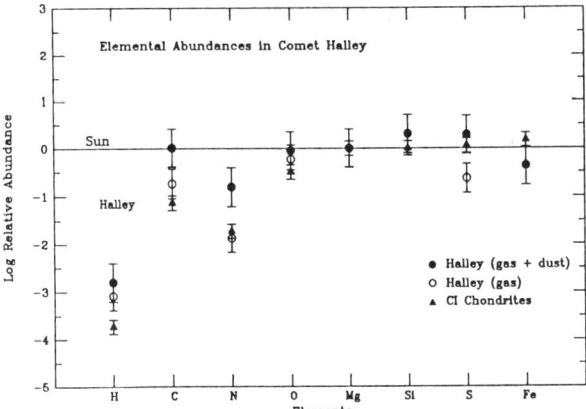

Fig. 1. Elemental abundances in the gas and the combined gas+dust components of P/Halley relative to solar photospheric abundances and normalized to Mg. After Wyckoff et al. 1991.

at cold temperature from homogeneous IS matter.

5. Some remarks on the modelling of the transition from ISM to PSN

To understand the chemical composition of solar sytem objects, especially little evolved objects like comets, one needs to model the evolution of the collapsing gas and its subsequent evolution in the outer fringes of the nebula. This is a quite complex task, as many competing phenomena are involved. However the rapid increase in the amount of data calls for such work, and makes it a very exciting subject. We comment here briefly on some aspects of this new field.

The density and temperature evolution could be taken from the numerical models of solar system formation (see Boss et al. 1989 for a summary). The results are however rarely given in Lagrangian (comoving) coordinates which are more convenient for the study of mostly local processes like chemical evolution. Such a comoving evolution of physical parameters can be more easily retrieved from approximate analytical models. The inside-out model of collapse (Shu 1977 ; Adams et al. 1987) although quite simplified, seems to include much of the relevant physics, according to its success (IR spectra of YSOs and spatial distribution of the matter around them : Adams et al. 1987 ; Butner et al. 1991). Regarding the PSN accretion disk, a quite manageable model is given by Wood and Morfill (1988).

Another question is that of the chemistry in the gas. Ion-molecule chemistry, dominant in the IS medium, will remain limited by its energy input, ionisation through cosmic rays, which has no reason to grow and will eventually decline in the innermost part of the protostellar core due to opacity effects (e.g. Ginzburg 1978). The increase in density will favor the recombination of preexisting ions and radical species, and, even with a modest increase in temperature, neutral-neutral reactions will take place preferentially. Shock chemistry may take place at the boundary of the disk, if there is an accretion shock as in the model of Hollenbach and Neufeld (1990) ; shocked gas will also be present around the bipolar flow which -in the present view -should accompany the early solar evolution. Other high-T chemistry is expected in the inner solar nebula (cf. Prinn and Fegley 1989), but the species produced there should not contaminate much the outer nebula through diffusion (Stevenson 1990). Flares have even been proposed as the origin of the rapidly heated and cooled inclusions found in meteorites (chondrules)(Levy and Araki 1990). Other effects may be produced by the early sun.

Grain surface chemistry, as was suggested in the study of sulfur species, may become the dominant chemistry. Condensation of molecules on grains is an increasingly efficient process as collapse proceeds. Molecule encounters with grains grow linearly with density ; in a cold medium with H_2 density higher than 10^9 cm^{-3}, molecules will condense on grains in less than a year in the absence of any reevaporation process. It appears clearly compulsory now to take into account ice mixture properties in order to discuss the condensation and evaporation of the ices (Schmitt et al. 1989, Sandford and Allamandola 1990, Yamamoto 1991). *Some* CO for example may be retained at temperatures as high as 50 K on an H_2O ice surface, contrasting with the 10-20K for a pure CO layer ; a comparison of solid versus gas phase CO abundances in dark clouds is given in Whittet and Duley (1991).

The surface available for molecule condensation and reaction is however decreased to an unknown extent by grain coagulation (e.g. Cassen and Boss 1988). This phenomenon, first and essential step in the formation of larger bodies, is extremely difficult to model. Grain-grain velocities depend on the (unknown) turbulent properties of the gas, and aggregated grains are likely to have a fractal structure as is common in such processes. Such structures have already been considered in the ISM (to explain dust emission properties, Wright 1987) and in a PSN context (e.g. Weidenschilling et al. 1989). Important also are : the temperature of the grains, which is expected to be different from the gas, the eventual ejection processes releasing molecular mantles into the gas phase, and the surface reaction rates.

In the recent years, some models have already taken into account part of the above complexity : Tielens and Hagen (1982), d'Hendecourt et al. (1985) (grain surface chemistry) ; Boland (1982), Tarafdar et al. (1985), Brown et al (1988), Rawlings et al. (1992) (chemistry in a collapsing cloud).

6. Conclusions

Our knowledge of cometary abundances has considerably grown recently. We now have good estimates or significant upper limits on critical molecular abundance ratios like CO/H_2O, CO/CH_4, NH_3/N_2, $H_2S/(SO+SO_2)$, H_2S/H_2O, ... We have also through the combination of ground based and space results indications for a significant departure from solar elemental abundances in the case of nitrogen. With all these new data, we begin to be able to constrain models of solar system formation. If one admits further, as was done above, the scenario in which cometary matter results from condensation of interstellar matter with little processing, these abundances even give new constraints to ISM chemical models. It is thus desirable that IS chemical models including grains provide mantle composition for comparison. Due to the large range of phenomena occuring, many molecular abundance ratios should be considered together to test the various theories ; among them, minor species such as S-bearing species are extremely useful diagnostic tools. Processes linking IS abundances and cometary/solar system abundances need to be investigated, like gas phase chemistry at densities higher than $10^7 - 10^8$ cm-3, and non-LTE condensation of ice mixtures. Grain surface chemistry plays probably a dominant role, but other possiblities (accretion shock chemistry,...) have also to be investigated. To go further, more data is needed. This is fortunately what the next years should bring, through high spatial and spectral resolution mm and IR investigation of low-mass star forming region by means of millimetric interferometers and infrared satellites, through on-going ground based observation of comets, and ultimately return to Earth of a comet sample by the Rosetta spacecraft.

Acknowledgements

This work owes much to my collaboration with J.Crovisier, D. Bockelee-Morvan, P.Colom and G.Paubert on radio observations of comets. I acknowledge very stimulating discussion with M.Walmsley and enlightening remarks by T.Owen and M.Guélin. I would also like to thank A.Baudry for detailed comments on the manuscript and P.D.Singh for his patience as an editor.

References

Adams, F.C., Lada, C.J., Shu, F.H. 1987, Ap.J., 312, 788-806.
A'Hearn, M.F. 1992, these proceedings.
Bockelee-Morvan, D., Crovisier, J. 1990, in Asteroids Comets Meteors III, eds. C.-I. Lagerkvist, H. Rickman, B.A. Lindblad and M. Lindgren, Uppsala Univ. Press, 263.
Bockelee-Morvan, D., Crovisier, J. 1992, A.A., submitted.
Bockelee-Morvan, D., Crovisier, J., Colom, P., Despois, D., Paubert, G. 1990, in Formation of Stars and Planets and the Evolution of the Solar System, ESA SP-315, 243.
Boland, W. 1982, Ph.D. Thesis, Univ. of Amsterdam.
Boss, A.P., Morfill, G.E., Tscharnuter, W.M. 1989, in Origin and Evolution of Planetary and satellite atmospheres, eds. S.K. Atreya,J.B. Pollack and M.S. Matthews, Univ. of Arizona Press:Tucson ,35-77.
Brown, P.D., Charnley, S.B., Millar, T.J. 1988 M.N.R.A.S. 231, 409-417.
Butner, H.M., Evans, N.J., II, Lester, D.F., Levrault, R.M., Strom, S.E., 1991, Ap.J., 376, 636.
Cassen, P., Boss, A.P. 1988, in Meteorites and the Early Solar System, eds. J.F. Kerridge and M.S. Matthews, Univ. of Arizona Press : Tucson, 305-327.
Clairemidi, J., Rousselot, P., Moreels, G. 1991, in IAU colloqium 126 : Origin and Evolution of Dust in the Solar System, ed. A.C. Levasseur-Regourd, Kluwer : Dordrecht, 217-220.
Colom, P., Bockelee-Morvan, D., Crovisier, J., Despois, D., Paubert, G. 1992, these proceedings.
Crovisier, J. 1991a, in Asteroids, Comets, Meteors conference, Flagstaff, Arizona, June 1991.
Crovisier, J. 1991b, in Infrared Astronomy with ISO, proceedings of a school held at Les Houches.
Crovisier, J., Despois, D., Bockele-Morvan, D., Colom, P.,Paubert, G. 1991, Icarus, accepted.
Donn, B. 1991, in Comets in the Post-Halley Era, eds. R.L. Newburn, Jr., M. Neugebauer, and J. Rahe, Kluwer : Dordrecht, 335-360.
d'Hendecourt, L.B., Allamandola, L.J., Greenberg, J.M. 1985, A.A., 152 ,130.
Delsemme, A.H. 1991, in Comets in the Post-Halley Era, eds. R.L. Newburn, Jr., M. Neugebauer, and J. Rahe, Kluwer : Dordrecht, 377-428.
Despois, D., Crovisier, J., Bockelee-Morvan, D., Colom, P. 1991, these proceedings.
Encrenaz, Th., Puget, J.L., d'Hendecourt, L. 1991, Sp. Sci. Rev., 56, 83-92.
Espinasse, S., Klinger, J., Ritz, C., Schmitt, B. 1991, Icarus, 92, 350.
Ginzburg, V. 1978, Physique Théorique et Astrophysique, Editions Mir, Moscou
Greenberg , 1985, Phys.Scripta 11 ,14.
Greenberg, J.M., Hage, J.I. 1990, A.A., 361, 260-274.
Hollenbach, D.J., Neufeld, D.A., poster presented at Protostars and Planets III conference, Tucson, March 1990.
Irvine, W.M., Knacke, R.F. 1989, Origin and evolution of Planetary and Satellite Atmospheres, eds S.K. Atreya, J.B. Pollack, and M.S. Matthews, Univ. of Arizona Press : Tucson, 3-34.
Jenniskens, P., de Groot, M., Greenberg, J.M., 1991, in Asteroids, Comets, Meteors conference, Flagstaff, Arizona, June 1991.
Jessberger, E.K., Kissel, J. 1991, in Comets in the Post-Halley Era, eds. R.L. Newburn, Jr., M. Neugebauer, and J. Rahe, Kluwer : Dordrecht, 1075-1092.
Kerridge, J.F., Matthews, M.S., eds. 1988, Meteorites and the Early Solar System, Univ. of Arizona Press : Tucson.
Kim, S.J.,A'Hearn, M.F. 1991, Icarus, 90, 79-95.
Léger, A., Puget, J-L. 1984, A.A., 137, L5.
Levy, E.H.,Araki, S. 1990, Icarus 81, 74-91.

McKinnon, W.B., Mueller, S. 1988, Nature, 335, 240.
Millar, T.J., Herbst, E. 1990, A.A., 231, 466-472.
Mumma, M., Blass, W., Weaver, H., Larson, H. 1988, Poster Book from the STScI workshop on the Formation and Evolution of Planetary Systems, eds. H.A. Weaver, F. Paresce, and L.Danly ,157-168.
Mumma, M.J., Stern, A., Weissman, P.R. 1992, in Protostars and Planets III, E.H. Levy, J.I. Lunine and M.S. Matthews eds., Univ. of Arizona Press : Tucson
Newburn, R.L., Jr., Neugebauer, M., Rahe, J., 1991, eds., Comets in the Post-Halley Era, Kluwer : Dordrecht, 361-376.
Owen, T. 1992, these proceedings.
Prinn, R.G., Fegley, B., Jr. 1989, in Origin and Evolution of Planetary and Satellite atmospheres, eds. S.K.Atreya, J.B. Pollack and M.S. Matthews,Univ. of Arizona Press:Tucson,78-136
Rawlings, J.M.C.,Williams, D.A. 1991, these proceedings.
Roettger, E.,E. 1991, Ph.D. dissertation, Johns Hopkins University.
Sandford, S.A., Allamandola, L.J. 1990, Icarus, 87, 188-192.
Schmitt, B., Espinasse, S., Grim, R.J.A., Greenberg, J.M., Klinger, J. 1989, in Physics and Mechanics of Cometary Materials, ESA SP-302, 65-69.
Shu, F.H., Adams, F.C. 1987, in IAU Symp 122 : Circumstellar Matter, eds. I. Appenzeller and C. Jordan, 7-22.
Shu, F.H. 1977, Ap.J., 214, 488-497.
Snyder, L.E. 1992, these proceedings.
Stevenson, D.J. 1990, Ap.J., 348, 730-737.
Tarafdar, S.K., Prasad, S.S., Huntress, W.T., Jr., Villere, K.R., Black, D.C. 1985, Ap.J., 289 ,220.
Tielens, A.G.G.M., Hagen, W. 1982, A.A., 114, 245.
Tielens, A.G.G.M., Allamandola, L.J. 1987, in Physical processes in Interstellar Clouds, eds. G.E. Morfill and M. Scholer, Reidel : Dordrecht, 333-376.
Walmsley, C.M., in Symposium I.A.U. 147 : Fragmentation of Molecular Clouds and Star Formation, E. Falgarone, F. Boulanger and G. Duvert eds., Kluwer : Dordrecht, 161-175.
Weaver, H.A., Mumma, M.J., Larson, H.P. 1991, in Comets in the Post-Halley Era, eds. R.L. Newburn, Jr., M. Neugebauer, and J. Rahe, Kluwer : Dordrecht, 93-106.
Weidenschilling, S.J., Donn., B., Meakin, P. 1989, in The Formation and Evolution of Planetary Systems, eds. H.A. Weaver and L. Danly, Cambridge Univ. Press : Cambridge, 131-150.
Weissman, P.R. 1991, in Comets in the Post-Halley Era, eds. R.L. Newburn, Jr., M. Neugebauer, and J. Rahe, Kluwer : Dordrecht, 463-486.
Williams, D.A. 1990, in Proceedings of the meeting "Molecular Clouds" held in Manchester, March 1990.
Wood, J.A., Morfill, G.E. 1988, in Meteorites and the Early Solar System, eds. J.F. Kerridge and M.S. Matthews, Univ. of Arizona Press : Tucson, 329-347.
Wright, E.L. 1987, Ap.J., 320, 818.
Wyckoff, S., Tegler, S.C., Engel, L. 1991, Ap.J. 367 ,641.
Yamamoto, T., Nakagawa, N., Fukui, Y. 1983, A.A., 122, 171-176.
Yamamoto, T. 1985, A.A., 142, 31-36.
Yamamoto, T. 1991, in Comets in the Post-Halley Era, eds. R.L. Newburn, Jr., M. Neugebauer, and J. Rahe, Kluwer : Dordrecht, 361-376.

QUESTIONS AND ANSWERS

B.Foing: In its early evolution in the T-Tauri and past T-Tauri phase, the Sun has an enhanced activity in terms of UV flux (~ 100 times higher than now), intense flaring and strong winds, as well as energetic particles. How do you expect this active early Sun to affect and alter grains before them being integrated into cometesimals?

D.Despois: We don't know yet. Several groups are investigating the effects of UV and energetic particles on ices. It is precisely the study of the chemical composition of comet nuclei which will inform us on the importance of these effects, and hence on the duration and relative starting time of the various phases of cometary matter formation, cometesimal accretion, clearing of the nebula and exposure to early Sun radiations.

FORMATION OF COMETS : CONSTRAINTS FROM THE ABUNDANCE OF HYDROGEN SULFIDE AND OTHER SULFUR SPECIES

D. DESPOIS[1], J. CROVISIER[2], D. BOCKELEE-MORVAN[2] and P. COLOM[2]

[1] *Observatoire de Bordeaux, B.P. 89, 33270 Floirac, France*
[2] *Observatoire de Paris, Section de Meudon, F-92195 Meudon, France*

Abstract. Recent determinations of H_2S and other sulfur compounds abundances in comets and in Orion KL bring new tests of the origin of cometary matter.

According to one of the two scenarios of cometary matter formation (Yamamoto 1991 ; Despois 1992), cometary composition should match closely the composition of interstellar medium (ISM). Recent observations of sulfur compounds provide new very useful tests ; sulfur, although not abundant, presents the advantage that most major S species are observed, whereas crucial C and N species like CH_4 and N_2 are at present still ill-known.

Comet results are summarized in Table I. Radio millimetric observations of the two bright comets Austin (1989c1) and Levy (1990c) have brought the first detection of H_2S (Colom et al. 1992 ; Crovisier et al. 1991). Optical observations of SO and SO_2 in Austin and other comets (Kim and A'Hearn 1991) have resulted in very low limits, corresponding to abundances lower than that of H_2S by 2 to 3 orders of magnitude. Also noticeable is the high variability of S_2 between comets - well detected in comet IRAS-Araki-Alcock (1983 VII), it is absent otherwise to very good limits (A'Hearn 1992) ; this contrasts with the limited variation encountered usually in mother molecules. CS_2 is the proposed parent for CS ; with H_2S this would be the main cometary sulfur species (Roettger 1991). The two last species are not well known : OCS limits need to be improved ; the lifetime of H_2CS is not constrained : if shorter than the assumed 3300 s at 1 AU, higher H_2CS abundances in the nucleus are possible.

Due to the paucity of available observation, the very complex Orion KL region is at present the main source of data for abundances ratios of S species in the ISM. It includes at least 4 (spatial as well as spectral) components : Hot Core, Plateau, Extended and Compact Ridges. Minh et al. (1990) have found very high H_2S abundances in both Hot Core and Plateau. However only in the Hot Core, where grain mantle evaporation has been suggested (e.g. Walmsley 1989), is H_2S much more abundant than sulfur oxides, whereas in the plateau, where shocks from outflows are believed to play an important role, their abundances are comparable. Regarding models of IS chemistry, the $[H_2S]/([SO]+[SO_2])$ ratio is usually between 0.1 and 10. Only grain surface chemistry models show a net tendancy for higher $[H_2S]$.

Is H_2S in other IS region frozen in grain mantles ? The tentative detection of solid H_2S towards W33A around 3.9 μ has raised some controversy as well on observational grounds (Smith 1991 ; Geballe 1991) as in the interpretation (Allamandola 1992). The level of present detection/upper limits is on the order of what is seen in comets, when compared to H_2O.

The resemblance of cometary and Hot Core ratios, together with results from state-of-the-art IS chemistry, tend to suggest a possible role of grain surface chemistry in the formation of cometary matter. A simple process could be the transformation of most atomic sulfur into H_2S when sticking on grains, through recombination with incoming H atoms ; atomic S is the dominant sulfur carrier in IS chemistry models (Millar and Herbst 1990).

	S_2	H_2S	SO	SO_2	CS (CS_2)	OCS	H_2CS
Austin (1989c1)	<0.05	2.7	<0.05	<4	0.5	–	–
Levy (1990c)	–	2	–	<2.5	–	<2	<1
P/Halley	<0.3	–	<0.2	<0.02	0.3-2.5	<7	–
IRAS-A.-A. (1983 VII)	0.2	–	<0.005	<0.001	–	–	–
Others	–	–	<0.2	<0.01	1	–	–
Orion Hot core	–	1000-5000	<20	<24	–	–	–
Orion Plateau	–	1000-4000	520	520	22	52	–
Orion Compact Ridge	–	–	–	–	–	3.3	1.6
Orion Extended Ridge	–	1.5	<0.9	<3.3	2.5	–	–
TMC1	–	0.7	5	<1	10	2	3

TABLE I

Abundances of S species. Comets : relative production rate Q/Q_{H_2O} in units of 10^{-3}. See Crovisier et al. 1991 for references. ISM : relative column densities N/N_{H_2} in units of 10^{-9}. Data from Minh et al. 1990, Blake et al. 1987, Millar and Herbst 1990 ; see also Guélin et al. 1990, Turner 1991.

In addition to chemistry, the presence of H_2S provides constraints on the history of cometary matter through its low sublimation temperature. For pure H_2S, at the typical density of 10^{13}, it is about 57 K (Yamamoto 1985). H_2S and CO are thus the most volatile parent species directly identified in comets. It is very important to study the more realistic case of ice mixtures ; laboratory work should precise temperature constraints deduced from the presence of H_2S when sticked on other ices.

The study of sulfur bearing species bring thus several important new abundance ratios to test theories of the origin of cometary matter : H_2S/H_2O , $H_2S /(SO+SO_2)$, H_2S/CS. More data is clearly needed both in comets to improve the statistics and in interstellar space towards regions forming sun-like stars.

References

A'Hearn, M.F. 1992, these proceedings.
Allamandola, L.J. 1992, these proceedings.
Blake, G.A., Sutton, E.C., Masson, C.R., Phillips, T.G. 1987, Ap.J., 315, 621-645
Colom, P., Bockelee-Morvan, D., Crovisier, J., Despois, D.,Paubert, G. 1992, these proceedings.
Crovisier, J., Despois, D., Bockelée-Morvan, D., Colom, P., Paubert, G. 1991, Icarus, accepted.
Despois, D. 1992, these proceedings.
Geballe, T.R. 1991, M.N.R.A.S., 251, 24p-25p.
Guélin, M., Rist, C., Cernicharo, J. 1990, in Proc. 7th Manchester Astronomical Conference, Cambridge University Press, in press.
Kim, S.J., A'Hearn, M.F. 1991, Icarus, 90, 79-95.
Millar, T.J., Herbst, E. 1990, A.A. 231, 466-472.
Minh, Y.C., Ziurys, L.M., Irvine, W.M., McGonagle, D. 1990, Ap.J., 360, 136-141
Mumma, M.J., Stern, A., Weissman, P.R. 1992, in Protostars and Planets III, E.H. Levy, J.I. Lunine and M.S. Matthews eds., Univ. of Arizona Press: Tucson.
Roettger, E.E. 1991, Ph.D. dissertation, Johns Hopkins University.
Turner, B.E. 1991, Ap.J., 76, 617-686.
Smith, R.G. 1991, M.N.R.A.S., 249, 172-176.
Walmsley, C.M. 1989, in I.A.U. Symposium 135 : Interstellar Dust, eds. L.J. Allamandola and A.G.G.M.Tielens, 263-273.
Yamamoto, T. 1985, A.A., 142, 31-36.
Yamamoto, T. 1991, in Comets in the Post-Halley Era, eds. R.L. Newburn, Jr., M. Neugebauer, and J. Rahe, Kluwer : Dordrecht, 361-376.

A DIAGNOSTIC SPECTRAL INDICATOR OF THE EXPOSURE AGE OF AN ASTEROIDAL SURFACE

Joseph A. Nuth III
Astrochemistry Branch, Code 691
NASA/Goddard Space Flight Center
Greenbelt, MD 20771

The energy of the SiH vibrational fundamental has been shown be extremely sensitive to the oxidation state of the silicon to which the hydrogen is bound, ranging from 4.4 microns in highly oxidized silicate grains to 4.74 microns in silicon carbide (Moore, Tanabé and Nuth, Ap. J. (Lett.) 373, L31-L34, 1991). Yin, Ghose and Adler (Appl. Spectrosc. 26, 355-7, 1972) have shown that the process of ion-sputtering a metal oxide results in chemical reduction, due to the high sputtering yield and volatility of oxygen relative to the metal. These authors have shown that solar wind ion-sputtering of lunar soils could provide an explanation for the observed solar wind darkening of the lunar surface. We hypothesize that a similar ion-reduction process could occur on asteroidal surfaces exposed to the solar wind. A second consequence of solar wind exposure would be the implantation of hydrogen ions into the asteroidal surface. Stein (J. Elec. Mat. 4, 159-174, 1975) has shown that hydrogen-ions implanted into silicon exhibit infrared absorptions between 4.5 and 5.5 microns: similar features should result from the ion-implantation of hydrogen ions into partially reduced silicate minerals and glasses. The position of such features would be indicative of the local chemical environment of the silicon atom to which the hydrogen was bound.

It may be possible to observe the SiH fundamental on the lunar surface, however, laboratory experiments have shown that exposure of grains containing-SiH groups to vacuum for several months at 370 K results in the elimination of hydrogen from the grains. Therefore, for lunar surfaces where the temperature might exceed 300 K for a considerable length of time each lunar day, it is possible that the SiH signature could be baked out if the rate of solar wind ion-implantation does not exceed the rate at which hydrogen is thermally driven from the grains. Nevertheless, as long as some SiH still exists in the lunar regolith, the wavelength of the SiH fundamental will be indicative of the oxidation state of the silicon in the surface of the grains.

Asteroidal regoliths will be considerably cooler than the lunar surface due to their greater distance from the sun and this should ensure that such regoliths will retain solar wind implanted hydrogen ions. Of course, increased distance of the asteroids from the sun decreases both the solar wind hydrogen ion implantation rate and the rate of ion-sputtering reduction of the asteroidal regolith. This will mean that surfaces require quite long exposure to the solar wind to become reduced. Exact correlation of the solar wind exposure ion-sputtering reduction age with the energy of the SiH fundamental requires careful laboratory studies, which have yet to be performed, of the reduction efficiency of solar wind ions for various meteorite types. These efficiencies would then be integrated over the orbit of the asteroid and corrected for variations in the solar wind flux through time in order to calculate absolute ages.

Despite the difficulty in obtaining absolute ages for asteroidal surfaces, it should be possible to obtain relative ages for asteroids of similar type in similar orbits quite easily if the peak frequency of the SiH fundamental can be determined. The surfaces of such asteroids should have received similar doses of solar wind irradiation over time and this radiation should have produced the same degree of chemical reduction in each. However, if due to collision a fresh surface on a chondritic-type asteroid were exposed, then this surface would be fully oxidized and one would expect the SiH fundamental to occur near 2300 cm^{-1} once a suitable quantity of hydrogen had been implanted into the surface to be observable. As the surface ages and becomes more reduced the energy of the SiH fundamental would gradually decrease towards 2100 cm^{-1}. The youngest surface is therefore the one with the highest energy SiH fundamental. For relatively slow rotating asteroids it may even be possible to determine the relative ages of different surfaces provided that sufficiently high signal-to-noise spectra can be obtained on a timescale which is short in comparison to the rotation period. Such data may not obtainable in this difficult spectral region until space-based infrared observatories become available.

Summary

D A WILLIAMS

Department of Mathematics, UMIST, PO Box 88, Manchester M60 1QD, England

August 29, 1991

1. Introduction

The quantity of scientific work in the area of astrochemistry is now enormous, and this meeting - in attempting to cover most of the aspects of our subject - has had problems in coping with it due to difficulty of reviewing very active, broad areas in a limited time. At this meeting we have had almost 40 hours of formal and informal discussion, and about 45 oral presentations at which I estimate that about 1500 transparencies and slides have been projected for our enlightenment. About 100 posters have been presented in three sessions. The posters were uniformly of a high standard and generated a great deal of interest and discussion. I think the prominence given in the programme to the posters has been particularly successful. The range of activity, and the enthusiasm with which it is pursued, demonstrate that astrochemistry is in a vigorous state. It has its own intrinsic interest, and in its present maturity is now making substantial contributions to understanding in astronomy.

The growth in the population of astrochemists is striking, and appears to be related to the number of molecular species identified in the interstellar medium. In 1965, the population was probably about a dozen, or several times the number of molecular species then identified. Today, the number of identified molecular species is about 100, and the number of astrochemists is approximately several hundred, of which about half are here. Given the advances in submillimetre and IR technology, and the imminence of several orbiting astronomical missions, we can confidently expect the number of identified interstellar species to rise dramatically. It will be interesting to observe whether the population of astrochemists rises in proportion.

There are few areas of present-day science which are so challenging and interesting as astrochemistry. We have, as the format of our meeting shows, the opportunity of addressing questions from the Early Universe to the Solar System, from the properties of tenuous quiescent gas to star formation, circumstellar envelopes, novae, and supernovae. To do this, we need acquaintance with a wide range of micro and macroscopic phenomena, from atomic and molecular physics to large scale dynamics. We need familiarity with the language of the quantum chemist, of laboratory chemical kinetics, of solid state physics and chemistry, of gas dynamics including hydrodynamic and magnetohydrodynamic shocks, flows, interfaces, and turbulence. We need to be aware of the value and limitations of observations made anywhere in the electromagnetic spectrum. We need especially to understand the astronomical context, and pose our questions appropriately.

Dalgarno, in his remarks at the IAU Symposium 120 on "Astrochemistry", Goa, 1985, gave what I believe is the first definition of astrochemistry. His opening remarks to this symposium illustrated the present state of knowledge and emphasised the developments since Goa. He also showed the first chemical equation at this meeting: exceptionally, it contained an error. This was surely not an accident, but a way of subtly reminding us that nothing is certain, and recommending caution and humility. This view was supported later in the meeting by Millar who, with disarming candour, said: "When I put the observational results in the left hand column, and calculated values in the right hand column, I want you to ignore both". Such scepticism was, however, short-lived in the enthusiasm of the meeting.

In this Summary it would be impossible to refer to every matter discussed at the Symposium. I can only give a personal view, and comment on some of the topics that have seemed to me to be particularly exciting. I conclude by highlighting important areas of activity, or themes, for the future.

2. Comments

It used to be a standard comment when discussing chemical networks that few of the relevant rate coefficients had been measured in the laboratory. This is no longer true: and is a testimony to the immense achievements of laboratory workers in Europe and the US over the last decade. Rowe emphasized especially the recent advances in low temperature ion molecular reactions, branching ratios and rate coefficients in dissociative recombination reactions. Recent results on the reaction of N^+ with H_2 and on CH_3^+ (via ternary reactions) radiation association have had immediate and significant impact on cloud models. A significant discrepancy between theory and experiment remains in ternary reactions and in dissociative recombination branching ratios. The most exciting and important single study reported by Rowe, was of the dissociative recombination of H_3^+, now believed to be fast in interstellar conditions.

Ellinger posed the question: Can Quantum Chemistry be a partner in Astrophysics Research? To which the answer was a resounding and emphatic: yes. The complexity of the systems that Quantum Chemistry is now confidently handling is very great. Ellinger illustrated this with work on IR intensity ratios in small PAH ions, and made the confident assertion that the DIBs arise from small ionized and partly dehydrogenated PAHs.

The chemistry of carbon molecules is an essential part of astrochemistry. Kroto discussed recent laboratory work and emphasised the discovery with which he has been associated of large closed structures such as C_{60} and C_{70}.

This is surely one of the most remarkable and important developments in chemistry today. Direct evidence of such structures may exist in observations of the Red Rectangle. The next few years should reveal whether these large molecules are widespread in the Galaxy.

The importance of both experiment and theory in determining data for astrochemistry cannot be overestimated. Great progress has been made. However, Tarafdar questioned whether enough effort is being put into the study of photoprocesses - echoed later by Glassgold in his discussion of circumstellar envelopes.

Good progress has been made in elucidating the role of chemistry in the Early Universe. The link between the details of the chemistry and the size of the collapsing region, to become a protogalaxy, a protocluster, or a protostar, was convincingly established by Shapiro. The discussions by Ferlet, Tielens and Owen of the abundance of deuterium in various locations are a useful pointer to the origin and evolution of matter. The concensus was that variations in the D:H ratio are real and are a consequence of different mixing processes. Owen presented data on deuterium abundances in the Solar System: these provided good evidence of both condensation and outgassing processes occurring in the protosolar nebula.

A large amount of information on External Galaxies is now available, as revealed in papers by Karawa, Guélin, Henkel and Millar. However, detail is missing - it is very large scale data. The scale size in studies of M82 discussed by Henkel and by Milllar is $\sim$ 100pc. One must assume, however, that the range of processes occurring in other galaxies is at least as complex as in our own.

Diffuse Clouds are one type of region in our own Galaxy which are still not fully understood, in spite of modelling efforts going back at least 40 years. van Dishoeck illustrated how the most precise and careful studies lead to some considerable successes; but there are also some failures in our understanding. The subject is, fortunately, receiving new observational input, particularly those involving high spectral resolution and high signal:noise, as described by Crane. Shocks are probably excluded as sources of CH^+. There seems to be a much greater variety of types of regions present in the diffuse interstellar medium than are normally considered (cf. Falgarone's remarks). The recent discovery of interstellar NH implies that dust is chemically active. Perhaps H_2 (high J) formation on dust will allow lower radiation fields to be possible. This area of research will become, with new instruments and techniques, once again observationally driven, and we can expect new developments and new understanding before the next astrochemistry symposium. Blitz reviewed the subject of high latitude clouds and created some order in what had been a very confusing situation. HLCs now include low extinction CO clouds, tiny little clouds, IRAS clouds without radio CO emission, and intermediate velocity clouds. Some of these are rich in molecules. How these objects are related to each other, and how they evolve is something to which astrochemistry will contribute much understanding in the next few years.

The study of Quiescent Clouds and regions of low mass star formation was by far the largest section, with 11 contributions. This is an area where much effort, both observational and theoretical, has been expended. Yet of all the areas discussed, this seems to be the area of least agreement between model and theory, or even between models. Falgarone's work illustrates the basis of the problem. Structures in the interstellar medium are generally found to be long and filamentary, with parameters that change rapidly with space and time. The description she proposes is one of fractal structure with which the tracers A_v, HI, $100\mu m$, CO (3 lines) and H_2 (1-0) all agree, giving the fractal dimension 1.4. Until we understand better what is driving this system, our understanding is bound to be limited. Some new aspects have been developed in recent years: gas-grain interaction is now agreed to be important (as emphasized by d'Hendecourt, Tielens, and Charnley) and mantle limitation must occur, perhaps both continuously and intermittently. Dynamics must not be ignored; Prasad discussed the collapse of single entities, while Charnley described a class of cyclic models in which gas moves between high and low density phases. Shocks may play a role in some regions but not all; Wu showed that some stable cores exist. The interpretation of some NH_3 observational data as representing quiescence was questioned by Rawlings. The applicability of pseudo-time-dependent models as represented, e.g., by the work of Suzuki and collaborators in interpreting chemical differences between TMC-1 and L134N, and the adoption of so-called early-time abundances, are areas of some discussion. Newly detected molecules in these sources (cf. Ohishi's presentation) may throw much light on this crucial comparison. Zuirys' talk on shocked regions in regions of high and low mass star formation and in SNRs (given by Snyder) presented observations that are interesting but which do not always provide very clear evidence of shocks. SNRs provide best direct evidence of the impact of a blast wave with molecular clouds; SiO seems to be shock enhanced. It is difficult to draw any clear conclusions on the effects of interstellar shocks. The next few years should be instructive, and our understanding of these regions will surely improve.

In discussion of regions of high mass star formation the question of the gas-dust interaction again received attention. Walmsley discussed a number of regions of high density in which the freeze-out time should be short, yet in which gaseous molecules are clearly present. Though freeze-out occurs, the evidence suggests that either some return of mantle material to the gas is occurring, or that the freeze-out is less efficient than often assumed. Studies in these regions are being influenced by beautiful new observational work. Maillard illustrated this with high resolution observations of gaseous and solid CO in the 4 μm band, towards young stellar objects. The solid:gas

CO fraction towards these sources is small, contrasting with the situation for H_2O. This suggests a differential mantle release mechanism is operating. Further work in the infrared was presented by Evans; high resolution observations at 13.4 μm of interstellar molecules such as C_2H_2 - previously thought to be undetectable - show what can now be achieved.

Interface regions were areas of some of the most successful activity and development in theory and observation described at this meeting. White, Stutzki and Jaffe in particular gave a coherent picture of interface regions with three nicely related talks. These showed that interfaces are clumpy, implying that clouds, too, are clumpy. The models of interfaces that have been developed by Tielens and Hollenbach, and by Sternberg and Dalgarno, are broadly successful. In the future, we shall expect significant refinements, such as time-dependence, shocks, dynamics, and realistic grain properties, to be included in the models. Stutzki's discussion of CI, CII, CO suggested that the brightest C^+ coincides with CS dense clumps so that C^+ follows molecular cloud material. The interpretation is that the C^+ observations delineate the cloud surface, if current models are correct. Stutzki also claims that CO surveys detect CO in PDRs around clumps, rather than the cool interiors, so the CO measurements merely count the clumps and hence give a measure of mass. Jaffe's conclusions also emphasized the general tenor of the meeting - that structure and dynamics are important. Warm gas is seen in many regions, and is heated in a variety of ways. Models must take account of this complexity. In the future we can expect observations in previously unexplored wavebands to enhance our knowledge and understanding. A sample of this enhancement was provided by Wright's description of the results obtained by the COBE satellite at wavelengths 200 μm to 0.5 cm. Positive first detections of several important interstellar species are reported, and whole-sky maps should reveal much about the energy balance in the interstellar medium.

The study of near stellar environments has been characterised by some beautiful new observational work, by good agreement between theory and observations, and by some successful exploratory calculations. Rawlings described recent work on novae and supernova. These models require a much greater degree of complexity than for any other type object discussed here: these regions are dense and hot, conditions are harsh, timescales rapid, yet molecules are observed. Models show conditions under which observed molecules can form, and constrain the picture we have of these dramatic events. Curiel, in his discussion of Herbig-Haro objects, presented some beautiful observations, together with very detailed modelling involving shocks, winds, jets, magnetism, etc. At last, an understanding of these fascinating objects is emerging. Glassgold and Lucas discussed observation and theory of circumstellar envelopes. In spite of some uncertainty concerning the rate of photoreactions, good agreement exists between observation and theory.

Interferometric work presented by Lucas illustrated very convincingly the contrast between pre-cursor and shell molecules.

The contribution of dust to interstellar chemistry now seems well established. Allamandola reviewed the large body of laboratory work on photo-induced reactions on ices, being carried out at Ames and at Leiden. The evidence that similar processes occur in mantles on interstellar dust is compelling.

There is enormous amount of recent data on cometary chemistry and A'Hearn, Snyder, Despois, and Delsemme gave us a flavour of this work. A'Hearn posed the interesting question: Can we put Humpty Dumpty together again? i.e. can we from reaction products infer the original pre-cursors and so define the chemistry in the cometary nucleus? Then, can we discuss the variations within cometary nucleus, and comet-comet variations. He showed that a tremendous amount is being achieved, and noted the similarity in many of the ideas with those popular in interstellar chemistry - especially grain-related phenomena. Snyder drew attention to the list of 15 molecular species identified in comets and compared it with the much larger list of identified interstellar species. He suggested that the cometary list incomplete, and his lists should stimulate further searches. Despois showed what can be done to explore the interstellar - Solar System connection through isotopic elemental and chemical abundances, the ortho:para ratio, and grain properties. The connection is indeed an important one. There are few astrochemists who have expertise in both areas, yet the benefits of strengthening the connection could be great.

3. Conclusions

After a week of intense activity one is forced to conclude that the subject of astrochemistry is in a very healthy state. There is a great deal of well-directed, purposeful activity. It is clear that the entire subject is so extensive that it cannot be covered in depth in a one-week meeting.

The subject is being observationally driven, and some excellent new work has been shown here, especially high spectral and angular resolution observations of high signal to noise. New wavebands, too, are becoming accessible. Future prospects for new discoveries and for detailed observations are most encouraging. To come, there will be ISO, SWAS, with more from HST, ASTRO, etc., and splendid ground-based developments. Theory is developing rapidly, too.

What are the lessons to take away from this meeting? Firstly, the importance of structure (fractals?) is indicated by all the measurements. Variations in density and temperature significantly affect the chemistry and all time scales. Secondly, dynamics is likely to be important in almost all situations. However the importance of shocks in chemistry in interstellar clouds is uncertain. One of the areas developing most rapidly is that of interfaces: Mixing, via turbulence, will need to be incorporated in our understanding. The importance of the gas-dust interaction is now confirmed; solid state chemistry does occur in the interstellar medium. Deposition and ejection (both continuous and intermittent) seem to be taking place. Much work on PAHs has been reported here. The role that they play, and their relation to interstellar dust will be clarified in the next few years. On data: astrochemists should challenge the quantum chemists with further problems. From the evidence presented here, any likely problem arising in astrochemistry is or will become amenable to theoretical attack. The response of laboratory workers to astrochemical needs has been magnificent. Yet, there is still more to do, with photo-reactions continuing to be an area for study. The value of making the connections between the branches of our subject is well illustrated in the discussion of cometary chemistry and the Solar System - interstellar dust relation. There, the benefits of such considerations are clear. There may be other connections whose study would bring similar rewards.

The Cosmic-Ray Ionization Rate

STEPHEN LEPP
University of Nevada, Las Vegas
Physics Department
4505 Maryland Parkway
Las Vegas, NV 89154 USA

ABSTRACT. A wide variety of molecules have been observed in the interstellar clouds. They are believed to be formed by reaction networks which begin with ionization by cosmic-rays. Cosmic-rays are also an important heating mechanism for many astrophysical regions. In this paper I shall review the methods used to infer the cosmic-ray ionization rate and the values which have been measured.

1 Introduction

Cosmic-rays are very high energy charged particles which stream throughout the galaxy. They are composed mostly of protons and alpha particles with about 1% heavier nuclei, electrons and positrons. The relatively high abundance of heavy elements suggests the origin of cosmic-rays are from regions with highly processed material such as supernova or pulsars (Cowsik and Price 1971).

There is a large excess of lithium,-beryllium-boron group elements compared to other cosmic objects. If these are assumed to be spallation products from collisions of the heavier nuclei cosmic-rays with normal matter then the cosmic-rays must pass through about 3 grams/cm^2 of material. This gives path lengths in the galactic disc of about $5 \times 10^5 pc$ (Spitzer 1968).

2 Chemistry of Cosmic Rays

In interstellar clouds cosmic-rays primarily ionize hydrogen and helium

$$\text{Cosmic} - \text{Ray} + H_2 \rightarrow H_2^+ + e \quad (1)$$
$$\rightarrow H^+ + H + e \quad (2)$$
$$\text{Cosmic} - \text{Ray} + He \rightarrow He^+ + e. \quad (3)$$

The electrons are energetic and cause additional ionizations and heating of the gas. The H_2^+ reacts quickly with molecular hydrogen to form H_3^+

$$H_2^+ + H_2 \rightarrow H_3^+ + H \quad (4)$$

which rapidly reacts with other atoms and molecules in the gas

$$H_3^+ + O \rightarrow OH^+ + H_2 \tag{5}$$
$$H_3^+ + CO \rightarrow HCO^+ + H_2 \tag{6}$$
$$H_3^+ + N \rightarrow NH_2^+ + H \tag{7}$$

usually by transferring a proton. These proton transfer reactions with H_3^+ are the initial reactions in the ion-molecule reactions networks which have been proposed to explain the molecular abundances. For example the OH^+ produced in reaction 3 goes on to form both OH and H_2O.

Prasad and Tarafdar (1983) pointed out that the fast electrons created by cosmic-ray ionization also produce ultraviolet photons by exciting the Lyman and Werner bands of H_2. Calculations of molecular dissociation and ionization rates by these cosmic-ray induced photons may be found in Sternberg, Lepp and Dalgarno (1987) and Gredel, Lepp and Dalgarno (1989).

3 Cosmic-ray Flux Measurements Near Earth

Measurements of the cosmic-ray flux at the earth allow one to infer a lower limit to the cosmic-ray ionization rate of $\xi = 6.8 \times 10^{-18} \, s^{-1}$ (Spitzer 1968). This represents a lower limit as the lowest energy cosmic-rays don't penetrate the solar wind into the earths orbit. More recently Cecchi-Pestellini and Aiello (1992) have used Voyager data and an some assumptions about the extrapolation to lower energies to estimate a cosmic-ray ionization rate of $4 \times 10^{-17} s^{-1}$ for the solar neighborhood.

4 Recombination Lines

Another method for inferring the cosmic-ray ionization rate is to compare high lying recombination lines with neutral hydrogen absorption measurements (Shaver 1976). The absorption measurements of neutral hydrogen allow one to identify cold gas along the line of sight. The ratio of recombination line optical depth to 21 cm neutral hydrogen optical depth is only a weak function of density and temperature times the cosmic-ray ionization rate. The measurements of ionization rate along various lines of sight by this method are summarized in table 1.

Table 1: Ionization rate: recombination lines

Authors	$\xi \, (s^{-1})$
Shaver (1976)	$\leq 2 \times 10^{-17}$
Shaver, Pedlar and Davies (1976)	$\leq 3 \times 10^{-17}$
Casse and Shaver (1977)	$< 2 \times 10^{-17}$
Payne, Salpeter and Terzian (1984)	$\leq 2 \times 10^{-17}$

The rates are upper limits when the recombination line is not observed. Hydrogen charge transfer to oxygen followed by the formation of molecules could lead to an effective recombination rate up to five times larger for high molecular fractions (Shaver et al 1976, Glassgold and Langer 1976). This means the rates listed in table 1 could be low by up to a factor of five.

5 Dense Clouds

In dense clouds the OH abundance is proportional to the H_3^+ abundance (Lepp, Dalgarno and Sternberg 1987). This could be combined with model assumptions about the H_3^+ destruction rate to determine an ionization rate. The fractionation of HCO^+ is a measure of the H_3^+ destruction rate because the fractionation in H_3^+ is passed along to HCO^+. The upper limit on the H_3^+ destruction allows one to infer a upper limit to the electron abundance for clouds with $DCO^+/HCO+$ measurments (Wooten et al 1982, Langer 1984 and Dalgarno and Lepp 1984). The combination of the fractionation measurment and the proportionality of H_3^+ and OH allow one to infer upper limits to the cosmic ray ionization rates for those clouds with both fractionation and OH measurments. The upper limits are $\xi > 4 \times 10^{-18} s^{-1}$ for L134N and $\xi > 8 \times 10^{-18} S^{-1}$ for B335 (Lepp, Dalgarno and Sternberg 1987). The rates are insensitive to the H_3^+ recombination rates as the H_3^+ is primarily removed by charge transfer reactions.

Each cosmic-ray ionization of a hydrogen molecule converts one molecule into two hydrogen atoms. The atoms reform into molecules on grain surfaces at a rate $k_g r$ so the cosmic-ray ionizion rate is given by

$$\xi n_{H_2} \times 2 \approx n_H n_{H_2} k_g r \qquad (8)$$
$$\xi \approx n_H k_g r / 2 \qquad (9)$$

Fukui and Hayakawa (1981) proposed using this to infer a cosmic-ray ionization rate for a measured hydrogen abundance, they infer the very low value of $\xi \approx 0.7 \times 10^{-18} s^{-1}$. Although this inferred rate is quite low in fact one might expect it to be an upper limit as atomic hydrogen may also be formed by photodissociation near the edges of the cloud.

6 Diffuse Clouds

In diffuse clouds OH is produced by reaction sequences initiated either by cosmic ray ionization of H or H_2 and the OH abundance is proportional to the cosmic-ray ionization rate (Black and Dalgarno 1973, Hartquist, Black and Dalgarno 1978). The OH is removed by photodissociation and so the rate inferred is dependent on the assumed rate. Recent quantum calculations (van Dishoeck and Dalgarno 1984) have shown the photodissociation cross section is larger then previously assumed and van Dishoeck and Black (1986) have rederived rates for several diffuse clouds. The inferred rates are also sensitive to the assumed rate constant for H_3^+ dissociative recombination. Table 2 shows the inferred cosmic ray ionization rates for some diffuse clouds for both slow ($< 10^{-8}\, cm^3\, s^{-1}$) and fast ($\approx 10^{-7}\, cm^3 s^{-1}$) values for the H_3^+ recombination rate.

Table 2: Ionization Rate: Diffuse Clouds

Cloud	ξ (s^{-1}) fast H_3^+ recombination	ξ (s^{-1}) slow H_3^+ recombination
ζ Per	2×10^{-16}	6×10^{-17}
ζ Oph	4×10^{-16}	1×10^{-16}
o Per	8×10^{-16}	2×10^{-16}

^{1} from van Dischoek and Black (1986)

7 Conclusions

The cosmic-ray ionization rate may be different for different parts of the interstellar medium. The average ionization rate is probably a $few \times 10^{-17}\, cm^3\, s^{-1}$. In some regions the infered rate is sensitive to the rates of reactions such as dissociative recombination of H_3^+ or the branching between OH and H_2O in the dissociative recombination of H_3O^+. In particular if the recent measurements of rapid dissociative recombination of H_3^+ stand up then their is a puzzle as to why the OH in diffuse clouds leads to such a high inferred rate.

8 References

Black, J. and Dalgarno A. 1973, *Ap. J. (Letters)* **184**, L101.
Casse, J.L. and Shaver, P.A. 1977, *Astr. Ap.* **61**, 805.
Cecchi-Pestellini, C. and Aiello, S. 1992 in preparation.
Cowsik, R. and Price, P. 1971 *Physics Today* 24, No. 9, 30.
Dalgarno A. and Lepp S. 1984, *Ap. J. (Letters)* **287**, L47.
Fukui, Y. and Hayakawa, S. 1981, International Cosmic Ray Conference p. 3.2-1.
Glassgold, A.E. and Langer, W.D. 1976, *Ap. J.* **206**, 85.
Gredel, R., Lepp, S. and Dalgarno A. 1989, *Ap. J.* **347**, 289.
Hartquist, T., Black, J. and Dalgarno, A. 1978, *M. N. R. A. S.* **185**, 643.
Langer, W.D. 1984, in "Protostars and Planets II", p 650.
Lepp, S., Dalgarno, A. and Sternberg, A. 1987, *Ap. J.* **321**, 383.
Prasad and Tarafdar 1983
Shaver, P.A. 1976, *Astr. Ap.* **49**, 149.
Shaver, P.A., Pedlar, A. and Davies, R.D. 1976, *M. N. R. A. S.* **177**, 45.
Spitzer, L. 1968 "Diffuse Matter in Space" (Interscience, New York).
Sternberg, A., Dalgarno A. and Lepp, S. 1987, *Ap. J.* **320**, 676.
van Dishoeck, E. and Black J. 1986, *Ap. J. Supp.* **62**, 109.
van Dishoeck, E. and Dalgarno, A. 1984, *Ap. J.* **277**, 576.
Wooten, A., Loren, R. and Snell, R. 1982, *Ap. J.* **255**, 160.

QUESTIONS AND ANSWERS

L.Blitz: Why can't you just use the gamma-ray data to get the cosmic ray ionization rate?

S.Lepp: The gamma-rays track the high energy portion of the cosmic-rays whereas the ionization is mainly from the low energy.

W.Langer: It has been suggested that cosmic rays scatter Alfven waves in clouds and could be attenuated somewhat. Wouldn't this lead to variations in CR ionization rate in dense cores versus diffuse clouds?

S.Lepp: That's right.

E.van Dishoeck: (comment) The cosmic ray ionization rates derived from diffuse cloud models quoted in your talk are very high and refer to the case that the H_3^+ dissociative recombination is high. If that rate is low, we find $\zeta \approx 7 \times 10^{-17}$ s^{-1}.

S.Lepp: Yes, the second column on my view graph had the rates assuming H_3^+ recombination is slow.

DENSE KNOTS IN HIGH LATITUDE MOLECULAR CLOUDS

LEO BLITZ
Astronomy Department, University of Maryland
College Park, MD, 20742 USA

Abstract. The high latitude molecular clouds are discussed in the context of the formation of very low mass condensations: protojupiters. The ability to identify such objects is intimately tied to the resolution of the current debate concerning the density of the condensations inferred from observations of high density tracers. At the heart of the debate is whether inclusion of electron excitation plays an important role in the determination of column densities. A brief review of molecular abundances shows that with the possible exception of H_2CO, the molecular abundances in the high latitude molecular clouds are not abnormal.

Key words: Interstellar:Molecules; Stars:Formation; Molecules: Abundances

1. Introduction

The high latitude molecular clouds (HLCs) are a diverse set of objects that encompass both dark and diffuse molecular clouds. On one end of the scale are the Barnard objects, classical dark clouds found at large angles from the Galactic midplane. Next, are the low extinction clouds found by means of their CO J = 1 − 0 emission (Blitz, Magnani, and Mundy 1984). They are synonymous with the transluscent molecular clouds of van Dishoeck and Black (1989). The mean visual extinction in these clouds is ∼ 1 magnitude, but there is a great deal of local variation. These clouds have been shown to have a mean distance from the Sun of about 100 pc, and they are, in general, not gravitationally bound. Furthermore, they are the high density cores of the infrared cirrus emission (See Blitz 1991 and references therein). A related set of objects are the "tiny little clouds," identified by Knapp and Bowers (1988) near Betelgeuse. Although they are observed closer to the plane than the HLCs identified by Magnani, Blitz and Mundy (1985), and are at the low end of the HLC size scale, their physical characteristics appear to be quite similar to the HLCs.

There are two other categories of high latitude molecular clouds identified in the literature. The first is a set of three intermediate velocity clouds (Mebold *et al.* 1985; Heiles, Reach and Koo 1988; Desert, Bazell and Blitz 1990) that have large negative velocities compared to the dispersion of the HLC sample. The most extensively studied of these is the Draco cloud (*e.g.* Mebold *et al.* 1985). All three may be related to material falling onto the galactic plane from the corona. The second category is a set of clouds identified from their IRAS emission to be molecular, but which contain no detectable CO J = 1 − 0 emission (Blitz, Bazell, and Désert 1990). These clouds have been argued to be the classical diffuse molecular clouds seen by means of their dust emission. Since the diffuse molecular clouds have lower CO abundances than dark or transluscent clouds, the radio CO emission would be much harder to detect. At least one of them appears now to have been detected in CO with a long integration (Boulanger, personal communication), confirming their molecular nature.

Because even giant molecular clouds can be found far from the plane, we require clouds to satisfy two criteria in order to be classified as HLCs. First, they must be at latitudes greater than $b = 25°$, and there must be no emission contiguous with a cloud at lower latitude. In this way, one can avoid contamination of the HLCs with the high latitude extensions of large clouds located at lower latitudes.

2. Protostars and Protojupiters

One of the outstanding characteristics of the HLCs is their proximity. Distances have been measured to only a handful of clouds, but one of them, MBM 12, has a distance of only 65 pc, which was determined from the presence and absence of strong interstellar absorption lines in the spectra of stars of known distance along the line of sight (Hobbs, Blitz and Magnani 1986). The HLCs thus may afford us the opportunity to observe the process of star formation at the highest possible angular resolution if stars do indeed form in these clouds. In fact, the HLCs may be the best place to look for substellar masses: objects with masses less than 0.08 $M_\odot$. Sensitive searches for such "brown dwarfs" have been carried out in the infrared, but the sensitivity is considerably increased when one looks for gravitationally bound *molecular* condensations that have not yet collapsed to form the brown dwarfs (Pound and Blitz in prepration). With conventional single dishes, it is possible to detect molecular condensations with masses as low as the mass of Jupiter in the nearby high latitude clouds. Such "protojupiters" should be seen in abundance if the star formation efficiency in the high latitude clouds is similar to that in GMCs and if the shape of the initial mass function (IMF) is similar to a Salpeter function. Even if the the power law index of the IMF is as low as 1.0, as indicated by Kroupa, Tout and Gilmore (1991), protojupiters should still be observable in the HLCs.

The reason that the detection sensitivity of substellar masses is so high in their protojupiter phase compared to their brown dwarf phase is that in the earlier phase, the objects are much bigger and therefore are radiating through a much larger surface area. However, the search for protojupiters in the HLCs supposes that they can indeed form stars, even though the clouds as whole are not gravitationally bound. There is little reason to look for protojupiters in the HLCs unless there is already some evidence for star formation in them. The direct evidence to date is rather ambiguous. At least one of the HLCs, MBM 20, is known to have several T Tauri stars in its nucleus. However, this cloud is the known dark cloud L1642. Sensitive searches for young stellar objects in the HLCs that are not in the Lynds (1962) compilation (Magnani, Caillault and Armus 1990) have failed to turn up any uncatalogued stars. We must therefore ask the more fundamental question: is it possible for clouds that are not themselves gravitationally bound, to form dense cores that are bound? The answer to this question leads to a controversy which has been brewing in the study of HLCs.

To put the controversy in context, it is necssary to first describe how one would identify a protojupiter. Such objects would surely appear as dense knots in molecular maps of HLCs. To show that the mass of a candidate protojupiter is less than a stellar mass it would be necessary to get two independent measurements of the mass, both would have to agree to within the uncertainties, and both would have to show that the mass is less than 0.08 $M_\odot$. On the one hand, it would be a relatively simple task to get a virial mass estimate, from the measured linewidth and size of the condensation (assuming that it was observed in a cloud of known distance). A virial measurement is not sufficient however, because the condensation may have a much lower true mass, and the clump might be supported by pressure rather than by gravity. It is therefore necessary to obtain another measurement of the mass of the clump from measurements of several molecules and molecular transitions to establish that the clump mass is approximately equal to the virial mass. But here is where we run into trouble.

3. Densities in the HLCs

In a first attempt to determine molecular abundances in HLCs, Magnani, Blitz, and Wouterloot (1988) observed a number of clouds in the 6 cm line of H_2CO, and made small maps in four of the clouds. Turner, Rickard and Xu (1989) reobserved most of these positions in the 2 cm line of H_2CO, as well as in two lines of C_3H_2; a few positions were also observed in HC_3N. These positions may be characterized as the high density cores of the HLCs. Turner et al. used the Magnani et al. results to do a multilevel analysis in both H_2CO and C_3H_2 for seven of the cores and found a median value for the density in the cores of 4×10^4 cm^{-3}. They concluded that the cores are gravitationally bound, and are very opaque with $A_v \sim 10$ mag.

If correct, this result is very striking, because it implies that the HLCs should be active sites of star formation. Furthermore, it raises the possibility that even though both the CO and IRAS results imply that the clouds are not massive enough to be gravitationally bound, there might be enough mass in the dense cores to bind many of the unbound clouds.

Another implication is that the appearance of the clouds on the Palomar Sky Survey prints is misleading. The HLCs are very difficult to see on the Palomar prints in most cases; that is why they were, for the most part, not catalogued by Lynds. Star counts (Magnani and de Vries, 1986) show that the clouds have mean visual extinctions of 1 magnitude or less, yet if the Turner *et al.* results are correct, they should have cores, sometimes extended by as much as 10', with extinctions an order of magnitude higher. There is certainly no hint from the optical appearance of the clouds that they have such dense knots, but extinction observations are on too coarse a scale to be able to make definitive judgements.

On the other hand work by van Dishoeck *et al.* (1991) and Gredel *et al.* (1991) suggests lower densities in the condensations based on multitransition studies of CO in a few HLC cores and observations of the important HD 210121 cloud. The latter source is particulary important because the star from which it gets is name is in the backgound and optical lines of C_2 and CN are observed in absorption. The cloud is, however, different from most of the HLCs in that it is gravitationally bound to within the uncertainties of the measurements. In any event, van Dishoeck *et al.* and Gredel *et al.* argue that the clouds are sufficiently diffuse and transparent that electron electron excitation is important, and must be taken into account in the multitransition studies. They find that the density of the cores is likely to be significanly smaller than the values derived by Turner *et al.*, who ignored electron excitation, by as much as a factor of about five. Unfortunately, there were no cores common to the various studies to make a direct comparison possible. The core densities derived when electron excitation is included is closer to $\sim 10^3\ cm^{-3}$; this is smaller than one would expect if the differences in the density estimates were due to electron impact alone. However, the differences between the two analyses is no longer so alarming. Furthermore, the implied extinctions are no longer at odds with the optical appearance of the clouds. Ongoing observations using different molecular tracers should resolve the issue definitively.

4. Abundances

It is worthwhile mentioning that at least 10 different molecules have been observed in the HLCs to date. Radio detections include CO (including the ^{13}CO isotope and the J = 2 - 1 and 3 - 2 transitions), OH, H_2CO, CH, C_3H_2, HC_3N(?), CS, and NH_3. Optically detected molecules include C_2, CN, and CH^+.

The abundance of CO apparently varies from its value in diffuse clouds of $\sim 10^{-6}$ to the dark cloud value of $\sim 10^{-4}$ depending on the cloud density and environment (Magnani *et al.* 1985; Blitz *et al.* 1988, van Dishoeck 1991). The OH abundances were found by Magnani *et al.* (1988) to be very high $\sim 2 \times 10^{-6}$, implying a substantial enhancement, perhaps as a result of shock formation. However, Magnani and Siskind (1990) have subsequently shown that with better resolution and sensitivity, the abundance anomalies of the original observations disappear, and the OH abundances are $\sim 10^{-7} - 10^{-8}$. Turner et al. (1989) concluded that the abundances of HC_3N and C_3H_2 were anomalously *low* compared to Galactic plane dark cloud values. However, if the H_2 column densities they derived are too high becuase electron excitation was ignored, then the abundances of these molecules might also be normal.

There seems to be considerable disagreement on the abundance of H_2CO. Abundance determinations vary by four orders of magnitude ranging from 1.6×10^{-6} (Meyerdierks et al. 1990) and $10^{-7} - 10^{-7.7}$ (Grossman et al. 1990), to $2 - 3 \times 10^{-9}$ (Magnani et al. 1988) and $2.5 \times 10^{-9} - 2.5 \times 10^{-10}$ Turner et al. (1989). The first three sets of observations are from the 6 cm transition alone and the last one from both the 6 cm and the 2 cm transitions taken together. Again, the Turner et al. observations are commensurate with the Magnani et al. observations if the derived H_2 column densities in the former observations are too high. It is unclear why the first two sets of observations obtain such high abundances of H_2CO compared to the latter two, but a consistent set of new multitransition studies sould resolve the issue.

The observational evidence therefore suggests that the chemical abundances in the HLCs are not abnormal, with the possible exception of H_2CO. As more molecules are observed, the physical conditions and the chemistry of these transitional objects will become more tightly constrained.

References

Blitz, L., Maganani, L, and Mundy, L. (1984), *Ap. J. (Letters)*, **282**, L9.
Blitz, L., Maganani, L, and Mundy, L. (1984), *Ap. J. (Letters)*, **282**, L9.
Blitz, L., Bazell, D., and Désert, F.X., 1988, *Ap. J. (Letters)*, **352**, L13.
Désert, F.X., Bazell, D., and Blitz, L., 1990, *Ap. J. (Letters)*, **355**, L51.
Gredel, R., van Dishoeck, E.F., de Vries, C.P., and Black, J.H., 1991, *Astr. Ap*, in press.
Grossman, V., Heithausen, A., Meyerdierks, H., and Mebold, U. 1990, *Astr. Ap*, **240**, 400.
Heiles, C., Reach, W.T., and Koo, B.-C., 1988, *Ap. J.*, **322**, 313.
Hobbs, L.M., Blitz, L. and Magnani, 1986, *Ap. J. (Letters)*, **306**, L109.
Knapp, G.R., and Bowers, P.F. 1988 *Ap. J.*, **331**, 974.
Kroupa, P., Tout, C.A., and Gilmore, G. 1991, *M.N.R.A.S.*, **251**, 293.
Lynds, B.T., 1962, *Ap. J. Suppl.*, **7**, 1.
Magnani, L., Blitz, L., and Mundy, L., 1985, *Ap. J.*, **295**, 402.
Magnani, L., and de Vries, C.P., 1986, *Astr. Ap*, **168**, 271.
Magnani, L., Blitz, L., and Wouterloot, J.G.A., 1988, *Ap. J.*, **326**, 909.
Magnani, L., Caillault, J.-P., an Armus, L. 1990 *Ap. J.*, **357**, 602.
Magnani, L., and Siskind, L. 1990, *Ap. J.*, **359**, 355.
Mebold, U., Cernicharo, J., Velden, L., Reif, K., Crezelius, C., and Georigk, W., 1985, *Astr. Ap*, , **151**, 427.
Meyerdierks, H., Brouillet, N., and Mebold, U. 1990, *Astr. Ap*, **230**, 172.
Turner, B.E, Rickard, L. J, and Xu., L-P., 1989, *Ap. J.*, **344**, 292.
van Dishoeck, E.F., and Black, J.H., 1989, *Ap. J.*, **340**, 273.
van Dishoeck, E.F., Black, J.H., Phillips, T.G., and Gredel, R., 1991, *Ap. J.*, **366**, 141.

QUESTIONS AND ANSWERS

B.E.Turner: We have included electron excitation now in our models of H_2CO in cirrus cloud cores, which was omitted in our earlier paper (Turner et al.1989,Ap.J.). The H_2CO abundance does not change much ($\lesssim 30\%$) by this inclusion, but the total densities derived are reduced by a factor 3 to 5, bringing them in good agreement with those derived from $C^{18}O$ observations, and in better agreement with those derived by van Dishoeck et al.(1991), although we have studied only two clouds in common with these authors. We still find extinctions rangiong from 2 to 8 magnitudes in 17 cirrus cloud cores studied in detail, but the central densities now do not exceed 10^4 cm^{-3}.

L.Blitz: It remains to be seen whether the extinctions of 2-8 magnitudes are consistent, with extinction estimates from IRAS maps. I expect that 8 magnitudes in a small area will be easily visible on the palomar prints. Lowering the densities by 3-5 should now make the clumps unbound by gravity.

R.Opher: You talked about the importance of whether clouds are bound or unbound. I would like to point out that thermal instability due to bremsstrahlung radiation (e.g. Opher and Valio, 1990) would make many of the clouds, that you say are unbound, bound.

L.Blitz: I am not familiar with this work, but I doubt that a thermal instability could bind by gravity the unbound clouds. Bremstrahlung is unimportant in these objects (the ionization is low) and in any event, it's hard to see how one can cool the clouds to lower temperatures than they already are.

SOURCE INDEX

AFGL 2136 : 285
AFGL 2688 : 47
R.R. Aql : 389
R Aqr : 409
AzV 211 : 109
AzV 221 : 109
AzV 456 : 109

B1 : 199,237
B335 : 193,217,471
BD-21.3873 : 409
BD +30°549 : 351

C 1003 : 403
α Caen A : 85
γ Cas : 85,319
Cas A : 85
λ Cep : 127
Cepheids : 327
Cen A : 111
α Cen B : 85
CIT 6 : 401
μ Col : 85
Comet,Bennett(1969i=1970II) : 443
    ,Austin(1989c1) : 439, 459
    ,Bradfield (1987s) : 443
    ,Brorsen-Metcalf(1989X) : 439,441
    ,Giacobini-Zinner(1984e): 447
    ,Hartley-Good(1985l) : 443,447
    ,Halley(1982i) : 97,441,443
    ,IRAS Araki-Alcock(1983d) : 459
    ,Levy(1990c) : 439,459
    ,Thiele(1985m) : 447
    ,West(1975n=1976VI) : 443
    ,Temple 2 : 449
CRL 618 : 363,389
CRL 2688 : 363,389
Cyg 12 : 65

DR 21-OH : 227

AFGL 2179 : 133
λ And : 85
κ Aql : 127
α Aur : 85
AzV 215 : 109
AzV 398 : 109

B5 : 193,199,205
BN : 259,265,285,387
BD +30 : 359
BM GEM : 403

Canopus : 85
C ans : 389
β Cen : 85
Cep A : 217,271
Cepheus A/GCD37 : 373
α Cen A : 85
Ceres : 421
α C Mi : 85

CRL 2591 : 217
α Cru : 85

Earth : 421
ER Del : 409

F H Ser : 365

Ganymede : 421
GL 437 : 217
GL 2136 : 259
G 10.47+0.03 : 251
G 34.3+0.15 : 191,227

HD 44179 : 47,65
HD 169454 : 143
HD 29647 : 143
Heiles Cloud : 199
Hen 2-99 : 359
HH1-2 : 373,217
HH7-11 : 373
HH24 : 217
HH26IR : 217
HH43 : 373
HH52/53/54 : 373

IC 10X : 285
IC 342 : 111
IC 443G : 181,237
IC 5146 : 351
ε Ind : 85
IRAS 063301+3057 : 237
IRAS 16293-2422 : 191
IRc 4 : 265
IRC +10216 : 23,31,181,285,363,379,389,399,401
IRC+10420 : 389
IRS 3 : 133
IRS 7 : 65,133,423

Jupiter : 97,435

K 3-50 : 225

L 43 : 217
L 134N : 1,171,193,205,217,319
L 183 : 217
L 778 : 217

ε Eri : 85

41.9+58 : 111

GC IRS7 : 65,133
GL 490 : 217,225,259
G 305.8-0.2 : 405
G 31.41+0.31 : 251
G 35.2N : 217

HD 154 368 : 143
HD 62542 : 143
HD 210121: 143,477
Hen 1 : 395
Hen 2-113 : 359
HH6 : 373
HH12 : 373
HH25-26 : 373
HH32 : 373
HH47 : 373
HR 1099 : 85

IC 63 : 319
IC 418 : 351
IC 443 : 319
IK Tau : 389
Io : 435
IRAS 15194-5115 : 401
IRc 2 : 1,237,265,297
IRc 7 : 265

IRS 1 : 199
IRS 5 : 217,237
IRS-LAS : 65

L 43B : 217

L 723 : 217
L 1450 : 373

L 1455 : 217
L 1582B : 217
$L_k H_\alpha$ 101 : 351,259

L 1515 : 217,237
L 1642 : 477
L 1709A : 217
L Ori : 85

βc Ma : 85
Mars : 97,421
Mercury : 421
MWC 1080 : 217
M 17 : 297
M-0.13-0.08 : 223
M 51 : 285

Maffei 2 : 111
MBM 12 : 143,477
Mon R 2 : 217
M 8E : 217,259
M 17SW : 205,271,303
M 42 : 333,297
M 82 : 111,121

N 19 : 111
N 5315 : 359
Neptune : 97
NGC 89 : 285
NGC 1068 : 103,329
NGC 1333 : 285,351,373
NGC 2023 : 303
NGC 2024 FIR3 : 251
NGC 2071(N) : 199,217
NGC 2264 IRS : 285
NGC 4945 : 111
NGC 6270 : 333
NGC 6857 : 225
NGC 6946 : 117
NGC 7293 : 363
NGC 7538 IRS1 : 259,347
NGC 7538 IRS9 : 65,217,227,259,285

N 159 : 111,121
N 7538-IRS1 : 251
NGC 253 : 111,121
NGC 1331 : 351
NGC 1977 : 303
NGC 2024 : 333
NGC 2024 FIR5 : 191
NGC 2024 IRS2 : 285
NGC 2261 : 217
NGC 4151 : 103
NGC 6240 : 103
NGC 6302 : 351
NGC 7027 : 333,363
NGC 7538 IRS2-3 : 347
NGC 7674 : 335
NQ Vul : 365

NML Cyg : 389

OH 231.8+4.2 : 389
ζ Ophiuchi : 127,143,157,193,285,309
ρ Ophiuchi A : 127,191,199,335

χ Oph : 127

Orion A : 193,227,297,303,335

Orion Molecular Cloud(OMC): 91,103,191,211,223,237,265,275,
281,349,317,395
ON3 : 225
ε Ori : 85
Ori B : 227,303
Orion KL : 1,91,181,193,223,217,227,237,251,271,285, 311,317,335,459
α Pav : 85
ψ Per : 127

α Ori : 285,389
δ Ori : 85

ε Per : 85
o Per : 127

ζ Per : 127
ζ Persei : 143
ζ Pup : 85

QU Vul : 365
RX-Boo : 389
R Leo : 389

Saturn : 97
S140 : 217,227,297,303
S 150 IRS1 : 259
R Scl : 389
δ Sco : 127
π Sco : 143, 285
S Sct : 389
Sgr A : 223
Sgr B2(M) : 181,187,227
68 Cyg : 127
SVS 3 : 351
Sw St-1 : 359

Taurus-Auriga-Perseus : 159

Titan : 421
23 Tau : 127
TT Cyg : 389
TW Hya : 395

U Aut : 389
Ursa Majoris : 159
UV Aur : 409

V 842 Cent(1986) : 365
Venus : 97,421
Ve 2 : 45
α Vir : 85
V Cyg : 389

W3(OH): 19,227,277,341
W 3 : 303
W 3IRS5: 65,191,259
W 33A : 65,259,459
W 49N : 349
W 51 : 349

ξ Persei : 143,157
δ Persei : 157

R Cas : 389
R LMi : 389

S 140 IRS1 : 65,217
S187 : 351
λ Sco : 85
β^1 Sco : 127
S CrB : 389
μ Sgr : 127
Sgr B2 : 179,181,187,191,223
67 Oph : 127
SN 1987A : 41,285,365,371,387
SVS 13 : 373, 379
SY Mus : 409

TMC-1 : 1,23,143,155,169,171,
181,199,211,223,237,285,319
20 Tau : 127
23 Ori : 127
TX Cam : 389

Uranus : 97
U Cam : 389

V 1370 Aql : 365
Vul : 365
γ Vel : 85
VII Zw31 : 335
VI Cygni # 12 : 389

W 28 : 237
W3 IRS4 : 191
W 31c : 251
W 44 : 237
W 49 : 321
W 51-e1,e2 : 345

W 51d : 251
W 51(MS) : 227
W 51S : 345

X Cyg : 389

W 51M : 181,191,227,345
W 51(N) : 227
W 58 : 225

INDEX OF CHEMICAL SPECIES

ONE ATOM

Al : 181, 411
Ar : 361
Ba : 409
C : 31,55,127,143,193,
199,249,251,281,
285,297,303,321,
329,333,335,357,
359,365,379,389,
421,459
^{13}C : 127
Ca : 181
e$^-$: 1,7,31,73,83,143,
211,329,357,379,
387,471
Fe^{++} : 281
H$^+$: 1,73,329,379,387,471

H$^-$: 73,83,365,379
D : 1,73,85,91,97
He$^+$: 7,73,193,359,365,471

^{4}He : 73,85
^{7}Li : 73,85,395,471
Mg$^+$: 181
N$^+$: 7,103,143,321,361
Na : 47,181,357
Nd : 409
O : 1,103,143,193,217,
245,249,251,297,
329,335,357,361,
365,373,379,415,
451,471
S : 1,181,217,357,361,
389,415,459
Si : 31,181,281,379
Sm : 409
Sr : 409

Al$^+$: 181
B : 471
Be : 471
C$^+$: 1,7,117,143,153,193,211,217,
285,297,303,309,321,329,359,
365,379,389
C^{++} : 281,359
C^{+++} : 359
C$^-$: 365
^{13}C$^+$: 193,303
^{56}Co : 365
e$^+$: 471
Fe : 39,181,281
Fe$^+$: 1,103
H : 1,7,31,47,55,65,73,83,85,117
143,153,159,211,217,285,309
329,357,361, 365,373 ,379,
387,389,415,471
He : 7,55,73,357,359,361,365
He^{++} : 73,471
^{3}He : 73,85
K : 181
Mg : 39,181
N : 1,13,211,251,357,361,365,
459,471
Na$^+$: 181
Ne : 361
O$^-$: 365
^{18}O$^+$: 335
O^{++} : 361
P : 181
Ps : 41
S$^+$: 1,373
S^{++} : 361
Si$^+$: 1,181,379
Si^{++} : 281
Ti : 181

Y : 409

TWO ATOMS

AlCl : 181,427
AlN : 181
AlO : 181
$^{12}C^{13}C$: 427

$^{13}CH^+$: 127,193

CD : 155

CN$^+$: 427
CO : 1,27,31,47,55,65,91,
   111,117,121,143,153,
   157,159,169,171,181,
   193,199,217,225,237,
   245,249,251,259,265,
   271,285,297,303,311,
   317,319,321,329,363,
   365,379,387,389,399,
   407,413,415,421,427,
   449,451,471,477
CP : 181,389,427
C^{34}S : 225,271

FeO : 181
H$_2$: 1,7,31,55,73,83,91,97,
   103,111,143,153,155,
   159,169,181,211,217,
   237,245,251,285,309,
   311,319,329,363,365,
   373,379,387,415,427,
   451,471
HD : 7,73,91,97,169,211
HCl : 7,55,427
HeH$^+$: 73,365
LiH : 73
MgO : 181
NaO : 181
N$_2$: 1,27,285,449,451,459
NH : 143,181,363,415,427,451

AlF : 181,427
C$_2$: 143,153,171,193,365,427,443,447,47

CH : 127,153,171,199,285,329,
   357,363,365,427,477
CH$^+$: 1,127,143,153,159,193,
   309,329,363,427,477
CN : 1,13,55,111,121,127,143,
   153,171,193,237,271,317,
   319,363,365,389,401,403,
   415,427,443,447,477
^{13}CN : 193
CO$^+$: 329,365,427
^{13}CO : 143,153,159,199,217,225,
   259,265,271,285,297,303,
   311,319,335,347,389,401,477

C^{18}O : 159,171,193,199,217,225
   251,265,271,285,297

$^{13}C^{18}$O: 193

CS : 31,111,121,171,193,199,
   217,225,237,251,271,311,
   319,389,401,415,477
FeS : 181,415
H$_2^+$: 73,83,365,379,387,471

D$_2$: 7
DCl : 97
KCl : 181,427
LiH$^+$: 73
MgN : 181
NaN : 181
N$_2^+$: 427,451
NH$^+$: 7

NO : 13,171,317,427
NaCl : 181,389,427

$^{16}O^{18}O$: 191,335

OH^+ : 427,471
PN : 181,389,427
PO : 181
SH : 1,217,357,415

SO^+ : 181,427
SiH : 39,461

SiN : 171,181,379,389,427
TiC : 181
TiN : 181

NS : 227,427
O_2 : 1,13,27,55,189,191,193,
      285,329,335,365,451
OH : 1,7,39,111,143,171,211,
     217,237,245,285,309,317,
     329,345,357,365,379,389,415,427,477
OD : 155
PS : 181
S_2 : 415,451,427,459
SO : 1,111,121,171,179,181,199
     237,271,319,357,389,415,
     427,451,459
SiC : 39,181,365,379,427
SiO : 1,39,111,181,199,223,225,
      237,251,365,371,379,389,401,427
SiS : 181,237,379,389,401,427
TiO : 181,409

THREE ATOMS

AlOH : 181
C_3^+ : 365
CaOH : 181
C_2H : 111,121,171,193,271,
        379,389,401,427
C_2S : 171,389,427
CO_2 : 27,65,97,285,435,449,
         427,451
CCH : 389
DCN : 1,211
H_3^+ : 1,7,91,143,169,211
         259,329,365,389,
         427,471
HCN : 111,121,171,181,199,
      211,225,231,237,251,
      265,271,317,319,347,
      363,379,389,401,439,
      449,427,451
HNC : 111,121,171,211,237,
      271,317,379,389,401,427
$HN^{13}C$: 401
HCO : 31,223,357,427
HCO^+ : 1,7,31,91,111,121,
         171,211,223,225,237,

C_3 : 31,139,389,443,447,427
CH_2 : 193,363
CH_2^+ : 329,363
C_2O : 171,181

C_2D : 211
CO_2^+ : 427

CS_2 : 415,449,451,459
DCO^+ : 91,155,169,211
DNC : 211
$H^{13}CO$: 223
$HC^{18}O^+$: 223
$H^{13}CN$: 271,401

251,271,319,329,347,
357,373,389,427,471
$H^{13}CO^+$: 169,171,225,271
HCS : 31
HNO : 187,317,357,427
H_2O : 1,7,27,29,65,97,157,
181,191,193,231,237,
249,259,275,285,317,
329,335,345,349,389,
413,415,421,427,435,
437,449,451,459
HDO : 1,97,211
H_2D^+ : 91,155,169,211

NH_2 : 55,317,415,427,451
MgOH : 181
MgNH : 181
NH_2^+ : 471
N_2H^+ : 171,211,237,357,111
271,427
OCH : 217
OCS : 7,27,65,171,223,265
415,427,439,451,459
$^{33}SO_2$: 179
Si_2C : 181,379
SiH_2 : 181
SiC_2 : 31,181,379,389,401
SiC_2^+ : 339

HCP : 181,389
HCS^+ : 7,31,171,223,427
HNSi : 181
H_2O^+ : 415,427
$H_2^{18}O$: 335

H_2S : 1,7,171,199,237,357,389
415,427,435,439,449,451,459

MgC_2 : 181
NaCN : 181
NaNH : 181
N_2D^+ : 155,211

OCN : 357
SO_2 : 1,7,171,179,199,237,319
389,421,427,435,439,451,459
$^{34}SO_2$: 179
c-SiC_2 : 427
SiCC : 399
SiO_2 : 365

FOUR ATOMS

AlCCH : 181
CH_3^+ : 7,211,363
C_2H_2 : 55,181,259,265,317,
389,427,449,451
CH_3D : 97, 211
1-C_3H: 427
c-C_3H: 171,427
C_3O : 171,427
D_2CO: 1,91,181,211
HC_2N : 181
$HCNH^+$: 171,211,427
$HNCD^+$: 211
H_2CO : 29,91,111,121,143,
171,181,193,237,285,

CH_3 : 317
CH_2D^+: 155,211
$C_2H_2^+$: 181,211,389
C_2HD^+: 155,211
C_3H : 171,181,211,389,401

C_3N : 171,401,427
C_3S : 171,389,427
F_3C : 181
$HCND^+$: 211

HCCN: 389
$H_2^{13}CO$: 193
HCO_2^+ : 223

317,319,347,415,427,
439,441,449,451,477
HDCO : 91,211
H_2CS : 171,415,427,439,451,459
HNCS : 227,427
$HOCO^+$: 179,427
H_3O^+ : 1,7,181,191,389,427
NaCCH : 181
NH_3 : 1,7,29,47,111,143,
169,171,181,199,205,
211,217,223,231,237,
251,317,345,357,373,
379,389,415,427,449,451
NH_2D : 1, 211
SiC_2H^+: 379

H_2DO^+: 155
H_2NC^+: 211
HNCO : 111,187,427
$HSiNH^+$: 379
H_3S^+ : 415

NH_3^+ : 31,415

$OCSH^+$: 7

FIVE ATOMS

Al_2O_3 : 17,181
CH_2C_2 : 171
CH_3D : 97,211

C_2H_3 : 31
C_3HD : 211
C_3H_2 : 111,121,143,171,181,
211,223,271,389,401,477
C_4D : 211
C_4Si: 427
CH_2CN: 1,171,427
C_5 : 139
DC_3N : 211
HCOOH : 171,223,427,451
HC_3N : 121,171,181,205,211,
237,271,317,363,379,
389,401,427,439,451,477
H_2CCC : 23,389
H_2CCO : 1,171,427
H_2NCN : 427
NH_4^+ : 7,31,415
NaCCH : 181
SiC_4 : 379

CH_2CH: 181
CH_2O_2 : 231
CH_4 : 27,65,97,199,231,259,
317,379,421,427,451,459
$C_2H_3^+$: 31

c-C_3H_2 : 427
C_4H : 171,181,211,389,401,427

C_4H^+: 211

C_4N : 31

$Fe(OH)_2$: 181

H_2CHN : 427
H_2CO_2 : 449
$MgSiO_3$: 181
SiH_4 : 181,379,427
$SiCl_4$: 389

SIX ATOMS

C_2H_3N : 55

C_2H_4 : 55,317,427,451

$C_3H_3^+$: 211
C_5H : 171,181,389,427
C_5N : 31

CH_3NC : 427
$C_3H_2D^+$: 211

CH_3OD : 1,211
CH_2C_3 : 171
HDC_3N^+ : 211
HC_3NH^+ : 211
H_2CCCC : 23,389
$HCONH_2$: 427
HC_2CHO : 171

$C_4H_2^+$: 211
C_5O : 31,181,427
CH_3CN : 111,171,181,311,317, 427,449,451
CH_3OH : 27,29,55,65,111,171, 179,181,199,223,251, 271,285,319,341,427, 437,439,449,451
CH_3SH : 427
H_2C_4 : 181
HC_3ND^+ : 211
$H_2C_3N^+$: 211
$H_3CO_2^+$: 449
NH_2CHO : 179,181

SEVEN ATOMS

C_7 : 139
C_6O : 181
$C_4H_3^+$: 389
C_3H_3N : 111
CH_3CCH : 171,223,271,427,451
DC_5N : 211
H_3CCCH : 237
HC_5N : 171,223,237,317,363,427
$H_2C_3H_2$: 449

C_6H : 31,171,181,389,427
C_2H_4O : 55
CH_3NH_2 : 427,449
CH_2CHCN : 171,317,427
CH_3CHO : 171,181
Fe_2SiO_4 : 181
$HCOCH_3$: 427

Mg_2SiO_4 : 181

EIGHT ATOMS

C_2H_6 : 317
C_7H : 31
CH_3OHCO : 181
H_3CCH_3 : 181

$C_4H_3^+$: 389
CH_3C_3N : 171,427
$HCOOCH_3$: 179,427
S_8 : 415

NINE ATOMS

CH_3C_4H : 171,427
C_2H_5CN : 179,237,427
$(CH_3)_2O$: 55,181,237,427
HC_7N : 171,317,363,427
H_3CCHCH_2 : 181

CH_3CH_2OH : 223
C_2H_5OH : 55,179,181,427
DC_7N : 211
H_3COCH_3 : 181

TEN ATOMS

C_6H_3N : 427 $\qquad$ H_3CCOCH_3 : 181,427

MORE THAN TEN ATOMS

HC_9N : 171,427
$HC_{13}N$: 31
$C_{10}H_6^+$: 31
$C_{10}H_7^+$: 31
$C_{10}H_8^+$: 25
$C_{14}H_{10}^+$: 31
$C_{16}H_{10}^+$: 31
C_{60}(Buckministerfullerene) : 47

$HC_{11}N$: 171,427
$C_{10}H_6$: 31
$C_{10}H_7$: 31
$C_{10}H_8$: 25,31
$C_{14}H_{10}$: 31
$C_{16}H_{10}$: 31

SUBJECT INDEX

Abundance : 21,31,39,65,111,137,143,153,155,157,169,181,
 191,193,199,211,251,281,319,329,357,363,395,
 401,409,411,425,439,451,459,471
 ,deuterium : 85,91,97,155,169,211
 ,molecular : 121,171,211,223,225,227,237
 ,gradient : 1,85,91,97,155,171,193,327,395
 ,fractional : 91,121,155,169,211,223,281,317,471
 ,isotopic : 85,91,127,169,193,211,451
Accretion : 47,91,181,189,199,251,285,317,421,451
Adsorption : 249,265,285,357
Albedo : 231,451
Approximation,Large velocity gradient(LVG) : 277
Association, radiative : 7,387
Asteroidal surfaces : 461
Attachment,radiative : 73,365,379
Atmosphere,cometary : 55,97,439,443,441,449
 ,jovian : 97,435,421
 ,planetary : 55,97,361,421
 ,stellar : 55,363
Bands,diffuse interstellar : 47,133,135,139,143,389
 ,unidentified IR : 47
 ,unidentified UV : 47
Brown dwarf : 477
Buckministerfullerene(C60) : 47
Calculation, ab-initio : 31
Carbon,amorphous : 65
 ,chain : 47
Center,galactic : 73
Chemical gradients : 271
Chemistry, circumstellar : 47,379,389
 ,gas-phase : 7,143,171,181,191,199,211,227,249,311,317,
 357,363,379,389,449,471
 ,grain-surface : 143,181,189,211,223,249,265,311,317,357,
 379,451,459,471
 ,shock : 143,153,181,199,223,237,249
 ,isotopic carbon : 193
 ,time dependent : 73,121,155,157,169,357
 ,premordial : 73
 ,stellar atmosphere : 285

Chondrites carbonaceous : 47,133,421
Clouds,contracting : 451
   ,dark : 171,205,249
   ,dense : 1,55,65,91,179,181,189,191,199,205,211,217,231,
       237,245,251,259,285,297,333,471
   ,diffuse : 1,65,127,143,157,159,181,231,245,285,309,321,471
   ,translucent : 1,127,143,153,159,193,285
   ,evolving : 55
   ,high-latitude : 1,143,159,285,477
   ,clumpy : 1,143,187,199,237,249,251,285,297,303,309,317,
       319,329,335,351,363,477
   ,interstellar : 1,97,143,193,205,211,245,285,401,471
   ,molecular : 1,143,155,159,169,171,179,181,191,193,
       205,211,217,237,251,259,265,297,303,
       311,319,329,333,335,341,395,407
   ,quiescent : 1,153,159,205,259,297,335
Cluster,globular : 73
   ,carbon : 1,139
   ,trapezium star : 297
Coagulation : 451
Collapse, gravitational : 73,317,357
Collisions,atom-electron : 41,303,333,365
   ,atom-molecule : 303,311
   ,atom-proton : 41,303
   ,grain-gas : 259,285
   ,grain-grain : 47
   ,ion-polar : 303
   ,molecule-molecule : 311
Comets : 55,415,421,427,439,441,443,447,449,451,459
Condensation : 1,39,83,251,357,451,477
Corona, solar : 425
Coronene,cation : 135
Cosmic background radiation : 73,127,321
Cyanopolyyenes : 171,211,389
Dark matter,decay(DDM) : 55
   ,cold : 73
Depletion : 1,121,251,285,357,395,451
Density,column : 111,121,143,153,157,159,223,251,399,447
Desorption : 249
Detachment, associative : 73

Deuterium fractionation : 169
Diamond : 47,65
Diffusion ambipolar : 1,155,379
Dissociation,collision-induced : 55
    ,collisional : 365
Dust : 15,17,39,47,65,91,109,133,143,159,231,251,259,285,
      321,365,371,379,405
    ,circumstellar : 47,259,285,403
    ,interstellar : 47,143,159,231,251,259,285
    ,attenuation : 55,65,285
    ,interplanetary : 97
    ,cometary : 427,451
Emission,chromospheric : 425
    ,thermal : 103
    ,non-thermal : 103
    ,quasi-thermal : 103
    ,interstellar : 137
Endohedral fullerene complexes : 47
Energy,excitation : 139
    ,internal : 157
Evaporation : 1,251,265,285,317,451
Evolution,chemical : 171,199
    ,galactic : 73,85,193
Excitation,collisional : 1,103,157,159,361,373
    ,cosmic rays : 1,357,389
    ,electron-impact : 1,333
    ,UV-radiation : 103
    ,shock : 103,285
Exohederal complexes : 47
Extinction: 21,47,109,133,143,153,231,251,265,285,329,379,477
Field,magnetic : 73,155,407
Flow,bipolar : 47,217,259,379,389,405,407
Fluorescence : 137,373,427
Fragmentation,electrostatic : 17
    ,gaseous-lumps : 73,47
    ,population III objects : 83
Galaxies : 73
    ,external : 103,109,117,121
    ,seyfert : 103
    ,SMC : 109,111,121,285
    ,LMC : 111,121,285
Gas,shock-heated : 1,103,285

,shocked : 73,91,103,285
Grains : 1,39,159,223,259,281,357,415
 ,amorphous : 47,309
 ,dirty ice : 65,189,415
 ,graphite : 47,285
 ,mantle : 91,189,199,249,251,265,285,317
 ,circumstellar : 1,39,285
 ,interstellar : 1,39,189,231,285,317
 ,CHON : 415,451
 ,sulfur-bearing : 415
 ,refractory : 415,451
 ,presolar : 451
 ,silicate : 39,65,223,365,371,403,415,451,461
Herbig-Haro objects(H-H) : 373
Hydrocarbons,aliphatic : 65,133
Ice : 27,29,65,181,189,199,249
 ,amorphous : 27,437
 ,dirty : 27,65,181,189,249,415,435,437
 ,feature : 27,65,181
Index,refractive : 27
Interaction,rotation-vibration : 1,103,371,373
Interface (transition) zones : 1,143,153,159,199,245,249,
       285,297,303,309,329,333,341,379,389
Ionization,associative : 365,387
 ,collisional : 121
Irradiation : 17
Kerogens type II : 423
Lifetime,atomic : 41
 ,molecular : 231
Lyman alpha line : 41,85,373
Masers,mega : 111
 ,OH : 111,277,329,345
 ,H_2O : 245,275,311,335,345,349,405
 ,H_2CO : 347
 ,CH_3OH : 341
 ,NH_3 : 345
Mass loss : 259,285,359,363,365,379,389,411,413
Mechanism,maser : 245,277
 ,pumping : 143,245,277,329
Medium,diffuse interstellar : 65,85,133,231,285,411
 ,dense interstellar : 65,231,285,411

,intergalactic : 73
Metamorphism : 39,47
Meteorite : 47,97,423
Model,big-bang : 73,83,85
    ,evolutionary : 85,143,205
    ,cold dark matter : 73
    ,MHD : 143,153
    ,chemical : 121,199,387
    ,thermal approximation : 137
    ,shock : 143,153,159,181,199,245,285,311,351
    ,steady-state : 143,365
Molecules,circumstellar : 23,181,245,285,379,389,399,403,405,413,427
    ,cometary : 29,427
    ,complex : 29,181
    ,destruction : 181,297,379
    ,formation : 121,181,217,297,379
    ,interstellar : 23,29,171,181,187,191,223,217,225,227,231,
            237,251,265,271,345,349,357,363,389,427,477
    ,novae : 365
    ,supernovae : 365
    ,stellar winds : 379
    ,parent : 415,439
    ,planetary : 435
    ,polar : 23
Nebulae,planetary : 47,333,351,359,361,363
    ,presolar : 97
    ,reflection : 351
    ,solar : 97
Novae : 47
Nucleosynthsis : 85,193
Nucleus,cometary : 415,451
Olivine,annealed : 365
Oort cloud : 451
Oscillator strength : 139
Oxidation : 39
Photochemistry : 55,379
Photodestruction : 55,73,143,181,357,363,379
Photodissociation : 1,55,73,143,181,231,297,329,351,363,389,415
Photoionization : 55,143,231,329,351,363,389,415
Photon-dominated region(PDR) : 1,91,103,117,121,193,281,285,
                   297,303,309,311,319,321,329,333,351

Planets : 97,421
Plasma : 387
    ,primordial : 83
    ,MHD : 155
Polarization : 47,109,275
Polycyclic Aromatic Hydrocarbons(PAHs) : 1,15,21,25,31,47,
                65,91,135,137,285,365,427,451
    ,deuterated(PADs) : 91
Polymer,polyoxymethylene(POM) : 29,449
    ,orgneil carbonaceous : 423
    ,meteorite : 423
Processes,destruction : 7,91
    ,photodestruction : 1,55,143,231
    ,photoionization : 1,55,143,231
    ,physico-chemical : 1,39,111,143
    ,s- : 403
Protojupiters : 477
Pulsar : 471
Pumping,UV : 157
    ,diamagnetic : 425
Radiation field,interstellar : 1,55,143,159,193,231,297,303,
                309,329,333,365,389
    ,transfer : 311
Rates,cosmic ray ionization : 1,249,259,471
    ,ionization : 55,249
    ,mass loss : 1,217,401,407,447
    ,photodissociation : 55,143,231,329,333
    ,photoionization : 55,143,231
    ,reaction : 7,387
Ratio,abundance : 1,29,31,55,65,73,91,97,127,133,143,
                155,157,159,169,181,193,211,249,251,259,
                265,281,285,297,335,341,347,359,361,
                365,389,403,439,451,459
    ,branching : 7
    ,dust-to-gas : 251
    ,isotopic : 15,73,85,91,97,127,143,193,155,157,169
Reactions,chemical : 91,179,193
    ,exothermic : 73,181
    ,grain-surface : 143
    ,ion-atom : 73
    ,ion-neutral : 73

,ion-molecule : 7,19,31,47,143,155,181,193,237,329,379,389,471
,ion-polar neutral : 7
,neutral-neutral : 13,19,91,171,193,329,317
,neutral-radical : 143,329,387
,radiative association : 143
,radical-radical : 389
,recombination : 193,329,333
,ternary : 365,379
Recombination,dissociative : 7,31,169,365,459,471
,dissociative electron : 193
,radiative : 7,41,245,359,471
,three body : 365
Regions, HII : 117,225,303,351,361,389
,HII ultra compact : 143,225,251,259,303,311,341,345,347
,shocked : 91,143,181,217,285,297,311,319,329,335
Shell,circumstellar : 1,181,285
Shielding,self : 1,285,363,379
Shocks : 1,73,143,199,237,309,311,335,363,373,389
,C-type : 155,181,199,237,285,311,373
,J-type : 181,199,237,285,311,373
Solar system : 421,425,435
Spectra,rotational : 23,91,159,271,303,373,427,439
,infrared(IR) : 15,25,27,31,39,47,65,117,159,189,259,265,285,303,
311,329,365,371,373,379,401,403,427,435,437
,ultraviolet(UV) : 25,47,65,427
,visible : 25,65,127,135,409,427
,mm & submm : 303,373,379,399,401
Spectroscopy,grain : 65
,absorption : 65,135,265,259
,emission : 65
Sputtering : 309,415
Star formation : 199,217,259,297,335,477
Stars,O-type : 85,117,329
,B-type : 85,103,117,329,351
,carbon : 285,363,379,401,403,411
,central of planetary nebulae : 285,359
,cool : 85
,late-type : 389
,M giant : 245,389
,Mira : 39,245,285
,OH/IR : 65,389

,oxygen rich : 39,389
,main sequence : 85
,proto : 1,217,245,259,277,285,311,357,379,387,405,407
,red giant : 359,379,409,413
,S-type : 389
,supergiant : 245,413
,T-Tauri : 379,395,477
,Wolf Rayet : 359
Supernovae : 1,47,65,319,365,371,387,471
Temperature,excitation : 111,127,143,157,159,285
,rotational : 127,143,157,159
,vibrational : 259
Transitions,collison induced : 245
,excitation : 139
,fine structure : 1,297,303,329,333,365
,hyperfine : 1,351
,quadrupole : 285
,radiative : 245,361
Turbulance : 1,143,153,159,217,407
Unidentified lines : 55,179
Universe, early : 1,73,83,85
,expanding : 73,83
Waves,shear Alfven : 155,407
Widths,equivalent : 127,359,395
Winds,BN : 387
,stellar : 85,97,359,363
,protostellar : 1,357,387
,solar : 425,461

AUTHOR INDEX

Abraham, Z. : 275, 349
Aiello, S. : 231
Aitken, D. K. : 371
A'Hearn, M. F. : 415
Allamandola, L. J. : 15, 25, 27, 29, 65, 133, 137, 435, 437
Amano, T. : 399
Auh, B. R. : 225
Avery, L. W. : 399
Badrinathan, C. : 139
Barlow, M. J. : 387
Batrla, W. : 347
Behar, F. : 423
Bell, M. B. : 399
Bergman, P. : 179
Bergin, E. A. : 271
Black, J. H. : 153, 319
Blake, D. F. : 437
Bockelee-Morvan, D. : 439, 459
Boice, D. C. : 447, 449
Booth, R. S. : 401
Breukers, R. : 189
Burdyzha, V. V. : 41
Butner, H. M. : 169
Carlstrom, U. : 401
Carpenter, J. : 271
Cecchi Pestellini, C. : 231
Cernicharo, J. : 23
Charnley, S. B. : 155, 199, 317
Colgan, S. W. J. : 281
Colom, P. : 439, 459
Costa, R. D. D. : 359, 361
Coyne, G. : 109
Crane, P. : 127
Crovisier, J. : 439, 459
Curriel, S. : 373
Dalgarno, A. : 1, 333
d'Hendecourt, L. : 135, 189, 423
de Almeida, A. A. : 443
de Araujo, F. X. : 359
de Araujo, J. C. N. : 83
de Freitas Pacheco, J. A. : 359, 361
de Pater, I. : 441

Delsemme, A. H. : 421
Defourneau, D. : 21
Defrance, A. : 13
Despois, D. : 439, 451, 459
Doel, R. C. : 277
Drew, J. E. : 387
Duley, W. W. : 309
Dufour, R. J. : 281
Ehrenfreund, P. : 135, 189, 423
Ellinger, Y. : 31
Erickson, E. F. : 281
Escalante, V. : 333
Evans II, N. J. : 265
Falgarone, E. : 159
Federova, A. V. : 413
Federova, O. V. : 413
Feldman, P. A. : 399
Ferlet, R. : 85
Field, D. : 277
Frantsman, Ju L. : 327, 411
Freund, F. : 437
Friberg, P. : 179, 223
Gangopadhaya, P. : 19
Geis, N. : 117
Genzel, R. : 117
Glassgold, A. E. : 285, 379
Goldsmith, P. F. : 271
Gomes Balboa, A. M. : 349, 405
Gottlieb, C. A. : 23
Gray, M. D. : 277
Gredel, R. : 153
Greenberg, J. M. : 189
Gregorio-Hetem, J. : 395
Grun, E. : 17
Guelin, M. : 23, 285
Hare, J. P. : 47
Hartquist, T. W. : 309, 357
Hass, M. R. : 281
Henkel, C. : 111
Heske, A. : 245
Herbst, E. : 121
Herrmann, F. : 117
Hjalmarson, A. : 179
Hollis, J. M. : 187

Howe, D. A. : 363
Hudgins, D. M. : 27, 437
Huebner, W. F. : 447, 449
Hyashi, S. S. : 399
Irvine, W. M. : 171, 179, 223, 227, 271
Jackson, J. : 117
Jaffe, D. T. : 285, 311
Jansen, D. J. : 319
Jatenco-Pereira, V. : 407
Joblin, C. : 21
Johansson, L. E. B. : 401
Johns, J. W. C. : 399
Kaifu, N. : 169
Kauts, V. L. : 41
Kawara, K. : 103
Keene, J. B. : 191
Killian, T. C. : 23
Kim, H. R. : 225
Konno, I. : 447, 449
Koo, B. C. : 225
Krishna Swamy, K. S. : 139
Kroto, H. W. : 47
Kuan, Y. -J. : 187
Kunoff, E. : 425
Landaberry, S. J. C. : 403, 409
Langer, W. D. : 193
Leger, A. : 21, 135
Lepine, J. R. D. : 245, 395
Lepp, S. : 471
Lovell, A. : 271
Lucas, R. : 389
MacLeod, J. M. : 399
Madden, S. C. : 117
Magalhaes, A. : 109
Maillard, J -P. : 259
Martin, P. : 21
Martins, S. L. : 403
Mathews, H. E. : 399
Mathur, D. : 139
Mauersberger, R. : 111
McGonagle, D. : 227, 271
Menten, K. M. : 345, 357
Millar, T. J. : 121, 179, 211, 363
Minh, Y. C. : 223, 225, 227

Mitchell, G. F. : 259
Moran, J. M. : 345
Morton, D. C. : 399
Nejad, L. A. M. : 249
Neufeld, D. A. : 285, 335
Nuth III, J. A. : 39, 461
Nyman, L. -A. : 401
Ohishi, M. : 171, 179
Olofsson, H. : 401
Opher, R. : 83, 407
Owen, T. : 97
Pacheco, G. M. : 349, 405
Padman, R. : 297
Palmer, P. : 441
Paubert, G. : 23
Paubert, G. : 439
Pendleton, Y. : 133
Pereira, C. B. : 409
Petrini, D. : 359
Phillips, T. G. : 191, 319
Piirola, V. : 109
Poglitsch, P. : 117
Prasad, S. S. : 19, 205
Pratap, P. : 345, 347
Quefflelec, J. L. : 13
Rawlings, J. M. C. : 357, 365, 387
Rebrion, C. : 13
Reid, M. J. : 345
Roberge, W. G. : 155
Robert, F. : 423
Roche, P. F. : 371
Rodrigues, C. : 109
Rodriguez, L. : 351
Roh, D. G. : 225
Rowe, B. R. : 7, 13
Rubin, R. H. : 281
Saito, S. : 179
Salama, F. : 25, 435
Sandford, S. A. : 27, 29, 133, 435, 437
Sanzovo, G. C. : 405
Scalise, Jr., E. : 349, 405, 447
Schilke, P. : 251
Schloerb, F. P. : 271
Schmidt, W. : 135

Schutte, W. A. : 29, 137
Sellgren, K. : 133
Shapiro, P. R. : 73
Shmeld, I. K. : 413
Simpson, J. P. : 281
Singh, P. D. : 447, 449
Smith, C. H. : 371
Snell, R. L. : 271
Snyder, L. E. : 187, 347, 427, 441
Stacy, G. J. : 117
Steinitz, R. : 425
Sternberg, A. : 309, 329, 333
Strelnitskij, V. S. : 413
Stutzki, J. : 303
Svestka, J. : 17
Tapia, M. : 133
Tarafdar, S. P. : 55, 139
Thaddeus, P. : 23
Tielens, A. G. G. M. : 27, 91, 133, 137, 285, 317
Tine, S. : 231
Townes, C. H. : 117
Travers, D. : 13
Turner, B. E. : 181, 399
Ungerechts, H. : 271
van Dishoeck, E. F. : 143, 153, 191, 285, 319
Verstraete, L. : 135
Vilas Boas, J. W. S. : 275, 405
Vrtilek, J. M. : 23
Wagenblast, R. : 157, 309
Walmsley, C. M. : 251, 285, 345
Watson, J. K. G. : 399
Watt, G. D. : 399
Webster, A. S. : 399
White, G. J. : 297
Williams, D. A. : 249, 309, 357, 363, 463
Wolstencroft, R. : 401
Wright, E. L. : 321
Wu, Y. : 217
Yudin, N. P. : 41
Zeng, Q. : 341
Ziurys, L. M. : 187, 237